# Process Grammar: The Basis of Morphology

Michael Leyton

# Process Grammar: The Basis of Morphology

Springer

Michael Leyton
DIMACS Center for Discrete Mathematics
and Theoretical Computer Science
Rutgers University
Busch Campus
New Brunswick
NJ 08854
USA
mleyton@dimacs.rutgers.edu

ISBN 978-1-4614-1814-6 e-ISBN 978-1-4614-1815-3
DOI 10.1007/978-1-4614-1815-3
Springer New York Dordrecht Heidelberg London

Library of Congress Control Number: 2011940966



Printed on acid-free paper

Springer is part of Springer Science+Business Media (www.springer.com)

# Preface

The Process Grammar has been applied by scientists and engineers in many disciplines, including medical diagnosis (e.g., cardiac diagnosis); geology (e.g., the analysis of volcanic islands); computer-aided design (to establish new CAD operators); meteorology (to analyze weather patterns); biological anatomy (e.g., MRI human brain scans, dental radiographs); engineering bridge design; chemical engineering; etc.

The Process Grammar is based on a theorem that I proved in the 1980s called the Symmetry-Curvature Duality Theorem. Researchers have shown that this theorem defines an enormous number of aspects of biology, e.g., anatomy with applications in radiotherapy, surgery, and psychiatry, the tracking of DNA molecules, musculoskeletal development, the morphology of leaves in botany, the morphology of fish, etc; also geology, e.g., for the analysis of drainage patterns, etc; also graphics, e.g., for interactive rendering and cartoon vectorization, etc. The considerable applications of this theorem demonstrate that the theorem is *fundamental to morphology*. Its importance to morphology is explained by the Process Grammar.

In fact, every rule in the Process Grammar is an instance of a rule in a much larger system, my New Foundations to Geometry, that I have elaborated in my book *A Generative Theory of Shape* (Springer-Verlag, 550 pages).

Chapter 2 of the present book gives a brief introduction to these New Foundations, so that the reader is then shown how the Process Grammar is an instance of these New Foundations.

The central proposal of my New Foundations to Geometry is that shape is equivalent to memory storage. Therefore, in the New Foundations, geometry is the mathematical theory of memory storage, invented by the New Foundations. This opposes the Standard Foundations to Geometry, which are based on the invariants program. My argument is that invariants are memoryless.

The New Foundations to Geometry, being a generative theory of shape, define any shape by a sequence of operations needed to create it. Furthermore, the New Foundations require that this sequence be intelligent. In fact, the New Foundations gives a mathematical theory of intelligence, and base the entire New Foundations on this mathematical theory. The two most basic principles of this mathematical theory of intelligence are Maximization of Transfer and Maximization of Recoverability of the generative operations. The New Foundations give a Mathematical Theory of Transfer, and a Mathematical Theory of Recoverability. In the Mathematical Theory of Transfer, transfer is modeled by a group-theoretic structure called a wreath product. In the Mathematical Theory of Recoverability, the fundamental claim is that the only recoverable operations are symmetry-breaking ones. Furthermore, the Mathematical Theory of Recoverability gives a mathematical theory of symmetry-breaking that is fundamentally opposed to the conventional theory of symmetry-breaking used in physics, chemistry, etc. In the New Foundations, the Mathematical Theory of Transfer and the Mathematical Theory of Recoverability are unified, by claiming that any memory store is structured by a group invented in the New Foundations, called a symmetry-breaking wreath product.

An important fact is that the New Foundations are fundamentally a theory of *complex* shape. This is used to solve problems in many disciplines. For example, it solves major well-known problems in object-oriented software. Furthermore, it solves the interoperability problem in Computer-Aided Design (CAD). For example, consider mechanical CAD, which is the design process in mechanical engineering – forming the basis, for example, of the aerospace and automotive industries. It is generally accepted that mechanical CAD proceeds by a process called *feature attachment*. This is the process of the successive addition of structural units and components. The New Foundations give a mathematical theory of feature attachment. Furthermore, as another example of complex shape, the New Foundations give a mathematical theory of musical composition. A remarkable result is that the New Foundations show that feature attachment in mechanical CAD is mathematically equivalent to musical composition.

The Process Grammar is a *component* of the New Foundations to Geometry. To help the reader understand this, note the following: In the New Foundations, Geometry is the Mathematical Theory of Memory Storage. According to the First Fundamental Law of Memory Storage, given by the New Foundations, all asymmetries are memory stores. The different components of the New Foundations are the application of the laws of memory storage, given by the New Foundations, to different asymmetries. The Process Grammar is one such component of the New Foundations. The asymmetry used as memory storage in the Process Grammar is *curvature variation*.

Chapter 3 begins by describing the inference rules invented in the Process Grammar for recovering process-history from curvature variation. This uses the type of differential symmetry axis called PISA, invented in the Process Grammar, that has very different properties from the other differential symmetry axes called the Medial axis, the Medial axis (MA) symmetry set, and the SLS. The properties of the PISA axes are crucial. To understand this, one should first observe that there are four types of curvature extrema: a positive maximum $M^+$, negative minimum $m^-$, positive minimum $m^+$, and positive maximum $M^-$. Using the PISA axis, the Process Grammar defines $M^+$ and $m^-$ as penetrative extrema, and defines $m^+$ and $M^-$ as compressive extrema. This is because, the PISA axis gives, to the four extrema, the following *strongly plausible causal explanations*: penetrative processes for the $M^+$ and $m^-$ extrema, and compressive processes for the $m^+$ and $M^-$ extrema. Thus the PISA axis fulfills a basic principle of the New Foundations to Geometry, that *shape is defined by causal explanation*.

The standard literature on shape fails to do this. The reason is that the standard literature fails to make the correct distinction between penetrative and compressive extrema, and does so because it fails to understand compressive extrema. The New Foundations show that compressive extrema are fundamentally important to understanding shape in biology, geology, meteorology, vehicle design, architecture, paintings, etc. However, the standard literature fails to understand this, and this is because it fails to understand compressive extrema. The reason is this:

The PISA axis is the only differential symmetry axis that describes the *correct history* of a compressive extremum, i.e., the trajectory of the curve point that became the compressive extremum as a result of the flattening process, and the PISA axis is the only axis that describes the force that created the flattening. It does so because it leads to the extremum from the concave side of the curve. In contrast, the MA symmetry set, being on the convex side and pointing away from the extremum, does not give the history of the extremum in the flattening process that created the extremum. Furthermore, the SLS axis, being on the convex side and pointing towards the extremum, does not give the history of the extremum in the flattening process that created the extremum.

The basis of the correctness of PISA is the following: In the region of a $m^+$ or $M^-$ extremum, the PISA symmetry axis *correctly* violates one of the fundamental properties of symmetry of the last 2500 years: that a symmetry axis must be on the *convex* side of an object. This violation of conventional symmetry makes the PISA symmetry axis the *correct* way of describing the history of compressive extrema, and is the reason why, *prior to the invention of the PISA axis, compressive extrema have always been failed to be described correctly.*

Chapter 3 also describes the six Level 3 Process Grammar operations: These operations are: $Cm^+$ squashing continues till it indents; $CM^-$ resistance continues till it protrudes; $BM^+$ shield-formation; $Bm^-$ bay-formation; $Bm^+$ breaking-through of a protrusion; $BM^-$ breaking-through of an indentation. The first two are called the $C$ process-continuations, and the other four are called the $B$ process-bifurcations.

Using the process-inference rules of the Process Grammar, these six operations have the following causal explanations: The $C$ process-continuations are caused by the continuation of a compressive process till it penetrates. The $BM^+$ and $Bm^-$ operations are caused by a compressive process that opposes a pre-existing penetrative process. The $Bm^+$ and $BM^-$ operations are caused by a penetrative process that opposes a pre-existing compressive process.

The Process Grammar claims that these operations are fundamentally important to morphology, and Chapter 3 demonstrates that the conventional theories of morphology completely fail to model these, because they completely fail to model compressive extrema, which are fundamentally important to the causal structure of all these operations.

That is, Chapter 3 shows that the PISA axis gives the correct causal explanation of the $C$ process-continuations, whereas the MA symmetry set and the SLS axis do not. Furthermore, the PISA axis gives the correct causal explanation of the $B$ process-bifurcations, whereas the MA symmetry set and SLS axis do not.

Chapter 3 also shows how the group theory invented in the New Foundations to Geometry, for complex shape, also defines the structure of these process-bifurcation operations. In fact, Chapter 3 shows how this same group theory, which models *mechanical CAD* and *musical composition*, in the New Foundations, also models these *process-bifurcations in morphology*.

Now note the following: The process-bifurcations elaborated in Chapter 3 are the 3-fold process-bifurcations of the Process Grammar. These operations are applied at an existing curvature extremum. Chapter 4 then elaborates the 2-fold process-bifurcations of the Process Grammar. These are called pair-creations. These operations are applied at a point away from an existing extremum, i.e., at a point on a spiral. The Process Grammar claims that such an operation simultaneously realizes two design intents.

When the spiral, on which a pair-creation operation is applied, is the *side* of a pre-existing extremum, the Process-Grammar calls the pre-existing extremum, an anchor-extremum.

In the case where the application-point of a pair-creation operation has non-zero curvature, the operation simultaneously produces a penetrative and compressive extremum. Therefore, the Process Grammar is the only morphology theory that correctly describes this situation, whereas other morphology theories cannot.

The Process Grammar elaborates six pair-creation operations, $\pi\emptyset^{\uparrow +}$, $[C(3)\pi]0^{\uparrow}$, $\pi\emptyset^{\uparrow -}$, $\pi\emptyset^{\downarrow +}$, $[C(3)\pi]0^{\downarrow}$, $\pi\emptyset^{\downarrow -}$, and elaborates their types of applications in relation to the various anchor-extrema.

Chapter 5 describes the theory of *parts* given by the Process Grammar. It follows from the theory of parts defined by the New Foundations to Geometry. Recall that the New Foundations are based on two basic principles: Maximization of Transfer and Maximization of Recoverability. Correspondingly, the New Foundations define *parts* as *phases* of the recovered causal history, such that they *maximize transfer and recoverability* in the recovered causal history.

This is fundamentally different from the standard research on geometry, because the standard research defines a part as a segment of the whole, where both the segment and the whole are defined independently of recoverable causal history. That is, standardly, a part is defined *non-causally*. For example, the literature in computer vision is full of diagrams of an animal body which is separated into rigid closed shapes corresponding to arms, legs, torso, head, etc., all floating independently in space. Thus, there is no conception that an arm actually grew out of the body.

The fact that the New Foundations to Geometry define *parts* as *phases* of the recovered causal history, such that they *maximize transfer and recoverability* in the recovered causal history, has the consequence that, in the Process Grammar, a part is defined as a recovered process and the recovered subsequent history of that process; i.e., its subsequent Process Grammar operations.

Notice therefore that, because the standard research literature fails to understand compressive extrema $m^+$ and $M^-$, i.e., fails to understand the processes that produced them, it fails to understand them as parts. This is a fundamental reason why the standard literature is stupid, in accord with the technical definition of intelligence given by the New Foundations.

The two extrema $m^+$ and $M^-$ define two parts, which the Process Grammar call respectively, *shields* and *bays*, which Chapter 5 demonstrates, in detail, are fundamentally important to morphology.

According to the Process Grammar, *shields* and *bays* should be regarded as extremely important to many disciplines such as biological morphology, automotive, aerospace and architectural design, computer vision, robotics, etc. Nevertheless, these disciplines tend not to notice shields and bays; and on the rare occasions when the disciplines notice them, the disciplines describe them incorrectly.

In contrast, the Process Grammar defines them correctly, and, in so doing, shows the fundamental importance of these structures.

Chapter 5 shows that examples of shields in biological morphology are the forehead, top and back of the human head, the chin and jaw, the chest, the tips of fingers, the knee-cap, the heel of the foot. Also, examples of shields in road vehicles are the front, top, back, and sides of cars, buses and vans. This unifies biological morphology and automotive design. For example, according to Chapter 5, the forehead of the human head is the same structure as the front of a van. In fact, both are what the Process Grammar calls bi-directional shields in the sense that they are shields both in the vertical and horizontal direction.

It is important to understand that conventional geometry, such as the highly popular Medial Axis of Blum, completely fails to describe shields. Blum invented and published the Medial Axis as an analysis of biological morphology. Yet it completely fails to describe what Chapter 5 demonstrates are fundamental and frequent in biological shape: i.e., shields. A major claim in Chapter 5 is that the fundamental importance of shields in biological shape is because biology requires shields for *protection*. In contrast to Blum's Medial Axis, the symmetry axis invented in the Process Grammar, the Process-Inferring Symmetry Analysis (PISA) not only recognizes the existence of shields but describes them correctly by revealing their protective structure. The same applies to shields in automotive, aerospace and architectural design.

Chapter 5 also shows that bays are important structures in architectural design, such as the design of bridges, doorways, ceilings, windows, etc. Again, Blum's Medial Axis completely fails to describe these, and PISA axis not only describes them but does so revealing their functional structure.

After Chapter 5 gives an extensive demonstration of the importance of shields in biological morphology, vehicle design, and architecture, Chapter 5 then gives an analysis of car design as the successive application of Process Grammar operations. A fundamental property of this analysis is that it demonstrates that the important functions of a car are processes inferred from the curvature extrema in the successive stages defined by the Process Grammar operations. This means that the Process Grammar recovers the *layers* of *design intents* in the car.

In fact, the recovery of design intent is one of the main achievements of the laws of recoverability given by the New Foundations to Geometry.

This is relevant for solving major problems that exist in the integration of large-scale engineering systems and product lifecycle management.

For example, the manufacturing of a major engineering product, such as an aircraft, involves supply chains that tend to use different design softwares that cannot exchange their model information. What would solve this interoperability problem is the recovery of design intent. Furthermore, over the product lifecycle, the recovery of design intent would also solve the reuse, adaptability, and maintenance problems.

A crucial fact is that the mathematics of recoverability, invented by my New Foundations to Geometry, solves these problems. This is corroborated by the fact that aspects of this mathematics have been applied by scientists and engineers in over 40 disciplines. Furthermore, certain members of the International Standardization Organization (ISO) are implementing my mathematics to establish international standards in engineering design.

Before discussing bays, Chapter 5 gives a geometric analysis of *grasping* and shows that the PISA axis is totally correct in describing the forces of grasping; and that the Medial Axis symmetry set is totally wrong in describing the forces of grasping.

Chapter 5 then goes on to demonstrating the important role of the parts that the Process Grammar calls bays in defining arches in architecture.

A crucial aspect of Chapter 5 is that it demonstrates the following:

**The Process Grammar gives a unification of biological morphology and vehicle design.**

In the analysis of parts, Chapter 5 also defines the crucial concept of *restricting* the *zone-of-influence* of a process, a concept that the standard literature completely fails to understand. Furthermore Chapter 5 also defines the Process Grammar operations that perform this zone-of-influence restriction.

Chapter 6 then gives an extensive analysis of what the Process Grammar calls 5-fold parts; that is, parts with 5 extrema. These include parts that standard morphological theories completely fail to understand and represent, for example, parts that the Process Grammar calls *triple-shields* and *triple-bays*. This book shows that triple-shields are fundamentally important in biological morphology and vehicle design.

Chapter 6 defines three scenarios that produce 5-fold parts, and the Process Grammar operations that realize those scenarios.

This book shows that the Process Grammar operations capture important *intended functions*, in both biological morphology and manufacturing design. Consequently, the Process Grammar operations define new CAD operations that achieve fundamentally important *design intents*. Therefore, since this book defines over 30 Process Grammar operations, there is this important consequence:

**This book defines over 30 important new CAD operations.**

Another aspect of the book is that it invents a new singularity theory of shape, called Interactive Singularity Theory. This singularity theory of shape is elaborated in Chapters 8 – 13.

A crucial fact is that Interactive Singularity Theory's definition and organization of singularities match the *causal* structures which the Process Grammar has proved are fundamentally important to biological morphology and manufacturing design. In contrast, conventional singularity theories of shape have completely failed to match the fundamentally important causal structures of biological morphology and manufacturing design. As an example, Chapter 7 shows that Catastrophe Theory is inadequate for modeling morphology.

A crucial aspect of Interactive Singularity Theory is that it gives the organization of relationships between the morphological operators, i.e. their Interactive Structures, in a way that matches the *causal* structure of the operators correctly given by the PISA axes.

For example, Chapter 12 shows that a crucial situation defined by the Process Grammar – a situation in which a process opposes and breaks-through a compressive process, resulting in a zone-restricted penetrative process – is given by what Interactive Singularity Theory calls Interactive Structure, in a way that matches the causal structure of those situations as correctly given by the PISA axes. In contrast, conventional singularity theories of shape cannot model this.

June 2011 Michael Leyton

# Contents

# 1

# Fundamental Theorem of Morphology

## 1.1 The Fundamental Theorem of Biological Morphology

The Process Grammar is based on a theorem I proved in the 1980s called the Symmetry-Curvature Duality Theorem (Leyton, [16]). This theorem states that, to every curvature extremum, there is a symmetry axis leading to, and terminating at, the extremum. My theorem has been applied by scientists in over 40 disciplines.

We shall begin by seeing that the theorem should be regarded as the fundamental theorem of biological morphology. First, in my book *Symmetry, Causality, Mind* (MIT Press, 1992), I showed that the theorem gives the basic structure of all biological limb-formation, as follows:

Consider the work of the biologists French, Bryant & Bryant, [7] [3]. Using tissue regeneration experiments, they have shown that, prior to the formation of a limb in a region, the region is homogeneous, e.g., chemically. It then becomes circularly organized around a point. In fact, the surface in the region takes on the organization of a polar coordinate system, as shown in the left diagram in Fig 1.1.

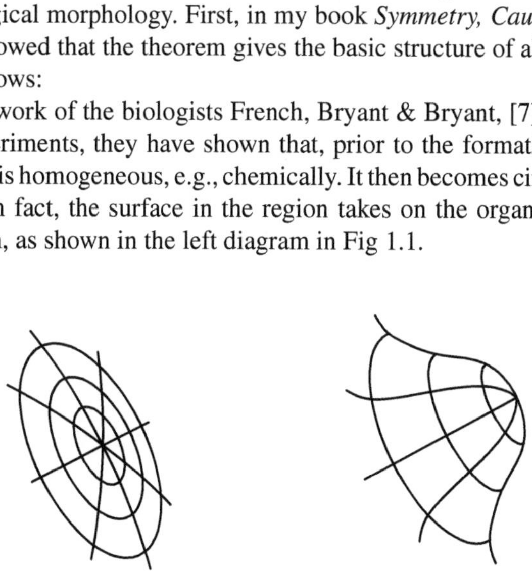

**Fig. 1.1.** Biological limb-formation.

Then, the development of the limb consists of "pushing out" this polar plot, as shown in the right diagram in Fig 1.1. The pole center itself becomes the tip of the limb. Viewing this within the framework of the Process Grammar, the tip is, in fact, a *curvature extremum*, a structure fully described for limbs in Chapter 5 of the present book. Furthermore, according to the Process Grammar, when we are given the final developed limb, the trajectory that the curvature extremum took over time, in creating the limb, is given by the symmetry axis which the Symmetry-Curvature Duality Theorem proves must exist and lead to the extremum. This means that the structure of the limb is determined by the theorem.

In fact, the theorem applies not only to biological limbs but also to an enormous number of other aspects of biological morphology.

For example, Shemlon [38] has shown that the theorem applies to the structure of the *human brain*, *neuronal growth*, *dentistry*, and *insect blood cells*. Lopéz [29] showed that the theorem applies to *cardiac* data. Pizer, Fritsch, Yushkevich, Johnson, & Chaney [36] showed that the theorem is crucial to the structure of *abnormal anatomy*, with applications in *radiotherapy, surgery, and psychiatry*. Parvin, Peng, Johnston, & Maestre [33] applied the theorem in tracking *DNA molecules*. De Sa, Radice, & Kerckhove [5] showed that the theorem is basic to *musculoskeletal development*. Costa [4] applied the theorem to the imaging of *neurons*. Ogniewicz [32] showed the fundamental role of the theorem in the morphology of *leaves in botany*. Torres & Falcão, [40], and many other researchers, have applied the theorem in the analysis of the morphology of *fish*.

In fact, surveying the literature, we see that the theorem is used endlessly in the analysis of biological morphology. We will see that the reason is this:

**FUNDAMENTAL THEOREM OF BIOLOGICAL MORPHOLOGY**

**The Symmetry-Curvature Duality Theorem is the fundamental theorem of biological morphology.**

The word *fundamental* is used here because the theorem is nearly always the *first step* in the morphological analyses carried out by modern researchers. This is because the remaining steps in the morphological analyses cannot proceed without first applying the theorem.

## 1.2 The Universal Power of the Symmetry-Curvature Duality Theorem

In fact, the Symmetry-Curvature Duality Theorem has been applied by researchers in many other disciplines besides the biological ones.

For example, Milios [31] used the theorem in *meteorology* in order to analyze *weather patterns*. Larsen [9] used the theorem in *geology* for the analysis of *volcanic islands*.

Lopéz [29] used the theorem in geology for the analysis of *drainage patterns*. Pernot, Guillet, Leon, Falcidieno, & Giannini [34] [35] used the theorem in *computer-aided design* to create *design tools* for the automotive and aerospace industry. Lee [10] used the theorem in *chemical engineering* to model *molecular dynamics*. Katz [8] used the theorem in *graphics* for *interactive rendering*. Zou & Yan [41] used the theorem in graphics for *cartoon vectorization*; etc.

The crucial fact is that several applications, e.g., applications in radiology, meteorology, geology, computer-aided design, chemical engineering, etc., were used as the basis of applying the *Process Grammar*.

This is conceptually important to understand, because the Process Grammar shows *why* the theorem is so frequently used in morphology.

**EXPLAINING THE SCIENTIFIC USEFULNESS OF THE SYMMETRY-CURVATURE DUALITY THEOREM**

**The deep reason why the Symmetry-Curvature Duality Theorem is so scientifically useful in morphology is due to the position that the theorem has in the Process Grammar.**

**That is, the Process Grammar explains why the theorem is useful in so many disciplines.**

In fact, every rule in the Process Grammar is an instance of a rule in my much larger system, called *Generative Geometry*, which I invented to encompass all the sciences and computational disciplines. This mathematical system is elaborated in my book *A Generative Theory of Shape* (Springer-Verlag, 550 pages). The next chapter gives a very brief introduction to Generative Geometry, so that the reader is given some deep foundations to the Process Grammar. Using this introduction, I will then show how the Process Grammar is an instance of that geometry.

# 2

# Generative Theory of Shape

## 2.1 New Foundations to Geometry

In my book, *A Generative Theory of Shape* (Springer-Verlag, 550 pages), I give New Foundations to Geometry. The present chapter gives a brief summary of the New Foundations because the Process Grammar is based on these foundations.

These New Foundations to Geometry directly oppose the Standard Foundations to Geometry that have existed for almost three thousand years. The central proposal in my New Foundations to Geometry is this:

**NEW FOUNDATIONS TO GEOMETRY**

**Leyton 1992**

**According to the New Foundations to Geometry:**
**Shape is equivalent to memory storage.**

**Therefore, in the New Foundations:**
**Geometry is the mathematical theory of memory storage,**
**invented by the New Foundations.**

Let us see how this opposes the Standard Foundations to Geometry. In the Standard Foundations, a geometric object consists of those properties of a figure that do not change under actions. These unchanged properties are called the *invariants* of the actions. Geometry began with the study of invariance, in the form of Euclid's concern with *congruence*, which is really a concern with invariance (properties that do not change). The full invariance program was explicitly defined in the 19th century by Felix Klein who defined geometry as the study of invariants under groups of transformations. Klein's invariance program became the basis of 20th century mathematics and physics.

My argument is that the problem with invariants is that they are *memoryless*. That is, if a property is invariant (unchanged) under an action, then one cannot infer, from the property, that the action has taken place. Thus I argue: *Invariants cannot act as memory stores*. In consequence, I conclude that the Standard Foundations to Geometry are concerned with *memorylessness*. In fact, since the Standard Foundations try to maximize the discovery of invariants, those foundations essentially try to maximize memorylessness.

A basic argument of mine is this: The Standard Foundations to Geometry are inappropriate for the computational age, because the computational age is based on memory storage, archival systems, etc., whereas the Standard Foundations are based on the invariants program, which tries to maximize memorylessness.

As a consequence, I embarked on a 30-year project to build up entirely New Foundations to Geometry. Rather than basing geometry on the *maximization of memorylessness* (the aim of the Standard Foundations), I base geometry on the *maximization of memory storage*. The result is a theoretical system that is profoundly different, both on a conceptual level and on a detailed mathematical level. Everything in the Standard Foundations is *inverted* in the New Foundations. For example, whereas in the Standard Foundations, groups are used to describe symmetries, in the New Foundations, groups are used to describe asymmetries. Again, whereas in the Standard Foundations, the goal is reference-frame independence, in the New Foundations, the goal is reference-frame dependence. Again, the relationship between hierarchical levels in the Standard Foundations, e.g., Euclidean to Affine to Projective, are entirely the opposite relationships in the New Foundations. These are all consequences of the fact that the Standard Foundations try to maximize memorylessness, and the New Foundations try to maximize memory storage.

The conceptual structure of the New Foundations is elaborated in my book *Symmetry, Causality, Mind* (MIT Press, 630 pages); and the mathematical structure is elaborated in my book *A Generative Theory of Shape* (Springer-Verlag, 550 pages). The latter book also gives extensive applications of the mathematics to computer-aided design, software engineering, computer vision, gestalt psychology, robotics, music, architecture, and physics. Besides these applications, parts of my mathematical theory have been applied by scientists in over 40 disciplines, such as chemical engineering, meteorology, radiology, geology, botany, structural engineering, mathematical control theory, etc.

The remainder of the present chapter will briefly describe some of the basic concepts and mathematical structures of my New Foundations to Geometry. This will give some useful background to the Process-Grammar.

First, being a generative theory of shape, the New Foundations define any shape by a sequence of operations needed to create it. Next, the New Foundations require that this sequence be *intelligent*. In fact:

**The New Foundations to Geometry elaborate a mathematical theory of intelligence, and base the entire New Foundations on this mathematical theory.**

The two most basic principles of this mathematical theory of intelligence are:

**(1) Maximization of Transfer.** Any agent is regarded as displaying intelligence and insight when it is able to *transfer* actions used in previous situations to new situations. In fact, the agent must maximize the transfer of parts of a generative sequence onto other parts of a generative sequence. Thus a basic part of the New Foundations to Geometry is giving a *Mathematical Theory of Transfer*.

**(2) Maximization of Recoverability.** Any intelligent agent must be able to infer the causes of its own current state, in order to identify why it failed or succeeded, and thereby edit its behavior. Notice that this is part of a still larger problem, which we call the problem of recoverability: Given the present state of an object, recover the sequence of operations which generated that current state. Thus a basic part of the New Foundations to Geometry is giving a *Mathematical Theory of Recoverability*.

## 2.2 Mathematical Theory of Transfer

Now let us begin to understand the Mathematical Theory of Transfer given by my New Foundations to Geometry. According to this theory, a situation of transfer involves two levels as illustrated in Fig 2.1: a group which the New Foundations call a **fiber group**, which is the group of actions *to be transferred*; and a group which the New Foundations call a **control group**, whose group elements will *transfer the fiber group*. The transferred versions of the fiber group are illustrated as the vertical copies in Fig 2.1, and the New Foundations calls these the **fiber-group copies**. The control group acts from above, and transfers the fiber-group copies onto each other, as illustrated by the arrow in the figure.

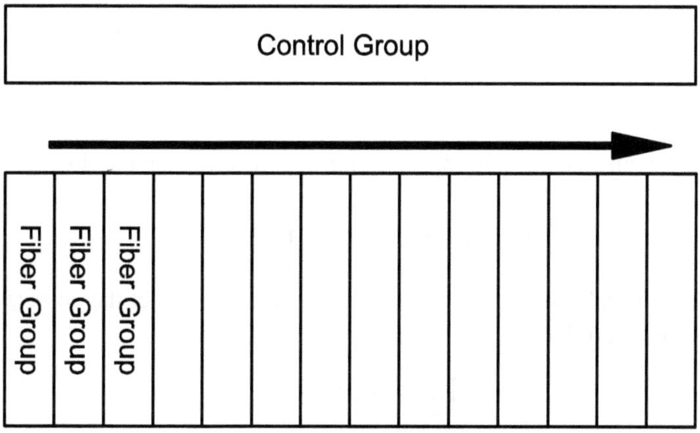


**Fig. 2.1.** The control group transferring the fiber-group copies onto each other.

Let us now describe the Mathematical Theory of Transfer in more detail. A claim in the theory is that a situation of transfer is built from two *group actions*. On the lower level, there is a group action of the fiber group $G(F)$, on a set $F$, and the New Foundations calls $F$ the **fiber set**. On the upper level, there is a group action of the control group $G(C)$, on a set $C$, and the New Foundations calls $C$ the **control set**.

Next, to model transfer, make the transferred versions of the fiber group, as follows: For each member $c$ in the control set, make a copy $G(F)_c$ of the fiber group $G(F)$. These will be the transferred versions, i.e., the fiber-group copies, which again are illustrated by the columns in Fig 2.1. Most crucially, the action of the control group $G(C)$ on the control set $C$ can therefore be *imitated* by an action of the control group $G(C)$ on the collection of fiber-group copies. It is this imitating action that will be regarded as the transferring action that the control group has on the fiber-group copies, i.e., this will be regarded as sending the fiber-group copies onto each other.

A fundamental property of the Mathematical Theory of Transfer is that it puts all these components together in a single encompassing structure. According to the theory, the encompassing structure is best given by a group-theoretic construct called a *wreath product*, which is defined as follows: Intuitively, a wreath-product is a group that contains the entire structure shown in Fig 2.1. The structure of this total group is as follows: In Fig 2.1, the entire lower block shown is the direct product $\prod_{c\in C} G(F)_c$ of the fiber-group copies. In the New Foundations, this is called the **fiber-group product**. Accordingly, the wreath product group is then built up by adding, to the fiber-group product, the control group $G(C)$ by a group-theoretic *semi-direct product*, explained as follows:

In any semi-direct product, the upper group (here the control group) sends the lower group (here the fiber-group product) to itself, by rearrangements that preserve the latter's group structure. Such rearrangements are called automorphisms. In a wreath product, this automorphic action is one in which the control group sends the fiber-group copies onto each other in a way that exactly imitates the action of the control group on the control set.

To state this rigorously: One constructs an **automorphism representation**

$$\tau : G(C) \longrightarrow Aut\{\prod_{c\in C} G(F)_c\}$$

such that, given an element $g$ in the control group, its effect on the fiber-group product is defined thus:

$$\tau(g) : \prod_{c\in C} G(F)_c \longrightarrow \prod_{c\in C} G(F)_{g^{-1}c}$$

From this automorphism representation, one can then construct the corresponding external semi-direct product:

$$\{\prod_{c\in C} G(F)_c\} \circledS_\tau G(C)$$

To understand this notation, notice that, to the left of the semi-direct product symbol $\circledS$ is the fiber-group product, i.e., the entire bottom block we diagramed in Fig 2.1. To the right of the $\circledS$ symbol is the control group $G(C)$, the upper block diagramed in Fig 2.1. Notice also that the subscript $\tau$ on the symbol $\circledS$ is the automorphism representation which defines what I call the *transfer* effect of the control group on the fiber-group copies.

It is this semi-direct product that is the *wreath product* of the fiber group and the control group, written like this:

$$\begin{aligned} \textbf{Fiber Group} \circledR \textbf{ Control Group} &= G(F) \circledR G(C) \\ &= \{\prod_{c \in C} G(F)_c\} \circledS_\tau G(C) \end{aligned} \tag{2.1}$$

Let us now understand how the New Foundations to Geometry model transfer within the wreath product. The claim is that transfer corresponds to what is algebraically called conjugation, $g\phi g^{-1}$ where $g$ is a member of the control group and $\phi$ is a member of the fiber-group product. Most crucially, let us understand its effect on the fiber-group copies. Notice that each fiber-group copy has an embedded version within the wreath product. We can call this, an *embedded fiber-group copy*; and, often, for convenience we will simply call it, a *fiber-group copy*. Similarly, the control group has an embedded version within the wreath product. Again, we can call this, the *embedded control group*; and, often, for convenience we will simply call it, the *control group*. The crucial fact is that, within the wreath product, the members of the control group send the fiber-group copies onto each other via *conjugation*. Therefore, we conclude:

**Transfer of the fiber-group copies is modeled by their** *algebraic conjugation,* **within the wreath product, by the members of the control group.**

Now a crucial role of the New Foundations to Geometry is that they give a comprehensive theory of **raw data representation** such that the data set is maximally useful throughout large-scale scientific and engineering systems. My book *A Generative Theory of Shape* [22] elaborates this theory in detail and comprehensively explains the use of this theory to represent **scientific data** and **manufacturing data**. The present chapter will give a summary of some parts of this theory, and the present book will show how the Process Grammar realizes this for morphology in both biology and manufacturing.

First we should note the following. Given each member $c$ of the control set $C$, let us call the set-theoretic Cartesian product $F \times \{c\}$ the corresponding **fiber-set copy**, which will also be notated as $F_c$. By the principle of the Maximization of Transfer, any data set will actually be the union $F \times C$ of the fiber-set copies given by a transfer structure, i.e., a wreath product $G(F) \circledR G(C)$. Thus, given a wreath product, we will refer to the union of the fiber-set copies as the **data set**. A crucial fact is that there is a group action of the wreath-product group $G(F) \circledR G(C)$ on the data set $F \times C$ as follows: Given an element in the wreath-product group, i.e., an ordered pair $\langle\, \phi \mid g \,\rangle$ where $\phi \in \prod_{c \in C} G(F)_c$, $g \in G(C)$, and given an element $(f, c)$ in the data set, define the effect of the former element on the latter, thus:

$$\langle\, \phi \mid g \,\rangle (f, c) \;=\; (\phi(gc)f, gc) \;\in\; F \times C \tag{2.2}$$

Notice that this relates the **data-set element** $(f, c)$ in the fiber-set copy $F_c$ to the **data-set element** $(\phi(gc)f, gc)$ in the fiber-set copy $F_{gc}$.

Now let us examine an example to illustrate the Mathematical Theory of Transfer. Later in this chapter, and the Process Grammar, we will consider much more complex examples. But to enable the reader to begin to understand the mathematical theory, we will initially study a simple example. This example is the way the theory structures a square. We will model the typical way in which a person draws a square on a sheet of paper – i.e., drawing the sides sequentially around the square. Notice that, in fact, this involves a crucial transfer structure as follows:

The first side is generated by starting with a corner point, and applying translations to trace out the side, as shown in Fig 2.2.

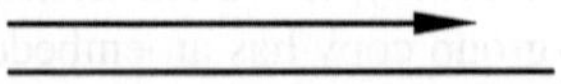

**Fig. 2.2.** The generation of a side, using translations.

Next, this translational structure is *transferred* from one side to the next – rotationally around the square. In other words, there is *transfer of translations by rotations*. This is illustrated in Fig 2.3.

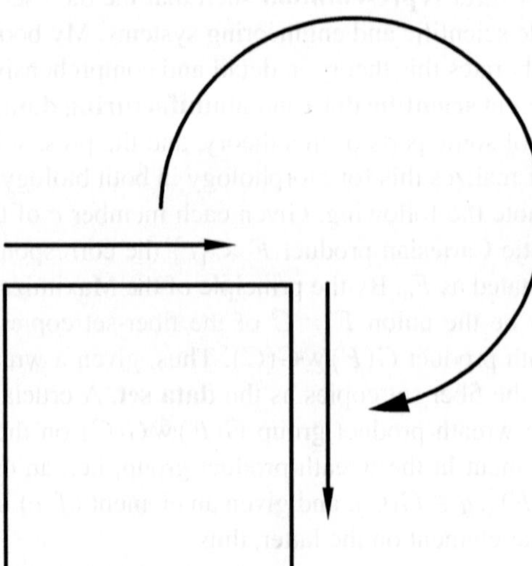

**Fig. 2.3.** Transfer of translation by rotation.

Therefore, according to our theory, the transfer structure, in drawing a square, is defined by the wreath product:

$$\text{Translations} \circledw \text{Rotations}$$

where Translations is the fiber group (generating the side) and Rotations is the control group *transferring* the translation operations that generated the first side, in order to generate the next side, and so on, around the square. This will now be defined rigorously, as follows:

The translation group will be denoted by the additive group $\mathbb{R}$. The rotation group is $\mathbb{Z}_4$, the cyclic group of order 4, which will be represented as

$$\mathbb{Z}_4 = \{ e, \quad r_{90}, \quad r_{180}, \quad r_{270} \}$$

where $r_\theta$ means clockwise rotation by $\theta$ degrees. We now construct our wreath product of these two groups.

The control group $G(C)$ will be $\mathbb{Z}_4$, and the control set $C$ will be the set of four side-positions around the square:

$$c_1 = \text{top}, \quad c_2 = \text{right}, \quad c_3 = \text{bottom}, \quad c_4 = \text{left}. \tag{2.3}$$

The effect of the control group $\mathbb{Z}_4$ on the control set $\{c_1, c_2, c_3, c_4\}$ will be the clockwise rotation of the four side-positions onto each other.

The fiber group $G(F)$ will be the translation group $\mathbb{R}$, and the fiber set $F$ will be the infinite line containing the finite side as a subset. The relationship between the infinite line $F$ and the finite side, that it contains, will be defined in our mathematical theory in a crucial way to be described later. First, however, we note that the action of the fiber group $\mathbb{R}$ on the fiber set $F$ will be the obvious translation of the infinite line along itself.

The fact that there are four elements in the control set $\{c_1, c_2, c_3, c_4\}$ implies that there are four fiber-group copies, which will be denoted as $\mathbb{R}_{c_1}, \mathbb{R}_{c_2}, \mathbb{R}_{c_3}, \mathbb{R}_{c_4}$. Also, it implies that there are four fiber-set copies, which will be denoted $F_{c_1}, F_{c_2}, F_{c_3}, F_{c_4}$. These are the four infinite lines that contain the four finite sides as subsets.

It is crucial to understand that each fiber-group copy (translation group $\mathbb{R}_{c_i}$) will act on its own "personal" copy of the fiber set (infinite line $F_{c_i}$). That is, for each member $c_i$ of the control set, we have the corresponding group action

$$\mathbb{R}_{c_i} \times F_{c_i} \longrightarrow F_{c_i}$$

Based on this, we can now define the wreath product:

$$\mathbb{R} \circledw \mathbb{Z}_4 \tag{2.4}$$

First observe that this is the semi-direct product:

$$[\mathbb{R}_{c_1} \times \mathbb{R}_{c_2} \times \mathbb{R}_{c_3} \times \mathbb{R}_{c_4}] \circledS_\tau \mathbb{Z}_4 \tag{2.5}$$

where $\tau$, the automorphism representation,

$$\tau : \mathbb{Z}_4 \longrightarrow Aut\{\mathbb{R}_{c_1} \times \mathbb{R}_{c_2} \times \mathbb{R}_{c_3} \times \mathbb{R}_{c_4}\}$$

is such that, given any element in the control group, i.e., a rotation $r_\theta$, its automorphic effect $\tau(r_\theta)$ on the fiber-group product, $\mathbb{R}_{c_1} \times \mathbb{R}_{c_2} \times \mathbb{R}_{c_3} \times \mathbb{R}_{c_4}$, corresponds to the effect of that rotation on the control set $\{c_1, c_2, c_3, c_4\}$. Therefore, the fiber-group copies are rotated around the square.

Now let us understand the data set $F \times C$, in this example. It is the *disjoint union* of the four fiber-set copies; i.e., the four infinite lines containing the four finite sides. Therefore it is important to understand that the fiber-set copies are independent sets, i.e., the four infinite lines do not intersect but *overlap*. To help understand this, one can think of them as *four infinite wires* overlapping each other.

Let us now model their relationship to the finite sides. According to the New Foundations two fundamental principles, Maximization of Transfer, and Maximization of Recoverability, the relationship is this: First, using the Theory of Recoverability of the generative operations (to be described later), the four finite sides are generated by cutting down the visibility of the four infinite lines at the end-points of the finite sides, by an extra generative operation that can switch visibility on or off. Furthermore, using the Theory of Transfer, the switching operation is incorporated as follows: First, it is defined by what our theory calls the **occupancy group**, $\mathbb{Z}_2$ (a cyclic group of order 2). The group switches between two states, "occupied" and "non-occupied", which, in the current example, determines whether a point is visible or not visible. Also, by the Maximization of Transfer principle, this group is *transferred* to each point along the infinite line, because the option of switching on and off the side-drawing program is available at any point along the infinite line.

Therefore, by the Mathematical Theory of Transfer, the group $\mathbb{Z}_2$ is placed as a fiber group below the group given in expression (2.4), thus:

$$\mathbb{Z}_2 ⓦ \mathbb{R} ⓦ \mathbb{Z}_4 \tag{2.6}$$

Notice that, with respect to the left wreath product symbol ⓦ, the occupancy group $\mathbb{Z}_2$ is the fiber group, and the subsequence $\mathbb{R}$ ⓦ $\mathbb{Z}_4$ is the control group. Therefore, the subsequence $\mathbb{R}$ ⓦ $\mathbb{Z}_4$ has the effect of *transferring* the occupancy group. Our theory states that transfer maps the fiber-group copies onto each other. In the present case, the fiber-group copies are the copies of the occupancy group, i.e., one copy at each point in the data set $F \times C$ of the group $\mathbb{R}$ ⓦ $\mathbb{Z}_4$. Therefore, the copies of the occupancy group can be identified with the points in the data set $F \times C$. Furthermore, since the group $\mathbb{R}$ ⓦ $\mathbb{Z}_4$ transfers the copies of the occupancy group onto each other, we can understand the group $\mathbb{R}$ ⓦ $\mathbb{Z}_4$ as *transferring* the points in the data set onto each other. Therefore, we have this crucial conclusion: There is only one point. The remaining points have been created by *transferring* that point. Therefore, the square was created purely from a single point. The other points are merely transfers of that single point. This is an example of the principle of the Maximization of Transfer.

Furthermore, as seen in expression (2.6), the transfer is hierarchical; i.e., it is *transfer of transfer*. The occupancy group $\mathbb{Z}_2$ is transferred by the translation group $\mathbb{R}$, and this *transfer is transferred* by the rotation group $\mathbb{Z}_4$. This illustrates the following statement:

**TRANSFER OF TRANSFER**

**An important property of the New Foundations to Geometry is hierarchical transfer, i.e., transfer is transferred.**

To give the reader another illustration of my Mathematical Theory of Transfer, we will see how the theory represents another object, a cylinder.

Note first that, in computer vision and graphics, cylinders are described as the sweeping of a circular cross-section in the direction of the axis. The group of this sweeping structure has never been given. In contrast, my Mathematical Theory of Transfer creates the following group theory of the structure of a cylinder.

By the principle of the Maximization of Transfer, we proceed as follows:

First consider the cross-section. This is given generatively by a circular rotation of a point, as illustrated in Fig 2.4. That is, it is given by the following structure of *transfer*:

$$\mathbb{Z}_2 \text{ \textcircled{w} } SO(2) \tag{2.7}$$

i.e., a *single point*, given by the occupancy group $\mathbb{Z}_2$, is transferred by the group $SO(2)$ which is the continuous rotation group in a plane.

Next, the sweeping of the cross-section, in the direction of the rotation axis, is given by the *transfer*, by translation, of the generative structure of the cross-section, as illustrated in Fig 2.5. Therefore, the wreath product in expression (2.7) is given as the fiber-group to which one adds, via an additional wreath product $\text{\textcircled{w}}$, the translation group $\mathbb{R}$ as the control group, thus:

$$\mathbb{Z}_2 \text{ \textcircled{w} } SO(2) \text{ \textcircled{w} } \mathbb{R} \tag{2.8}$$

Notice therefore that, as a result of this, the cylinder is decomposed into a structure of rotational fibers, as illustrated in Fig 2.6.

With respect to notation, we now make the following comment: When we need to help the reader concentrate on *spatial* levels of a structure, the occupancy level will be omitted from the notation. Thus for example, the structure of the square will be given as

$$\mathbb{R} \text{ \textcircled{w} } \mathbb{Z}_4 \tag{2.9}$$

and the structure of the cylinder will be given as

$$SO(2) \text{ \textcircled{w} } \mathbb{R} \tag{2.10}$$

Notice that, in both these cases, the fiber group is the movement of a point, and our theory defines this as the *transfer* of the generation of a point, and this implies the existence of the occupancy group as a fiber of the fiber. Thus, even though expressions (2.9) and (2.10) help one concentrate on the *spatial* levels of the generative structure, the presence of the occupancy group as an additional fiber is implied, because, by the Maximization of Transfer, the lowest *spatial* fiber must be transferring the occupancy group that generates the initial point.

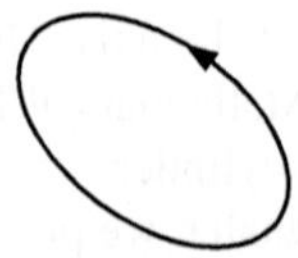

**Fig. 2.4.** A point is transferred by rotations, producing a circle.

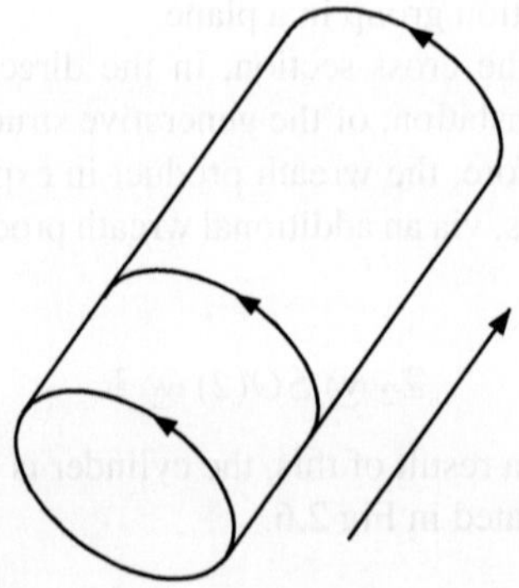

**Fig. 2.5.** The rotation is then transferred by translations, producing a straight cylinder.

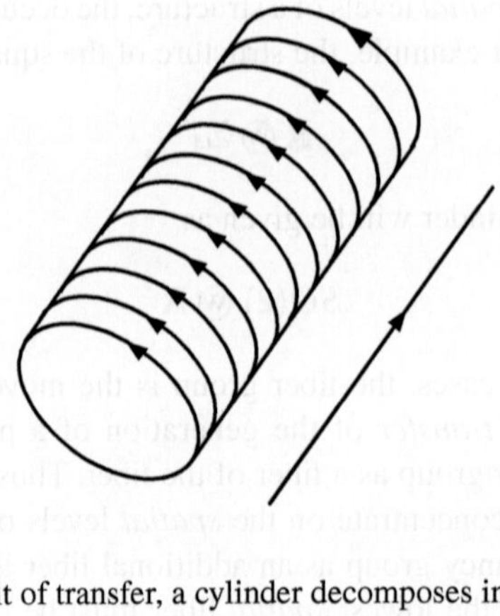

**Fig. 2.6.** As a result of transfer, a cylinder decomposes into rotational fibers.

## 2.3 Mathematical Theory of Recoverability: 1

Recall from section 2.1 that the New Foundations to Geometry are based on two fundamental principles: Maximization of Transfer and Maximization of Recoverability. Furthermore, the New Foundations give a Mathematical Theory of Transfer and a Mathematical Theory of Recoverability. So far, we have been looking at the Mathematical Theory of Transfer. We will now turn to the Mathematical Theory of Recoverability. This will take a number of sections to elaborate.

First, note that by *recovery*, we mean the following:

**THE RECOVERY PROBLEM**

**Given a data set $\mathcal{D}$, infer from $\mathcal{D}$ the generative history that produced it.**

According to the New Foundations, when one has recovered, from a data set $\mathcal{D}$, its generative history, one has *converted the data set into a memory store*. The New Foundations give a massive mathematical theory of memory storage. This theory begins with proposing the following two laws:

**FIRST FUNDAMENTAL LAW OF MEMORY STORAGE**
**(Leyton, 1992)**
**Memory is stored only in asymmetries.**

**SECOND FUNDAMENTAL LAW OF MEMORY STORAGE**
**(Leyton, 1992)**
**Memory is erased by symmetries.**

Let us begin with a simple example. Consider the sheet of paper shown on the left in Fig 2.7. Even if one had never seen that sheet before, one would conclude that it had undergone twisting. I claim that this is because the asymmetry in the sheet yields the information about the generative history. In other words, from the asymmetry, one can *recover* the past history. That is, the *asymmetry acts as a memory store for the past action* – as stated in my First Fundamental Law of Memory Storage (above).

Now let us un-twist the paper, thus obtaining the straight sheet given on the right in Fig 2.7. Suppose we show this straight sheet to any person on the street. They would not be able to infer from it the fact that it had once been twisted. I claim that this is because the symmetry of the straight sheet has wiped out the ability to recover the preceding history. This means that the *symmetry erases the memory store* – as stated in my Second Fundamental Law of Memory Storage (above).

In fact, from the symmetry, one concludes that the straight sheet had always been like this. For example, when you take a sheet of paper from a box of sheets of paper

**Fig. 2.7.** A twisted sheet is a source of information about the previous deformation history. A non-twisted sheet is not.

you have just bought, you do not assume that the sheet of paper had once been twisted or crumpled. Its very straightness (symmetry) leads you to conclude that it had always been straight.

The two diagrams in Fig 2.7 illustrate the two fundamental laws of memory storage given above. These two laws are the basis of the theory of recoverability in the New Foundations to Geometry. As *inference rules*, the two laws are stated in the following way:

**ASYMMETRY PRINCIPLE**
**(Leyton, 1992)**

**Given a data set $\mathcal{D}$, a sequence of operations for generating $\mathcal{D}$ is *recoverable from* $\mathcal{D}$ only if asymmetries in $\mathcal{D}$ go back to symmetries in the previously generated states; i.e., only if, in the forward-time direction, the sequence of operations was symmetry-breaking on each of the successively generated states.**

**SYMMETRY PRINCIPLE**
**(Leyton, 1992)**

**Given a data set $\mathcal{D}$, a sequence of operations for generating $\mathcal{D}$ is *recoverable from* $\mathcal{D}$ only if symmetries in $\mathcal{D}$ are preserved backwards through the previously generated states.**

At first it might seem as if there are exceptions to these two principles. However, my books show that the apparent exceptions are due to the incorrect description of data sets. For example, the Asymmetry Principle states that the only recoverable operations are symmetry-breaking ones. The reader might think that there are exceptions to this law because one is aware of many processes in the world that are not symmetry-breaking, but are symmetry-increasing; e.g., a tank of gas settling to equilibrium under the standard entropy-increasing process. Concerning such situations, my theory says this:

**SYMMETRY-INCREASING PROCESSES.** *A symmetry-increasing process is recoverable only if it is symmetry-breaking on successive data sets.*

So, for example, you can recover the fact that the tank of gas was entropy-increasing over time, if you kept a set of records (e.g., photographs) and the records are linearly ordered, e.g., they are laid out from left to right on a table, in which case the sequence of photographs breaks the left-right symmetry of the table. That is, the increase in spatial symmetry in the tank of gas is made to correspond to a decrease in spatial symmetry of the record structure. This issue is related to the concept of external and internal inference to be described in section 2.6. Most crucially, one must explicitly define the data set, and understand the detailed theory of how these laws are applied to data sets. To understand this theory, the reader should read Chapter 2 of my book *A Generative Theory of Shape*.

The Asymmetry Principle and Symmetry Principle lead to very powerful mathematics, for example a new algebraic theory of symmetry-breaking, and an extensive mathematical theory of memory storage.

However, before going into some of this mathematics, let us begin to develop a familiarity with the Asymmetry Principle and Symmetry Principle. The use of the two principles requires that we go through the following procedure: First partition the presented situation into its asymmetries and its symmetries. Then use the Asymmetry Principle on the asymmetries, and the Symmetry Principle on the symmetries. Note that the application of the Asymmetry Principle will return the asymmetries to symmetries. And the application of the Symmetry Principle will preserve the symmetries.

What does one obtain when one applies this procedure to a situation? The answer is this: One obtains the *past*!

Since this procedure is a basic component of the New Foundations, it will be stated succinctly as follows:

**PROCEDURE FOR RECOVERING THE PAST**

**(1) Partition the situation into its asymmetries and symmetries.**
**(2) Apply the Asymmetry Principle to the asymmetries.**
**(3) Apply the Symmetry Principle to the symmetries.**

An extended example will now be considered that will illustrate the power of this procedure, as follows: In a set of psychological experiments that I carried out in the psychology department in Berkeley in 1982, I found that, when subjects are presented with a rotated parallelogram, as shown in Fig 2.8a, they refer it in their minds to a non-rotated parallelogram, Fig 2.8b, which they then refer in their minds to a rectangle, Fig 2.8c, which they then refer in their minds to a square, Fig 2.8d. It is important to

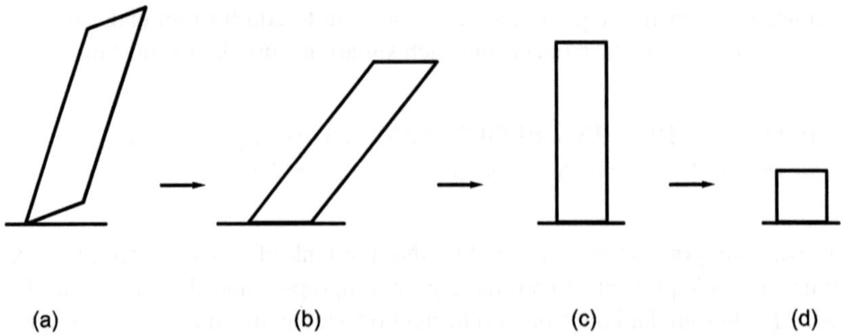


**Fig. 2.8.** The history inferred from a rotated parallelogram.

understand that the subjects are presented with only the first shape. The rest of the shapes are actually created by their minds, as a response to the presented shape.

Close examination reveals that what the subjects are doing is *recovering the history* of the rotated parallelogram. That is, they are saying that, prior to its current state, the rotated parallelogram, Fig 2.8a, was non-rotated, Fig 2.8b, and prior to this, it was a rectangle, Fig 2.8c, and prior to this, it was a square, Fig 2.8d.

The following should be noted about this sequence. The sequence from *right to left* – that is, going from the square to the rotated parallelogram – represents the direction of *forward time*; i.e., the history starts in the *past* (the square) and ends with the *present* (the rotated parallelogram). Conversely, the sequence from *left to right* – that is, going from the rotated parallelogram to the square – represents the direction of *backward time*. Thus, what the subjects are doing, when their minds create the sequence of shapes from the rotated parallelogram to the square, is this: They are **running time backwards!**

We shall now see that the subjects create this sequence by using the Asymmetry Principle and the Symmetry Principle. Recall that the way one uses the two principles is to apply the simple three-stage Procedure for Recovering the Past, given above: (1) Partition the presented situation into its asymmetries and symmetries, (2) apply the Asymmetry Principle to the asymmetries, and (3) apply the Symmetry Principle to the symmetries.

Thus to use this procedure on the rotated parallelogram, let us begin by identifying the asymmetries in that figure. It is important first to note that *asymmetries are the same as distinguishabilities*. In the rotated parallelogram, there are three distinguishabilities:

(1) The distinguishability between the orientation of the shape and the orientation of the environment – indicated by the difference between the bottom edge of the shape and the horizontal line which it touches.

(2) The distinguishability between adjacent angles in the shape: they are different sizes.

(3) The distinguishability between adjacent sides in the shape: they are different lengths.

It is clear that what happens in the sequence, from the rotated parallelogram to the square, is that these three distinguishabilities are removed successively backwards in time. The removal of the first distinguishability, that between the orientation of the shape and the orientation of the environment, results in the transition from the rotated parallelogram to the non-rotated one. The removal of the second distinguishability, that between adjacent angles, results in the transition from the non-rotated parallelogram to the rectangle, where the angles are equalized. The removal of the third distinguishability, that between adjacent sides, results in the transition from the rectangle to the square, where the sides are equalized.

Therefore, each successive step in the sequence is a use of the Asymmetry Principle, which says that an asymmetry must be returned to a symmetry backwards in time.

Having identified the asymmetries in the rotated parallelogram and applied the Asymmetry Principle to each of these, we now identify the symmetries in the rotated parallelogram and apply the Symmetry Principle to each of these. First we need to note that *symmetries are the same as indistinguishabilities*. In the rotated parallelogram, there are two indistinguishabilities:

(1) The opposite angles are indistinguishable in size.

(2) The opposite sides are indistinguishable in length.

The Symmetry Principle requires that these two symmetries in the rotated parallelogram must be preserved backwards in time. And indeed, this turns out to be the case. That is, the first symmetry, the equal size of the opposite angles in the rotated parallelogram, is preserved backwards through the entire sequence; i.e., each subsequent shape, from left to right, has the property that opposite angles are equal in size. Similarly, the other symmetry, the equal length of the opposite sides in the rotated parallelogram, is preserved backwards through the entire sequence; i.e., each subsequent shape, from left to right, has the property that opposite sides are equal in length.

Furthermore, what can also be seen is that, after any symmetry has been recovered, it is also preserved backward through the remaining sequence. For example, in going from Fig 2.8b to Fig 2.8c, the adjacent angles, which are initially different, are made the same size. This symmetry is then preserved in all the remaining figures. Again, this is a use of the Symmetry Principle.

Thus what we have seen in this entire example is this: The sequence from the rotated parallelogram to the square is determined by two rules: the Asymmetry Principle which returns asymmetries to symmetries, and the Symmetry Principle which preserves the symmetries. These two rules allow us to *recover the past*, i.e., *run time backwards*.

We have just given an illustration of the use of the two basic inference rules, in the New Foundations, for recovering generative history: the Asymmetry Principle and Symmetry Principle. My books also instantiate these two basic rules as an enormous number of inference rules for recovering history in many scientific and artistic domains; from general relativity and software engineering to painting and music.

Now recall that, according to the New Foundations, a data set becomes a memory store when one has *inferred* its generative history. Thus, the theory says that the inference has *converted* the data set into a memory store.

As an example, return to the psychological results in Fig 2.8. What these results show is that the human mind *converts* the first shape, the rotated parallelogram, into a memory store, by recovering its generative history backwards in time, via the fundamental inference rules, the Asymmetry Principle and Symmetry Principle. The crucial fact is that the first shape is then *defined* by its generative history.

This fundamentally contrasts with the Standard Foundations to Geometry, in which a *geometric object* is defined as an *invariant* with respect to its history; i.e., a property that is *unchanged* through the history. In the present example, a geometric object, i.e., an invariant, would be the existence of four sides and their straightness. But this would not include the lengths of the sides. That is, the invariant would be blind to the side-lengths. Consequently, as our theory argues generally, the invariant is *memoryless* with respect to the history.

This illustrates the fact that a fundamental difference between the Standard Foundations to Geometry and the New Foundations to Geometry is that the two systems define a *geometric object* in the opposite way, as follows:

**DEFINITION OF A GEOMETRIC OBJECT**

**STANDARD FOUNDATIONS TO GEOMETRY**
**(Klein)**
**A geometric object is an** *invariant;* **i.e.,** *memoryless.*

**NEW FOUNDATIONS TO GEOMETRY**
**(Leyton)**
**A geometric object is a** *memory store.*

It is important to recall that the above definition given by the New Foundations arises from the Mathematical Theory of Intelligence elaborated in the New Foundations, as follows: Recall from page 7 that one of the two fundamental principles of this theory of intelligence is the Maximization of Recoverability. In fact, let us see how recoverability is related to intelligence, as follows: First recall from page 15 that recovery means this: Given a data set $\mathcal{D}$, infer from $\mathcal{D}$ the generative history that produced it. Thus, recovery means inferring, from the data set, an *explanation* of the data set. The next important thing to understand is that, according to the New Foundations, there should be no separation between the representation of the data set and the explanation given of the data set. This is a fundamental principle introduced in my book *Symmetry, Causality, Mind* (MIT Press), where it is called the *Representation is Explanation Principle*. In fact, in that book, there is an entire chapter called *Representation is Explanation.*

**REPRESENTATION IS EXPLANATION**
**(Leyton, 1992)**

**In the New Foundations to Geometry:**
**The representation of a data set is an explanation of the data set.**

In fact, in the New Foundations, I argue that it is the explanation that defines the data set as a *shape*. That is:

**SHAPE $\equiv$ EXPLANATION**

**According to the New Foundations to Geometry:**
**All explanation is shape and all shape is explanation.**
**That is: shape and explanation are equivalent terms.**

Thus, putting the above two claims together, it should be understood that, according to the New Foundations, we have this:

**DEFINITION OF ANY OBJECT**

**Since, according to the New Foundations to Geometry, the representation of any object is an explanation of the object, the New Foundations conclude this:**

**Every object is defined as a shape; i.e., an explanation. That is:**

**Every object is a geometric object.**

Again, this all relates to the issue of intelligence. The following contrasts intelligence and stupidity:

**INTELLIGENCE vs. STUPIDITY**

**An intelligent representation of an object is one that explains it.**

**A stupid representation of an object is one that does not explain it.**

The reader should understand that, as a result of the fact that the New Foundations are an extensive mathematical theory of explanation, the terms *intelligence* and *stupidity* are given a highly technical meaning in the New Foundations. As a result of this technical meaning, we have:

**DEFINITION OF SHAPE**

**Definition of shape in the Standard Foundations to Geometry:**

**Shape = Configuration**

**This definition is independent of an explanation of the shape. Therefore:**

**The Standard Foundations to Geometry are *stupid.***

**Definition of shape in the New Foundations to Geometry:**

**Shape = Explanation**

**This definition equates shape with its explanation. Therefore:**

**The New Foundations to Geometry are *intelligent.***

## 2.4 Combining Transfer and Recoverability

Now let us look further at the mathematics of the New Foundations to Geometry. Recall, from page 7, that the two fundamental principles of the New Foundations are (1) Maximization of Transfer, and (2) Maximization of Recoverability. It is important to understand that these two principles are deeply integrated in the theory, conceptually as well as mathematically.

First, conceptually, a fundamental argument in the New Foundations is that the *recovered past state* is *transferred* onto the *present state*. For example, in Fig 2.8 (page 18), the recovered past state, the square, is understood as *transferred* onto the present state, the rotated parallelogram. That is, having *recovered* the square from the parallelogram, via the inference rules, we then represent the parallelogram as a deformed and rotated version of the square.

Mathematically, this relationship between transfer and recoverability is realized in the following way. Recall from section 2.2 that the New Foundations model transfer by the group

**Fiber Group ⓦ Control Group**

where the fiber group is the group that is being transferred, the control group is the group that is doing the transfer, and the wreath product ⓦ models the relation *is transferred by*. That is, the above expression should be understood as saying *the fiber group is transferred by the control group.*

Now, combining this with the theory of recoverability, the fiber group gives the recovered past state, and the control group gives the recovered generative history that transfers the recovered past state onto the present state. In this way, the wreath product of the fiber group and control group gives the structure of the *present state*.

For example, in the case of the rotated parallelogram (Fig 2.8 page 18), the fiber group and control group must be as follows: The fiber group is the structure of the past state, the square. On page 11, we defined the structure of the square as $\mathbb{R}$ ⓦ $\mathbb{Z}_4$. Thus the fiber group must be $\mathbb{R}$ ⓦ $\mathbb{Z}_4$. The control group, which gives the generative operations that created the parallelogram from the square, must be the group of linear transformations, i.e., the general linear group $GL(2,\mathbb{R})$. Therefore, taking the wreath product of the fiber group $\mathbb{R}$ ⓦ $\mathbb{Z}_4$ and the control group $GL(2,\mathbb{R})$, we get:

$$\mathbb{R} \circledS \mathbb{Z}_4 \circledS GL(2,\mathbb{R})$$

This group produces, from the square, a space of parallelograms, each of which is defined as a *transferred version* of the square. Thus the group captures the psychological result that, given a parallelogram, one sees it as a distorted version of a square. That is, the group captures the fact that the parallelogram acts as a *memory store* for the past state, the square.

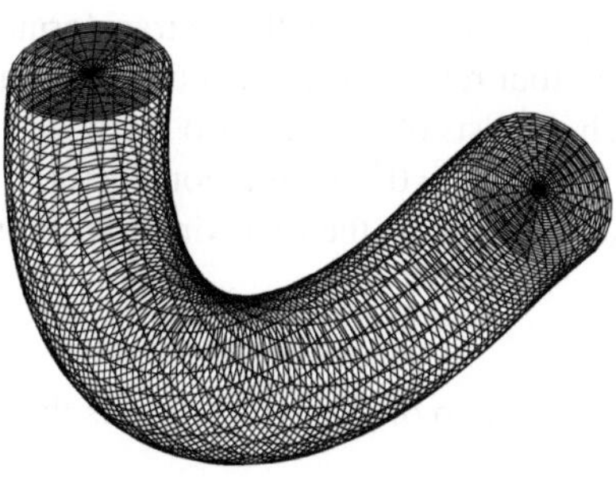

**Fig. 2.9.** A bent cylinder.

As another example, consider a bent cylinder, e.g., as shown in Fig 2.9. It is clear that, by describing this as *bent*, we are inferring that, originally, it must have been a straight cylinder. Furthermore, we see the past state, the straight cylinder, as *transferred* onto the present state, the bent cylinder. On page 13, the structure of the straight cylinder was defined to be the group $SO(2)$ Ⓦ $\mathbb{R}$. Consequently, $SO(2)$ Ⓦ $\mathbb{R}$ must be the fiber group that is transferred onto the bent cylinder by the control group, which we can take to be the group of diffeomorphisms $Diff$. Therefore, the present state, the bent cylinder, is given by the wreath product of the fiber group $SO(2)$ Ⓦ $\mathbb{R}$ and the control group $Diff$, thus:

$$SO(2) \text{ Ⓦ } \mathbb{R} \text{ Ⓦ } Diff$$

The wreath product captures the psychological effect that, given a bent cylinder, one sees it as a distorted version of a straight cylinder; i.e., the bent cylinder acts as a *memory store* for the past state, the straight cylinder.

## 2.5 New Theory of Symmetry-Breaking

Our way of combining the theory of transfer and the theory of recoverability leads to a profound set of conclusions, as follows: Recall first that our theory of recoverability says that the only recoverable operations are symmetry-breaking ones. Thus, combining this with the fact that the past state must be transferred by the generative history onto the present state, we are lead to the following conclusion: *The control group must be symmetry-breaking on its fiber*. This, in turn, leads the New Foundations to the invention of a crucial structure which I call *symmetry-breaking wreath products*.

Close examination reveals that this gives a far more powerful theory of symmetry-breaking than the conventional one that underlies physics and chemistry.

**CONVENTIONAL THEORY OF SYMMETRY-BREAKING**

**Symmetry-breaking is a reduction of symmetry group.**

As an example, consider the transition from a square to a parallelogram, which is a symmetry-breaking one. The conventional theory says that the symmetry group of a square is $D_4$ which consists of the eight Euclidean transformations that map the square to itself: four rotations and four reflections. In contrast, a parallelogram is given by the symmetry group $\mathbb{Z}_2$ which consists of the only two Euclidean transformations that map a parallelogram to itself: rotation by $0^0$ and rotation by $180^0$. Thus, the transition from a square to a parallelogram is given by the following transition of groups:

$$D_4 \quad \longrightarrow \quad \mathbb{Z}_2 \tag{2.11}$$

In fact, the group $\mathbb{Z}_2$ is a subgroup of $D_4$, which means that the transition is given by reduction of symmetry group.

This theory of symmetry-breaking has dominated physics and chemistry for nearly a century. However, according to our theory, this is inherently weak because it means a loss of algebraic structure. That is, in the conventional theory, as one goes from a simpler object (such as a square) to a more complex object (such as a parallelogram), one is reducing the size of the description – which is logically absurd.

In contrast, we have just illustrated our theory of symmetry-breaking using the same example, the transition from a square to a parallelogram. We modeled this transition as follows: One takes our symmetry group of the square, and *adds* the group of linear transformations, *via a wreath product*. Therefore, in our approach, symmetry-breaking actually preserves the original group, and in fact increases it. Thus generally:

### NEW THEORY OF SYMMETRY-BREAKING

**The breaking of a symmetry group $G_1$ is given by its extension by another group $G_2$ via a wreath product thus: $G_1 \circledS G_2$, where $G_2$ is the symmetry group of the asymmetrizing action.**

The fundamentally important feature of this theory is as follows: In the conventional theory of symmetry-breaking, the past symmetry is *lost* in the present broken symmetry. In contrast, in the new theory, the past symmetry is *transferred* onto the present broken symmetry. That is, one actually sees the past symmetry in the broken symmetry; i.e., the past symmetry is *stored* in the broken symmetry.

Let us give another illustrating example of this concept: Consider a bent pipe that one comes across in the street. Merely by the fact that one understands it as a bent pipe means that one sees the past symmetric version as transferred onto the present asymmetric version. That is, the past symmetric version is stored in the present asymmetric version.

This illustrates what the New Foundations mean by saying that *shape is equivalent to memory storage*. That is, the present shape of the pipe equals the history recovered from it.

The approach we have been describing, in fact, leads to a general theory of the mathematical structure of memory stores:

### FUNDAMENTAL STRUCTURE OF MEMORY STORES

**According to the New Foundations to Geometry:**
**Any memory store is structured as a symmetry-breaking wreath product.**

Since, according to the New Foundations, any intelligent description of an object is the representation of the object as a memory store, the above principle gives the structure of objects in all scientific, computational, and artistic domains, when they are *intelligently* represented.

## 2.6 External vs. Internal Inference

This section gives more concepts from the theory of recoverability in the New Foundations, and therefore more of the theory of memory storage. First, the theory makes a fundamental distinction between two types of inference, as follows:

**Definition 2.1.** *In* **external inference***, also called the* **single-state assumption***, the observer assumes that the data set contains a record of only a single state of the generative process (i.e., a single snap-shot). Any inferred previous state therefore does not have a record in the data set. Therefore, in this case, we say that any inferred previous state is* **external** *to the data set.*

An example of external inference is the psychological experiment where subjects were presented with a rotated parallelogram, and inferred its past state to be a square (Fig 2.8 page 18). Although the square is inferred from the data set, the rotated parallelogram, it is not present in the data set. Thus the subjects have inferred a past state, the square, that is *external* to the data set. Therefore, by the above definition, we say that the inference is *external*; i.e., the inference goes outside the present data set.

Another example of external inference is the fact that, when one comes across a bent pipe in the road, one understands it to have originated as a straight pipe. This inference is made despite the fact that a straight pipe is not visible in the present data set. That is, the inferred past state, a straight pipe, is *external* to the present data set. Thus, by the above definition, we say that the inference is *external*; i.e., the inference goes outside the present data set.

The next definition is this:

**Definition 2.2.** *In* **internal inference***, also called the* **multiple-state assumption***, the observer assumes that the data set contains records of multiple states of the generative process (i.e., multiple snap-shots taken over time). A state, recorded in the data set, can therefore have a past state that also has a record* **internal** *to the data set.*

An example of internal inference is where one is presented with a trace of states, e.g., a scratch on a table. Each point along the trace was created at a different moment in time. Therefore, the present data set, the trace, contains records from all those previous moments in time. Thus, according to the above definition, we say that the past states have records that are *internal* to the present data set. Correspondingly, we say that the inference from the present data set, the complete trace, to the past states, the points inside the trace, is *internal inference*.

A detailed theory of external and internal inference is given in my book *Symmetry, Causality, Mind* (MIT Press, 630 pages). Remarkably, although external and internal inference are used in very different types of situations, they are both carried out by using the Asymmetry Principle. The difference is as follows:

**APPLICATION OF RULES.** *In external inference, the Asymmetry Principle is applied to* **intra-record** *asymmetries (asymmetries within the record of a state). In internal inference, the Asymmetry Principle is applied to* **inter-record** *asymmetries (asymmetries between the records of states).*

We are now going to describe one of the most powerful laws in the New Foundations. To do so, we first have to introduce one of the basic types of groups that were invented in the New Foundations:

**Definition 2.3.** *The New Foundations define an* **iso-regular group** *to be a group that satisfies the following three conditions:*

*(1) It is an $n$-level wreath product, $G_1 Ⓦ G_2 Ⓦ \ldots Ⓦ G_n$; i.e., a structure of hierarchical transfer.*
*(2) Each level $G_i$ is generated by a single generator (i.e., it is either a cyclic group or a connected 1-parameter Lie group).*
*(3) Each level $G_i$ is an isometry group on its space of action.*

In fact, two examples of iso-regular groups were given earlier in this chapter:

Square: $\mathbb{R}$ Ⓦ $\mathbb{Z}_4$

Cylinder: $SO(2)$ Ⓦ $\mathbb{R}$

The reader can easily see that they are iso-regular groups by comparing their structure with each of the three properties listed in Definition 2.3.

Given the sequence of definitions in this section, we are now ready to state a fundamental law from the New Foundations, a law that turns out to be one of the most powerful laws of memory storage:

**EXTERNALIZATION PRINCIPLE**

**To maximize recoverability, any generative sequence, inferred by external inference, must lead back to a starting state whose internal structure is given by an iso-regular group.**

As an example, consider the psychological experiment where subjects were presented with a rotated parallelogram, and inferred its past state to be a square (Fig 2.8 page 18). We have seen that this is an example of *external* inference. Observe that the inferred starting state, the square, is given by the iso-regular group $\mathbb{R}$ Ⓦ $\mathbb{Z}_4$. Thus the inference, made by the subjects in the experiment, accords with the Externalization Principle; i.e., the starting state, inferred by the external inference, is given by an iso-regular group.

As another example, consider the case of the bent cylinder. Again, we note that the fact that it is called "bent" means that one infers its starting state to have been a straight cylinder. We have seen that this is an example of *external* inference. Observe that the inferred starting state, the straight cylinder, is structured by the iso-regular group $SO(2)$ Ⓦ $\mathbb{R}$. Therefore, the inference again accords with the Externalization Principle; i.e., the starting state, inferred by the external inference, is given by an iso-regular group.

## 2.7 Algebraic Theory of Inheritance

The term *inheritance*, in object-oriented programming, refers to the passing of properties from a parent to a child. The child incorporates these parent properties, but also adds its own [30].

This kind of structure covers two types of situation. The first is class inheritance, which is a static software concept. The second is a type of dynamic linking created at run-time. My book *A Generative Theory of Shape* (Springer-Verlag) gives an algebraic theory of both types of inheritance; but in the present book we will require the theory of only the second type of inheritance.

This *run-time* created inheritance is fundamental to all computer-aided design, assembly, robotics, animation, etc. A typical example is a child object inheriting the transform of a parent object, and adding its own. For example, in architectural CAD, a door is defined as a child of a wall, and moves with the wall if the designer decides to change the position of the wall. However, the door can also open and close with respect to its attached position in the wall. This means that the door inherits the movement of the wall, but adds its personal movement with respect to the latter. In Leyton [24], I refer to this type of inheritance as *object-linked inheritance*.

The mathematical theory of object-linked inheritance, in the New Foundations, states this:

**ALGEBRAIC THEORY OF OBJECT-LINKED INHERITANCE.** *Object-linked inheritance arises from a wreath product:*

$$\begin{array}{ccc} \textit{Parent} & \longleftrightarrow & \textit{Control group} \\ \textit{Child} & \longleftrightarrow & \textit{Fiber group} \end{array}$$

Notice that this means that the basis of inheritance is *transfer*. The enormous power of this theory is that it explains inheritance in all of CAD, robotics, assembly, animation, etc. This is fully elaborated in my book *A Generative Theory of Shape* (Springer-Verlag).

Now let us return to the architectural example, where the door is a child of the wall. This example can be extended further: The designer can design a door-handle and attach it to the door. Standardly, the designer would define the door-handle as a child of the door. Thus, the door-handle would move with the door when the latter is opened or closed. However, the door-handle could also rotate with respect to its attached position in the door. This means that the door-handle would inherit the movement of the door, but also add its personal movement with respect to the latter. Notice that, because the door-handle is a child of the door, and the door is a child of the wall, the door-handle would also be inheriting the movement of the wall, if the designer decided to move the wall.

It will be helpful for the reader to consider Fig 2.10, which illustrates the diagrammatic method used in the animation program *3D Studio Max* to represent an inheritance hierarchy. Here, inheritance is represented by indentation – i.e., an indented object is a child of the next object above with respect to which it is indented. Each object, except the World object, has a transform shown just below it. The transform relates the coordinate

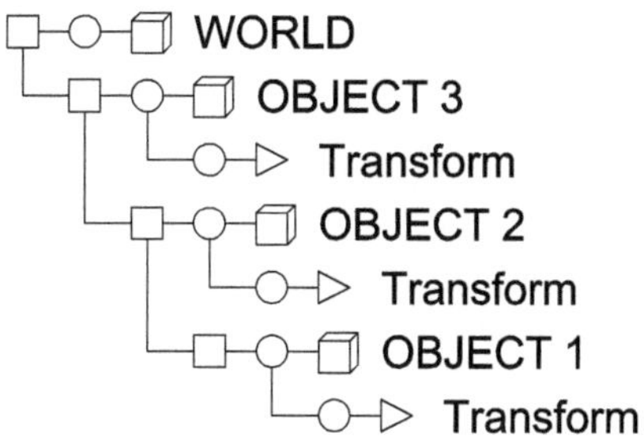


**Fig. 2.10.** The representation of parent-child relations in a program like *3D Studio Max.*

frame of the object to the coordinate frame of its parent. This transform is the "personal" transform of the object. In addition, the object inherits the transform of its parent. The object therefore adds its personal transform to its inherited transform. This means, of course, that via its parent, the object inherits the transform of its parent's parent, and so on.

Thus, return to the architectural example. In this case, Object 3 in Fig 2.10 would be the wall; Object 2 would be the door in the wall; and Object 1 would be the door-handle in the door. Thus, in the diagram, each of these three objects is shown with its personal transform below it; and the inheritance hierarchy is given by the successive indentations.

The New Foundations define the full group of such structures as follows:

**GROUP OF ENTIRE TRANSFORM STRUCTURE.** *Consider a set of $n+1$ objects: Object 1 to $n$, and the World. Suppose that they are linked such that Object $i$ is the child of Object $i+1$, and Object $n$ is the child of the World. Then the group of the entire transform structure is the wreath product:*

$$ {}_{F_1}G_1^{F_2} \circledR \; {}_{F_2}G_2^{F_3} \circledR \; \ldots \; \circledR \; {}_{F_n}G_n^{W} $$

*where:*

*(1) Object $i$ has personal transform group $G_i$ and frame $F_i$.*
*(2) Personal transform group $G_i$ relates frame $F_{i+1}$ of the parent, upper index, to the personal frame $F_i$, lower index. (The world frame $F_{n+1}$ is written as $W$.)*

Clearly this statement gives a new algebraic theory of *relative motion*, which is explained in detail in my book *A Generative Theory of Shape* (Springer-Verlag).

## 2.8 Complex Shape

The New Foundations are fundamentally a theory of *complex* shape. It is now necessary to understand a major way in which the New Foundations represent complex shape.

Consider the main problem in providing a generative theory of complex structure. As stated in section 2.3, according to the New Foundations, recoverability is possible only if the generative operations are symmetry-breaking. Notice that, in the Standard Foundations to Geometry, this would mean that, as one proceeds forward in the generative sequence, the symmetry group of the structure would quickly reduce to nothing. Thus, there would be a loss of algebraic information. In the New Foundations, this problem is solved using an entirely opposite theory of symmetry-breaking, as described in section 2.5. In this theory, the group describing the symmetric past state is actually increased in the symmetry-broken state. This is done by making it the fiber group of a wreath product in which the control group is the group of the asymmetrizing action. Thus, in using this wreath product, the group of the past state is *transferred* onto the symmetry-broken state.

In the previous sections, we have seen how this theory of symmetry-breaking gives a formulation of *deformation* that does not have the problems of the Standard Foundations to Geometry. That is, as illustrated by the examples of deforming a square into a parallelogram and deforming a straight cylinder into a bent one, we saw that, according to the New Foundations, the inferred symmetric past state is given by what the New Foundations call an iso-regular group, and the deformation is modeled by extending the iso-regular group, as a fiber group, by the deforming group, as a control group, in a wreath product.

Having shown how our theory of symmetry-breaking solves the problems of the Standard Foundations in modeling *deformation*, we now need to show how our theory solves the problems in modeling *concatenation*. Consider Fig 2.11. Each of the two objects *individually* has a high-degree of symmetry. However, the *combined* structure, shown, looses much of this symmetry. Thus, the Standard Foundations would encode the concatenated structure by a reduced group. In contrast, the New Foundations develops the opposite kind of group theory. In this, the group of the concatenated structure not only preserves the symmetry groups of the individual objects, but adds the extra information of the concatenation.

This is how the New Foundations proceed: The generative history starts with the two independent objects, and therefore the symmetry of this starting situation is given thus:

$$G_{cylinder} \times G_{cube}$$

which is the *direct product* of the *iso-regular groups*, $G_{cylinder}$ and $G_{cube}$, that describe the two independent objects.

Now, by the Maximization of Transfer, the starting group, i.e., this direct product group, must be transferred onto subsequent states in the generative history, and therefore it must be the fiber of the wreath product in which the control group creates the subsequent generative process.

Let us take the control group to be the *affine group* $AGL(3, \mathbb{R})$ on three-dimensional real space.The full structure, fiber plus control, is therefore the following wreath product:

$$[G_{cylinder} \times G_{cube}] \circledw AGL(3, \mathbb{R})$$

Now, it is necessary to fix the *group representation* of this wreath product. First, by our theory of recoverability, the control group must have an asymmetrizing action with respect to the fiber. Thus proceed as follows: The particular fiber-group copy

$$[G_{cylinder} \times G_{cube}]_e$$

corresponding to the identity element $e$ in the affine control group, i.e., the starting state, must be the most symmetrical configuration possible. This exists only when the cube and the cylinder are coincident, with their symmetry structures (axes, etc.) *maximally aligned*. The New Foundations call this configuration the **alignment kernel**.

Next, choose one of the two objects to be a reference object. This will remain fixed at the origin of the world coordinate frame. Let us choose the cube as the referent. Given this, now describe the action of the affine control group as providing an affine motion of the cylinder *relative* to the cube. Each fiber-group copy

$$[G_{cylinder} \times G_{cube}]_g$$

for some member $g$, of the control affine group, is therefore an arrangement of this system. In fact, any fiber copy will be called a **configuration** of the system. For example, Fig 2.11 corresponds to a configuration. The crucial concept is this: The role of the affine control group is to **transfer configurations onto configurations**.

The wreath product we have presented:

$$[G_{cylinder} \times G_{cube}] \circledS AGL(3, \mathbb{R})$$

gives the *complete* symmetry group of the concatenated situation. It has all the internal symmetries of the objects individually, as well as their relationships.

Let us now understand how to add a further object, for example a sphere. First of all, the fiber becomes the following, with the added *iso-regular group* $G_{sphere}$ of the sphere:

$$G_{sphere} \times G_{cylinder} \times G_{cube}$$

Let us define the cube is the referent for the cylinder-sphere pair, and the cylinder is the referent for the sphere.

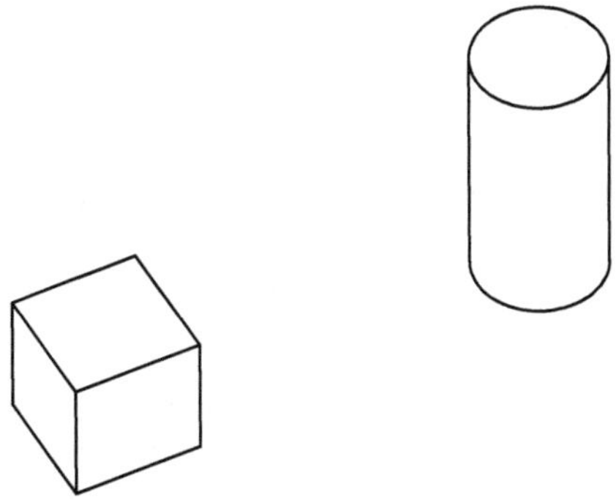

**Fig. 2.11.** Concatenation of cylinder and cube.

Accordingly, there are now two levels of control, each of which is the affine group $AGL(3,\mathbb{R})$, and each of which is added via a wreath product. Thus we obtain the 3-level wreath product:

$$[G_{sphere} \times G_{cylinder} \times G_{cube}] \circledS AGL(3,\mathbb{R}) \circledS AGL(3,\mathbb{R})$$

This is interpreted in the following way: Initially, the three objects (sphere, cylinder, cube) are coincident with their symmetry structures maximally aligned. This corresponds to the fiber-group copy that the New Foundations call the *alignment kernel*. The higher affine group moves the cylinder-sphere pair in relation to the cube. The lower affine group moves the sphere in relation to the cylinder.

Recall that, in the above situation, the cube is fixed at the origin of the world-frame. In fact, in our theory, its symmetries are maximally aligned with the symmetries of the world frame. Now, if we also allow the cube to move with respect to the world-frame, then we add the group $G_W$, defining the symmetries of world frame, into the alignment kernel, and add a third level of control above the two control levels that have already been included, thus:

$$[G_{sphere} \times G_{cylinder} \times G_{cube} \times G_W] \circledS AGL(3,\mathbb{R}) \circledS AGL(3,\mathbb{R}) \circledS AGL(3,\mathbb{R})$$

The new top control level will move the cube-cylinder-sphere triple with respect to the world-frame. Notice that the full control group corresponds to the group of the entire transform structure given on page 29 for our algebraic theory of inheritance. In my book *A Generative Theory of Shape*, there is a chapter devoted to giving an algebraic theory of reference frames, and I show that the appropriate group $G_W$, for the world frame, is the iso-regular group that is the maximal normal subgroup of the hyperoctahedral group.

The crucial point here is that, initially, the four objects (sphere, cylinder, cube, world-frame) are coincident with their symmetry structures maximally aligned. This corresponds to the fiber-group copy that I call the *alignment kernel*. The hierarchy of control groups move these objects hierarchically out of alignment with each other, in correspondence with the inheritance hierarchy.

The above discussion has been illustrating a class of groups which were invented in the New Foundations, called **unfolding groups**. The basic idea is that any complex structure, such as a design in CAD, or a musical composition, is *unfolded* from a maximally collapsed form, which I call the alignment kernel. Two main properties characterize unfolding groups:

**Selection:** The control group acts selectively on only part of its fiber.

**Misalignment.** The control group is symmetry-breaking by misalignment.

The New Foundations show that unfolding groups are a fundamental structure in all complex shape generation. The remainder of this section will give a brief introduction to this. First, it is necessary to give more of the algebraic theory of object-oriented software developed by the New Foundations. Within that theory, there is an analysis of class structure which says that each geometric class consists of (1) an *internal symmetry group*, which gives the correct structure of what are conventionally, but incorrectly, called the invariants clauses of the software text for the class; and (2) an *external group* consisting of command operations, such as deformations, specified in the feature clauses of the class text.

A principal claim of the theory of software, in the New Foundations, is that the relation between the internal symmetry group and the command structure, in the software text, is a wreath product, thus:

$$G_{sym} \circledR G(C)$$

where $G_{sym}$ is the internal symmetry group and $G(C)$ is the group of command operations. Thus *transfer* is made to be the basic structure of a class.

This solves major well-known problems in object-oriented software. For example, what is standardly called an invariance clause of the class which is known as the *square* – the specification that it has four equal sides – is violated by the stretching deformation operator applied to the square. This illustrates the fact that object-oriented software does not accord with the Standard Foundations to Geometry, but instead accords with my New Foundations to Geometry: That is, as illustrated earlier in this chapter, according to the New Foundations, the internal group of a square is *transferred* onto the stretched version, a rectangle. The way that this corrects major problems in software-engineering is shown in detail in my paper *Interoperability and Objects* [24].

Now let us turn to *cloning* in object-oriented programming. It is important to notice that, when one clones an object, one is producing a copy with the same instance values. This means that one is essentially creating a copy that is *aligned* with the original, as can be seen in such programs such as 3D Studio Max and Viz. The copy can then be manipulated via its command operations, which will pull the clone out of alignment, i.e., break the symmetry of the object-clone pair. For example, a designer might wish to clone a cylinder that already exists in the design. The clone command creates an exact copy of the cylinder that is completely coincident with the existing one, and, after its creation, the designer will then move the copy out of alignment with the existing one.

Now let us consider a fundamental process in design. Consider first mechanical CAD, which is the design process in mechanical engineering – forming the basis, for example, of the aerospace and automotive industries. It is generally accepted that mechanical CAD proceeds by a process called *feature attachment*. This is the process of the successive addition of structural units and components.

The New Foundations give a mathematical theory of feature attachment in mechanical design, and also shows that it is mathematically equivalent to other design processes such as musical composition. The basic proposal is this:

**THEORY OF FEATURE ATTACHMENT**

**When one creates objects and attaches them in the design structure, one is entering new instances into the alignment kernel, and positioning the command group for each new instance in the appropriate wreath level within the unfolding group corresponding to the inheritance hierarchy of the structure.**

If the reader wishes to see how this shows that mechanical CAD is equivalent to musical composition, then the reader should consult my paper *Musical Works are Maximal Memory Stores* [23], which fully explains this equivalence.

# 3

# Process Grammar

## 3.1 Curvature Variation as Memory Storage

The Process Grammar is a *component* of the New Foundations to Geometry. Let us first explain what this means, using the following sequence of statements:

**THE ROLE OF THE PROCESS GRAMMAR**

**In the New Foundations to Geometry:**
**Geometry is the Mathematical Theory of Memory Storage.**

**According to the First Fundamental Law of Memory Storage,**
**in the New Foundations:**
**All asymmetries are memory stores.**

**The different components of the New Foundations are**
**the application of the laws of memory storage to different asymmetries.**

**The Process Grammar is one such component of the New Foundations.**

**The asymmetry used as memory storage in the Process Grammar is**
*curvature variation.*

To illustrate, consider the smooth closed curve shown in Fig 3.1. This may be the outline of an embryo, tumor, cloud, etc. Observe that the curvature, i.e., the amount of bend, *varies* around the curve. According to the New Foundations, the variation

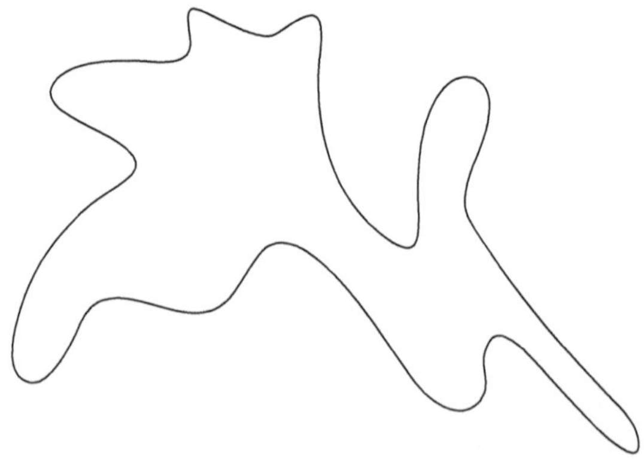

**Fig. 3.1.** According to the New Foundations to Geometry, the curvature variation is memory storage of a past process-history.

in curvature is *memory storage* of the processes that created the shape. The Process Grammar applies the laws of memory storage, from the New Foundations, to *recover* this past process-history.

Using these laws, the Process Grammar shows that a significant aspect of the recovery of process-history from curvature variation is given by the curvature extrema. The following sections will show how the New Foundations to Geometry explain this. We shall begin by considering planar curves, but will later see that the concepts are easily extendable to surfaces.

## 3.2 Differential Symmetry Axes

By our fundamental laws of recoverability (page 16), symmetry has an important role with respect to the recovery of process-history. With respect to curvature variation as memory storage, we shall see that it is important to be able to define reflectional symmetry with respect to complex shape.

Defining reflectional symmetry on a simple shape is, of course, trivially easy, as follows: Consider the triangle shown in Fig 3.2. It is a simple shape. One establishes symmetry in this shape merely by placing a mirror in such a position that it reflects one half of the shape onto the other. The line, along which the mirror lies, is called the symmetry axis. It is shown as the vertical dashed line in the figure.

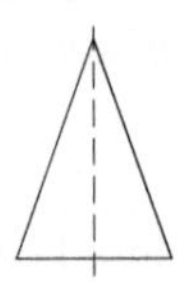

**Fig. 3.2.** A simple shape having a straight mirror symmetry.

In contrast, consider a complex shape like that shown earlier in Fig 3.1 (p36). We cannot place a mirror on it so that the mirror will reflect one half onto the other. Nevertheless, we shall now see that such a shape does contain what I call a *differential* form of reflectional symmetry, as follows:

Consider the two smooth curves, $c_1$ and $c_2$, shown in Fig 3.3. Observe that one cannot take a mirror and reflect one curve onto the other. For example, the upper curve is more curved than the lower one. Therefore a mirror would not be able to send the upper one onto the lower one.

However, now consider the next figure, Fig 3.4, which shows the same two curves. In this figure, a particular point A was selected on the upper curve. Note the following crucial property: There is a point B, on the lower curve, where the *tangent* at B is reflectionally symmetric to the *tangent* at A. The mirror, about which these two tangents are symmetric, is shown as M in the diagram. The fact that, with the same mirror M, there is no neighborhood of the curve, at A, that is reflectionally symmetric to a neighborhood of the curve, at B, means that this reflectional symmetry is limited only to the tangents at the respective points. This is the kind of symmetry that, in Leyton [16], I named **differential symmetry**. I argue that the term is appropriate because differential structures concern tangent properties of manifolds, and these properties correspond to what, in conventional calculus, one calls *infinitesimal* properties. Thus the differential symmetry that exists between points A and B can be re-phrased by saying that, at points A and B, the curves are *infinitesimally* symmetric. The mirror M, that reflects the two tangents onto each other, will be called a **differential mirror**.

Now observe that, because there are no actual neighborhoods of A and B that are symmetric about the mirror M, the instant we move from point A along the upper curve, the mirror will rotate in order to capture a differential symmetry between subsequent

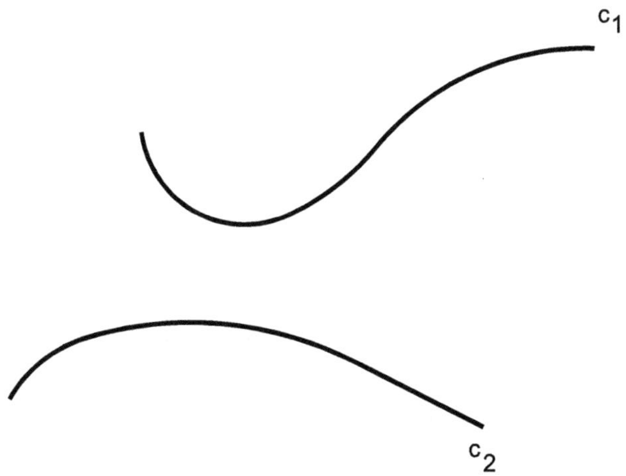


**Fig. 3.3.** How can one construct a symmetry axis between these to curves?

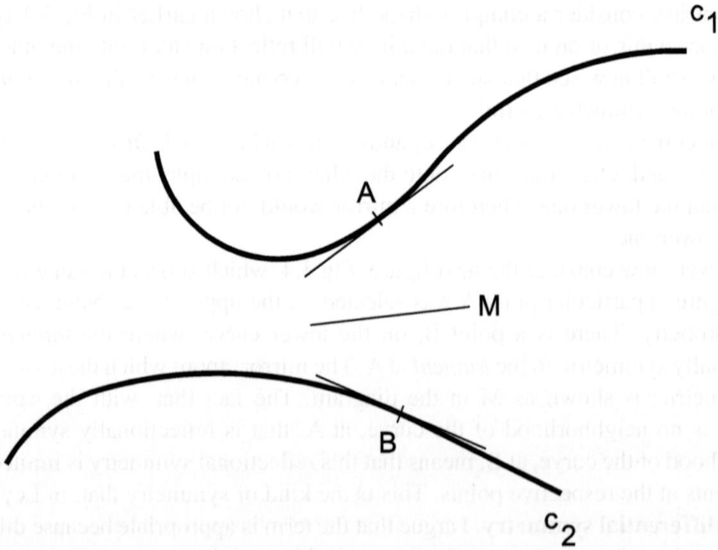


**Fig. 3.4.** A differential symmetry exists between points A and B.

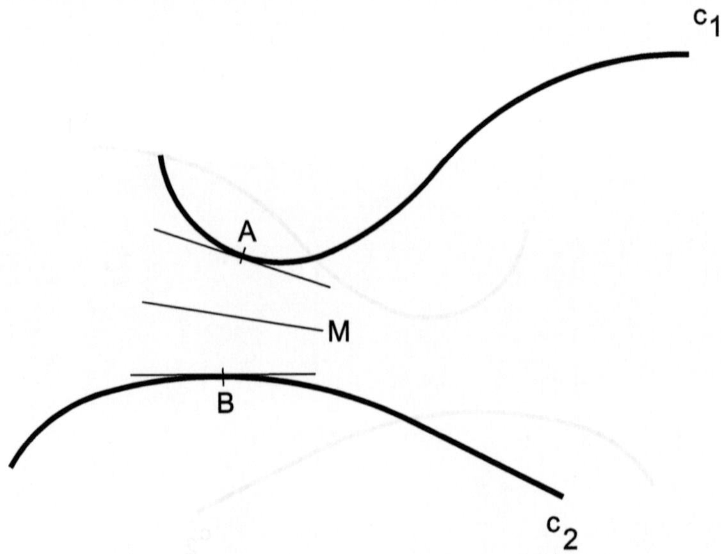


**Fig. 3.5.** A differential symmetry exists also between another pair of points A and B.

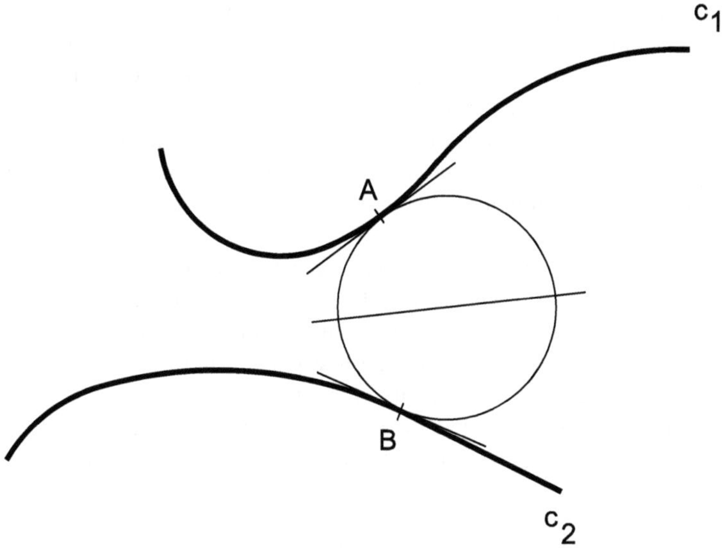


**Fig. 3.6.** A differential symmetry can be discovered by using a bitangent circle.

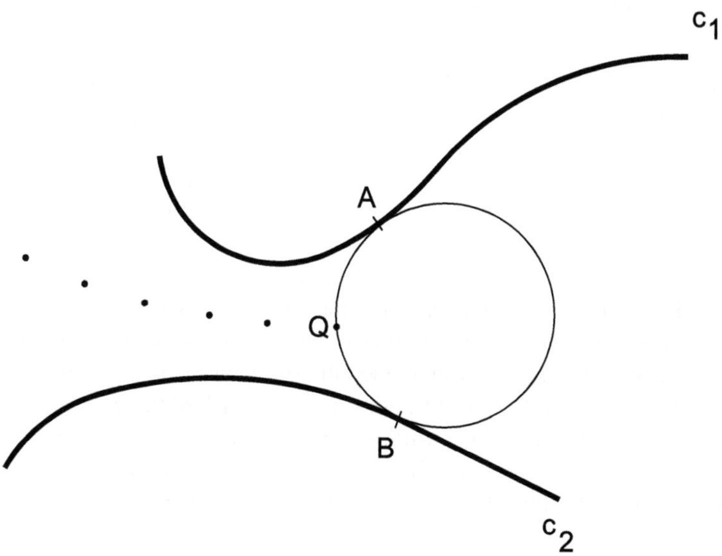


**Fig. 3.7.** The points Q define the PISA differential symmetry axis.

pairs of points. For example, consider another point A on the upper curve, as shown in Fig 3.5. Its tangent is reflectionally symmetric to the tangent at another point B on the lower curve, about the mirror labeled M in that diagram. Observe that this mirror is a different orientation from the previous mirror.

Note that, as we moved from the first pair of points, A and B, to this second pair of points, A and B, the mirror must have been smoothly rotating, and capturing the successive differential symmetries that exist along the curve-segments between these two pairs.

Now, there is a convenient way of picking out the pairs of points, A and B, that are differentially symmetric. This is done by inserting a circle between the two curves such that the circle is tangential to the two curves simultaneously. The circle is called a *bitangent* circle. For example, given the first pair, A and B, that we discussed, the corresponding bitangent circle is shown in Fig 3.6.

The next thing to observe is that, on a mirror, any point can be an incidence point at which a light ray can define the reflection. This lead me to the following definition:

**DIFFERENTIAL SYMMETRY AXIS**

**In Leyton [16], I defined a *differential symmetry axis* to be a rule-determined selection of an incidence point from each of the successive differential mirrors.**

**In order to fully understand the concept of a differential symmetry axis, the reader should read pages 376 - 410 in my book *Symmetry, Causality, Mind* (MIT Press, 1992), which explains this concept in detail.**

In the particular differential symmetry axis, which I invented in Leyton [18], the chosen incidence point is the point $Q$ shown in Fig 3.7, which is on the circle, half way between the two tangent-points $A$ and $B$. Notice that it is on the differential mirror. Thus, the trajectory of differential mirrors yields a trajectory of points Q, as indicated by the dotted line in Fig 3.7. I call this axis the **Process-Inferring Symmetry Analysis (PISA)** axis.

Notice that this axis contrasts with the two other symmetry axes which I have also classed as differential symmetry axes: (1) the Medial Axis (MA) of Blum [1], who defined the symmetry axis as the trajectory of circle centers, and (2) the Smooth Local Symmetry (SLS) of Brady [2], who defined the symmetry axis as the trajectory of the chord midpoints between the tangent-points A and B. Each of the three axes, MA, SLS, and PISA, captures different *configurational* properties of curves. My argument is that PISA is more appropriate for capturing the *application of forces* and *processes* to the *boundary* of an object. This will be shown throughout the book. It should also be noted that I have defined a number of *formats* for PISA-axes: those that are calculated *directly* from the presented shape, and those that are *transferred* from versions of the shape (e.g., past versions) inferred by additional rules from my New Foundations to Geometry. I call these two types of formats, the *direct* and *transferred* PISA-axes.

## 3.3 Symmetry-Curvature Duality

As we have said, according to the fundamental laws of recoverability (page 16), in the New Foundations to Geometry, symmetry has an important role with respect to the recovery of process-history. In relation to this, we shall see that curvature extrema are fundamentally important.

In the 1980s, I proved a theorem which shows that there is a very strong relationship between differential symmetry axes and curvature extrema – in fact, a duality. I called the theorem, the Symmetry-Curvature Duality Theorem. This is the theorem which, in Chapter 1, was called the Fundamental Theorem of Biological Morphology, because of the enormous number of biological areas to which scientists have applied the theorem. Furthermore, as we saw, the theorem has also been applied by scientists in many other disciplines, including meteorology, geology, chemical engineering, computer-aided design, etc.

**SYMMETRY-CURVATURE DUALITY THEOREM.**
**Leyton (1987)**

**Any segment of a smooth planar curve, with one and only one curvature extremum, has one and only one differential symmetry axis of the type being used.**

**Furthermore, the PISA axis terminates at the extremum.**

**Also the SLS axis terminates at the extremum**

**However, the Medial axis does not.**

To illustrate this theorem, consider the curve shown in Fig 3.8. It can be understood as a segment of a much larger curve. The segment shown here has only one curvature extrema, the point at the far left.

The curve segment shown can be understood as the beak of a bird, the extremum being the sharpest point on the beak.

The question to be asked is this: How many symmetry axes does this curve segment possess? The above theorem gives us the answer. It says: Any curve segment with only one curvature extremum has only one differential symmetry axis, of the type chosen. In Fig 3.8, the type of differential symmetry axis chosen was the PISA type. The theorem proves that, because this curve segment has only one extremum, it has only one PISA axis. Similarly, if the chosen type of differential symmetry axis was a SLS axis or a Medial axis, the curve segment would have only one of each of those types.

The next question to be asked is this: Where does this symmetry axis go? Could it, for example, hit the upper side or lower side of the curve segment? Again, the theorem provides us with the answer. It says that the PISA axis is forced to terminate at the

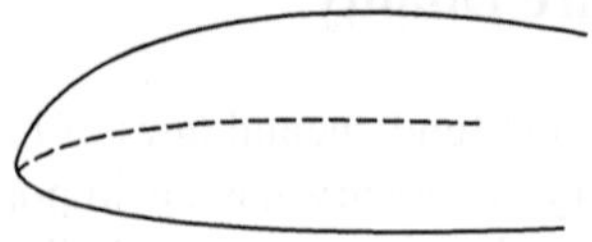

**Fig. 3.8.** Illustration of the Symmetry-Curvature Duality Theorem on the beak of a bird.

curvature extremum, i.e., the tip of the beak, as shown in Fig 3.8. The same would be true of the SLS axis, but not the Medial axis.

A crucial part of my proof of the theorem is this. Note that a *spiral* is a segment of curve with only increasing, or only decreasing, curvature. The curve on each side of the extremum is a spiral.

**SPIRAL THEOREM**

**In Leyton [16], I proved a theorem that** *a spiral cannot have a differential symmetry axis*. **This has important consequences for a curve segment that contains only one extremum: It forces any two symmetry-related tangents of the curve segment to come from the two opposite spiral sides of the extremum rather than the same spiral side. This also proves that a differential symmetry axis cannot terminate on either side of the extremum.**

Now, the Symmetry-Curvature Duality Theorem gives an enormously valuable understanding of the structure of any complex curve. This is done by what I call the **decompositional procedure**: Break down the curve into segments, each with only one curvature extremum. The theorem then tells us that each of these segments has only one symmetry axis, and that the axis terminates at the extremum.

Fig 3.9 illustrates this decompositional procedure. The curve has sixteen extrema. Thus, the theorem says that there must be sixteen symmetry axes associated with and terminating at those extrema. These axes are shown as the dashed lines on the figure.

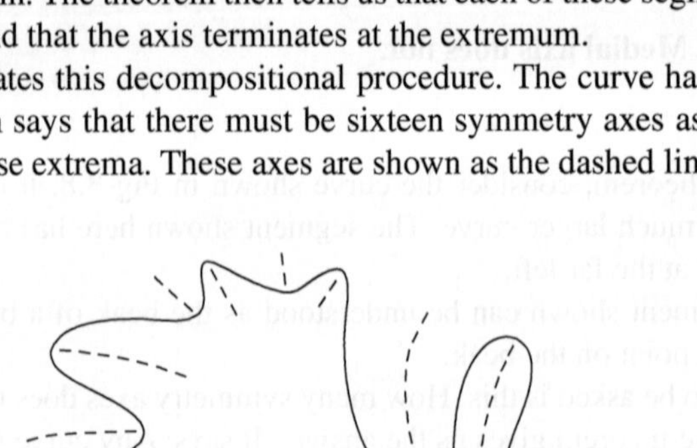

**Fig. 3.9.** By the theorem: sixteen extrema imply sixteen symmetry axes.

## 3.4 Curvature Extrema and the Symmetry Principle

Recall that the problem that the New Foundations defines as solving is this: When presented with a shape like Fig 3.1, how can one recover from it the process-history that produced it? On page 16, we gave the two fundamental inference rules that the New Foundations give for the recovery of process-history. These rules are the Asymmetry Principle, which states that asymmetries in the present shape should be assumed to have arisen from past symmetries; and the Symmetry Principle, which states that symmetries in the present shape should be assumed to have always existed. Both principles must be applied to the shape. Let us first use the Symmetry Principle.

The Symmetry Principle requires that one must preserve symmetries in the shape, backwards in time. What are the symmetries? The previous section established significant symmetries in the shape: the symmetry axes illustrated in Fig 3.9, predicted by the Symmetry-Curvature Duality Theorem; i.e., those axes corresponding to the curvature extrema.

Applying the Symmetry Principle requires that the symmetry axes must be preserved backwards in time.

There is, in fact, only one way that this preservation of axes can be accomplished: As one runs time backwards, the past processes must move backwards along the axes. Therefore, this implies that, in the forward-time direction, the processes must have moved forwards along the axes. Thus, according to the New Foundations:

**INTERACTION PRINCIPLE**
**Leyton (1984)**

**The symmetry axes in the present are the directions along which past processes are most likely to have acted.**

This rule is analyzed in depth in my book *A Generative Theory of Shape*. With respect to the Process Grammar, this rule is very significant when combined with the considerations of the next section.

## 3.5 Curvature Extrema and the Asymmetry Principle

According to the New Foundations, the recovery of process-history from shape requires that one apply *both* the Symmetry Principle *and* the Asymmetry Principle (page 16). The previous section applied the Symmetry Principle. The present section applies the Asymmetry Principle.

The Asymmetry Principle states that asymmetries in the present must be understood as having arisen from past symmetries. It is now necessary to define the first asymmetry that is basic to the Process Grammar. To understand it, let us look at a curve such as the outline of the human hand, shown in Fig 3.10. The purpose is to examine the curvature at different points along the curve.

It is important to fully appreciate this, as follows: Consider the hand outline, shown in Fig 3.10, as parameterized by arc length, starting at the far right, and moving towards the left. It is clear that, as one moves along the curve, the curvature, i.e., the rate of rotation of the tangent, undergoes considerable variation. For example, observe that, in going through point A, the tangent rotates anti-clockwise, but not very much; i.e., the curvature is not large here. However, observe that, in going through point B, although the tangent also rotates anti-clockwise, it rotates considerably faster than at point A; i.e. the curvature at B is much larger than at A. But then, in going through point C, the tangent does not rotate at all, i.e., there is no curvature at C. Then, in contrast, in going through point D, the tangent rotates considerably, and in fact clockwise, in contrast to points A and B where the rotation was anti-clockwise.

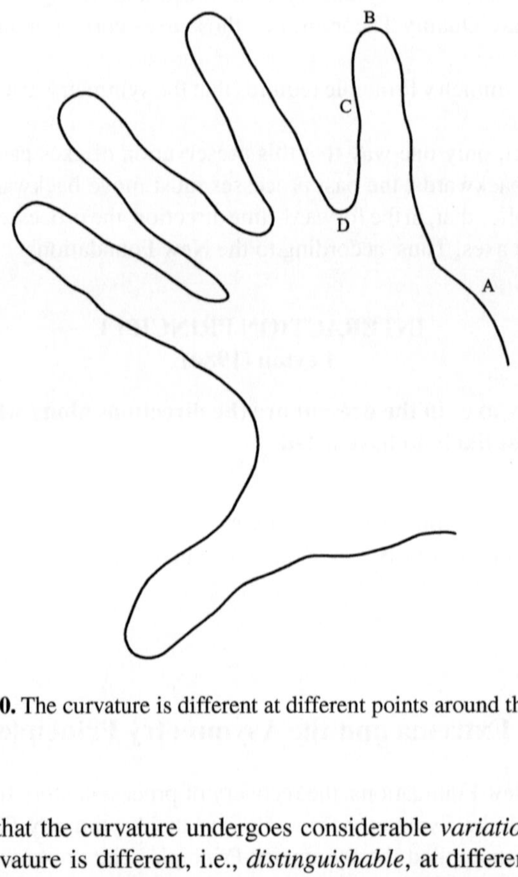


**Fig. 3.10.** The curvature is different at different points around this curve.

Thus we see that the curvature undergoes considerable *variation* along the curve. Therefore the curvature is different, i.e., *distinguishable*, at different points along the curve. The reader will recall that the term *asymmetry* really means distinguishability. The distinguishability in curvature, at different points along the curve, is one of the basic asymmetries of the Process Grammar, as follows:

**FIRST FUNDAMENTAL ASYMMETRY**

**The first fundamental asymmetry that is used in the Process Grammar, for the recovery of past history, is curvature variation; i.e., the distinguishability in curvature at different points along a curve.**

Let us now see what happens when one applies the Asymmetry Principle to this distinguishability. The principle says that distinguishability in the present must go back to indistinguishability in the past. This means that the differences in curvature on the curve must be removed backwards in time, leaving a curve in which the curvature is the *same* (i.e., *indistinguishable*) at *all* points along the curve. There is, in fact, only one type of smooth curve in which the curvature is the same at all points: a *circular* curve.

**THE PAST OF ANY SMOOTH CURVE**

**According to the Asymmetry Principle, the past of any smooth curve with curvature variation is a curve that has constant curvature – i.e., a curve that is circular. In particular:**

**The past of any closed smooth curve is a circle.**

Consider an example. We have been looking at the outline of a hand. This curve is part of a closed smooth shape: the outline of the entire body. Now, it is a remarkable fact that the past of the body is circular; i.e., the body grows from a circular egg.

The extraordinary thing is that one need know nothing about biology to arrive at this conclusion. The Asymmetry Principle gives us this conclusion immediately. The Asymmetry Principle removes the need for a biological science. It is a basic argument of my New Foundations to Geometry that the different laws of the different sciences can be replaced by the single set of memory laws given by the New Foundations; i.e., the laws for recovering the past from shape. See the theory of science given in my book *A Generative Theory of Shape* (Springer-Verlag).

## 3.6 General Shapes

In most of the illustrations so far, we applied the inference rules to smooth, closed curves. However, the rules apply equally to any piece-wise smooth curve; i.e., a curve that can contain sharp corners and can be non-closed. As I have shown in my previous books, the past of any such curve is either a (1) circle, (2) straight line or (3) regular polygon. Notice that, like a circle, a straight line has the same curvature at each point (in fact, the curvature is zero at each point). Notice also that, except at the corners, a polygon is straight everywhere, and therefore also has constant curvature everywhere. According to my laws of recoverability, constant curvature is a basic symmetry of past shapes.

## 3.7 The Three Rules

Let us now put together the rules established in sections 3.3 – 3.5, for recovering process-history from curvature variation. There are a total of three rules, as follows.

**Rule 1.**
This is the Symmetry-Curvature Duality Theorem. It says that, to each curvature extremum, there is a unique symmetry axis leading to, and terminating at, the extremum.

As an *inference rule*, this theorem works in the following way: Suppose one is presented with a curve such as that shown in Fig 3.11. Then the first thing one should do is identify the curvature extrema. The theorem then says that, from each curvature extremum, one can *infer* the presence of a symmetry axis leading to the extremum. Therefore, the use of this rule inserts symmetry axes into the shape, thus producing the diagram given in Fig 3.12.

**Rule 2.**
The second inference rule is the Symmetry Principle, which states that recoverability requires symmetries in the present to be preserved backwards in time. This means, in particular, that any symmetry *axis* must be preserved in the backward history, which means that, in the backward history, the process must run back along the axis. But this means that, in the forward history, the process went forward along the axis. I call this inference, the Interaction Principle.

**Rule 3.**
The final inference rule is the Asymmetry Principle, which states that recoverability requires asymmetries in the present to go to symmetries backwards in time. The asymmetry being considered here is curvature variation, i.e., distinguishability in curvature at the different points along the curve. The Asymmetry Principle implies that this distinguishability must be removed backwards in time. Thus the past is a curve that has the same curvature at each point.

## 3.8 Process Diagrams

Together, the three rules imply that the shape was created by processes that *pushed the boundary along the axes*. For example, the protrusions, in Fig 3.12, were created by pushing the boundary out along the axes, and the indentations were created by pushing the boundary in along the axes. In fact, the processes actually created the isolated curvature extrema. This is a crucial conclusion from the above rules. The original curve, in the past, had no isolated extrema; i.e., although a circular curve, having constant curvature, can be regarded as having an extremum at every point, it has no isolated extrema! On a curve with isolated extrema, the extrema are separated by curvature variation. Thus, according to our three rules, we conclude:

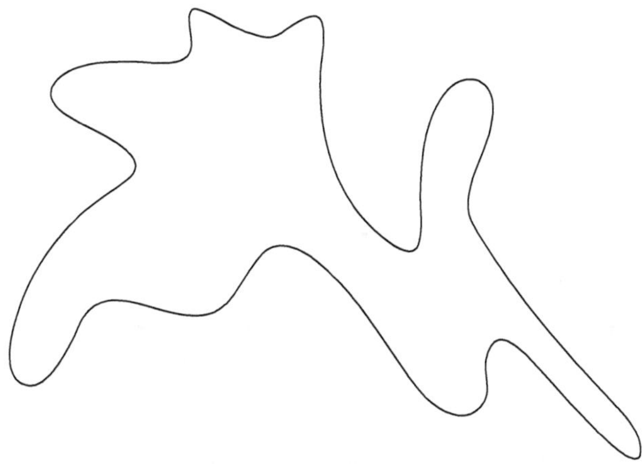

**Fig. 3.11.** A closed smooth shape.

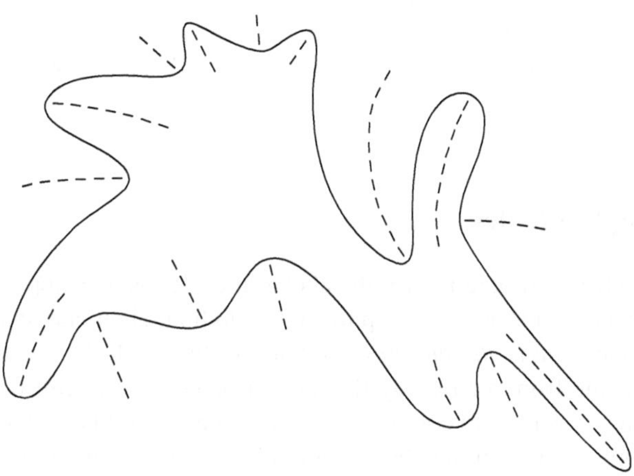

**Fig. 3.12.** The inferred axes.

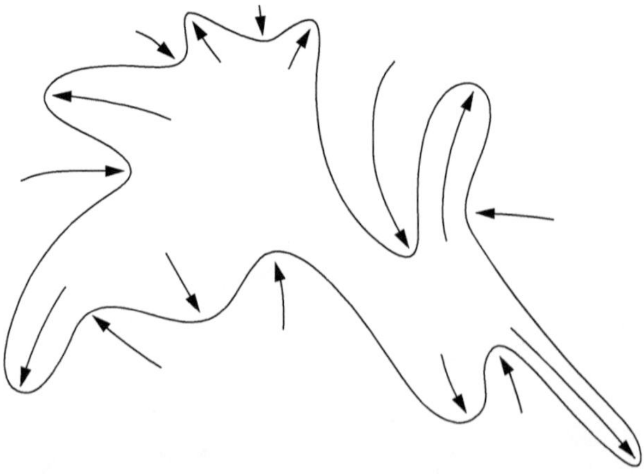

**Fig. 3.13.** The processes inferred by the rules.

**EXTREMA-BASED PROCESSES**

**According to the above three rules: In a curve with isolated extrema, each process (e.g., force) that created the shape, went along a differential symmetry axis and created the isolated extremum at the end of the axis.**

**Each such process will be called an** *extremum-based process.* **According to the New Foundations, such processes are fundamental to biological growth and mechanical design.**

The processes will be represented by putting arrows along the axes, as illustrated in Fig 3.13. The arrows lead to the isolated extrema, and indicate that the extrema were created by the processes. A diagram like this will be called a **process-diagram.**

## 3.9 Applying the Rules

To obtain extensive corroboration for the above rules, let us now apply them to a large catalogue of shapes: all shapes with up to, and including, eight curvature extrema. The catalogue provides *purely the outlines* exhibited in Fig 3.14, 3.15, and 3.16. Most of these outlines come from a paper by Richards, Koenderink & Hoffman [37], and the Process Grammar was used to complete the catalogue. What I have done is taken these outlines and applied to them the above three rules for the recovery of process-history. These rules put the arrows on each shape, indicating how the shapes were formed over time. As the reader can see, the inferred histories accord very strongly with one's sense of how these shapes were formed.

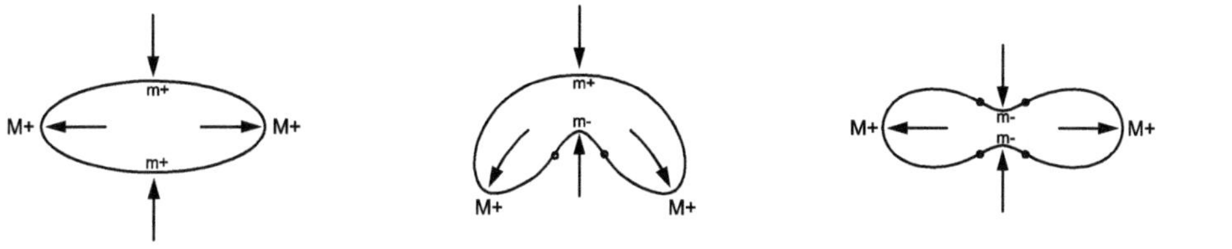


**Fig. 3.14.** The inferred histories on the shapes with 4 extrema.

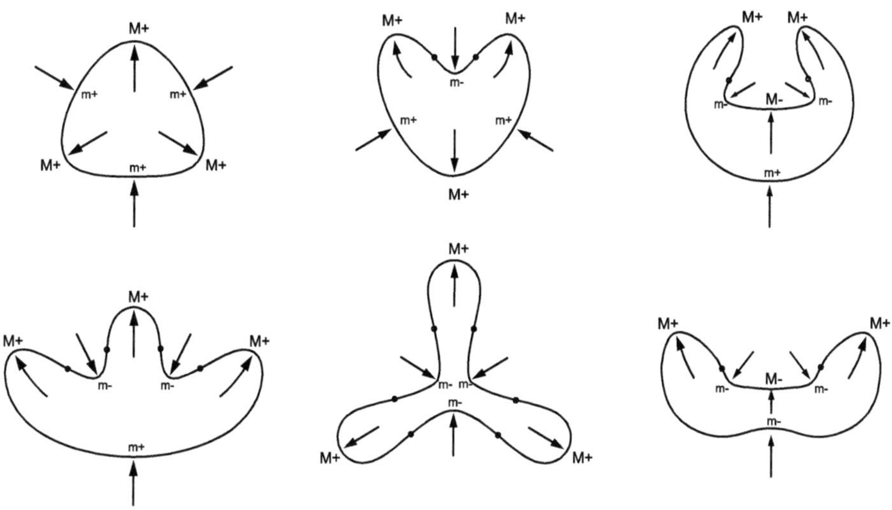


**Fig. 3.15.** The inferred histories on the shapes with 6 extrema.

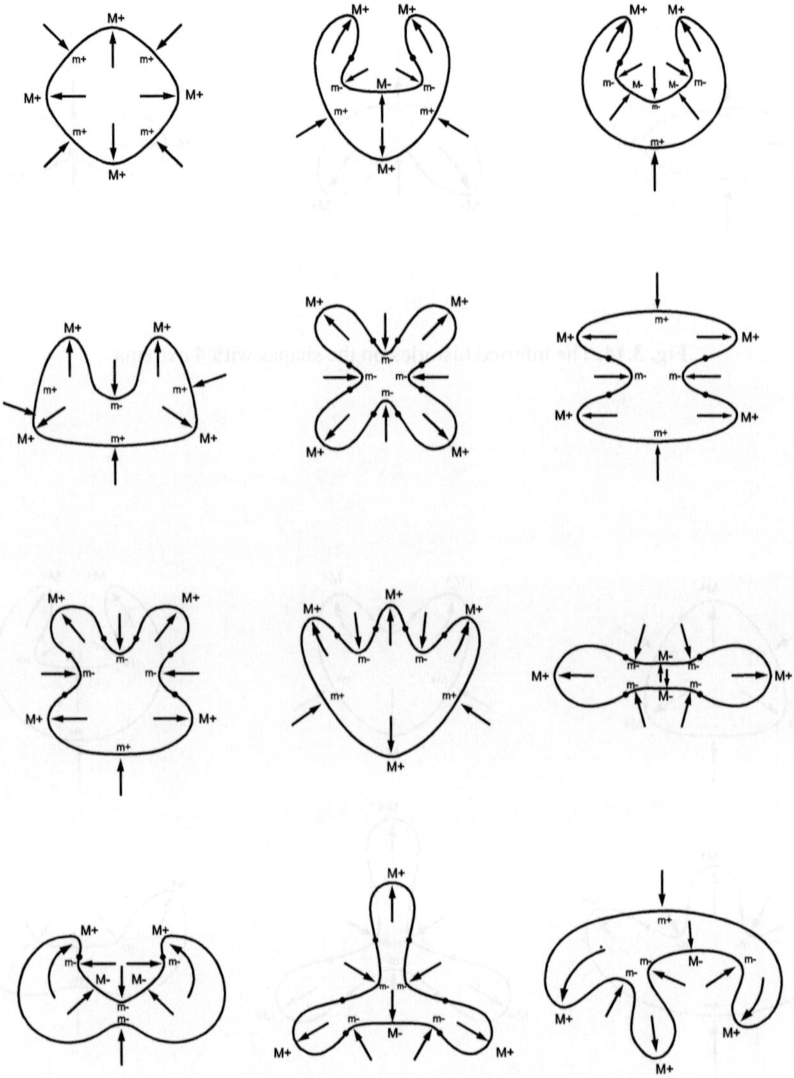


Fig. 3.16. The inferred histories on the shapes with 8 extrema.

## 3.10 How the Rules Accord with the Procedure for Recovering the Past

The three rules are clearly an example of the three-part procedure given on page 17 for recovering the past. The procedure was this:

**PROCEDURE FOR RECOVERING THE PAST.**

**(1) Partition the situation into its asymmetries and symmetries.**
**(2) Apply the Asymmetry Principle to the asymmetries.**
**(3) Apply the Symmetry Principle to the symmetries.**

The correspondence between the three rules and the three parts of the procedure is this: Recall that Rule 1 is the Symmetry-Curvature Duality Theorem, which states that, to each curvature extremum, there is a symmetry axis leading to the extremum. This theorem, in fact, corresponds to part 1 of the above procedure: It gives the partitioning of the shape into asymmetries and symmetries. The particular asymmetries it chooses are defined by the curvature extrema, which decompose the curvature variation into what we can call *fluctuations*, which are violations of the symmetry of constant curvature. The particular symmetries it chooses are the symmetry axes. The theorem describes a partitioning of the shape into asymmetries and symmetries, in this way: It states that, for each unit of asymmetry – i.e., each fluctuation defined by a curvature extremum – there is a unit of symmetry – i.e., a symmetry axis. The Asymmetry Principle then removes the curvature variation, while the Symmetry Principle preserves the symmetry axes.

## 3.11 The Four Extremum-Types

It is important now to observe that, on each of the process-diagrams in Figs 3.14–3.16 (pages 49–50) a letter-label was placed at each extremum (the end of each arrow). There are four alternative labels, $M^+$, $m^-$, $m^+$, $M^-$, and these correspond to the four alternative types of curvature extrema. These are explained as follows:

First, we will understand the curve to be the *boundary* of an object. Correspondingly, we will refer to the object side of the curve as "solid" and the other side as "empty". Next, it is necessary to choose the direction for traveling along the curve; i.e., the direction of the increasing arc-length curve parameter $t$. The direction chosen will be that which keeps the solid on the *left* side of the curve. Then define curvature as the rate of *anti-clockwise* rotation of the tangent. Denote a curvature maximum and minimum by $M$ and $m$, respectively; and denote positive and negative curvature by $+$ and $-$, respectively (i.e., anti-clockwise and clockwise rotation). Then, in the *curvature function* on the curve, the four kinds of extrema are as illustrated by the graph in Fig 3.17.

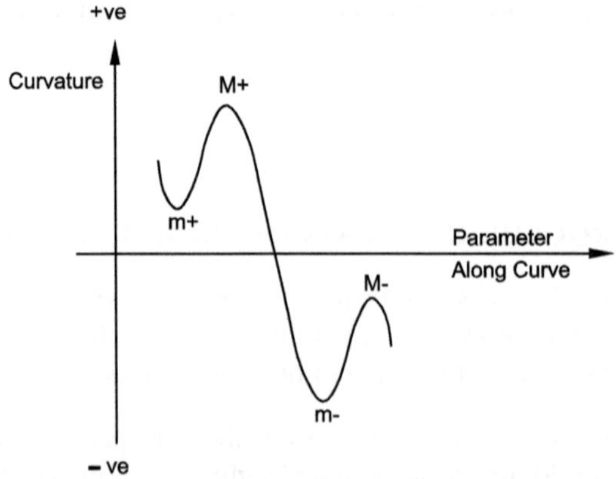


**Fig. 3.17.** A curvature function showing the four types of extrema.

The four types of extrema will be called the four **extrema-types**. While Fig 3.17 has shown how the four types are defined in the curvature function, we now need to understand carefully how the four types look qualitatively on an actual shape. This is illustrated in Fig 3.18. The four types, given by the four labels in the figure, are the *top* points on the four shapes shown. The following qualitative facts should be observed:

The first thing to note about Fig 3.18 is that the *shading* represents the *solid* side of the curve, and the *non-shading* represents the *empty* side of the curve.

Now observe the following: We can see that the *curves* for the first two kinds of extrema, $M^+$ and $m^-$, have the *same shape*. They are the *sharpest points* on their respective curves (the points at the top). Their difference is that they change the side on which the solid (shaded) and empty (non-shaded) occur; i.e., they are *figure/ground reversals* of each other. They will be called *duals* of each other.

Observe also that the *curves* for the other two extrema, $m^+$ and $M^-$, are also the same shape as each other. They are the flattest points on their respective curves (the points at the top). Again, they are *figure/ground reversals* of each other. They will be called *duals* of each other.

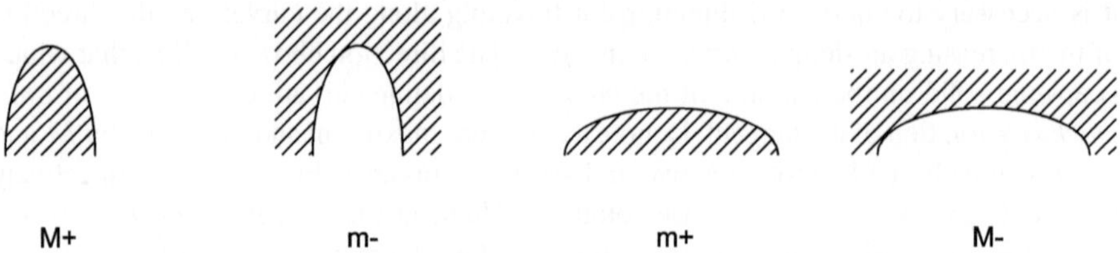


**Fig. 3.18.** The four extremum-types are the top points on the four shapes shown.

## 3.12 The Inferred Process-Types

Since the New Foundations argue that shape is equivalent to inferred process-history, we now come to a fundamental concept of the Process Grammar: the *types of processes* that are inferred from the *four extrema-types*. We will refer to these as the *process-types*.

According to the New Foundations, the process-types are given by the differential symmetry axis invented in the New Foundations, called PISA, that is, *Process-Inferring Symmetry Analysis*. The process-types are not correctly given by the other forms of differential symmetry axes. We will now see the following:

> **While it is easy to understand the differential symmetry axes leading to the first two types of extrema, $M^+$ and $m^-$, the differential symmetry axes leading to the other two types of extrema, $m^+$ and $M^-$, are much more difficult to understand, but must be deeply understood for profound reasons that are totally ignored in the standard literature.**

To begin our analysis, we should first note that, while a differential symmetry axis is sensitive to the shape of the curve, it is independent of which side of the curve is assigned the property of solid or empty. Thus the differential symmetry axis for the first two types of extrema, $M^+$ and $m^-$, must be the same as each other; and the differential symmetry axis for the other two types of extrema, $m^+$ and $M^-$, must be the same as each other.

It is important to notice a crucial fact about the differential symmetry axis for the first two types of extrema, $M^+$ and $m^-$, as follows: Fig 3.19 shows a curve with a $M^+$ or $m^-$ extremum at the top. What one must observe is that the bitangent circles are *between* the two sides of the curve.

Now, since the particular differential symmetry axis that is necessary for the process-inference is PISA, Fig 3.19 shows the PISA symmetry point for each circle as the dot on the circle; i.e., it is the midpoint on the circle arc between the two tangent-points. In accord with the Symmetry-Curvature Duality Theorem, the PISA-axis, i.e., the trajectory of PISA-points, lead to the extremum, as shown.

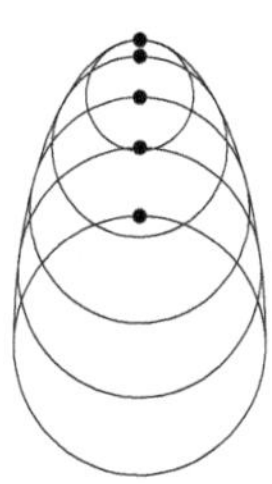

**Fig. 3.19.** The trajectory of bitangent circles, and PISA-points, leading to the top extremum, which is either $M^+$ or $m^-$.

Now let us turn to the remaining two extrema, $m^+$ and $M^-$. To understand the differential symmetry axis leading to these extrema, consider Fig 3.20. The sequence of diagrams in this figure shows a bold curve whose top extremum is either $m^+$ or $M^-$. The sequence illustrates the trajectory of bitangent circles that lead to this extremum. Observe that, in the top diagram, the two tangent-points of the circle are far apart from each other on the curve. Then, in the second diagram, the two tangent-points have moved a little towards each other along the curve. Then, in the remaining sequence of diagrams, the tangent-points continue to move further towards each other. Finally, in the bottom diagram, the two tangent points have converged at the extremum at the top of the curve; i.e., the circle has reached tangency to the extremum.

Fig 3.20 illustrates another theorem that I proved in relation to the Symmetry-Curvature Duality Theorem: Given a $m^+$ or $M^-$ extremum of a curve, there is curve segment that is a neighborhood of the extremum, such that all the bitangent circles to the curve segment are *outside* (exscribe) the curve segment, Leyton [16].

There are many deep consequences of this fact, as will be seen throughout the book. First consider the PISA-axis leading to the extremum. In Fig 3.20, the PISA symmetry point, for each circle, is given by the dot at the top of the circle; i.e., this is the midpoint on the circle arc between the two tangent-points. Observe that, in the succession of circles, the PISA-point moves successively downwards towards the extremum, reaching the extremum in the final circle.

Now let us put together the information we have gained so far in this section about the PISA axes of the four extrema. Fig 3.21 shows the four extrema; and the four *arrows* are the PISA axes for the four extrema. We see the following extraordinary fact: The PISA axes for the first two extrema, $M^+$ and $m^-$, are on the *convex* side of the curve, and the PISA-axes for the other two extrema, $m^+$ and $M^-$, are on the *concave* side of the curve. The reason is this: For each of the first two extrema, $M^+$ and $m^-$, the PISA axis is necessarily *between* the two sides of the curve, because the bitangent circles are *between* the two sides of the curve. For each of the other two extrema, $m^+$ and $M^-$, the PISA axis is *not between* the two sides of the curve, because the bitangent circles are *outside* the curve, and therefore the bitangent circles put the PISA axis on the concave side of the curve.

We can now turn to the fundamental purpose, in the New Foundations, of establishing the PISA-axes for the extrema: the *inference of the processes that created the extrema*. We will demonstrate that the processes went along the PISA-axes leading to the extrema. Therefore the first important conclusion will be that the arrows shown in Fig 3.21 actually give the process-history of the extrema. We can refer to the arrows as the *process-arrows*, and can often think of them as the *forces* that created the extrema.

It is now necessary to define the following concept:

**RECIPIENT SPACE**

**A process-arrow lies on one side of the curve. The space on the other side of the curve will be called** *the recipient space.* **That is, the recipient space is the space towards which the process-arrow points.**

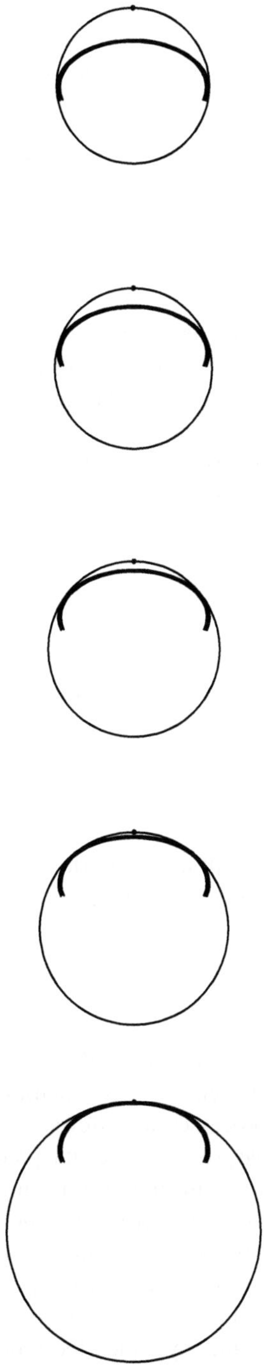

**Fig. 3.20.** The trajectory of bitangent circles, and PISA-points, leading to the top extremum, which is either $m^+$ or $M^-$.

**Fig. 3.21.** The four types of extrema, and their process-arrows inferred by PISA.

Now observe the following facts about the arrows in Fig 3.21. Observe that, because, in the first two cases, $M^+$ and $m^-$, the arrow is between the two sides of the extremum, the action of the arrow has been to *penetrate* the recipient space. We shall therefore refer to these two cases as the *penetrative* cases.

Now consider the other two cases, $m^+$ and $M^-$. Because, in these two cases, the arrow is not between the two sides of the extremum, the arrow has not penetrated the recipient space. In fact, what it has done is to *compress* the recipient space. We shall therefore refer to these two cases as the *compressive* cases.

Accordingly, we will give the following classification of curvature extrema:

**PENETRATIVE vs. COMPRESSIVE EXTREMA**

**The $M^+$ and $m^-$ extrema will be called the *penetrative* extrema, because the inferred processes, given by the PISA axes, caused a penetration of the recipient space.**

**The $m^+$ and $M^-$ extrema will be called the *compressive* extrema, because the inferred processes, given by the PISA axes, caused a compression of the recipient space.**

It is at this inference stage, that we are going to take into account the positions of the solid vs. empty space. Because this creates the distinction between all four extrema, it also creates a distinction between the two penetrative processes, and a distinction between the two compressive processes, as follows:

First consider the *penetrative* processes; i.e., the processes corresponding to the first two extrema in Fig 3.21. Observe, from this figure, the following: Because the penetrative process for the first extremum $M^+$ comes from the solid space, it is a *protruding* process. In fact, the PISA axis gives the *trajectory*, over time, by which the process *protruded* into the recipient space. Next observe that, because the penetrative process for the second extremum $m^-$ comes from the empty space, it is an *indenting* process. In fact, the PISA axis gives the *trajectory*, over time, by which the process *indented* into the recipient space.

Now consider the *compressive* processes; i.e., the processes corresponding to the remaining two extrema in Fig 3.21. Observe, from this figure, the following: Because the compressive process for the third extremum $m^+$ comes from the empty space, its effect is that of *squashing* the recipient space. In fact, the PISA axis gives the *trajectory*, over time, by which the process *squashed* the recipient space. To fully understand the process at the fourth extremum, we need considerations to be explained later. In fact, it will be shown that this process corresponds to *internal resistance*; i.e., resistance within an object. For example, we will see that, in geology, the fourth extremum represents a *bay*, and this process corresponds to a ridge of mountains that *resisted* the invasion of the water.

As a result of the above discussion, we see that, using the PISA-method of defining symmetry, the four extremum-types correspond to four types of processes defined as listed in Table 3.1.

| **EXTREMUM-TYPE** | $\longleftrightarrow$ | **PROCESS-TYPE defined by PISA axis** |
|---|---|---|
| $M^+$ | $\longleftrightarrow$ | **protrusion** |
| $m^-$ | $\longleftrightarrow$ | **indentation** |
| $m^+$ | $\longleftrightarrow$ | **squashing** |
| $M^-$ | $\longleftrightarrow$ | **internal resistance** |

**Table 3.1.** Correspondence between extremum-type and process-type inferred by the PISA axis.

In fact, what we have shown is that the PISA-method gives, to each of the four extrema, a **strongly plausible causal explanation**. Thus the PISA-method fulfills the basic requirement of the New Foundations to Geometry, that *shape is defined by causal explanation*. Throughout this book, we will see that the Process Grammar fulfills this requirement in a powerful way.

The standard literature on shape fails to do this. The reason is that the standard literature fails to make the distinction between penetrative and compressive extrema, and does so because it fails to understand compressive extrema. The New Foundations show that compressive extrema are fundamentally important to understanding shape in biology, geology, meteorology, vehicle design, architecture, paintings, etc. However, the standard literature fails to understand this, and this is because it fails to understand compressive extrema. We will see why, as follows:

The following facts concerning the three types of differential symmetry axes should be observed.

Consider first the Medial Axis, defined by Blum [1]. It is crucial to note that Blum only allowed maximal *inscribed* circles to the shape. This means that he would not have allowed the bitangent circles shown in Fig 3.20, which lead to the top extremum; i.e., these circles *exscribe* (surround) the shape; i.e., they are not *inscribed*. Therefore, we see that Blum could not have identified the $m^+$ or $M^-$ extremum in a shape, and therefore could not have identified compressive extrema. Thus, as I show in Chapter 5, although Blum created his Medial Axis to describe biological shape, his axis fails to describe many of the most important structures in a biological shape.

Now, in the research literature, there are sometimes discussions of the centers of *all* the bitangent circles; i.e., not just the maximally inscribed ones. This set of centers is standardly referred to as the "symmetry set". We will refer to it as the *MA symmetry set* since it is standardly viewed as an extension of Blum's Medial Axis (MA). Nevertheless, we will see that the MA symmetry set fails to give a correct description of compressive extrema. To see this, let us consider the MA symmetry set for Fig 3.20. Observe, that the center of the top circle is on the *convex* side of the curve; i.e., below the curve. Furthermore, the center, of each of the subsequent circles, moves successively *downward*. In fact, the centers of these successive circles are shown as the downward succession of dots in the left-most diagram in Fig 3.22. A crucial fact we see from this diagram is that, for a compressive extremum, the MA symmetry set *does not lead to the extremum* – it moves *away* from the extremum. Another crucial fact is this: Observe that the MA symmetry set of the curve is on the *convex* side of the curve. Later we will see that this fact, together with the fact that the axis moves away from the extremum, makes the MA symmetry set completely wrong for describing fundamental morphological structures.

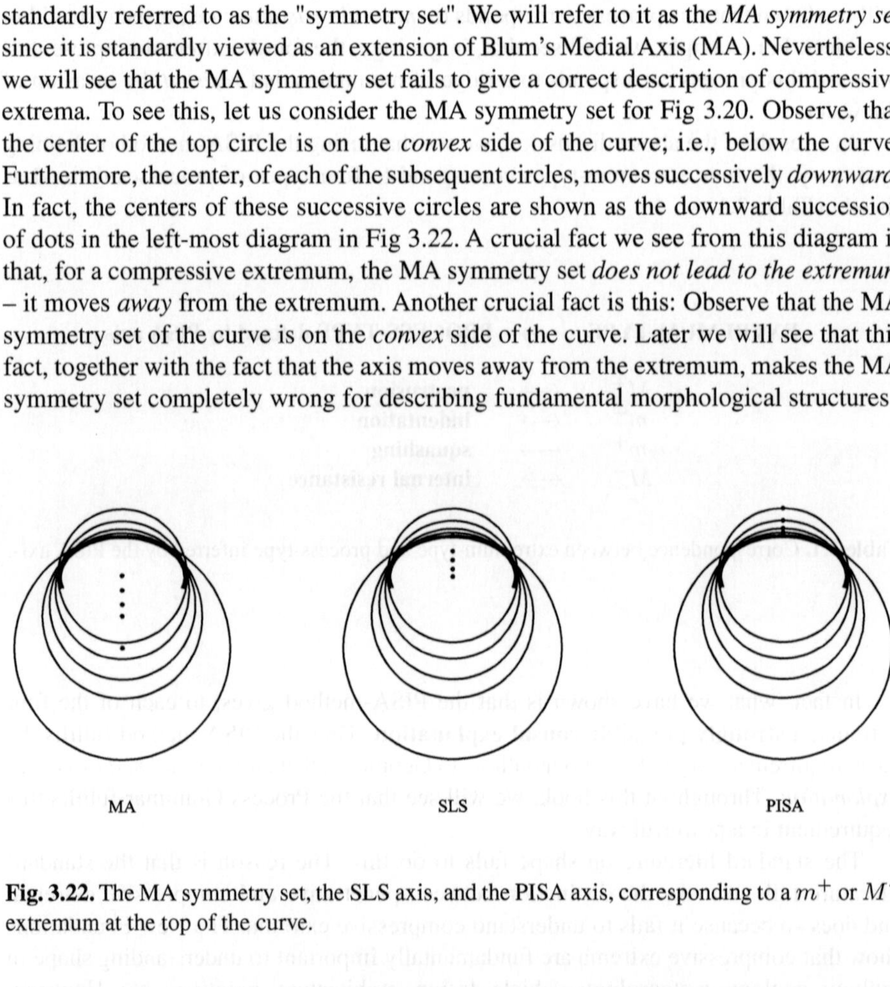


**Fig. 3.22.** The MA symmetry set, the SLS axis, and the PISA axis, corresponding to a $m^+$ or $M^-$ extremum at the top of the curve.

Next consider the Smooth Local Symmetry (SLS) axis defined by Brady [2]. This is defined as the chord midpoints of the bitangent circles. Return to Fig 3.20. Observe that the chord midpoint of the top circle (i.e., the midpoint of the straight line between the two tangent points) is on the *convex* side of the curve. Furthermore, the chord midpoints, given by the subsequent circles, move successively *upward*. Most crucially, a chord midpoint never occurs above the curve. In fact, because the two tangent points, of each of the successive circles, converge, in the final circle, onto the extremum itself, the trajectory of chord midpoints terminates at the extremum, as proved by the Symmetry-Curvature Duality Theorem. Thus, for Fig 3.20, the SLS axis is an upward axis that terminates at the extremum. In fact, the chord-midpoints of these successive circles are shown as the upward succession of dots in the middle diagram in Fig 3.22. The final dot is shown on the curve itself. Therefore, generally, for a $m^+$ or $M^-$ extremum, the SLS axis is on the *convex* side of the curve and leads to the extremum.

Next consider the PISA axis. Return again to Fig 3.20. In this diagram, we saw that the PISA points, i.e., the circle arc mid-points between the tangent points, started from above the compressive extremum and moved down to the extremum; i.e., for a compressive extremum, the PISA axis is on the concave side of the curve. This downward succession of points is also shown in the right-most diagram in Fig 3.22.

Therefore, Fig 3.22 gives a comparison of the differential symmetry axes for what the Process Grammar calls a compressive extremum. That is, the left-most figure shows the MA symmetry set, which is the downward set of dots on the *convex* side of the curve, and leading away from the extremum. The middle figure shows the SLS axis, which is the upward set of dots on the *convex* side of the curve, and leading to the extremum. The right figure shows the PISA axis, which is the downward set of dots on the *concave* side of the curve, and leading to the extremum.

**FUNDAMENTAL FACT**

**As shown in Fig 3.22, one can see:**

**The PISA axis is the only differential symmetry axis that describes the** *correct history* **of a compressive extremum, i.e., the trajectory of the curve point that became the compressive extremum as a result of the flattening process, and the PISA axis is the only axis that describes the force that created the flattening. It does so because it leads to the extremum from the concave side of the curve.**

**In contrast, the MA symmetry set, being on the convex side and pointing away from the extremum, does not give the history of the extremum in the flattening process that created the extremum.**

**Furthermore, the SLS axis, being on the convex side and pointing towards the extremum, does not give the history of the extremum in the flattening process that created the extremum.**

Therefore, we see that a major failure of the MA symmetry set and SLS axis, in describing a compressive extremum, is that those axes are on the convex side of the curve. This is why they fail to give the history of the extremum in the flattening process. In contrast, the PISA axis, being on the concave side (and leading to the extremum) correctly gives the history of the extremum in the flattening process.

We therefore see the following crucial fact:

**PISA CORRECTLY VIOLATES**
**THE CONVENTIONAL VIEW OF SYMMETRY**

**In the region of a $m^+$ or $M^-$ extremum, the PISA symmetry axis violates one of the fundamental properties of symmetry of the last 2500 years: that a symmetry axis must be on the *convex* side of an object.**

**This violation of conventional symmetry makes the PISA symmetry axis the *correct* way of describing the history of compressive extrema, and is the reason why, prior to the invention of the PISA axis, compressive extrema have always been failed to be described correctly.**

Thus, return to Fig 3.21 which shows the PISA axes of the four types of curvature extrema, i.e., the first two being the penetrative extrema and the other two being the compressive extrema. Observe the following crucial fact: In the entire 2500 year history of symmetry, the symmetry axis of each of the four curves shown would be on the *convex* side of the curve, i.e., below each of these curves. In contrast, the PISA axis is on the *convex* side of the penetrative extrema, but on the *concave* side of the compressive extrema. We are going to see that this violation of the conventional view of symmetry is fundamentally important for understanding biological morphology and manufacturing design.

## 3.13 Process-Continuations and Bifurcations

The rules, given so far, infer what our system calls a *process-diagram*. Such a diagram is a curve together with a set of arrows leading to its extrema. The figures on pages 49 – 50 exhibited many examples of process-diagrams.

We will now show how additional historical information can be inferred with respect to the extrema: The information concerns the *temporal order* in which the extrema were created – e.g., which extrema were created first, and whether some extrema arose out of other extrema, etc.

Our procedure for establishing the order information will be as follows: Suppose we have records of two stages in the history of the shape. For example, imagine being a doctor looking at two X-rays of a tumor taken a month apart. Observe that any doctor examines two such X-rays (e.g., on a screen) in order to assess what has happened in the intervening month. If one considers the way the doctor's thinking proceeds, one realizes that there is a basic inference rule that is being used: The doctor tries, as much as possible, to explain processes seen in the later shape as *extrapolations* of processes already seen in the earlier shape. We will show how to discover these extrapolations.

Recall that the processes we have defined are those that move along symmetry axes, creating extrema. As a simple first distinction, we can say that extrapolations have one of two forms:

> **(1) Process-Continuation:** The process simply continues along its symmetry axis, maintaining that single axis.

> **(2) Process-Bifurcation:** The process branches two processes of the same type.

Later chapters will look at more complex forms of continuation and bifurcation. However, the present chapter will concentrate on the two forms just listed.

Now recall, from section 3.11, that there are four types of extrema $M^+$, $m^-$, $m^+$, $M^-$. It is necessary therefore to look at what happens when one continues the process at each of the four types, and at what happens when one branches (bifurcates) the process at each of the four types. This means that there are *eight* possible events that can occur: four continuations and four bifurcations.

## 3.14 Continuation at $M^+$ and $m^-$

Let us start by considering continuations, and then move on to bifurcations. It turns out that, when one continues the process at either of the first two extrema, $M^+$ or $m^-$, nothing significant happens, as follows:

First consider $M^+$. Recall from Table 3.1 (p57) that the $M^+$ extremum corresponds to a protrusion. The shape in Fig 3.23 has three examples of $M^+$, the three protrusions. We want to understand what happens when any one of the $M^+$ processes is continued. For example, what happens when the protruding process at the top $M^+$ continues pushing the boundary further along the direction of its arrow?

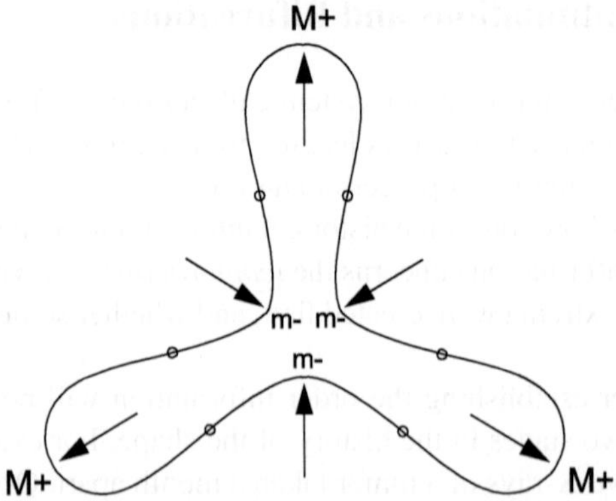


**Fig. 3.23.** Continuation at $M^+$ and $m^-$ does not change the extremum-type.

The answer is simple: The boundary would remain a $M^+$ extremum, despite being extended further upwards. Intuitively, this is obvious: A protrusion remains a protrusion if it continues. For example, a finger remains a finger if it continues to grow. Therefore, from now on, we will ignore continuation at $M^+$ as structurally trivial.

Now observe that exactly the same result occurs with respect to continuation at any $m^-$ extremum. For example, notice that the same shape, Fig 3.23, has three $m^-$ extrema. Observe also that, in accord with Table 3.1 (p57), each of these corresponds to an *indentation*. It is clear that, if the process continues at a $m^-$, the boundary would remain $m^-$. Again, this is intuitively obvious: An indentation remains an indentation if it continues. As a consequence, we will also ignore continuation at $m^-$ as structurally trivial.

In summary, the two cases considered in this section, continuation at $M^+$ and at $m^-$, are structurally trivial. It will now be seen that continuations at the remaining two extrema, $m^+$ and $M^-$, induce *much more interesting effects* on a shape.

## 3.15 Continuation at $m^+$

According to Table 3.1 (p57), a $m^+$ extremum is always associated with a *squashing* process. Consider the $m^+$ at the top of the left shape in Fig 3.24. Notice that, in accord with our theory, its process, the bold downward arrow, explains the flattening at this extremum, relative to the greater bend at each end of the top.

Our goal is to understand what happens when the process at this $m^+$ extremum is continued forward in time; i.e., the bold downward arrow pushes further downward. Clearly, a continuation of the process can result in the indentation shown within the top of the right shape in Fig 3.24. That is, the continuation of the bold downward squashing arrow at the top of the left shape has resulted in it becoming the bold downward indenting arrow in the right shape.

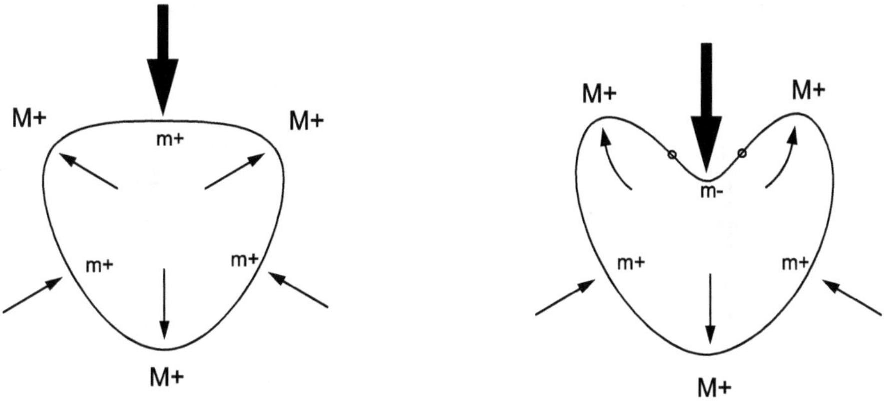


**Fig. 3.24.** Continuation at $m^+$.

The structural change, in going from the left to the right shape, in Fig 3.24, should be understood as follows: First, the $m^+$ extremum at the top of the left shape changes to the $m^-$ extremum within the top of the right shape. Notice that the $m^-$ extremum corresponds to an indentation, as predicted by Table 3.1 (p57).

An extra feature should be observed: On the curve on each side of the $m^-$ extremum, in the right shape, a small circular dot has been placed. Such a dot marks a position where the curvature is zero; i.e., the curve is locally straight.

With these facts, one can now describe exactly what occurred in the transition from the left shape to the right shape: The $m^+$ extremum at the top of the left shape has changed into a $m^-$ extremum in the right shape, and two points of zero curvature, 0, have been introduced, one on each side of the $m^-$. One can therefore say that the transition from the left shape to the right shape is the replacement of $m^+$ (left shape) by the triple, $0m^-0$ (right shape). The transition is therefore:

$$m^+ \longrightarrow 0m^-0$$

This transition will be labeled $Cm^+$ meaning *Continuation at* $m^+$. Thus the transition is given fully as:

$$Cm^+ \;:\; m^+ \longrightarrow 0m^-0$$

This mathematical expression is easy to translate into English. Reading the symbols, from left to right, the expression says:

*Continuation at* $m^+$ *takes* $m^+$ *and replaces it by the triple* $0m^-0$

It is worth having a simple phrase defining the transition in Fig 3.24. Notice that, since the extremum $m^+$ in the left shape is a squashing, and the extremum $m^-$ in the right shape is an indentation, the transition can be described as:

*A squashing continues till it indents.*

**Fig. 3.25.** Continuation at $M^-$.

## 3.16 Continuation at $M^-$

We will now show what happens when the process at the fourth type of extremum $M^-$ is continued forward in time. As an example, consider the $M^-$ in the center of the left shape in Fig 3.25. In accord with Table 3.1 (p57), the process at this extremum is an *internal resistance*. In order to understand this process, let us suppose that the left shape represents an island in the ocean. Initially, this island was circular. Then, there was an inflow of water at the top (creating an inward dip). This flow continued inward, but was eventually *resisted* by a ridge of mountains along the center of the island, thus forming the *bay* shown in the left shape. In the center of the bay, the point labeled $M^-$ is a curvature extremum, the flattest point on the bay.

Now return to the continuation issue: What happens when the upward resistance arrow, the bold upward arrow shown as terminating at the $M^-$ extremum, is continued along the direction of the arrow. This could happen for example, if there were a volcano in the mountains, that erupted, sending lava down into the sea. The result would therefore be the shape shown on the right in Fig 3.25. That is, a protrusion would be formed into the sea. Thus, the continuation of the bold upward resistance arrow in the left shape has resulted in it becoming the bold upward protruding arrow in the right shape.

The structural change, in going from the left shape to the right shape, should be understood as follows: First, the $M^-$ in the center of the bay (left shape) changes into the $M^+$ in the central region of the right shape, the protrusion.

An extra feature should be observed: On the curve on each side of the new $M^+$ extremum, in the right shape, a small circular dot has been placed. Such a dot again marks a position where the curvature is zero; i.e., the curve is locally straight.

Thus we can describe what has happened in the transition from the left shape to the right shape: The $M^-$ extremum in the bay of the left shape has changed into the new $M^+$ extremum in the right shape, and two points of zero curvature, 0, have been introduced, one on each side of the new $M^+$. In other words, the $M^-$ in the left shape has been replaced by the triple $0M^+0$ in the right shape. The transition is therefore:

$$M^- \longrightarrow 0M^+0$$

This transition will be labeled $CM^-$ meaning *Continuation at* $M^-$. Thus the transition is given fully as:

$$CM^- \ : \ M^- \longrightarrow 0M^+0$$

This mathematical expression is easy to translate into English. Reading the symbols, from left to right, the expression says:

*Continuation at $M^-$ takes $M^-$ and replaces it by the triple* $0M^+0$

It is worth having a simple phrase defining the transition in Fig 3.25. Notice that, since the extremum $M^-$ in the left shape is a resistance, and the extremum $M^+$ in the right shape is a protrusion, the transition can be described as:

*A resistance continues till it protrudes.*

## 3.17 The Continuation Format

In sections 3.14 to 3.16, we have gone through each of the four extrema, and defined what happens when the process at each extremum is allowed to continue. Continuation at the first two extrema, $M^+$ and $m^-$, the penetrative extrema, involved no structural change, so we have chosen to ignore their continuations. However, continuation at the other two extrema, $m^+$ and $M^-$, the compressive extrema, did involve a structural change. Each compressive extremum changed into a penetrative extremum. Therefore it is now possible to understand the structure of what I call continuation, thus:

---

**THE STRUCTURE OF CONTINUATION**

**A compressive process continues till it becomes penetrative.**

---

We have also seen that continuation is coded by the following transition between extrema: A compressive extremum is replaced by the penetrative extremum of the opposite sign, with a zero on each side. The logic of the encoding is as follows:

A positive curvature sign means that the curve bends out, and a negative curvature sign means that the curve bends in. Therefore, when a compressive process continues till it becomes penetrative, it is clear that the sign of the extremum must change. Now, for obvious reasons in calculus, the sign of the initial compressive extremum ($m^+$ or $M^-$) must be the same sign as the next extremum on each side of it. Therefore, when the compressive extremum undergoes the transition to the penetrative extremum of the opposite sign, a zero must appear between the penetrative extremum and each of those two side extrema. This explains the format of the two continuation operations.

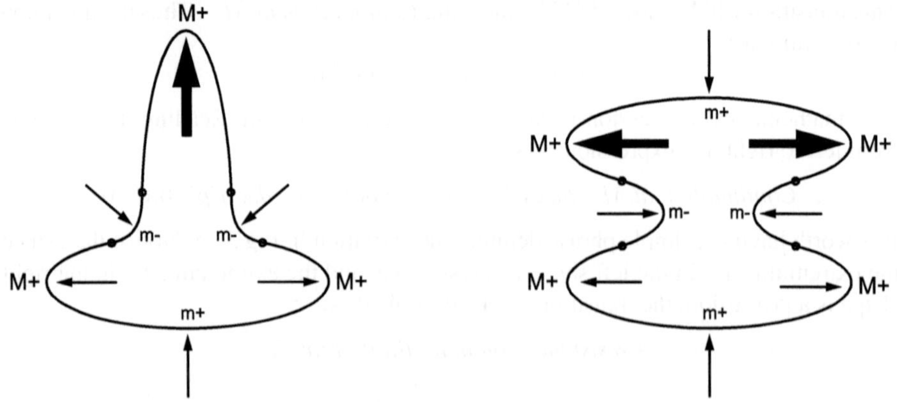


**Fig. 3.26.** Bifurcation at $M^+$.

## 3.18 Bifurcation at $M^+$

We now turn from continuations to bifurcations at extrema. Again, each of the four extrema will be investigated in turn.

First we examine what happens when the process at a $M^+$ extremum bifurcates forward in time. As an example, consider the $M^+$ at the top of the left shape in Fig 3.26. In accord with Table 3.1 (p57), the process at this extremum is a *protrusion*, shown by the upward bold arrow in the left shape in Fig 3.26. The result of bifurcating this process is shown in the right shape. The upward protrusion process in the left shape branched into *two protrusion processes* in the right shape: one given by the left-ward pointing bold arrow and the other given by the right-ward pointing bold arrow in the right shape.

The structural change, in going from the left shape to the right shape, should be understood as follows: First observe that the single $M^+$, at the top of the left shape, splits into two copies of $M^+$, shown at the ends of the two branches in the right shape.

There is also another feature. In the center of the top of the right shape, a new extremum has been introduced, $m^+$. The $m^+$ is a minimum, and is mathematically required because the two branching extrema are maxima $M$, and two maxima cannot exist without a minimum between them. Notice that the process at the $m^+$ extremum is a *squashing*, as predicted in Table 3.1 on p57. And now for this fundamental claim:

**According to the New Foundations,**
**the squashing process *caused* the bifurcation of the protruding process.**

Now let us describe the transition in terms of the extrema involved: The $M^+$ extremum at the top of the left shape bifurcated into three extrema in the right shape. It split into two copies of itself, and introduced a new extremum $m^+$. That is, the transition from the left shape to the right shape is the replacement of $M^+$ (left shape) by the triple, $M^+m^+M^+$ (right shape). The transition is therefore:

$$M^+ \longrightarrow M^+m^+M^+$$

This transition will be labeled $BM^+$, meaning *Bifurcation at* $M^+$. Thus the transition is given fully as:

$$BM^+ \; : \; M^+ \longrightarrow M^+m^+M^+$$

This mathematical expression is easy to translate into English. Reading the symbols, from left to right, the expression says:

*Bifurcation at* $M^+$ *takes* $M^+$ *and replaces it by the triple* $M^+m^+M^+$

It will also be worth having a simple phrase to summarize the effect of the transition in Fig 3.26. The top structure formed on the right shape has the shape of a *shield*, and therefore the transition will be referred to thus:

*Shield-formation.*

## 3.19 Bifurcation at $m^-$

Next we examine what happens when the process at a $m^-$ extremum bifurcates forward in time. As an example, consider the $m^-$ in the center of the left shape in Fig 3.27. In accord with Table 3.1 (p57), the process at this extremum is an *indentation*, shown by the downward bold arrow in the left shape in Fig 3.27. The result of bifurcating this process is shown in the right shape. The downward indentation process in the left shape branched into *two indentation processes* in the right shape: one given by the left-ward pointing bold arrow in the right shape, and the other given by the right-ward pointing bold arrow in the right shape. A crucial fact of this bifurcation is that a *bay* has been formed! Notice, therefore, that one can regard the transition from the left shape to the right shape as the stage preceding Fig 3.25 on p64.

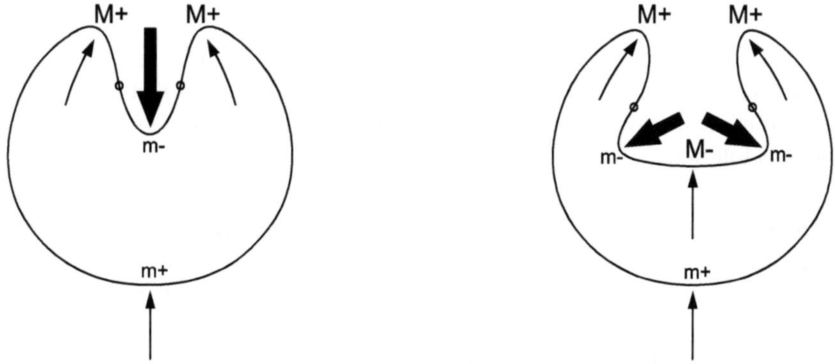


**Fig. 3.27.** Bifurcation at $m^-$.

The structural change, in going from the left to the right shape in Fig 3.27, should be understood as follows: First observe that the single $m^-$, in the center of the left shape, splits into two copies of $m^-$, shown at the ends of the two branches in the right shape.

There is also another feature. In the center of the right shape, a new extremum has been introduced, $M^-$. Notice that the process at the $M^-$ extremum is a *resistance*, as predicted in Table 3.1 on p57. And now for this fundamental claim:

**According to the New Foundations,**
**the resistance process *caused* the bifurcation of the indenting process.**

Now let us describe the transition in terms of the extrema involved: The $m^-$ extremum in the center of the left shape bifurcated into three extrema in the right shape. It split into two copies of itself, and introduced a new extremum $M^-$. That is, the transition from the left shape to the right shape is the replacement of $m^-$ (left shape) by the triple, $m^-M^-m^-$ (right shape). The transition is therefore:

$$m^- \longrightarrow m^-M^-m^-$$

This transition will be labeled $Bm^-$ meaning *Bifurcation at* $m^-$. Thus the transition is given fully as:

$$Bm^- \ : m^- \longrightarrow m^-M^-m^-$$

This mathematical expression is easy to translate into English. Reading the symbols, from left to right, the expression says:

*Bifurcation at $m^-$ takes $m^-$ and replaces it by the triple $m^-M^-m^-$*

It will also be worth having a simple phrase to summarize the effect of the transition in Fig 3.27. The structure formed on the right shape has the shape of a *bay*, and therefore the transition will be referred to thus:

*Bay-formation.*

## 3.20 The Bifurcation Format

The previous two sections established the first two bifurcations: those at $M^+$ and $m^-$. The next two sections will describe the remaining two bifurcations that have the same format. From the first two bifurcations, we can see that the format of this type of bifurcation is as follows:

$$E \longrightarrow EeE$$

An extremum $E$ is sent to two copies of itself, and a new extremum $e$ is introduced between the two copies. A crucial fact is this: Although the new extremum $e$ must be of the opposite type from $E$, that is, Max ($M$) vs. min ($m$), it must be the *same sign* as $E$.

**My version of singularity theory will show that this *same sign condition* captures fundamentally important aspects of morphology.**

## 3.21 Bifurcation at $m^+$

Next we examine what happens when the process at a $m^+$ extremum bifurcates, forward in time. As an example, consider the $m^+$ at the top of the left shape in Fig 3.28. In accord with Table 3.1 (p57), the process at this extremum is a *squashing*, shown by the downward arrow at the top of the left shape in Fig 3.28.

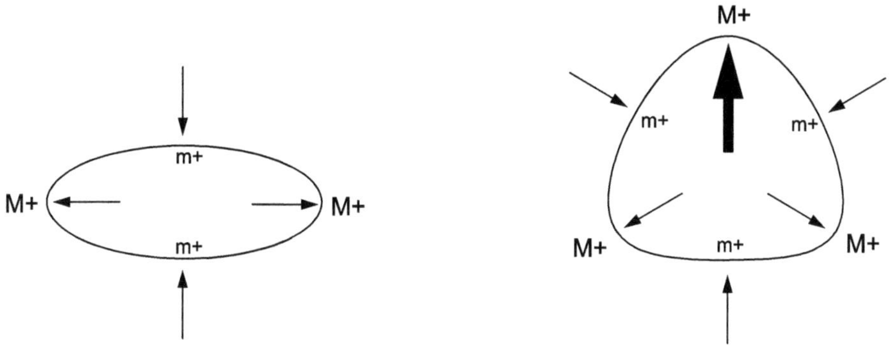


**Fig. 3.28.** Bifurcation at $m^+$.

The result of bifurcating this process is shown in the right shape. The downward squashing process in the left shape branched into *two squashing processes* in the right shape: one given by the arrow leading to the left $m^+$, in the right shape, and the other given by the arrow leading to the right $m^+$, in the right shape. One should imagine the two copies as *sliding* away from the top till they reached their current positions. Thus the single $m^+$, at the top of the left shape, split into two copies of $m^+$, shown at the ends of the two branches in the right shape.

The other crucial event is the introduction of a new extremum $M^+$ in the top of the right shape. Notice that the process here, the upward bold arrow, conforms to Table 3.1 on p57, which says that a $M^+$ extremum always corresponds to a *protrusion*. And now for this fundamental claim:

**According to the New Foundations,**
**the protrusion process *caused* the bifurcation of the squashing process.**

Now let us describe the transition in terms of the extrema involved: The $m^+$ extremum at the top of the left shape bifurcated into three extrema in the right shape. It split into two copies of itself, and introduced a new extremum $M^+$. That is, the transition from the left shape to the right shape is the replacement of $m^+$ (left shape) by the triple, $m^+M^+m^+$ (right shape). The transition is therefore:

$$m^+ \longrightarrow m^+M^+m^+$$

This transition will be labeled $Bm^+$ meaning *Bifurcation at* $m^+$. Thus the transition is given fully as:

$$Bm^+ \; : m^+ \longrightarrow m^+M^+m^+$$

This mathematical expression is easy to translate into English. Reading the symbols, from left to right, the expression says:

*Bifurcation at* $m^+$ *takes* $m^+$ *and replaces it by the triple* $m^+M^+m^+$

It will also be worth having a simple phrase to summarize the effect of the transition, as follows: Notice that the main effect in Fig 3.28 is that the initial squashing process is pushed to each side by the breaking-through of a new protrusion. Thus the transition can be summarized by the following phrase:

*Breaking-through of a protrusion.*

This protrusion, that breaks through, is given by the bold upward arrow in the right shape in Fig 3.28.

## 3.22 Bifurcation at $M^-$

Now we establish the final bifurcation of this type. We examine what happens when the process at a $M^-$ extremum bifurcates, forward in time. As an example, consider the $M^-$ in the center of the bay, in the left shape, in Fig 3.29. In accord with Table 3.1 (p57), the process at this extremum is an *internal resistance*, shown by the upward arrow in the center of the left shape in Fig 3.29.

The result of bifurcating this process is shown in the right shape. The upward resistance in the center of the bay of the left shape branched into *two resistance processes* in the right shape. These two resistance processes are shown as the two diagonal upward processes on the sides of the resulting bay in the right shape. The Process Grammar calls this structure, a **deepened bay**.

The structural change, in going from the left to the right shape in Fig 3.29, should be understood as follows: First observe that the single $M^-$, in the center of the left shape, splits into two copies of $M^-$, shown at the ends of the two resulting, upward diagonal, branches on the deepened bay of the right shape.

The other crucial event is the introduction of a new extremum $m^-$ in the bottom of the deepened bay. Notice that the process here, the downward bold arrow, conforms to Table 3.1 on p57, which says that a $m^-$ extremum always corresponds to an *indentation*. And now for this fundamental claim:

**According to the New Foundations,**
**the indentation process *caused* the bifurcation of the resistance process.**

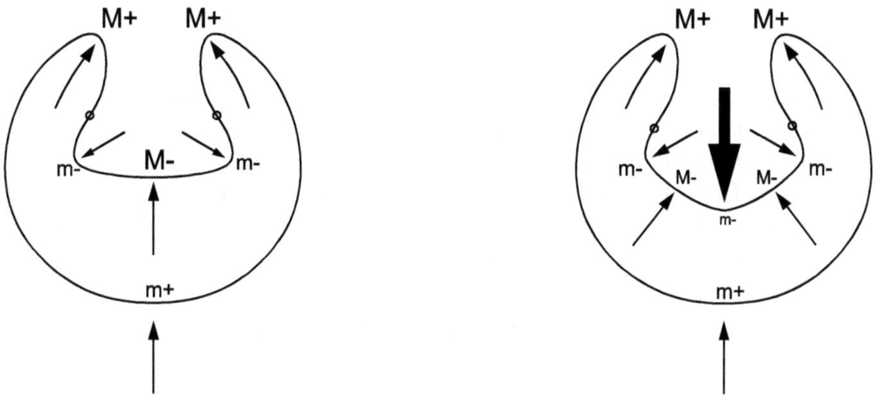


**Fig. 3.29.** Bifurcation at $M^-$.

Now let us describe the transition in terms of the extrema involved: The $M^-$ extremum in the center of the left shape bifurcated into three extrema in the right shape. It split into two copies of itself, and introduced a new extremum $m^-$. That is, the transition from the left shape to the right shape is the replacement of $M^-$ (left shape) by the triple, $M^-m^-M^-$ (right shape). The transition is therefore:

$$M^- \longrightarrow M^-m^-M^-$$

This transition will be labeled $BM^-$ meaning *Bifurcation at* $M^-$. Thus the transition is given fully as:

$$BM^- \ : \ M^- \longrightarrow M^-m^-M^-$$

This mathematical expression is easy to translate into English. Reading the symbols, from left to right, the expression says:

*Bifurcation at $M^-$ takes $M^-$ and replaces it by the triple $M^-m^-M^-$*

It is also worth having a simple phrase to summarize the effect of the transition, as follows: Notice that the main effect in Fig 3.29 is that the initial resistance process is pushed to each side by the breaking-through of a new indentation. Thus the transition can be summarized by the following phrase:

*Breaking-through of an indentation.*

This indentation, that breaks through, is given by the bold downward arrow in the right shape in Fig 3.29.

## 3.23 The Process Grammar

We have thus finished elaborating the types of process-continuation and process-bifurcation that accord with the forms of extrapolations defined in section 3.13. We showed that there are *two* types of continuation and *four* types of bifurcation. Therefore, this defines a total of *six* types of process-transition. We saw that each transition can be encoded by a transition of extrema.

A crucial fact is that the six transitions correspond to conceptually powerful real-world situations. The following table lists the extremum-coding of each transition, together with the real-world situation to which it corresponds.

**PROCESS GRAMMAR (Level 3)**

| | | |
|---|---|---|
| $Cm^+$ : | $m^+ \longrightarrow 0m^-0$ | squashing continues till it indents |
| $CM^-$: | $M^- \longrightarrow 0M^+0$ | resistance continues till it protrudes |
| $BM^+$: | $M^+ \longrightarrow M^+m^+M^+$ | shield-formation |
| $Bm^-$ : | $m^- \longrightarrow m^-M^-m^-$ | bay-formation |
| $Bm^+$ : | $m^+ \longrightarrow m^+M^+m^+$ | breaking-through of a protrusion |
| $BM^-$: | $M^- \longrightarrow M^-m^-M^-$ | breaking-through of an indentation |

After I published this component of the Process Grammar, in the 1980s, it was applied by scientists in many disciplines: For example, Milios [31] applied it in meteorology to analyze weather patterns. Larsen [9] applied it in geology to analyze the formation of volcanic islands. Deguchi & Furukawa [6] applied it in medicine for cardiac diagnosis. Lin, Liang, & Chen [28] applied it to analyze general radiological structure. Lee [10] applied it in chemical engineering to model molecular dynamics. Pernot, Guillet, Leon, Falcidieno, & Giannini [34] [35] applied it in computer-aided design to create design operators for the automotive and aerospace industry. Shemlon [38] applied it to analyze morphology in the human brain, neuronal growth, dentistry, and insect cells.

The six operations, listed above, constitute what I call Level 3 of the Process Grammar, because, in each operation, an extremum bifurcates into a *triple* of singularities. The letters $C$ and $B$ denote these Level 3 operations. Later chapters in the book will give other levels of the Process Grammar.

## 3.24 The Bifurcation-Causing Process

Recall that a fundamental principle of my New Foundations to Geometry is that shape is equivalent to causal explanation. As an example, we saw in sections 3.18 – 3.22 that the power of the extrema-based process-inference rules, in the New Foundations, is that they *causally explain* the *bifurcation* operations. They do so as follows:

**THE BIFURCATION-CAUSING PROCESS**

**By our process-inference rules, each of the $B$ bifurcations of the Process Grammar is causally explained by the process inferred from the central extremum that appears between the copies of the bifurcating extremum.**

**This central process will be called the *bifurcation-causing process.***

**Notice that the bifurcation-causing process *opposes* the pre-existing process that splits in the bifurcation.**

**Furthermore, the bifurcation-causing process causes that split of the pre-existing process.**

In sections 3.18 – 3.22, we saw that the explanations for the four $B$ bifurcation operations are:

**THE EXPLANATIONS**

**In the shield-formation operation $BM^+$, the process that causes the protrusion to split is the squashing force that creates the flattening in the center of the resulting shield.**

**In the bay-formation operation $Bm^-$, the process that causes the indentation to split is the resistance force that creates the flattening in the center of the resulting bay.**

**In the breaking-through operation $Bm^+$, the process that causes the squashing to split is the protruding force that breaks through the squashing.**

**In the breaking-through operation $BM^-$, the process that causes the resistance to split is the indenting force that breaks through the resistance.**

## 3.25 Causal Plausibility

The concepts of this section are fundamentally important for understanding shape. The reader who fails to understand these concepts will never be able to understand spatial morphology.

In section 3.12, we saw that PISA gives a *plausible causal explanation* of each of the four extrema. In contrast, we saw that the Medial Axis (MA) symmetry set and SLS do not give a plausible causal explanation of two of the four extrema, $m^+$ and $M^-$. In our system, these two extrema are called the *compressive* extrema, since this is their plausible causal explanation.

We will see, throughout this book, that the inability of conventional symmetry analyses, such as the Medial Axis, to give a plausible causal explanation of compressive extrema is a major problem for those analyses.

In the present section, we will show something that is fundamentally important to understand with respect to causal plausibility:

**CAUSAL PLAUSIBILITY**

**Our theory claims that compressive extrema are fundamentally important in morphology. This is demonstrated with many examples in Chapter 5.**

**Recall that, in section 3.12, we have shown that the PISA axis gives the correct causal explanation of compressive extrema, whereas the MA symmetry set and SLS axis do not.**

**Furthermore, since compressive forces are involved in the causal explanations we gave in section 3.24 for all the $B$ Process-Bifurcation operations, defined by the Process Grammar, the PISA axis gives the correct causal explanation of these bifurcations, whereas the MA symmetry set and SLS axis do not. This also captures fundamental aspects of morphology.**

**Our theory claims that the Process-Continuation operations, $Cm^+$ and $CM^-$, defined by the Process Grammar, i.e., the continuation of a compressive process till it penetrates, are also fundamental in morphology.**

**We will now show that the PISA axis gives the correct model of these Process-Continuations, whereas the MA symmetry set and the SLS axis do not.**

**Thus we see that: In the important real-world situations modeled by the Process Grammar operations, PISA is causally plausible, whereas the MA symmetry set and SLS are not causally plausible.**

**The reader who fails to understand this will never understand spatial morphology.**

Let us therefore now study the causal modeling of the Process-Continuation operations, $Cm^{+}$ and $CM^{-}$, of the Process Grammar. In both these operations, a compressive process continues till it becomes penetrative. Correspondingly, a compressive extremum changes into a penetrative extremum of the opposite sign with two zeros on the sides of the penetrative extremum. Fig 3.30 serves as an illustration of both of these operations. The choice of which one of the two $C$ operations is used to describe the shape transition in this figure is given by which side of the curves (above or below) is chosen to be solid vs. empty. However, the crucial fact about this figure is this: The central extremum of the first shape is compressive. This extremum is replaced, in the second shape, by a central penetrative extremum of the opposite sign. Necessarily, two zeros of curvature (flat points) also appear on the sides of the new penetrative extremum.

**Fig. 3.30.** Transition from a central compressive extremum to a central penetrative extremum of the opposite sign, as given by both of the $C$ operations of the Process Grammar.

Although Fig 3.30 illustrates the extremum-transition coded by the two $C$ operations of the Process Grammar, the figure by itself does not give the *causal explanation* of this transition. We will now show that PISA gives a highly *plausible* explanation of this transition, whereas the MA symmetry set and SLS are not causally plausible.

We are going to discuss the three cases, MA symmetry set, SLS, and PISA, on three separate pages, so that the reader can easily view the diagrams that illustrate the three respective cases.

Let us begin by examining the MA symmetry set of the transition. Recall that the left-hand diagram in Fig 3.22 (page 58) showed the MA symmetry set for the compressive extremum. In that figure, we saw that the MA symmetry set is below the extremum and goes *away* from the extremum. Therefore, the MA symmetry set for the compressive extremum, in the shape transition we are now considering, is shown as the arrow in the left diagram of Fig 3.31.

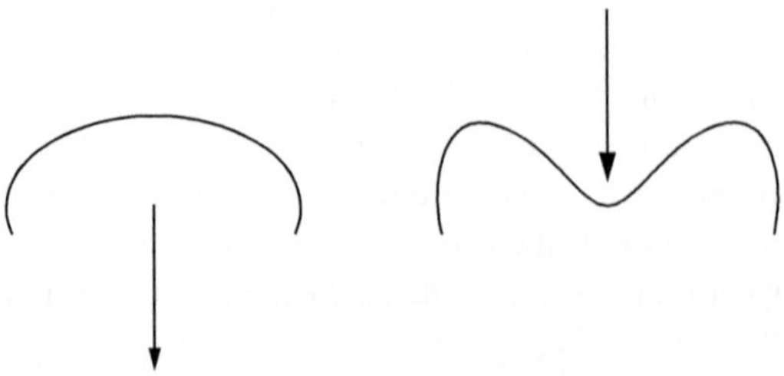

**Fig. 3.31.** The transition from compressive to penetrative, given by the MA symmetry set, does not have a plausible causal structure.

For the penetrative extremum, in this shape transition, the MA symmetry set is given by the arrow in the right diagram in Fig 3.31. Notice that, because the bitangent spheres leading to a penetrative extremum are between the sides of the penetrative extremum, the MA symmetry set is *above* this penetrative extremum. As a result of this, Fig 3.31 shows the following:

**FAILURE OF THE MA SYMMETRY SET TO DESCRIBE IMPORTANT PROCESS-CONTINUATIONS**

**The MA symmetry set fails to describe the following important real-world situation defined by the Process Grammar: the continuation of a compressive process till it penetrates.**

**The reason is that, as illustrated in Fig 3.31, the MA symmetry set of the compressive extremum is on the convex side of the extremum, and the MA symmetry set of the penetrative extremum is on the convex side of the penetrative extremum, and therefore, as shown in Fig 3.31, the MA symmetry sets of these two extrema are on opposite sides of the two curves in the transition. Therefore, the MA symmetry set at the penetrative extremum cannot be a *continuation* of the MA symmetry set at the compressive extremum.**
**Furthermore, as illustrated in Fig 3.31, because the MA symmetry set of the compressive extremum goes away from the extremum, and the MA symmetry set of the penetrative extremum goes towards the extremum, the MA symmetry set fails to describe the process continuation because the MA symmetry set of the compressive extremum is *beyond* the MA symmetry set of the penetrative extremum.**

Let us now examine the SLS of the shape transition. Recall that the middle diagram of Fig 3.22 (page 58) showed the SLS axis for the compressive extremum. In that figure, we saw that the SLS axis is below the extremum and goes *upward towards* the extremum. Therefore, the SLS axis for the compressive extremum, in the shape transition we are now considering, is shown as the arrow in the left diagram of Fig 3.32.

**Fig. 3.32.** The transition from compressive to penetrative, given by the SLS axis, does not have a plausible causal structure.

For the penetrative extremum, in this shape transition, the SLS axis is given by the arrow in the right diagram in Fig 3.32. Notice that, because the bitangent spheres leading to a penetrative extremum are between the sides of the penetrative extremum, the SLS axis is *above* this penetrative extremum. As a result of this, Fig 3.32 shows the following:

**FAILURE OF THE SLS AXIS**
**TO DESCRIBE**
**IMPORTANT PROCESS-CONTINUATIONS**

**The SLS axis fails to describe the following important real-world situation defined by the Process Grammar: the continuation of a compressive process till it penetrates.**

**The reason is that, as illustrated in Fig 3.32, the SLS axis of the compressive extremum is on the convex side of the extremum, and the SLS axis of the penetrative extremum is on the convex side of the penetrative extremum, and therefore, as shown in Fig 3.32, the SLS axes of these two extrema are on opposite sides of the two curves in the transition. Therefore, the SLS axis at the penetrative extremum cannot be a *continuation* of the SLS axis at the compressive extremum.**

**Furthermore, as illustrated in Fig 3.32, because the SLS axis of the compressive extremum goes towards the extremum, and the SLS axis of the penetrative extremum goes towards the extremum in the opposite direction to the SLS axis of the compressive extremum, the SLS axis fails to describe the process continuation because the SLS axis of the compressive extremum and penetrative extremum are in *opposite directions* from each other.**

Let us now examine the PISA of the transition. Recall that the right-hand diagram in Fig 3.22 (page 58) showed the PISA axis for the compressive extremum. In that figure, we saw that the PISA axis is above the extremum and goes *downwards towards* the extremum. Therefore, as we said, the PISA axis gives the correct causal explanation of the compressive extremum. Notice that the PISA axis for the compressive extremum, in the shape transition we are now considering, is shown as the arrow in the left diagram of Fig 3.33.

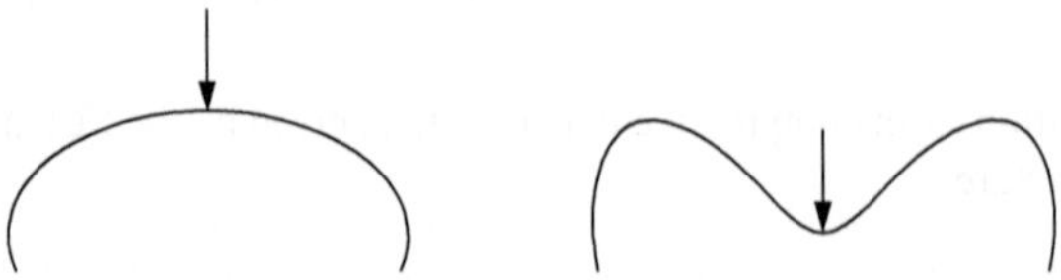

**Fig. 3.33.** The transition from compressive to penetrative, given by the PISA axis, has a plausible causal structure.

For the penetrative extremum, in this shape transition, the PISA axis is given by the arrow in the right diagram in Fig 3.33. Observe that, because the bitangent spheres leading to a penetrative extremum are between the sides of the penetrative extremum, the PISA axis is *above* this penetrative extremum. As a result of this, Fig 3.33 shows the following:

**CORRECTNESS OF THE PISA AXIS**
**IN DESCRIBING**
**IMPORTANT PROCESS-CONTINUATIONS**

**The PISA axis correctly describes the following important real-world situation defined by the Process Grammar: the continuation of a compressive process till it penetrates.**

**The reason is that, as illustrated in Fig 3.33, the PISA axis of the compressive extremum is on the *concave* side of the extremum, and the PISA axis of the penetrative extremum is on the *convex* side of the penetrative extremum, and therefore, as shown in Fig 3.33, the PISA axes of these two extrema are on the *same* sides of the two curves in the transition.**

**Furthermore, as illustrated in Fig 3.33, because the PISA axis of the compressive extremum goes towards the extremum, and the PISA axis of the penetrative extremum goes towards the extremum in the *same* direction as the PISA axis of the compressive extremum, on the *same* side of the curve, the PISA axis correctly describes the process continuation because the PISA axis of the penetrative extremum has moved further in that same direction, i.e., the PISA axis of the penetrative extremum correctly describes the continuation of the PISA axis of the compressive extremum.**

We therefore have this profoundly important conclusion:

**FUNDAMENTAL FACT OF SPATIAL MORPHOLOGY**

**PISA gives the correct causal explanation of compressive extrema; i.e., it describes the extremum in terms of a compressive force.**
**The MA symmetry set and the SLS axis do not.**

**Furthermore PISA gives the correct description of the continuation of a compressive force till it penetrates.**
**The MA symmetry set and the SLS axis do not.**

**This book shows that this is fundamentally important to understanding spatial morphology.**

**The standard literature on shape fails to understand this issue, and therefore fails to understand spatial morphology.**

## 3.26 Causal Effect on the Curvature Functions

It is important to understand that the terms *continuation* and *bifurcation*, in the Process Grammar, mean **process-continuation** and **process-bifurcation**. We will now see that *both* of these can be coded by bifurcations in the corresponding curvature functions.

To fully understand this, we first need to consider the shape as a parameterized curve, $z(t)$, where $t$ is the parameter along the curve. The curvature along the curve will be given by the *curvature function* $\kappa(t)$.

Now let us consider the first continuation operation of the Process Grammar:

$$Cm^+ : m^+ \longrightarrow 0m^-0$$

This operation codes the situation, *squashing continues till it indents*, which is an effect on the curve itself, and was illustrated in Fig 3.24 (page 63).

What this operation does to the *curvature function*, $\kappa(t)$, is illustrated in the sequence of three curvature functions in Fig 3.34. That is, the effect is to lower the positive minimum $m^+$ in the first function till it becomes the negative minimum $m^-$ in the third function. Necessarily, two zeros appear on the sides of the negative minimum. Clearly, this lowering of the minimum requires the minimum to be moved through the $t$-axis of the graph.

It is important to carefully understand the concepts and terminology involved. Generally, we will regard a *singularity* of the curvature function as a point on the parameter $t$, where the curvature function has zero value or zero gradient. If it has zero gradient, then it would be what mathematicians would call a *critical-point* of the curvature function.

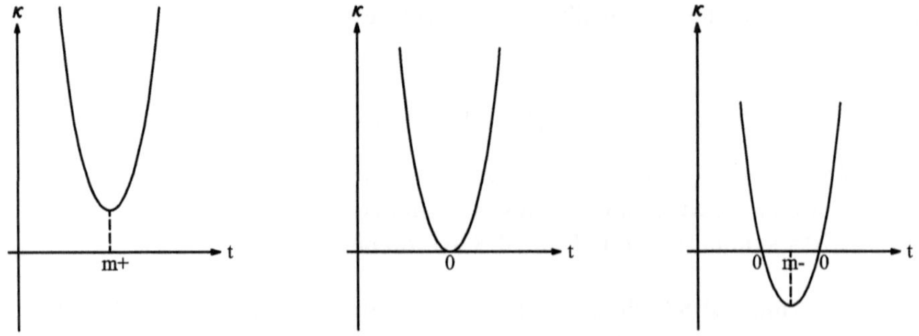


**Fig. 3.34.** How the $Cm^+$ operation alters the curvature function.

In a function, a critical point can be an extremum, i.e., a minimum or maximum, or it can be an inflection point where the gradient is zero.

A curvature function can have a sequence of singularities along the parameter $t$. This sequence will be called the **singularity-configuration** of the function. Thus let us return to Fig 3.34. The first function in this figure has a singularity-configuration consisting of one extremum, a positive minimum $m^+$. We can also see that the third function has a singularity-configuration consisting of two zeros and a negative minimum $m^-$. That is, the singularity-configuration of the third function is $0m^-0$. Therefore, the Process Grammar operation $Cm^+ : m^+ \longrightarrow 0m^-0$ gives the transition from the singularity-configuration of the first function to the singularity-configuration of the third function.

The singularity-configuration $m^+$ of the first function will be called the **domain** of the Process Grammar operation; and the singularity-configuration $0m^-0$ of the third function will be called the **codomain** of the Process Grammar operation.

Now, in Fig 3.34, we see the obvious fact that, since the transition lowers the graph of the first curvature function to obtain the graph of the third curvature function, there must be an intervening curvature function, whose graph has the property that the minimum is tangential to the $t$-axis, as illustrated by the second graph in Fig 3.34. We will call this intervening function, the *transition-function* of the Process Grammar operation. The singularity-configuration of the transition-function will be called the *transition-state* of the Process Grammar operation.

The way we will understand the transition-state will be as follows: Consider the *backward-time* direction from the third function to the second function. In this direction, the three singularities of the third function, i.e., the two zeros and the minimum, *coalesce* to become the singularity of the second function; i.e., the zero that is a minimum. This coalescence will mean that we can view the zero-minimum in the second function as the *coincidence* of two zeros and a minimum. Therefore, in the *forward-time* direction from the second function to the third function, we can understand the three *coincident* singularities of the zero-minimum in the second function as *bifurcating* into the three *separate* singularities in the third function.

Generally, we will define the **transition-state** of a Process Grammar operation to be the singularity-configuration, between the domain and codomain, where the singularities that will bifurcate in the codomain are coincident. The **transition-function** of the operation will then be defined as the function that contains the transition-state.

In singularity theory, a singularity that consists of two or more coincident singularities is called *degenerate*. Conversely, a singularity that does not consist of two or more coincident singularities is called *non-degenerate*. We can therefore have the following rule: Generically, each singularity in the domain and codomain of a Process Grammar operation is non-degenerate.

Now, by the Asymmetry Principle (page 16), of the New Foundations, a Process Grammar operation must *increase* the number of singularities in a curvature function, because this increases the curvature variation. The consequence of this is that the operation must have a transition curvature function in which there is a degenerate singularity. The coincident singularities in the degenerate singularity will then bifurcate to become separate non-degenerate singularities in the codomain curvature function.

For $n \geq 2$, when $n$ zeros bifurcate from a degenerate zero, we will say that the function has undergone an *$n$-fold zero-bifurcation*. Thus, we can see that a basic fact of the continuation operation $Cm^+ : m^+ \longrightarrow 0m^-0$ is that it involves a 2-fold zero-bifurcation. This is illustrated in Fig 3.34, where we see that the 2-fold zero-bifurcation occurs in going from the transition-function to the codomain function.

Let us now turn from the process-continuation operations, $C$, to the process-bifurcation operations, $B$. For example, consider the operation:

$$Bm^+ : m^+ \longrightarrow m^+M^+m^+$$

This operation codes the situation, *breaking-through of a protrusion*, whose effect on a curve was illustrated in Fig 3.28 (page 69).

What this operation does to the *curvature function*, $\kappa(t)$, is illustrated in the sequence of three functions in Fig 3.35. We see that the first function has the singularity-configuration $m^+$, which is the domain of the above operation $Bm^+$. And the third function has the singularity-configuration $m^+M^+m^+$, which is the codomain of the operation.

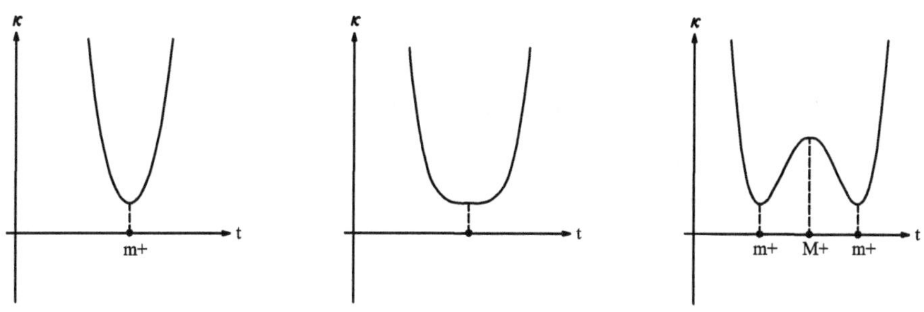


**Fig. 3.35.** How $Bm^+$ alters the curvature function.

Now let us use the rule that, generically, each singularity in the domain and codomain is non-degenerate, i.e., not the coincidence of multiple singularities. However, since the Process Grammar operation $Bm^+$ describes a bifurcation of processes, this must mean that the three singularities of the final function must have bifurcated from one singularity. Therefore, going *backward in time*, the three singularities of the final function, i.e., the minimum, maximum, minimum, must *coalesce* into one singularity that is 3-fold degenerate, i.e., is the coincidence of three singularities. This degenerate singularity is the one shown in the second function in Fig 3.35. Therefore, in the *forward-time* direction from the second function to the third function, we can understand the three *coincident* singularities in the second function as *bifurcating* into the three *separate* singularities in the third function.

Thus the second function is the *transition-function* of this Process Grammar operation, and the singularity in the second function is the *transition-state* of the operation.

Now consider the following important fact: In both the first and second function, the singularity is a positive minimum. However, in the first function the positive minimum is non-degenerate (i.e., not the coincidence of multiple singularities); but in the second function the positive minimum is degenerate (i.e., the coincidence of multiple singularities).

Recall, from section 3.24, that each $B$ operation of the Process Grammar is causally explained by the process inferred from the central extremum that appears in the final shape. In the present operation, $Bm^+$, the causing force is the protrusion that breaks-through.

Thus we see that the *initial* effect, that the causing force has on the curvature function, is to convert the *non-degenerate positive minimum* of the first function into the *degenerate positive minimum* of the second function. The subsequent effect is to cause the bifurcation of the degenerate positive minimum into the three non-degenerate extrema of the final function.

**Conclusion:** In this section, we have been examining the effects of the process-continuations $C$ and the process-bifurcations $B$ on the curvature functions corresponding to the shapes. We have seen the following:

**PROCESS-CONTINUATIONS & PROCESS-BIFURCATIONS**

**The process-continuations $C$ and the process-bifurcations $B$, of the Process Grammar, both *cause* a bifurcation of singularities in the curvature functions, as follows:**

**The process-continuations $C$ cause a bifurcation of zeros.**

**The process-bifurcations $B$ cause a bifurcation of extrema.**

## 3.27 Unfolding Groups and Process-Bifurcation

As we have said, the Process Grammar is a component of the New Foundations to Geometry. It is therefore based on the fundamental principles of the New Foundations: Maximization of Transfer and Maximization of Recoverability. For example, we have seen that the recoverability property of the Process Grammar is based on the inference rules given by the two Fundamental Laws of Memory Storage in the New Foundations – the Asymmetry Principle and Symmetry Principle. The asymmetry, to which the Asymmetry Principle is applied in the Process Grammar, is curvature variation. The symmetry, to which the Symmetry Principle is applied in the Process Grammar, is the PISA axis. In accord with the Asymmetry Principle, the Process Grammar operations define bifurcations of singularities that are extrema and zeros of the curvature, because these bifurcations increase curvature variation.

**So far we have been seeing how the particular singularity theory defined by the Process Grammar is set up to realize the Laws of Recoverability in the New Foundations to Geometry. Much more of this will be shown in later chapters.**

**We will invent a singularity theory which we will call** *Interactive Singularity Theory*, **that realizes fundamentally important causal aspects of morphology, that conventional singulary theories of shape completely fail to realize.**

**We will now see how the singularity theory of the Process Grammar realizes the Mathematical Theory of Transfer in the New Foundations.**

**The first example we will show of this is how the role of degeneracy in spatial morphology is given by the Mathematical Theory of Transfer in the New Foundations to Geometry. (A later example will be given in our theory of surface morphology.)**

First recall from section 2.2 that, according to the Mathematical Theory of Transfer, given by the New Foundations, the transfer relation is defined by the following structure:

**Fiber Group Ⓦ Control Group**

where the fiber group is the group of actions that is being transferred, the control group is the group of actions that is doing the transfer, and the wreath product symbol Ⓦ models the relation "is transferred by". Therefore, the above expression is interpreted thus:

**Fiber Group** *is transferred by* **Control Group**

In section 2.6, we saw that a basic class of wreath products, invented in the New Foundations, is called *iso-regular groups*, which are defined as follows: An iso-regular group is a group that satisfies three conditions: (1) It is an $n$-level wreath product

$G_1 ⓦ G_2 ⓦ \ldots ⓦ G_n$; i.e., an $n$-level hierarchy of *transfer*. (2) Each level $G_i$ is generated by a single generator; i.e., it is either a cyclic group or a connected 1-parameter Lie group. (3) Each level $G_i$ is an isometry group on its space of action (e.g., reflections, rotations, translations).

As an example, we saw that, in the New Foundations, the structure of a straight cylinder is defined as the wreath product $SO(2) ⓦ \mathbb{R}$. That is, the fiber group is the rotation group $SO(2)$ of the cross-section, and the control group is the translation group $\mathbb{R}$ along the axis. Since both $SO(2)$ and $\mathbb{R}$ are 1-parameter isometry groups, this wreath product is an iso-regular group.

The importance of iso-regular groups is given by one of the New Foundations' fundamental principles of recoverability, the *Externalization Principle* (page 27), which states that, to maximize recoverability, any generative sequence, inferred by external inference, must lead back to a starting state whose internal structure is an iso-regular group.

Next recall, from section 2.5, that the New Foundations gives a new theory of symmetry-breaking that is fundamentally different from conventional mathematics. This new theory states that the breaking of a symmetry group $G_1$ is given by its extension by another group $G_2$ via a wreath product thus: $G_1 ⓦ G_2$, where $G_2$ is the symmetry group of the asymmetrizing action. That is, whereas in the conventional theory of symmetry-breaking, the past symmetry is *lost* in the present broken symmetry; in contrast, in the new theory, the past symmetry is *transferred* onto the present broken symmetry.

One of the main symmetry-breaking actions that can be applied to the starting iso-regular group is deformation. Another is misalignment.

We saw that this leads to another crucial class of groups, invented in the New Foundations, called *unfolding groups*. Recall that the following two properties are basic to unfolding groups: (1) The control group acts *selectively* on only part of its fiber. (2)The control group is symmetry-breaking by *misalignment*.

It is necessary now to recall, from section 2.8, also the following details of unfolding groups, because these details will be used when we elaborate the mathematical theory that the New Foundations give of the bifurcation of singularity degeneracy.

Recall that the types of unfolding group illustrated in section 2.8 have the following structure: The fiber is the direct product of the iso-regular groups corresponding to the objects which are used in the New Foundations generation of the complex shape. Crucially, in this generation, one defines a reference hierarchy for those objects. Notice that the reference hierarchy is an example of what our theory calls *object-linked inheritance* in object-oriented programming. Recall, from section 2.7, that, according to the *algebraic theory of object-linked inheritance*, invented by the New Foundations, this type of inheritance is modeled by a wreath product, in which the personal transform group of the parent is the control group, and the personal transform group of the child is the fiber group. Therefore, in an unfolding group, one takes the direct product defining the fiber and one adds to it, by a hierarchy of wreath product operations, levels of control corresponding to the object-linked inheritance between the objects given in the fiber.

The entire group represents the following: Initially, the objects given in the fiber are coincident with their symmetry structures maximally aligned. I call this the *alignment kernel*. This configuration is given by the fiber-group copy that corresponds to the simultaneous selection of the identity elements from all the levels of control. The alignment of the objects is then broken hierarchically by the levels of control; i.e., in correspondence with the inheritance hierarchy.

Now, if one also allows the highest parent object to move with respect to the world-frame, one then adds the iso-regular group $G_W$, defining the symmetries of world frame, into the alignment kernel. Furthermore, one extends the control group by adding an extra level of control above the control levels that have already been included. The new top control level will move the highest parent object with respect to the world-frame.

Recall also that the New Foundations give a theory of *cloning* in object-oriented programming, as follows: This theory says that, because cloning an object produces a copy with the same instance values, the clone must be *aligned* with the original, and the use of its command group pulls it out of alignment, i.e., breaks the symmetry of the object-clone pair.

The theory of complex shape generation, in the New Foundations, was then illustrated by the theory of feature attachment in mechanical CAD, given by the New Foundations. Recall (page 33) that this theory says that when one creates objects and attaches them in the design structure, one is entering new instances into the alignment kernel, and positioning the command group for each new instance in the appropriate wreath level within the unfolding group corresponding to the inheritance hierarchy of the structure.

As a result of this, we have the following crucial consequence: My paper [23] shows that this theory of *mechanical CAD* in the New Foundations is equivalent to the theory of *musical composition* in the New Foundations. We will now see that this theory is also equivalent to the theory of *singularity bifurcation* in the New Foundations.

As an example, consider the process-bifurcation operation

$$Bm^+ : m^+ \longrightarrow m^+M^+m^+$$

which codes the situation, *breaking-through of a protrusion*, illustrated in Fig 3.28 (page 69).

In the last section, we discussed the structure of the curvature functions corresponding to this operation. The discussion used the sequence of three curvature functions in Fig 3.35 to illustrate the structure. In the first function, corresponding to the domain of the operation $Bm^+$, there is one singularity, a *non-degenerate* minimum. In the second function, the transition function, this minimum becomes 3-fold degenerate. In fact, its three coincident singularities will bifurcate as the three *non-degenerate* singularities, the minimum, maximum, minimum, in the third function, which corresponds to the codomain of the operation $Bm^+$.

Using this example, let us illustrate the theory of singularity bifurcation in the New Foundations.

First we need to consider the Symmetry-Curvature Duality Theorem. Recall that an important aspect of my *proof* of this theorem was the fact that there must be a neighborhood of the extremum in which the curve on each side of the extremum is a spiral, and my Spiral Theorem proved that a spiral cannot have a differential symmetry, and this forces any two symmetry-related tangents in the neighborhood of the extremum to come from the two opposite spiral sides of the extremum rather than the same spiral side. Now, consider one of the circles in the trajectory of bitangent circles leading to the extremum. As we have just said, the circle must be tangential to the curve at two points, A and B, on the two sides of the curvature extremum. Furthermore, the circle defines a reflection between the tangent-line at A and the tangent-line at B. Next, as the successive bitangent circles go towards the curvature extremum, the two tangent-lines, one on each side of the extremum, successively *converge*, and become coincident at the extremum.

Therefore, according to the New Foundations, the extremum is defined by the following iso-regular group

$$\mathbb{R} \circledS \mathbb{Z}_2 \tag{3.1}$$

which is explained as follows: In this wreath product, there are two copies of the fiber group, $\mathbb{R}$, corresponding to the two elements of the reflection control group $\mathbb{Z}_2$. These two copies of $\mathbb{R}$ correspond to the two tangent-lines that have become coincident at the extremum. The control group $\mathbb{Z}_2$ *transfers* the two copies of $\mathbb{R}$, i.e., the two tangent-lines, onto each other.

Notice that this wreath product group also describes the differential symmetry that relates the two tangent-lines for any bitangent circle in the trajectory. It is only at the extremum, that the two tangent-lines become coincident with each other, and their PISA-point becomes an actual point on the curve, i.e., the extremum. Thus, the extremum is the only individual curve-point that has the symmetry given by the above wreath product.

With respect to this, let us now apply our theory of feature attachment. The first thing our theory says is this: In creating and attaching objects into the structure, one enters them into the alignment kernel.

Therefore, in the transition from the non-degenerate minimum to the 3-fold degenerate minimum, one is adding two new objects into the alignment kernel, as follows: (1) One is *cloning* the existing minimum, and (2) one is adding a maximum. Thus, the alignment kernel is given as follows:

$$(\mathbb{R} \circledS \mathbb{Z}_2)_{m^+} \times (\mathbb{R} \circledS \mathbb{Z}_2)_{m^+} \times (\mathbb{R} \circledS \mathbb{Z}_2)_{M^+} \tag{3.2}$$

where each of the three components is the iso-regular group $\mathbb{R} \circledS \mathbb{Z}_2$, given in expression (3.1) for an extremum, and each is labeled by the non-degenerate extremum that it will become *after* bifurcation.

Next, our theory of feature attachment says this: In entering new instances into the alignment kernel, one positions the command group for each new instance in the appropriate wreath level within the unfolding group corresponding to the inheritance hierarchy of the structure.

To understand the command groups added into the control group, in the situation we are currently discussing, first observe the following: A valuable way of describing the bifurcation is to define the two minima as diverging *relative* to the central maximum. This means that, in the inheritance hierarchy, *both* minima are children of the central maximum.

Thus, taking the alignment kernel in expression (3.2), and adding the control structure that defines the movement of the two minima, we obtain the following unfolding group:

$$\begin{gathered}[(\mathbb{R} \circledS \mathbb{Z}_2)_{m^+} \times (\mathbb{R} \circledS \mathbb{Z}_2)_{m^+} \times (\mathbb{R} \circledS \mathbb{Z}_2)_{M^+}] \\ \circledS \;\; [\, AGL(2,\mathbb{R})_{m^+} \; \times \; AGL(2,\mathbb{R})_{m^+} \,]\end{gathered}$$

To help understand this structure, the group has been written on two lines. The first line is the fiber, which is the alignment kernel. The second line is the control group. This is seen from the fact that the wreath-product symbol is at the beginning of the second line. The important thing to observe is that the control group is the *direct product* of two affine groups $AGL(2,\mathbb{R})$, corresponding to the two $m^+$ extrema. These two affine groups will move the two $m^+$ extrema *relative* to the $M^+$ extremum. That is, they will move the two $m^+$ iso-regular groups, in the alignment kernel, relative to the $M^+$ iso-regular group, in the alignment kernel. The fact that the relation between the two affine groups is a direct product captures the fact that they are on the *same level* in the control group, which captures the fact that the two $m^+$ copies are on the same level in the inheritance hierarchy; i.e., they are both children of the $M^+$ extremum.

Thus, in the above unfolding group, the two copies of the affine group will *misalign* the *symmetries of the two minima* with respect to the *symmetries of the maximum*.

Now, the above structure defines the world-frame as fixed to, and aligned with, the maximum. There are some situations where this assumption can be useful; e.g., if the observer is traveling on the central extremum.

In other cases, where we wish to understand the maximum as moving relative to the world-frame, we expand the alignment kernel to contain the group $G_W$ of the world-frame, thus:

$$(\mathbb{R} \circledR \mathbb{Z}_2)_{m^+} \times (\mathbb{R} \circledR \mathbb{Z}_2)_{m^+} \times (\mathbb{R} \circledR \mathbb{Z}_2)_{M^+} \times G_W \quad (3.3)$$

In fact, according to the New Foundations, the group $G_W$ of the 2D world-frame is itself also a reflection structure $\mathbb{Z}_2 \circledR \mathbb{Z}_2$, as described in detail in the chapter on reference frames in my book *A Generative Theory of Shape*. Therefore, one of the fiber reflection axes of the reference frame must initially be coincident with the reflection axis of the extremum.

By adding the world-frame as an extra object, we have expanded the *inheritance* hierarchy thus: Both minima remain children of the maximum, but the maximum has now become a child of the world-frame.

Thus, taking the alignment kernel in expression (3.3), and adding the control structure that corresponds to this inheritance hierarchy, we obtain the following unfolding group:

$$\begin{aligned}
&[(\mathbb{R} \circledR \mathbb{Z}_2)_{m^+} \times (\mathbb{R} \circledR \mathbb{Z}_2)_{m^+} \times (\mathbb{R} \circledR \mathbb{Z}_2)_{M^+} \times G_W] \\
&\circledR \ [\, AGL(2,\mathbb{R})_{m^+} \ \times \ AGL(2,\mathbb{R})_{m^+} \,] \\
&\circledR \ AGL(2,\mathbb{R})_{M^+}
\end{aligned}$$

This group is now a wreath product of three levels, which have been put on three lines, to help the reader understand the structure. The first line is the fiber, which is the alignment kernel. The second line is the first level of the control group, which, as previously, is the direct product of the two affine groups that misalign the two minima with respect to the maximum. The third line is the second level of the control group, which has been added to misalign the maximum with respect to the world-frame.

Notice that both the second and third lines begin with a wreath-product symbol, which indicates that these are levels in the control group, and also indicates the *inheritance hierarchy*, which, according to the New Foundations, is coded by the wreath hierarchy.

The above discussion illustrates the following:

**ONE OF THE MATHEMATICAL THEORIES INVENTED IN THE NEW FOUNDATIONS TO GEOMETRY HANDLE:**
**BIFURCATION THEORY,**
**MECHANICAL CAD,**
**AND**
**MUSICAL COMPOSITION**

**The theory of singularity bifurcation, in the New Foundations, is equivalent to the theory of mechanical CAD, in the New Foundations, which is equivalent to the theory of musical composition, in the New Foundations.**

**In our singularity theory, bifurcation corresponds to the addition of iso-regular groups into the alignment kernel, which will undergo misalignment via command groups that are added into the wreath product at levels corresponding to the inheritance hierarchy.**

**In our theory of mechanical CAD, the design process corresponds to the addition of iso-regular groups (e.g., cylinders) into the alignment kernel, which will undergo misalignment via command groups that are added into the wreath product at levels corresponding to the inheritance hierarchy.**

**In our music theory, composing corresponds to the addition of iso-regular groups (e.g., meters) into the alignment kernel, which will undergo misalignment via command groups that are added into the wreath product at levels corresponding to the inheritance hierarchy.**

## 3.28 Operations and Operators

In the Level 3 Process Grammar (page 72), there are two continuation operations $Cm^+$, $CM^-$, and four bifurcation operations $BM^+$, $Bm^-$, $Bm^+$, $BM^-$. The symbol $C$, in the continuation operations, and the symbol $B$, in the bifurcation operations, will be called **operators**. Thus, the concept is that, when an *operator*, $C$ or $B$, is applied to a particular type of extremum, it becomes an *operation*. Thus we will say that an operator is **realized by** a certain set of operations. For example, the $C$ operator is realized by two operations, and the $B$ operator is realized by four operations.

As an intervening level between an operator and its operations, we will also distinguish between whether an operator is applied to a minimum or a maximum. Thus, when the $B$ operator is applied to a minimum ($m^+$ or $m^-$), it will be called the $B_3$ operator; and when the $B$ operator is applied to a maximum ($M^+$ or $M^-$), it will be called the $B_{-3}$ operator. The number 3, in the operators $B_3$ and $B_{-3}$, will refer to the fact that both of these operators induce a 3-fold extremum bifurcation. Also, when the $C$ operator is applied to a minimum ($m^+$), it will be called the $C_2$ operator; and when the $C$ operator is applied to a maximum ($M^-$), it will be called the $C_{-2}$ operator. The number 2, in the operators $C_2$ and $C_{-2}$, will refer to the fact that both of these operators induce a 2-fold zero-bifurcation. The form of this notation and terminology will be useful when we define the larger collection of Process Grammar operations, later in the book.

## 3.29 Parts of Shape

One of the purposes of this book is to describe additional *causal operations* in the Process Grammar. We shall see that the operations given so far can be viewed as defining the *parts* of a shape. Furthermore, several of the operations to be introduced later can also be viewed defining the parts of a shape.

**THEORY OF PARTS**

**The current literature on *parts* of shape is *stupid* by the technical definition of intelligence given by the New Foundations to Geometry.**

**The reason is that the current literature gives a *non-causal* theory of parts, and therefore is not based on *understanding* what a part is.**

**The theory of parts given by the New Foundations is the opposite.**

**Throughout all my books, parts are defined as:**
*phases of causal scenarios.*

**This is the only *intelligent* definition of a part.**

Many of the Process Grammar operations, to be described later in the book, will represent causal scenarios that create what we will call *three-fold parts* and *five-fold parts*. These operations will be applied to understand morphology in biology (e.g., chins, jaws, animal-heads), morphology in geology (e.g., different types of bays, harbors, etc), as well as architectural design (bridges, balconies, entrances, etc.), and the design of vehicles (cars, vans, and buses).

## 3.30 Interactive Singularity Theory

After we have elaborated several additional operations of the Process Grammar, in Chapters 4 – 6, there is then a sequence of chapters that explain the corresponding singularity theory, which I invented, called *Interactive Singularity Theory*.

Interactive Singularity Theory had to be invented because the standard singularity theories of shape are not sufficient to understand morphology. For example, we will see that the Catastrophe Theory of René Thom was not sufficient. Let me explain why:

Catastrophe theory is one of the most powerful and valuable branches of mathematics. It is a branch of mathematics I respect enormously. There have been many successful applications of Catastrophe Theory; e.g., in physics. However, despite these successful applications, one of the main areas to which René Thom intended the application of Catastrophe Theory was *spatial* morphology. For example, a principle goal of his was to explain the development of *biological* shape. Unfortunately, he did not succeed in producing a theory of spatial morphology, and, in particular, biological morphology. I argue that there are a number of reasons for this, as follows:

First, Thom did not have the Symmetry-Curvature Duality Theorem. I have shown that it is impossible to have a theory of biological morphology without this theorem.

Second, there is an important problem with respect to what Thom regarded as generic functions. A generic function in Thom's theory is a function $f$, whose *first derivative* $Df$ has a graph that is *transversal* to the domain axis $t$. However, Thom did not consider whether the graph of the function $f$ itself is transversal to the domain axis. In fact, most of the functions that Thom considered generic are non-transversal to the domain axis. In contrast, I argue that there are profound reasons why, in spatial morphology, such functions should be considered *non-generic*. The consequences of this are enormous, as we will see:

I invented what I call *Interactive Singularity Theory* as defining the singularity structure of spatial morphology. A fundamental rule of Interactive Singularity Theory is that generic functions are required to satisfy *both* of the above forms of transversality, i.e., not only the transversality required by Thom. The first major consequence of this is that the space of functions required to model morphology is considerably larger than the space of functions used by Thom. Furthermore, the structure of this larger space is considerably more complicated than the structure of the space used by Thom.

Nevertheless, the separatrix structure, which partitions this larger space of functions, involves *two* of the separatrix structures defined by Thom – and therefore I consider Thom's contribution to this theory as extremely valuable. However, the fact that *two* of

Thom's structures are involved, and cut across each other, causes the space of functions to be structured in a more complex way than he envisioned for spatial morphology.

The Process Grammar operations correspond to movements between the regions of generic functions in this more complex space. These movements are therefore organized in a much more complicated way than the movements defined by Thom in his smaller less complicated spaces.

The complexity of the space is profoundly important as follows: We will see that the two types of Process Grammar operations, process-continuations and process-bifurcations, correspond, respectively, to the two types of transversality that are involved in this larger space.

The crucial fact is this: *Because the two types of Process Grammar operations are modeled in the same space of functions, they structure each other's uses.* In particular, we will see that, to understand spatial morphology, it is important to recognize that the system of process-continuations structure the system of process-bifurcations. In Thom's Catastrophe Theory, this structural organization of bifurcations does not exist, and we shall see that this is a fundamental reason why his theory fails to model spatial morphology.

There is an additional basic issue: The Process Grammar not only encodes the structure of this more complex space of functions, in terms of a larger system of singularities than that used by Thom, but it achieves a further fundamental goal that Thom's Catastrophe Theory does not: It gives the powerful *causal* structure on which that space is founded. This is provided by the laws of recoverability in the New Foundations. That is, it is the laws of recoverability that lead to the Symmetry-Curvature Duality Theorem, the Interaction Principle, the PISA axes, etc. In fact, the laws of recoverability, in the New Foundations, provide the following important fact:

**FUNDAMENTAL PROPERTY**
**OF**
**INTERACTIVE SINGULARITY THEORY**

**A fundamental property of Interactive Singularity Theory is that the singularity bifurcations, defined in Interactive Singularity Theory, match the causal structure given by the PISA axes that correctly explain those singularity bifurcations; e.g., explaining any $B$ extremum-bifurcation as caused by a compressive process opposing a penetrative process, and explaining any $C$ zero-bifurcation as caused by the continuation of a compressive process till it penetrates.**

**We shall see that this fundamental property is crucially important for biological morphology and manufacturing design.**

## 3.31 Appendix: Notes on Presentation

In order to help the reader understand concepts in this book, we will use certain formats in presenting diagrams and texts, as follows:

### Extra Issues Concerning Direction of Curve Parameter

As we have already seen illustrated, an important aspect of understanding the Process Grammar operations is to understand the effects of the operations on the curvature functions. We know that, obviously, to be able to correspond a curvature function with a parameterized curve which has that curvature function, one needs to understand the direction of the curve parameter $t$ along the curve. This will involve certain *extra issues* that we will now explain.

Several figures in the book will contain *both* a curvature function and its corresponding (computed) parameterized curve. To illustrate, Fig 3.36 gives an example of the curvature extremum $M^+$, showing both its curvature function, and its corresponding parameterized curve. The upper diagram is the *graph* that plots the curvature (vertical axis) against curve parameter $t$ (horizontal axis). In contrast, the lower diagram shows the $(x, y)$ plane, into which the corresponding parameterized curve has been placed.

In such a pair of diagrams, we use the following obvious convention: The curve parameter $t$, which is the horizontal axis in the graph of the curvature function, is increasing from *left to right* along that axis. Correspondingly, this parameter $t$ goes from *left to right* along the parameterized curve. Thus, the direction of the parameter $t$, in *both diagrams*, is *left to right*.

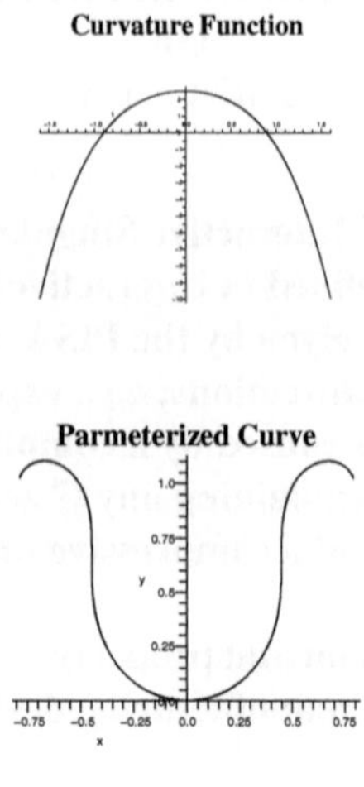


**Fig. 3.36.** When the curve parameter $t$ goes from left to right on a parameterized curve, a $M^+$ extremum points downwards on the parameterized curve.

This convention will be used whenever we show, within a figure, a curvature function together with its corresponding parameterized curve. The reason is that this allows us to easily match the features of the curvature function with features of the parameterized curve.

There is however a conceptual problem. Because the extremum in Fig 3.36 is $M^+$, it defines, on the parameterized curve, a *protrusion*. Usually, people conceive of protrusions as pointing upwards; e.g., the largest type of protrusion that people encounter is a *mountain peak*, which indeed points upwards. However, we see that, on the parameterized curve in Fig 3.36, the fact that the parameter $t$ goes from left to right, makes the protrusion point downwards. A downward pointing penetrative extremum is more like the way people conceive of a *valley*, rather than a mountain peak.

A valley corresponds to the opposite type of penetrative extremum, $m^-$. With respect to this issue, now consider Fig 3.37. It gives an example of $m^-$, showing both its curvature function, and its corresponding parameterized curve. Again, we use the convention that the curve parameter $t$ goes from left to right along the horizontal axis of the graph of the curvature function, and from left to right along the parameterized curve.

Here, we see another example of the conceptual problem. Because the extremum in Fig 3.37 is $m^-$, it defines, on the parameterized curve, an *indentation*. Usually, people conceive of indentations as pointing downwards; e.g., the largest type of indentation that people encounter is a valley, which indeed points downwards. However, we see that, on the parameterized curve in Fig 3.37, the fact that the parameter $t$ goes from left to right, makes the indentation point upwards. An upward pointing penetrative extremum is more like the way people conceive of a mountain peak, rather than a valley.

Despite these conceptual difficulties, we will not give up the convention of making the parameter $t$, on both the function and the parameterized curve, go from left to right, when both the function and curve are shown in the same figure.

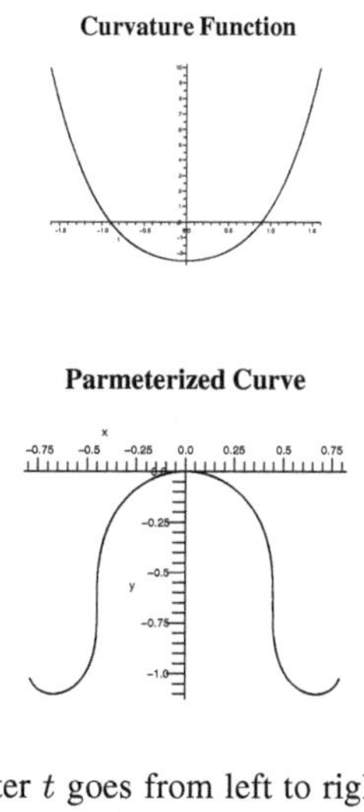


**Fig. 3.37.** When the curve parameter $t$ goes from left to right on a parameterized curve, a $m^-$ extremum points upwards on the parameterized curve.

However, when showing the parameterized curve *separately*, we will usually show a $M^+$ extremum (protrusion) as pointing upwards, and a $m^-$ extremum (indentation) as pointing downwards, since the figure will be conceptually easier for the reader to understand. This will imply that the curve parameter is going from *right to left*.

Now observe the following: Many of the illustrating curves will be closed planar curves (without self-intersection). Furthermore, in most cases, the "solid" will be understood as within the curve. Note that all the closed curves used in the present chapter, to illustrate the Process Grammar operations, were of this type. On such a closed curve, the curve parameter goes in the *anti-clockwise* direction. This is because, in the standard curvature convention, the direction of traveling along the curve keeps the solid to the left of the traveler.

Furthermore, to provide illustrations of the Process Grammar operations, it is perceptually easiest to view the operations when they are applied to the *top* of the closed curve. However, since the curve parameter $t$ is anti-clockwise, it moves from *right to left* along the top of the curve. Again, this goes in the opposite direction of the parameter $t$ in the graph of the curvature function, where the parameter goes from *left to right* along the horizontal axis.

In some chapters however, where the Process Grammar operations are complicated and need to be carefully matched with the curvature function, the operations will be applied to the bottom of the closed curves. This is due to the fact that, because the curve parameter goes anti-clockwise on the closed curve, it goes from *left to right* along the bottom of the closed curve.

### Text and Diagrams

We now describe a text format in relation to diagrams. Because each diagram will be a valuable tool in helping the reader to understand several details in the text, it will be important to keep a diagram as near as possible to its corresponding text. To do so, it will be necessary sometimes to stop the text at some point on a page and leave the rest of the page blank. For example, this will help the reader if that text concerns a full-page diagram on the next page, and the subsequent text concerns the next diagram. Therefore the subsequent text will be given after that full-page diagram, and will be as near as possible to the diagram it discusses, rather than preceding the previous diagram.

### Mathematical Expressions and Punctuation

Another format, we will use, concerns punctuation and any text line that consists of a single mathematical expression, e.g., a mathematical expression placed like this:

$$G_1 = G_2 \circledS_{\tau} G_3$$

We will create the following convention in this book. In most books, a text line, consisting of a single mathematical expression, conventionally contains a punctuation that gives the grammatical relation of the mathematical expression to the surrounding text. For instance, at the end of the mathematical expression, there can be a dot if the mathematical expression is the end of a sentence; for example:

$$G_1 = G_2 \circledS_{\tau} G_3.$$

Or there can be a comma, at the end of the mathematical expression, if the mathematical expression is at the end of a clause; for example:

$$G_1 = G_2 \circledS_{\tau} G_3,$$

However, we will adopt a new convention. We will never have a text punctuation in a text line consisting of a single mathematical expression. The reason is that a dot can be interpreted as a algebraic operation, and a comma can be interpreted as a mathematical comma from a mathematical set. Therefore, a text punctuation in a text line consisting of a single mathematical expression might be confused as having a mathematical role in the expression. It is for this reason that we will never include a text punctuation in a text line consisting of a single mathematical expression. In fact, I argue that, without the punctuation, it is nevertheless obvious to the reader if the mathematical expression is at the end of a sentence or the end of a clause, and therefore it is generally not necessary to include the punctuation.

# 4

# Pair-Creation

## 4.1 Introduction

A crucial aspect of the Process Grammar is that it captures *design intent* in morphology. This is shown in considerable detail in the next chapter. Other theories of morphology fail to capture design intent.

In the present chapter, we will describe a Process Grammar operator that defines an important morphological transition: that which *simultaneously* realizes *two design intents* on a curve. This morphological transition is created at a curve point that is *away from an existing extremum*; i.e., on a spiral.

An example is shown in Fig 4.1. The first shape (the left shape) has only one extremum $M^+$, and its single corresponding process is given by the arrow shown in that shape.[1] Notice that, since this extremum is the only one on that shape, then if we divide the shape at the extremum, the two halves would each be a *spiral.*

In the morphological transition that we are now going to define, the two processes that will simultaneously produce the *two design intents*, occur initially opposite each other at a single point on one of the two spiral-sides. In Fig 4.1, these two processes are shown as the two opposing arrows at point $I$ on the second shape. This shape also shows their initial effect, which is that they have made the curve less spiral at $I$, that is, more constant curvature locally, as seen in the neighborhood of $I$ on the second shape. Thus they made the curve below $I$ more flat, and the curve above $I$ more sharp. Then, in the creation of the two design intents, they bifurcate to become the forces shown at the $m^+$ extremum and the upper $M^+$ extremum in the third shape.

The third shape looks somewhat like an animal head. We will see important examples of this morphological transition in vehicle design in the next chapter.

> **Whereas the Process-Grammar models the transition in Fig 4.1 correctly, other morphological theories would fail to do so because they would not correctly model the resulting compressive extremum.**

---

[1] The fact that the first shape in Fig 4.1 has only one extremum is shown by the curvature function shown on page 416 for a curve that is congruent to this one, with the solid and empty side reversed.

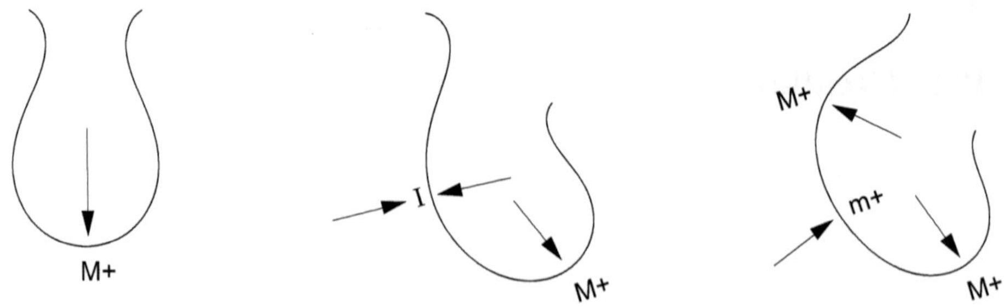


**Fig. 4.1.** Simultaneous realization of two design intents.

## 4.2 The Pair-Creation Operator

I will now develop a theory of what is going on in such situations. First, we introduce a new term: The existing extremum $M^+$ in the first shape of Fig 4.1 will be called the **anchor-extremum**. The process inferred from the anchor-extremum is a *previous* process that was responsible for creating the "body" of the shape

Now consider the left spiral in the first shape in Fig 4.1. Its curvature function is qualitatively illustrated by the first graph in Fig 4.2. Notice that the horizontal axis of the graph is $t$, the parameter along the curve; and movement along this axis, from *left to right*, corresponds to movement along the curve in Fig 4.1, from left to right. The position marked $M^+$ on the $t$-axis of the graph is the position on the curve parameter where the anchor-extremum occurs. Obviously, this is the point where the graph reaches the highest point. Observe that another point has been marked on the $t$-axis of this graph: the point labeled $\emptyset$. This is the point on the curve parameter, where the new causal action will later be applied. The reason why it is labeled with the empty-set symbol $\emptyset$ is because, in the initial stage (the first graph), it is not a critical point, i.e., not a point where the graph has a horizontal tangent. In particular, it is not an extremum. Notice that, at the point $\emptyset$, the tangent-line to the graph is shown as the dashed line.

Now, in going from the first to the second shape in Fig 4.1, the curvature function changes from the first graph to the second graph in Fig 4.2. Observe the following crucial feature about this change: The tangent-line shown in the first graph in Fig 4.2 is *rotated clockwise* to become *horizontal* in the second graph. This means that the graph at this point has become a horizontal *inflection*. The position on the $t$-axis is now labeled $I$ for inflection.

Finally, the third graph in Fig 4.2 illustrates the curvature function corresponding to the third shape in Fig 4.1. What has happened is that the tangent-line has rotated even further, and is therefore no longer horizontal. Thus: Within the region of the causal action we have described as applied away from the pre-existing extremum, the curvature gradient undergoes a rotation that changes the sign of the gradient.

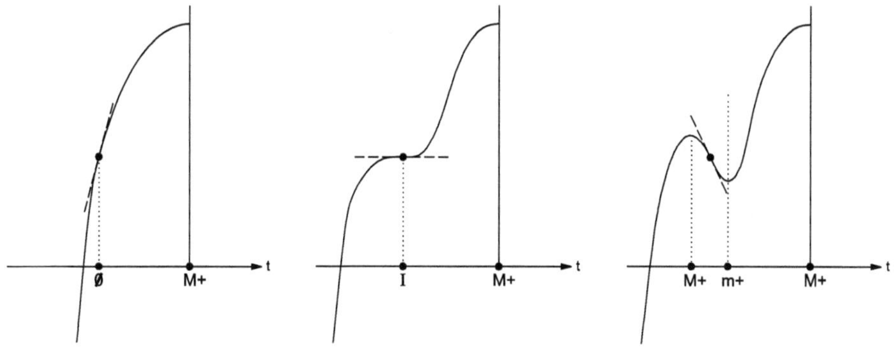


**Fig. 4.2.** Rotating tangent to the graph of the curvature function.

The important consequence of the rotation is this: Two new extrema are introduced into the third curvature function: a positive maximum $M^+$ and a positive minimum $m^+$. In fact, these two extrema bifurcated from the inflection point $I$ in the second graph.

I will call this type of bifurcation, **pair-creation**. Correspondingly, the operator describing the bifurcation will be called the **pair-creation operator**, and will be represented by the symbol $\pi$. This will be the new operator to be defined in the Process Grammar.

## PAIR-CREATION OPERATOR

**The pair-creation operator $\pi$ is applied to a point $\emptyset$ that is not a critical point, and causes a bifurcation of the point into two extrema of opposite types (Max and min).**

Now, just as the *operators* $B$ and $C$ each become a set of *operations* when applied to specific types of cases, the *operator* $\pi$ becomes a set of *operations* when applied to specific types of cases. To illustrate, let us return to the example in Fig 4.1. We observed that the first shape is divided into two spirals at the anchor-extremum $M^+$. It is important now to notice that, because a spiral is a curve with only monotonically increasing curvature, or monotonically decreasing curvature, the two spirals in the first shape in Fig 4.1 are distinguished by the fact that the left spiral has monotonically increasing curvature, and the right spiral has monotonically decreasing curvature (i.e., following the curve parameter $t$ from left to right along the curve).

We can see the fact that the left spiral has monotonically increasing curvature by the fact that the first graph in Fig 4.2 has an *upward* gradient. Thus, the particular pair-creation operator $\pi$, that is to be applied here, will be called the **upward pair-creation operator** $\pi^\uparrow$, where the upward arrow $\uparrow$ indicates the upward gradient of the region of application.

The next thing to observe about the first shape in Fig 4.1 is that each spiral has positive curvature in a neighborhood of the extremum, i.e., the shape bends out. However, further away from the extremum, the curvature becomes negative, i.e., the curve bends in. Therefore, the first shape in Fig 4.1 has a curvature function which is qualitatively illustrated by the graph in Fig 4.3. The two halves of the graph correspond to the two spirals in the shape. Furthermore, as shown, the graph is divided into *four* regions, labeled as follows:

| | |
|---|---|
| $\uparrow +$ | upward positive curvature |
| $\uparrow -$ | upward negative curvature |
| $\downarrow +$ | downward positive curvature |
| $\downarrow -$ | downward negative curvature |

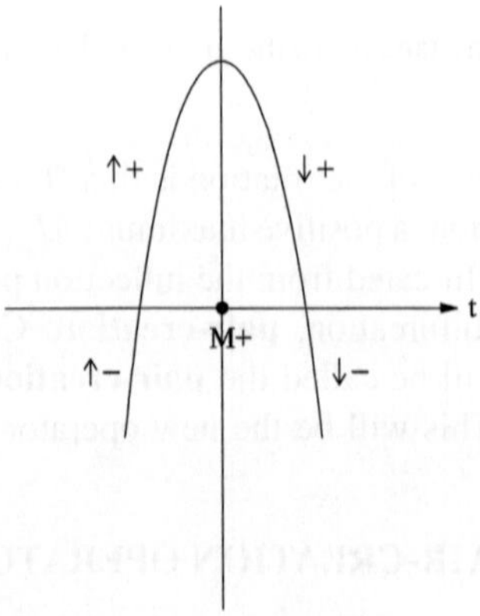

**Fig. 4.3.** Curvature function corresponding to a $M^+$ extremum.

Because, in the graph, the upper two regions $\uparrow +$ and $\downarrow +$ have the same curvature sign (positive) as the extremum $M^+$, they will be called the **same-sign components** of the extremum. Conversely, because the lower two regions $\uparrow -$ and $\downarrow -$, the "tails" of the graph, have the opposite curvature sign (negative) from the extremum $M^+$, they will be called the **opposite-sign tails** of the extremum.

Now, by applying the pair-creation operator $\pi$ to the four regions, we obtain four pair-creation operations, one for each region.

To illustrate, let us first return to the example shown in the graphs of Fig 4.2. It is clear that the pair-creation operator was applied to a point $\emptyset$ in the $\uparrow +$ region of the first graph. Thus, the pair-creation operation that describes this case will be written as follows:

$$\pi\emptyset^{\uparrow+} \; : \; \emptyset^{\uparrow+} \quad \longrightarrow \quad M^+m^+ \tag{4.1}$$

where the symbol $\emptyset^{\uparrow+}$ means that the application-point $\emptyset$ has an upward gradient and +ve curvature. Observe that the codomain $M^+m^+$, in this operation, is the pair that enters the third function in Fig 4.2.

This operation is an application of the *upward* pair-creation operator $\pi^{\uparrow}$. It is clear that the same operator can also be applied to a point in the *negative tail* in the first graph in Fig 4.2. That is, whereas the clockwise rotating tangent occurred previously at a point on the graph above the $t$-axis, it now occurs at a point on the graph below the $t$-axis. In this case, it is easy to see that the operation is:

$$\pi\emptyset^{\uparrow -} \ : \ \emptyset^{\uparrow -} \quad \longrightarrow \quad M^{-}m^{-} \tag{4.2}$$

where the symbol $\emptyset^{\uparrow -}$ means that the application-point $\emptyset$ has an upward gradient of -ve curvature.

The effect on the shape is shown in Fig 4.4. The left shape is the starting shape which, again, has a single $M^{+}$ extremum. On this left shape, two dots have been placed, which show where the curvature is zero on the two spirals. Consider the left spiral. Previously, we had applied the pair-creation operator to a point in the same-sign component of the extremum, i.e., the positive curvature component below the left zero-point. However, now we are applying the pair-creation operator to a point in the opposite-sign tail of the extremum, i.e., the negative curvature component above the left zero-point. The consequence is shown in the right shape; i.e., the pair of extrema, $M^{-}$ and $m^{-}$, are introduced into the *tail*. The resulting tail, in the right shape, will be called the **dragon's tail**.

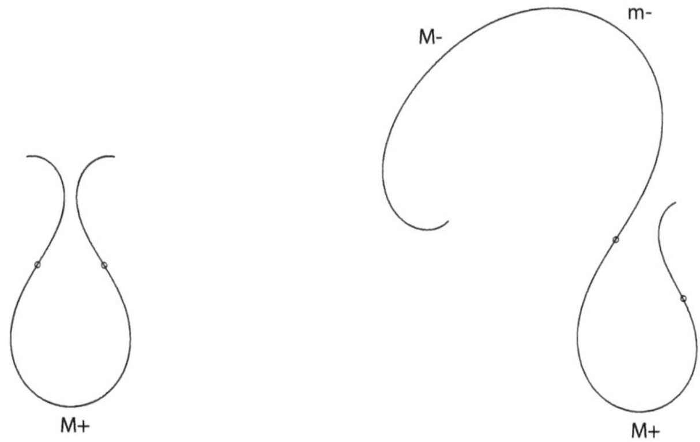


**Fig. 4.4.** Application of the pair-creation operator within the negative tail of the left spiral of the left diagram produces the *dragon's tail* in the right diagram.

Expressions (4.1) and (4.2) give the two operations that realize the upward pair-creation operator $\pi^{\uparrow}$. They are the operations that would be applied within the positive and negative regions on the left in the graph shown in Fig 4.3. For obvious reasons we will call them, respectively, **the positive and negative pair-creation operations that realize the upward operator** $\pi^{\uparrow}$.

On the right side of Fig 4.3, the graph is downward and correspondingly the applicable pair-creation operator will be called the **downward pair-creation operator** $\pi^{\downarrow}$. The downward side is again divided into two regions, positive and negative, and therefore the application of the operator within these two regions will be called, respectively, **the positive and negative pair-creation operations that realize the downward operator** $\pi^{\downarrow}$. We now show how they are defined:

When one applies the downward pair-creation operator to a point on the positive region $\downarrow +$, it is clear that this will correspond to rotating the tangent on the graph in the *anti-clockwise* direction. This would result in the following pair-creation operation:

$$\pi\emptyset^{\downarrow +} \; : \; \emptyset^{\downarrow +} \quad \longrightarrow \quad m^{+}M^{+} \tag{4.3}$$

where the symbol $\emptyset^{\downarrow +}$ means that the application-point $\emptyset$ has a downward gradient of +ve curvature. Observe that the codomain sequence $m^{+}M^{+}$ is the *reverse* of the pair produced by the previous positive operation in expression (4.1). Obviously, this new operation would be illustrated by Fig 4.1, except that the second and third shape would each be vertically reflected.

Next observe that, when one applies the downward pair-creation operator to a point on the negative region $\downarrow -$, it is clear that this will *also* correspond to rotating the tangent on the graph in the *anti-clockwise* direction. Since this rotation would be below the $t$-axis, it is easy to see that the operation would be:

$$\pi\emptyset^{\downarrow -} \; : \; \emptyset^{\downarrow -} \quad \longrightarrow \quad m^{-}M^{-} \tag{4.4}$$

where the symbol $\emptyset^{\downarrow -}$ means that the application-point $\emptyset$ has a downward gradient of -ve curvature. Observe that the codomain sequence $m^{-}M^{-}$ is the *reverse* of the pair produced by the previous negative operation in expression (4.2). Obviously, this new operation would be illustrated by Fig 4.4, except that the right shape would be vertically reflected.

Let us finally provide some extra clarification of the terminology introduced in this section; and we will use Fig 4.5 to illustrate: The overall operator that we have introduced is the pair-creation operator $\pi$. This has two cases: the upward pair-creation operator $\pi^{\uparrow}$, and the downward pair-creation operator $\pi^{\downarrow}$. Each of these is realized by a positive and negative pair-creation operation. Thus, the overall pair-creation operator $\pi$ is realized by four pair-creation operations: $\pi\emptyset^{\uparrow +}$, $\pi\emptyset^{\uparrow -}$, $\pi\emptyset^{\downarrow +}$, $\pi\emptyset^{\downarrow -}$. For later review purposes we summarize the structure in the following bold statements.

**UPWARD AND DOWNWARD PAIR-CREATION OPERATORS**

**When the pair-creation operator $\pi$ is applied to a point of upward curvature gradient, it will be called the** ***upward pair-creation operator*** $\pi^{\uparrow}$.

**When $\pi$ is applied to a point of downward curvature gradient, it will be called the** ***downward pair-creation operator*** $\pi^{\downarrow}$.

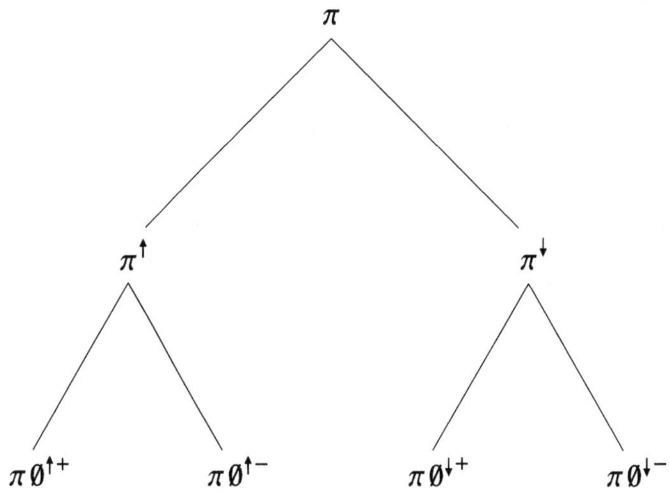


**Fig. 4.5.** The pair-creation operator $\pi$ realized as the four non-zero pair-creation operations.

## NON-ZERO PAIR-CREATION OPERATIONS

**When the upward and downward pair-creation operators, $\pi^{\uparrow}$ and $\pi^{\downarrow}$, are applied to points of non-zero curvature, they will be called the** *non-zero pair-creation operations.*

**There are a total of four non-zero pair-creation operations: Two of them have an application-point of positive curvature, and will therefore be called the** *positive pair-creation operations.* **The other two have an application-point of negative curvature, and will therefore be called the** *negative pair-creation operations.*

**The four non-zero pair-creation operations are therefore:**

$$\pi\emptyset^{\uparrow+} : \emptyset^{\uparrow+} \longrightarrow M^{+}m^{+}$$

$$\pi\emptyset^{\uparrow-} : \emptyset^{\uparrow-} \longrightarrow M^{-}m^{-}$$

$$\pi\emptyset^{\downarrow+} : \emptyset^{\downarrow+} \longrightarrow m^{+}M^{+}$$

$$\pi\emptyset^{\downarrow-} : \emptyset^{\downarrow-} \longrightarrow m^{-}M^{-}$$

**Notice that they all produce a penetrative and compressive extremum, and therefore the Process Grammar is the only morphology theory that would correctly describe them, whereas other morphology theories would not.**

## 4.3 The Application-Types

We will now see that each of the four non-zero pair-creation operations has three types of *application*, obtained by *changing the anchor-extremum*.

The logic of this is as follows: Consider Fig 4.6, which shows curvature functions corresponding to the four types of curvature extrema. The first two extrema are the *penetrative* extrema; and the other two extrema are the *compressive* extrema. For the first two extrema, the spiral on each side of the extremum has been extended to cross the $t$-axis, which means that the extrema have been given *opposite-sign tails*. This is impossible with the spirals of the other two extrema.

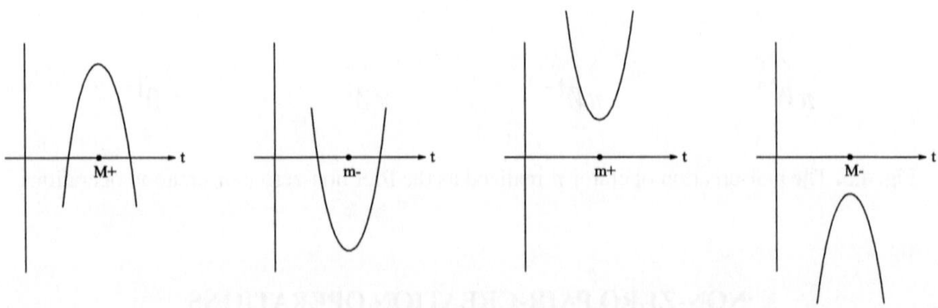


**Fig. 4.6.** Curvature functions corresponding to the four types of extrema.

This means that the first two extrema have all four possible regions, $\uparrow +$, $\uparrow -$, $\downarrow +$, $\downarrow -$, and therefore can allow all four non-zero pair-creation operations. However, the other two extrema have only two possible regions, either both positive, or both negative. Therefore, a consequence is that each of the operations can be applied to a side of *three* of the extrema: It can be applied to a side of both of the first two extrema, and to a side of only one of the last two extrema. These will be called the **three extrema side-applications** of the operation, or simply the **three applications** of the operation. Table 4.1 gives the three applications of each of the positive and negative pair-creation operations.

The previous section gave diagrams illustrating the application of the two *upward* operations for the first extremum $M^+$; and we noted that the diagrams for the two *downward* operations are obviously the reflections of these. It is clear that the same set of diagrams illustrate the application of the four operations with respect to the second extremum $m^-$, except that we simply change the extrema labels to their duals. Thus, the four operations have been illustrated with respect to the two *penetrative* extrema, $M^+$ and $m^-$. It remains therefore to illustrate the operations with respect to the two *compressive* extrema, $m^+$ and $M^-$.

Let us begin with the left side of the extremum $m^+$. From Fig 4.6, the reader can see that this side of $m^+$ is only one region, $\downarrow +$, positive curvature of downward gradient, and therefore, the only operation for this side is:

| Operations | Applications | | |
|---|---|---|---|
| $\pi\emptyset^{\uparrow +}$ | $+$ve left side of $M^+$ | $+$ve right side of $m^-$ | right side of $m^+$ |
| $\pi\emptyset^{\uparrow -}$ | $-$ve left side of $M^+$ | $-$ve right side of $m^-$ | left side of $M^-$ |
| $\pi\emptyset^{\downarrow +}$ | $+$ve right side of $M^+$ | $+$ve left side of $m^-$ | left side of $m^+$ |
| $\pi\emptyset^{\downarrow -}$ | $-$ve right side of $M^+$ | $-$ve left side of $m^-$ | right side of $M^-$ |

**Table 4.1.** Each of the positive and negative pair-creation operations has three applications obtained by changing the anchor-extremum. The words "left" and "right", in the table, refer to a side of the anchor-extremum as it appears in the curvature function.

$$\pi\emptyset^{\downarrow +} \; : \; \emptyset^{\downarrow +} \quad \longrightarrow \quad m^+M^+$$

The effect on an actual *shape* is illustrated in Fig 4.7. The left shape contains only one extremum $m^+$, which will be the anchor-extremum. The right shape shows the result of applying the pair-creation operator to the left side of this extremum; i.e., the pair $m^+M^+$ has been introduced into the left side.

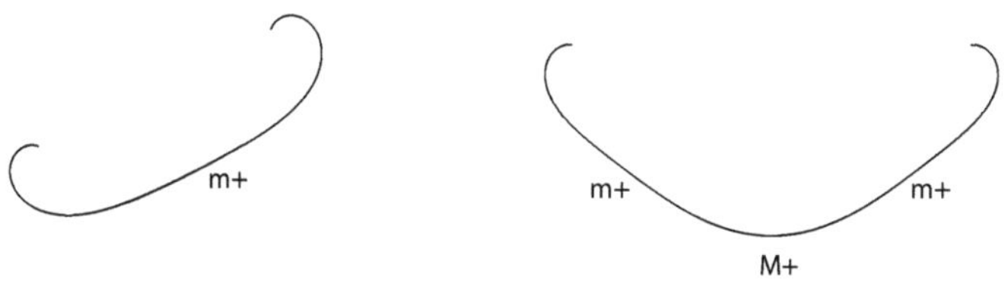


**Fig. 4.7.** Pair-creation on the left side of a $m^+$ extremum.

Now consider the right side of the graph of extremum $m^+$ in Fig 4.6; i.e., the side of upward gradient. It is clear that the application of the pair-creation operator to this side is simply the reflection of the application to the left side; i.e., it is the operation

$$\pi\emptyset^{\uparrow +} \; : \; \emptyset^{\uparrow +} \quad \longrightarrow \quad M^+m^+$$

Furthermore, considering the two sides of the final extremum $M^-$ in Fig 4.6, we obtain simply the duals of the two operations used for $m^+$, that is, the same-shaped curves with the extrema-labels replaced by their duals.

An important consequence of the above discussion is this: Recall first that each of the four operations has three applications. However, the illustrating shapes for the two upward operations are reflections of the illustrating shapes for the two downward operations. In addition, the illustrating shapes for the two positive operations are exactly the same as the illustrating shapes for the two negative operations, except the extrema-labels have been replaced by their duals. This means that the shapes that illustrate the three applications of one operation are the same as the shapes that illustrate the three applications of any other operation (i.e., up to reflection and duality). Therefore, we will refer to them as the *three application-types* of any non-zero pair-creation operation. They are listed in Table 4.2.

**THE 3 APPLICATION-TYPES**
**OF A NON-ZERO PAIR-CREATION OPERATION**

**Each of the non-zero pair-creation operations is realized by the three application-types listed in Table 4.2.**

| Anchor-Extremum | Application-Type |
|---|---|
| Compressive | Application to a point in one of the two spirals. |
| Penetrative | Application to a point in one of the two opposite-sign tails. |
| Penetrative | Application to a point in one of the two same-sign components. |

**Table 4.2.** The 3 application-types of a non-zero pair-creation operation.

The *diagrams* for the three application-types are shown in Fig 4.8. That is, the same three diagrams (up to reflection) hold for each of the operations. Notice that both, Table 4.2 and Fig 4.8, list the three types in the opposite order from that which they were introduced in this chapter. The reason for the new order is as follows: In the Process Grammar, the continuation operator $C$ goes from a compressive extremum to a penetrative extremum; i.e., by continuing a compressive force till it penetrates. It is for this reason that the first line of the table is an anchor-extremum given by a compressive force, which can be viewed as continuing till it creates the penetrative extremum in the second line of the table. However, even though the sign of the anchor-extremum changes, in this way, one can nevertheless preserve the sign of the application-point to apply the same pair-creation operation as in the previous case. This is achieved by locating the application-point within the *tail* of that penetrative anchor-extremum, as shown in the second line of Table 4.2. However, in the third case, to keep the same pair-creation operation as that in the previous two cases, one would have to consider its penetrative anchor-extremum to be the dual of the penetrative anchor-extremum of the second case.

Type 1: Compressive extremum: Application to a point in one of the two spirals.

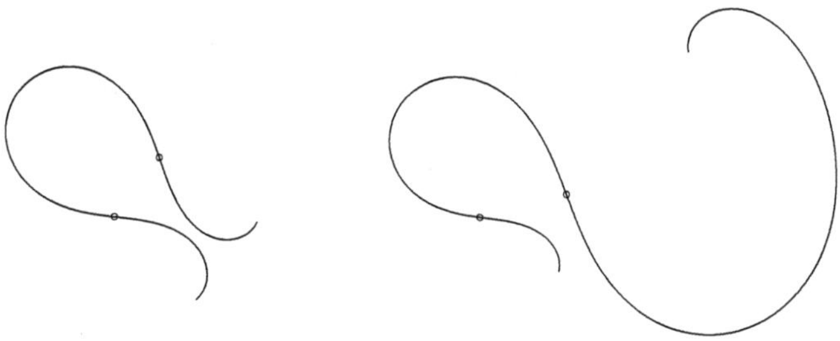

Type 2: Penetrative extremum: Application to a point in one of the two opposite-sign tails.

Type 3: Penetrative extremum: Application to a point in one of the two same-sign components.

**Fig. 4.8.** These three diagrams (up to reflection) illustrate the three application-types of a non-zero pair-creation operation.

## 4.4 The Adding-On Structure of Pair-Creation

In Fig 4.1 (page 98), which illustrated the pair-creation operation $\pi\emptyset^{\uparrow+}$, we placed the anchor-extremum $M^+$ downward because this allowed an easy visual match between the shapes and the curvature functions; i.e., it gave the curve parameter $t$ as going from left to right in the shapes and their curvature functions. However, as stated in section 3.31, the downward orientation of a protrusion is counter-intuitive because people generally think of protrusions as upward, like mountains, and indentations as downward like valleys. Thus Fig 4.9 gives exactly the same diagrams as in Fig 4.1, except that the anchor-extremum now has been oriented upward. This will make the discussion, in this section, easier to follow. Observe now that the curve parameter $t$ starts at the right end of each shape, and therefore the strings of extrema are read in that order.

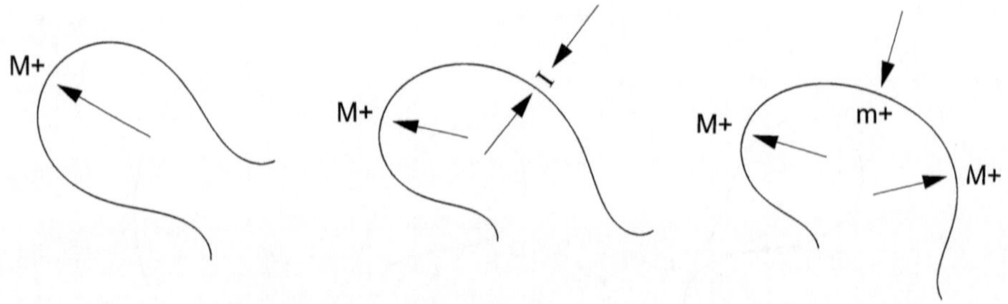


**Fig. 4.9.** Simultaneous realization of two design intents.

Observe, in Fig 4.9, which illustrates the pair-creation operation $\pi\emptyset^{\uparrow+}$, that the transition which describes the *full* set of *three* extrema involved, *including the anchor-extremum*, is given as follows:

$$\pi\emptyset^{\uparrow+} : M^+ \quad \longrightarrow \quad M^+m^+M^+$$

where the extremum $M^+$ at the *right-end* of both, the domain-string and the codomain-string, is the anchor-extremum. Notice that this expression is unambiguous because the symbol $\emptyset^{\uparrow+}$, on the $\pi$ operation, means that pair-creation must act on the left side of the extremum $M^+$ in the domain-string.

Now observe that this transition between strings has some similarity to the transition defined by the following bifurcation operation given in section 3.18:

$$BM^+ : M^+ \longrightarrow M^+m^+M^+$$

However, the notation informs us that there is a fundamental difference between the two situations. In the pair-creation case, we see that the pair $M^+m^+$, in the codomain, arises as a 2-fold bifurcation from the non-extremum $\emptyset^{\uparrow+}$ in the domain. That is, the bifurcation is *away* from the anchor-extremum. This means that the pair $M^+m^+$ is independent of, and *added to*, the starting extremum $M^+$. In contrast, in the case of the operation $BM^+$, all three extrema, in the codomain, arise out of the starting extremum $M^+$.

Nevertheless, both situations result in the string $M^+m^+M^+$, which our theory calls the *shield*, the dual of the *bay*. With respect to the pair-creation in Fig 4.9, this is very recognizable by continuing the right process in the manner shown in Fig 4.10.

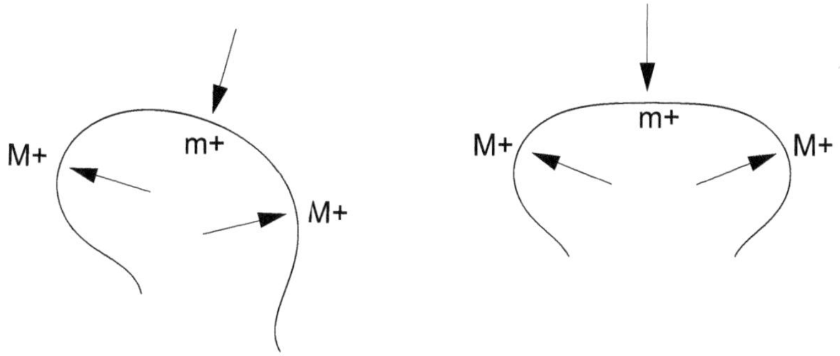


**Fig. 4.10.** Continuing the right process, creates a structure that is highly recognizable as the dual of the bay, the shield.

It is important to observe this: The string-transition given above in the pair-creation operation $\pi\emptyset^{\uparrow+}$ was for Application-Type 3 of that operation. Generally, Application-Type 3 produces a symmetric 3-fold extrema-string, that also results from the application of the 3-fold bifurcation operator $B$ to the same starting extremum. However, the two cases differ in the way described above. We can therefore refer to the pair-creation history of the symmetric string as the **asymmetric creation** of that string, and refer to the 3-fold bifurcation history of that string as its **symmetric creation**.

This kind of situation also arises in the case of Application-Type 1 of a pair-creation operation. For example, consider Fig 4.7 (page 105), which illustrated the pair-creation operation

$$\pi\emptyset^{\downarrow+} \;:\; m^+ \quad\longrightarrow\quad m^+M^+m^+$$

where the extremum $m^+$ at the right-end of both, the domain-string and the codomain-string, is the anchor-extremum. This expression, again, is unambiguous because the symbol $\emptyset^{\downarrow+}$, on the $\pi$ operation, means that pair-creation must act on the left side of the extremum $m^+$ in the domain-string.

The above would be the *asymmetric creation* of the symmetric string $m^+M^+m^+$. In contrast, the 3-fold bifurcation operation, given in section 3.21

$$Bm^+ \;:\; m^+ \;\longrightarrow\; m^+M^+m^+$$

would be the *symmetric creation* of that string.

Finally, observe that, whereas a 3-fold symmetric result arises in the case of Application-Types 1 and 3 of a pair-creation operation, it does not arise in the case of Application-Type 2. This is because, in the latter case, the operation is applied to the opposite-sign tail, and therefore does not have the 3-fold symmetric result; i.e., the created pair are of the opposite sign from the anchor-extremum. The reader can see an illustration of this by returning to Fig 4.4 (page 101).

## 4.5 Zero Pair-Creation Operations

All the applications given so far, of the pair-creation operator $\pi$, have been to points of positive or negative curvature. One additional case however remains: the application of the pair-creation operator to points of *zero* curvature. This produces different results whose importance will be shown a number of times in the book.

The reader will recall that Fig 4.2 (page 99) illustrated what happens to the curvature function when the pair-creation operator is applied to a (regular) point $\emptyset$ of *non-zero* curvature on a spiral: That is, the tangent to the graph at $\emptyset$ rotates till it becomes a horizontal inflection $I$, and then rotates further till the sign of the gradient becomes the opposite of what it was initially.

This rotating-tangent structure also occurs in the case which we will now be considering; i.e., application to a point $\emptyset$ of *zero* curvature. This effect on the curvature function is illustrated in Fig 4.11.

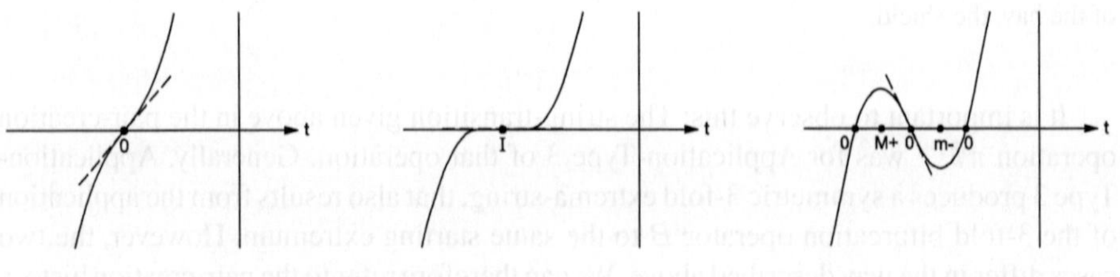


**Fig. 4.11.** Rotating tangent to the graph of a curvature function, at a point of zero curvature.

In the first graph, the application-point is now a position where the graph actually cuts transversally through the $t$-axis. The tangent at this point then rotates till it becomes a horizontal inflection in the second graph. At this stage, the tangent is actually *coincident* with the $t$-axis. Finally, the tangent rotates further, which results in the situation shown in the third graph, where, again, a pair of extrema of opposite types (Max and min) have been created.

However, there are fundamental differences between this situation and the non-zero situations discussed previously. The first difference is this: Previously, the two extrema of the created pair had the same sign; therefore, one *penetrative* and one *compressive*. In contrast, in the present case, the two extrema of the created pair are of opposite signs: a *positive* maximum $M^+$ and a *negative* minimum $m^-$; therefore, *both penetrative*. The second difference is this: Previously, no zeros were created in the bifurcation. In contrast, in the present situation, *three* zeros are created in the bifurcation.

Thus we see that, via the zero inflection-point $I$ in the middle graph, the zero in the first graph has undergone two *simultaneous* bifurcations in the final graph: a bifurcation into two extrema, and a bifurcation into three zeros. These two bifurcations are produced by two operators in the Process Grammar: The bifurcation into two extrema is produced by the pair-creation operator $\pi$, which we have already defined. The bifurcation into three zeros is produced by a new operator $C(3)$, which we will now add to the Process Grammar. The reader will recall that the operator $C$, given in Chapter 3, produces a bifurcation into *two* zeros. It can also be notated as $C(2)$. In contrast, the operator $C(3)$, introduced here, produces a bifurcation into *three* zeros. Notice also the following important difference: Whereas the operator $C$ creates a 2-fold zero-bifurcation from a point that is *not* initially a zero, the operator $C(3)$ creates a 3-fold zero-bifurcation from a point that *is* initially a zero, as illustrated in Fig 4.11.

Based on what we have just said, the transition from the first graph to the third graph, in Fig 4.11, is given by the simultaneous application of the two operators, $C(3)$ and $\pi$. This simultaneous application will be denoted by $[C(3)\pi]$, which is a new operator to be also added to the Process Grammar.

**Notation 4.1** *Square brackets $[Q_1Q_2]$ around two operators, $Q_1$ and $Q_2$, will mean that the operators are applied simultaneously to the same point.*

Notice that the new operator $[C(3)\pi]$ must necessarily be the simultaneous application of operators $C(3)$ and $\pi$ to *a point of zero curvature*, because *one* of its two components, $C(3)$, is defined as the 3-fold bifurcation of a zero. This means that $[C(3)\pi]$ must describe the transition from the zero, in the first graph in Fig 4.11, to the *five* singularities $0M^+0m^-0$ in the third graph; i.e., it is a *5-fold bifurcation*. Notice that the zero in the first graph is at a point of *upward* slope in the graph. Such a point will be denoted by $0^\uparrow$. Therefore the application of the new operator $[C(3)\pi]$ to this point, to produce the singularity-configuration in the third graph, will be the following operation:

$$[C(3)\pi]0^\uparrow \,:\, 0^\uparrow \quad\longrightarrow\quad 0M^+0m^-0 \tag{4.5}$$

Arrows need not be placed on the zeros in the codomain because their gradient-directions (upward or downward) can be deduced from their positions in the string. Similarly, to apply this operation to a general singularity-configuration, consisting of several zeros and extrema, one searches, in the string, for a zero whose gradient-direction is deduced, from its immediate neighbors, to be upward.

The operator $[C(3)\pi]$ will be called the *zero pair-creation operator*, which means that it is the application of the pair-creation operator $\pi$ to a point of zero curvature. However, we must also remember that, unlike the application of the pair-creation operator $\pi$ to a positive or negative point, it necessarily involves, besides $\pi$, the additional simultaneous use of the operator $C(3)$.

When the operator $[C(3)\pi]$ is applied to a zero of upward gradient, as in expression (4.5), it will be called the *upward zero pair-creation operation*; and when the operator $[C(3)\pi]$ is applied to a zero of downward gradient, it will be called the *downward zero pair-creation operation*, which is defined thus:

$$[C(3)\pi]0^\downarrow \,:\, 0^\downarrow \quad\longrightarrow\quad 0m^-0M^+0 \tag{4.6}$$

Notice that, in this case, the codomain is simply the reverse of the codomain of the upward zero pair-creation operation in expression (4.5).

### ZERO PAIR-CREATION OPERATOR $[C(3)\pi]$

**The Process Grammar operator $[C(3)\pi]$ is the simultaneous application of two operators: the 3-fold zero-bifurcation operator $C(3)$ and the 2-fold extremum-bifurcation operator $\pi$. It is realized by the following two operations:**

**Upward Zero Pair-Creation Operation**

$$[C(3)\pi]0^{\uparrow} \,:\, 0^{\uparrow} \quad\longrightarrow\quad 0M^{+}0m^{-}0$$

**Downward Zero Pair-Creation Operation**

$$[C(3)\pi]0^{\downarrow} \,:\, 0^{\downarrow} \quad\longrightarrow\quad 0m^{-}0M^{+}0$$

Fig 4.12 gives an illustration of the effect of the upward zero pair-creation operation on an actual shape. The dots on the curves indicate the positions of the zeros of curvature. There is one on the first curve, and three on the other curve. The curve parameter $t$ is understood as going from right-to-left on each curve. The two curves correspond to the first and last functions in Fig 4.11.

**Fig. 4.12.** The effect of a zero pair-creation operation on a curve.

Now, we saw that each of the *non-zero* pair-creation operations has *three* applications, corresponding to *three different anchor-extrema*. These were deduced by examining the curvature functions shown in Fig 4.6 (page 104) for the four types of extrema. From this same figure, we can deduce that each of the *zero* pair-creation operations has *two* applications, corresponding to *two different anchor-extrema*, as follows: Obviously, the last two extrema $m^{+}$ and $M^{-}$ cannot be anchor-extrema for a zero pair-creation operation, because, as seen from the figure, their spirals cannot contain a zero; i.e., they cannot be next to a zero in a singularity-configuration. However, from this figure, we see that the upward zero pair-creation operation $[C(3)\pi]0^{\uparrow}$ can be applied to the zero on the left side of $M^{+}$ and the zero on the right side of $m^{-}$. Conversely, the downward zero pair-creation operation $[C(3)\pi]0^{\downarrow}$ can be applied to the zero on the right side of $M^{+}$ and the zero on the left side of $m^{-}$. Thus, the two *penetrative* extrema provide anchor-extrema for the zero pair-creation operations, but the two *compressive* extrema do not.

Notice that, while each zero pair-creation operation, $[C(3)\pi]0^{\uparrow}$ and $[C(3)\pi]0^{\downarrow}$, has two applications, there is only one application-type: i.e., the curves that are produced in the applications are all the same, up to reflection and duality. This contrasts with the positive and negative pair-creation operations, where the three applications correspond to three application-types; i.e., because, besides the type provided by a compressive extremum, the two penetrative extrema provided two different types, one based on the *same-sign component* of a penetrative extremum, and the other based on the *opposite-sign tail* of the other penetrative extremum. The types on the same-sign component and the opposite-sign tail cannot be reflections or duals of each other. In contrast, because a zero is *between* two such components, it does not provide different types.

## 4.6 Misalignment of Processes

The Symmetry-Curvature Duality Theorem concerns *reflectional* symmetry. One of the proven results in the theorem is that a curvature extremum is a point about which a curve is locally reflectionally symmetric. The proof of the theorem also included my Spiral Theorem that a curve cannot be locally reflectionally symmetric about a point on a spiral.

In contrast, let us now consider *central* symmetry. Clearly, a curve is not locally centrally symmetric about a curvature extremum – because the curve bends the same way on both sides of the point. Furthermore, for the same reason, it is not locally centrally symmetric about a point of positive or negative curvature on a spiral. However, the curve can be regarded as locally centrally symmetric about a point of zero-curvature on a spiral, because it bends in opposite directions about such a point. This is illustrated in Fig 4.13, which illustrates the transition in a zero pair-creation operation. We see that each of the three curves is centrally symmetric about its central zero-point.

Therefore, a crucial property of a zero pair-creation operation is that it preserves the local central symmetry about the point of bifurcation. Notice that one of the contributing factors to the central symmetry of the final curve is the fact that the two extrema produced by the operation are both penetrative and of opposite sides.

Observe that the transition of the arrows shown in Fig 4.13 can be interpreted as representing the simultaneous realization of two design-intents, the concept stated, at the beginning of this chapter, as an important use of pair-creation. Furthermore, the situation could be causally explained by two physical forces that grip the curve and accidently slip passed each other, as shown in the final curve. In fact, the slipping of a grip can also be non-accidently due to design intent.

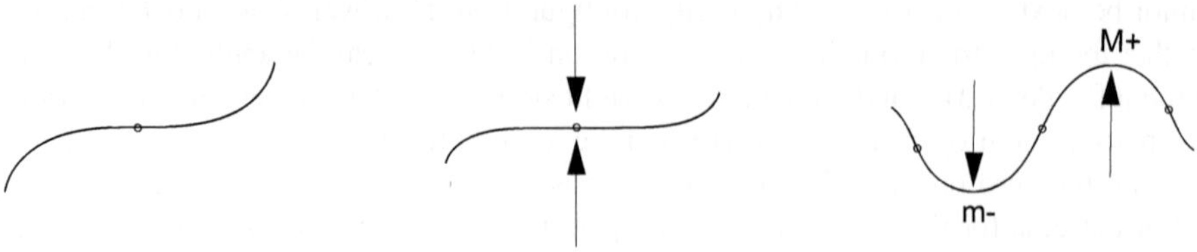


**Fig. 4.13.** Misalignment of forces.

A crucial fact of the pair-creation operator is that it involves the **misalignment of two opposing processes**. The existence of the two processes is given by the two extrema in the final state. The fact that the two processes in pair-creation were initially opposing each other is given by observing, backwards in time from the final state, the differential mirrors located at the two extrema, each of these mirrors being identified with its direction to its extremum on the PISA side of the extremum, these directional mirrors being viewed as the forces that created the extrema. One finds that, in the transition state, the back-ward time *limit*, of those directional differential mirrors, oppose each other at the bifurcation point in the transition state.

Let us now consider additional properties of the zero pair-creation operator. Recall first that the zero pair-creation operator not only involves the pair-creation operator $\pi$ but also the 3-fold zero-bifurcation operator $C(3)$. We are now going to look more closely at $C(3)$. Recall that the 2-fold zero-bifurcation operator $C$, introduced in Chapter 3, describes the *continuation of a process*. The new operator $C(3)$, which is the bifurcation of three zeros, can also be viewed as a *process-continuation* operator, as we will now see. We compare the operators $C$ and $C(3)$ as follows:

**CONTINUATION OPERATORS $C$ AND $C(3)$**

**The continuation operator $C$ expresses the continuation of a *single process*, as the bifurcation of *two zeros*, on each side of that single process.**
**In this case, the continuation is that of a compressive process that continues till it becomes penetrative.**

**The continuation operator $C(3)$ expresses the continuation of *two processes*, as the bifurcation of *three zeros*, on the sides of the two processes. Furthermore, the two processes go in the *opposite directions*.**
**In this case, each of the two processes can be viewed as having initially a compressive effect in the sense that the region surrounding the flat point, to which they are initially applied, becomes flatter (as shown in the second curve in Fig 4.13), due to their application, and then becomes penetrative on each of the two sides of that flat point, due to their bifurcation.**

A crucial factor in the comparison of $C$ and $C(3)$ is this: We have seen that, whereas the 2-fold zero-bifurcation operator $C$ does not involve the bifurcation of a process, we have seen that the 3-fold zero-bifurcation operator $C(3)$ is impossible without the simultaneous 2-fold bifurcation of a process; i.e., it impossible without the pair-creation operator. Thus, its full effect is given by the operator $[C(3)\pi]$.

Let us now say more about the $[C(3)\pi]$ operator. $[C(3)\pi]$ is an example of what we will call an *$n$-fold radial-undulation operator,* to be defined later in Definition 4.15 (page 122). For the moment, we will say that $[C(3)\pi]$ is the particular case that will be called the *2-fold radial-undulation operator,* to be explained thus:

**2-FOLD RADIAL-UNDULATION OPERATOR**
**i.e., ZERO PAIR-CREATION OPERATOR $[C(3)\pi]$**

**The zero pair-creation operator $[C(3)\pi]$, which is the *simultaneous* application of the 3-fold zero-bifurcation operator $C(3)$ and the 2-fold extremum-bifurcation operator $\pi$, to the *same application-point,* will also be called the *2-fold radial-undulation operator,* for the following reasons:**

**The term *2-fold undulation* refers to the created configuration, which consists of 3 zeros and 2 extrema that alternate with the zeros. Notice that, because of this alternation, the extrema must necessarily be *penetrative,* and in fact of opposite signs, $M^+$ and $m^-$; i.e., implying penetrative processes of opposite directions; i.e., an *undulation.***

**The term *radial-undulation* will mean that, under the operator, all 5 singularities of the undulation (3 zeros and 2 extrema) *radiate* out simultaneously from a *single* central application-point.**

**Therefore, $[C(3)\pi]$ describes the *opposition* and *misalignment* of two processes that become *penetrative.***

Now recall, from Chapter 3, that each of the $B$ bifurcations of the Process Grammar, i.e. the 3-fold extremum-bifurcations, is causally explained by the *opposition* of two processes, one that splits the other into two copies. In the current chapter, we have seen that the pair-creation operations are also due to the *opposition* of two processes. In this case, the two processes slip passed each other.

These facts are examples of the following important principle:

**FUNDAMENTAL CLAIM:**

**PROCESS-BIFURCATION IS DUE TO PROCESS-OPPOSITION**
**AND**
**ONLY THE PROCESS-GRAMMAR**
**CORRECTLY DESCRIBES THIS**
**WHEREAS**
**THE OTHER MORPHOLOGY THEORIES**
**DO NOT**

**The process-bifurcation operations of the Process Grammar give the different types of consequences that can occur as a result of** ***process-opposition.***

**For example, in the application of the Process Grammar operator $B$, a** ***pre-existing process*** **is** ***opposed*** **by a** ***new process*** **that breaks-through it and causes the pre-existing process to bifurcate to each side of the opposing process.**
**A crucial fact is that these two opposing processes are of opposite types, compressive and penetrative, and therefore the Process Grammar is the only morphology theory that correctly describes these situations, whereas other morphology theories do not.**

**In the application of the Process Grammar operator $\pi$, there are two opposing processes that slip passed each other.**
**Again, a crucial fact is that, all the non-zero pair creation operations produce a penetrative and compressive extremum, and therefore, again, the Process Grammar is the only morphology theory that correctly describes these situations, whereas other morphology theories do not.**

## 4.7 Erupting Effect on an Extremum-Side

In this chapter, the description we gave of the pair-creation operator has been in its application to a *side* of an extremum, i.e., the anchor-extremum. As was stated, this important concept is that the anchor-extremum corresponds to a process that was responsible for creating the "body" of the extremum on whose side the pair-creation operator is applied.

We shall now see that one of the two processes resulting from the pair-creation operator can be so strong that it can become more prominent than the process at the anchor-extremum. An illustration is shown in Fig 4.14. Consider the triple $0m^-0$ in the top of the first shape (the left shape), where the two zeros are given by the two dots. Let us suppose that we know that this triple was created by previously applying the Process Grammar operation $Cm^+ : m^+ \longrightarrow 0m^-0$. That is, originally, the central extremum was $m^+$, produced by a downward squashing process; and this process continued till it indented, thus producing the triple $0m^-0$ that we see in the first shape in Fig 4.14. This means that the two zeros (dots), in this shape, "belong" to the central extremum, because the zeros bifurcated from that extremum.

Now, let us consider the left zero in the first shape: We will apply the zero pair-creation operation

$$[C(3)\pi]0^{\uparrow} \ : \ 0^{\uparrow} \quad \longrightarrow \quad 0M^+0m^-0$$

to this zero, to produce the second shape. Note that the curve parameter is anti-clockwise on both of these closed curves and therefore the anti-clockwise direction gives the order in which the singularities in the codomain of the above operation are read on the second curve. That is, the succession of codomain singularities, $0M^+0m^-0$, are read, starting with the dot above the right diagonal arrow on the second curve, and moving from this dot, from right to left, along the top of that curve. Notice also that the terms left and right in the current text will refer to diagram positions in Fig 4.14 instead of the curvature functions.

It is conceptually very valuable to understand now what has happened in applying the above operation, as follows:

(1) The right zero (dot) in the first shape has become the right-most zero (dot) in the second shape.

(2) The central $m^-$ in the first shape has become the right $m^-$ in the second shape.

(3) The left zero (dot) in the first shape has bifurcated to become the *three* zeros (dots) to the left of the right $m^-$ in the second shape, as well as the top $M^+$ and the left $m^-$ in the second shape.

That is, in the second shape, the entire quintuple $0M^+0m^-0$, going leftwards from the right $m^-$, has bifurcated from a single point, the left zero in the first shape. This 5-fold bifurcation is the zero pair-creation operation given above.

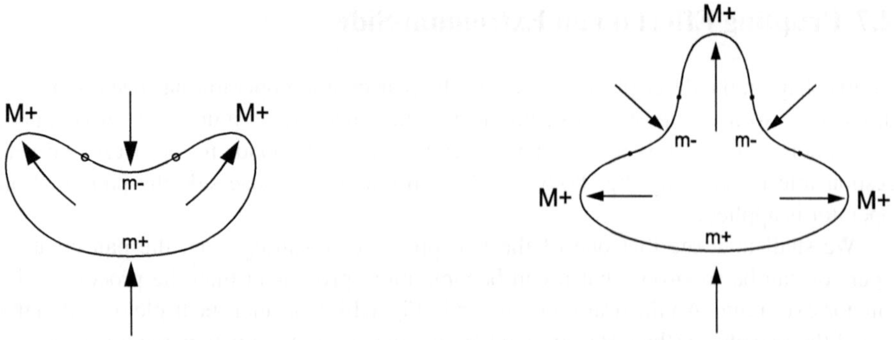


**Fig. 4.14.** Applying the pair-creation operator at a zero point.

Thus we see that the quintuple $0M^{+}0m^{-}0$ was created in the *side* of the initial indentation. In particular, the protrusion $M^{+}$, at the top of the second shape, *erupted* out of the side of the initial indentation. We can therefore refer to this as a **side-eruption**. Notice that it is this eruption that pushed the initial indentation around to the right in the second shape.

Finally, it is important to emphasize that the term *application*, defined in section 4.3, means application to a *side* of an extremum, the anchor-extremum. A side is a spiral connected to the extremum. We defined the different alternative applications of a pair-creation operation to be given by the different alternative anchor-extrema, to whose sides the operation can be applied. We saw that, for a positive or negative pair-creation operation, there are three alternative applications, i.e., anchor-extrema; and for a zero-pair creation operation, there are two alternative applications, i.e., anchor-extrema.

There is, of course, an extra alternative for each operation: applying it in a spiral that is not connected to an extremum; i.e., a non-side. Since this extra case is trivial to classify, i.e., there is no variation dependent on changing an anchor-extremum, we will call it the **anchor-independent use** of the operation. Thus, for pair-creation, we will reserve the term *application* to mean *application to a side of an extremum*; i.e., as stated in section 4.3, the term **application** is an abbreviation of **extremum side-application**.

## 4.8 The Complete Set of Pair-Creation Operations

We can now list all the pair-creation operations.

## THE COMPLETE SET OF PAIR-CREATION OPERATIONS

**There are six pair-creation operations.**

**As shown in Fig 4.15, they are grouped as follows:**

**The operator $\pi$, from which they are all derived, has two cases:**
**the upward pair-creation operator $\pi^{\uparrow}$**
**the downward pair-creation operator $\pi^{\downarrow}$**

**These two operators are each realized by three operations:**
**a positive, zero, and negative, pair-creation operation.**

**The positive and negative operations are *pure*, in the sense that they do not involve more than the creation of a pair of extrema.**

**The zero operation, between a positive and negative operation, involves, besides the creation of a pair of extrema, a simultaneous bifurcation into 3 zeros.**

**The six operations are given as follows:**

**Upward Pair-Creation Operations**

$$\pi\emptyset^{\uparrow+} \ : \ \emptyset^{\uparrow+} \quad \longrightarrow \quad M^{+}m^{+}$$

$$[C(3)\pi]0^{\uparrow} \ : \ 0^{\uparrow} \quad \longrightarrow \quad 0M^{+}0m^{-}0$$

$$\pi\emptyset^{\uparrow-} \ : \ \emptyset^{\uparrow-} \quad \longrightarrow \quad M^{-}m^{-}$$

**Downward Pair-Creation Operations**

$$\pi\emptyset^{\downarrow+} \ : \ \emptyset^{\downarrow+} \quad \longrightarrow \quad m^{+}M^{+}$$

$$[C(3)\pi]0^{\downarrow} \ : \ 0^{\downarrow} \quad \longrightarrow \quad 0m^{-}0M^{+}0$$

$$\pi\emptyset^{\downarrow-} \ : \ \emptyset^{\downarrow-} \quad \longrightarrow \quad m^{-}M^{-}$$

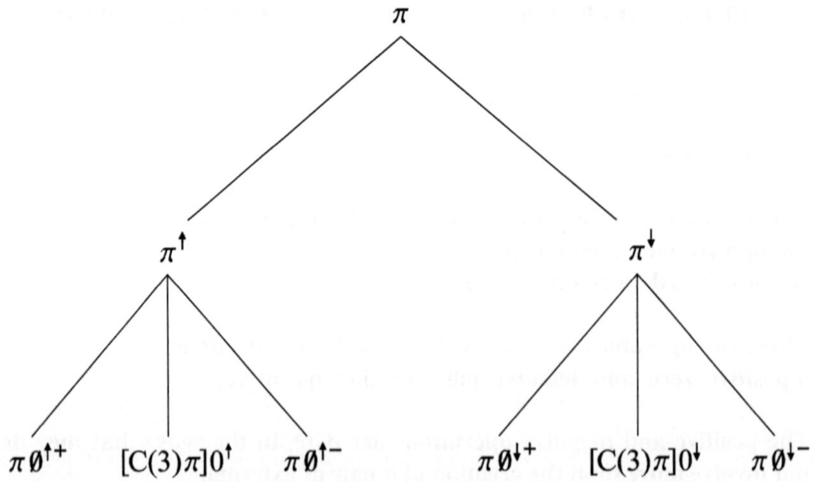


**Fig. 4.15.** The bottom row shows the complete set of pair-creation operations.

## 4.9 Notation

In the chapters so far, we have introduced the reader to a number of operators from the Process Grammar, and it is time now to explain the notation that we use generally for the operators of the grammar.

First observe that there are two kinds of operators: (1) process-bifurcation operators, and (2) process-continuation operators. The notations for these two kinds of operators will now be specified.

### Process-Bifurcation Operators

**Notation 4.2** Let $n \geq 2$ be a natural number. A bifurcation into $n$ processes ($n$ extrema) will be denoted by $B(n)$. The operators $B(n)$ will be called the **process-bifurcation operators**.

Since the two *lowest-order* process-bifurcation operators, $B(3)$ and $B(2)$, will be used the most frequently, they will be given by short-hand notations, thus:

**Short-hand 4.3** $B(3)$ will usually be denoted by $B$.

**Short-hand 4.4** $B(2)$ will usually be denoted by $\pi$.

We now distinguish between the cases where $n$ is odd and $n$ is even:

**Causal Interpretation 4.5** When $n$ is odd, $B(n)$ is interpreted as the application of a process in opposition to one that has already taken place. The application-point is therefore a curvature extremum.
When $n$ is even, $B(n)$ is interpreted as the application of a process to a point at which a process has not yet been applied; i.e., a regular point in the curvature function.

**Notation 4.6** When $n$ is odd, $B(n)$ is applied to a curvature extremum that is non-degenerate. If the extremum is a *minimum*, then $B(n)$ can be denoted by $B_n$. If, instead, the extremum is a *maximum*, then $B(n)$ can be denoted by $B_{-n}$.

**Notation 4.7** When $n$ is even, $B(n)$ is applied to a point that has non-horizontal gradient in the curvature function, i.e., a regular point in the curvature function. If, at this point, the gradient of the curvature function is positive, then $B(n)$ can be denoted by $B_n$ and also $B(n)^{\uparrow}$. If, instead, at this point, the gradient of the curvature function is negative, then $B(n)$ can be denoted by $B_{-n}$ and also $B(n)^{\downarrow}$.

**Short-hand 4.8** $B(2)^{\uparrow}$ will usually be denoted by $\pi^{\uparrow}$; and $B(2)^{\downarrow}$ will usually be denoted by $\pi^{\downarrow}$.

## Process-Continuation Operators

**Causal Interpretation 4.9** A **process-continuation operator** is interpreted as a process breaking-through a zero, causing the zero to bifurcate into a zero on each side of that process as well as on each side of any other process that simultaneously bifurcates from that process.

**Notation 4.10** Each process-continuation operator will be denoted by $C(n)$, where $n \geq 2$ is the number of zeros that bifurcate from the application-point.

Since the *lowest-order* continuation operator, $C(2)$, will be used the most frequently, it will be given by a short-hand notation:

**Short-hand 4.11** $C(2)$ will usually be denoted by $C$.

We now distinguish between the cases where $n$ is odd and $n$ is even.

**Notation 4.12** When $n$ is odd, $C(n)$ is applied to a zero of the curvature function, where the zero is a regular point. If the zero has upward gradient, it is denoted by $0^{\uparrow}$, and $C(n)$ can be denoted by $C_n$ and also $C(n)^{\uparrow}$. If, instead, the zero has downward gradient, it is denoted by $0^{\downarrow}$, and $C(n)$ can be denoted by $C_{-n}$ and also $C(n)^{\downarrow}$.

**Notation 4.13** When $n$ is even, $C(n)$ is applied to a compressive extremum, either $m^{+}$ or $M^{-}$. If that extremum is $m^{+}$, then $C(n)$ can be denoted by $C_n$. If, instead, that extremum is $M^{-}$, then $C(n)$ can be denoted by $C_{-n}$.

**Notation 4.14** Necessarily, when $n \geq 3$, the operator $C(n)$ is simultaneous to the operator $B(n-1)$ at the same point of application; i.e., bifurcation into $n$ zeros simultaneously produces bifurcation into $n-1$ extrema that alternate with those zeros. This simultaneous bifurcation can be explicitly notated as the operator $[C(n)B(n-1)]$. There are four cases:

When $n$ is odd, and the point of application is $0^{\uparrow}$, then $[C(n)B(n-1)]$ can be notated as $[C_n B_{n-1}]$.

When $n$ is odd, and the point of application is $0^{\downarrow}$, then $[C(n)B(n-1)]$ can be notated as $[C_{-n} B_{-(n-1)}]$.

When $n$ is even, and the point of application is $m^+$, then $[C(n)B(n-1)]$ can be notated as $[C_n B_{n-1}]$.

When $n$ is even, and the point of application is $M^-$, then $[C(n)B(n-1)]$ can be notated as $[C_{-n} B_{-(n-1)}]$.

**Definition 4.15. Radial-Undulation:**
The operator $[C(n+1)B(n)]$, which is the *simultaneous* application of the $n+1$-fold zero-bifurcation operator $C(n+1)$ and the $n$-fold extremum-bifurcation operator $B(n)$, to the *same application-point*, will be called the ***n*-fold radial-undulation operator**. This reasons for choosing this name are given as follows: The term ***n*-fold undulation** corresponds to the created singularity-configuration, which consists of $n+1$ zeros and $n$ extrema that alternate with the zeros. Notice that, because of this alternation, the extrema must necessarily be *penetrative*, and in fact of opposite signs, $M^+$ and $m^-$; i.e., implying *penetrative processes* of *alternating opposite directions*; i.e., an **undulation**. The term **radial-undulation** will mean that, under the operator, all $2n+1$ singularities of the undulation ($n+1$ zeros and $n$ extrema) *radiate* out simultaneously from a *single* central application-point. We have seen that, when $n$ is odd, the application-point is a compressive extremum, $m^+$ or $M^-$; and when $n$ is even, the application-point is a curvature zero that is regular, $0^{\uparrow}$ or $0^{\downarrow}$.

### Operations

**Notation 4.16** The application of an operator $Q$ to a point of curvature singularity $x$, will be called an **operation**, *which will be denoted by* $Qx$.

The notations given above allow for a number of alternative notations for an operation. For example, the operations

$Bm^+$
$Bm^-$
$BM^+$
$BM^-$

can also be denoted, respectively, by

$B_3m^+$
$B_3m^-$
$B_{-3}M^+$
$B_{-3}M^-$

As another example, the operations

$[C(3)\pi]0^\uparrow$
$[C(3)\pi]0^\downarrow$

can also be denoted, respectively, by

$[C_3B_2]0^\uparrow$
$[C_{-3}B_{-2}]0^\downarrow$

### Combinations

**Notation 4.17** Let $J_1$, $J_2$, ..., $J_n$ be a set of operators, or a set of operations. Then:
$J_1 \circ J_2 \circ \ldots \circ J_n$ denotes their successive application.
$J_1J_2 \ldots J_n$ denotes their simultaneous application to a distributed set of $n$ points.
$[J_1J_2 \ldots J_n]$ denotes their simultaneous application to a single point.

## PROCESS GRAMMAR SO FAR

### Process-Continuation

| | | | | |
|---|---|---|---|---|
| $Cm^+$ : | $m^+$ | $\longrightarrow$ | $0m^-0$ | (squashing continues till it indents) |
| $CM^-$: | $M^-$ | $\longrightarrow$ | $0M^+0$ | (resistance continues till it protrudes) |

### 3-Fold Process-Bifurcation

| | | | | |
|---|---|---|---|---|
| $BM^+$: | $M^+$ | $\longrightarrow$ | $M^+m^+M^+$ | (shield-formation) |
| $Bm^-$ : | $m^-$ | $\longrightarrow$ | $m^-M^-m^-$ | (bay-formation) |
| $Bm^+$ : | $m^+$ | $\longrightarrow$ | $m^+M^+m^+$ | (breaking-through of a protrusion) |
| $BM^-$: | $M^-$ | $\longrightarrow$ | $M^-m^-M^-$ | (breaking-through of an indentation) |

### Pair-Creation

Pure 2-fold Process-Bifurcation

| | | | | |
|---|---|---|---|---|
| $\pi\emptyset^{\uparrow+}$: | $\emptyset^{\uparrow+}$ | $\longrightarrow$ | $M^+m^+$ | (application-point: upward positive) |
| $\pi\emptyset^{\uparrow-}$: | $\emptyset^{\uparrow-}$ | $\longrightarrow$ | $M^-m^-$ | (application-point: upward negative) |
| $\pi\emptyset^{\downarrow+}$: | $\emptyset^{\downarrow+}$ | $\longrightarrow$ | $m^+M^+$ | (application-point: downward positive) |
| $\pi\emptyset^{\downarrow-}$: | $\emptyset^{\downarrow-}$ | $\longrightarrow$ | $m^-M^-$ | (application-point: downward negative) |

2-fold Radial-Undulation

| | | | | |
|---|---|---|---|---|
| $[C(3)\pi]0^{\uparrow}$: | $0^{\uparrow}$ | $\longrightarrow$ | $0M^+0m^-0$ | (application-point: upward zero) |
| $[C(3)\pi]0^{\downarrow}$: | $0^{\downarrow}$ | $\longrightarrow$ | $0m^-0M^+0$ | (application-point: downward zero) |

# 5

# Process-Theory of Parts (1)

## 5.1 The Definition of Parts in the New Foundations to Geometry

The reader will recall that My New Foundations to Geometry define shape as the causal history *recovered* (i.e., inferred) from a data set, i.e., the history needed to create that data set. That is, contrary to the Standard Foundations to Geometry, the New Foundations state that it is impossible to define shape independently of its recovered history.

It is important now to introduce the reader to the fact that, in my book *A Generative Theory of Shape*, a significant term is the word *understanding*. In fact, a fundamental aspect of the book is the development of a mathematical theory of how *understanding* can be formed of a complex shape. On the next page, we introduce the reader to some of the basic concepts related to the term *understanding* in the New Foundations, and show how this is crucially important to the definition of *parts* given by the New Foundations.

**THEORY OF UNDERSTANDING
IN THE
NEW FOUNDATIONS TO GEOMETRY**

**A fundamental theory invented in my book *A Generative Theory of Shape* is an extensive theory of *understanding* complex shape.**

**This theory is based on the Mathematical Theory of Intelligence invented in my New Foundations to Geometry.**

**Thus, according to the New Foundations, *understanding* a complex shape, means giving it an *intelligent* representation; i.e., a representation based on the Mathematical Theory of Intelligence in the New Foundations.**

**Recall that the two basic principles of my Mathematical Theory of Intelligence are: Maximization of Transfer and Maximization of Recoverability. Furthermore, my New Foundations give a Mathematical Theory of Transfer and a Mathematical Theory of Recoverability.**

**Crucially, the New Foundations *unifies* the Mathematical Theory of Transfer and the Mathematical Theory of Recoverability. This is crucial because this *unification* gives causal explanations of data sets.**

**In the world, any observer is presented with a situation without structure, i.e., a data set. According to the New Foundations, a data set is *understood* when it is causally explained. Furthermore, using the theory of understanding in the New Foundations, the causal explanation provided by the New Foundations gives the intelligent way to *structure* the data set.**

**Recall that, according to the New Foundations, the causal explanation that structures the data set is the correct meaning of the word *shape*, as opposed to the meaning of the word *shape* in the Standard Foundations.**

**DEFINITION OF PARTS:**
**A crucial consequence of the fact that the New Foundations give an extensive mathematical theory of the intelligent structure of complex shape, is that the New Foundations give an intelligent theory of the *parts* of shape.**

**That is, the New Foundations define *parts* as *phases* of the recovered causal history, such that they *maximize transfer and recoverability* in the recovered causal history.**

In contrast to the New Foundations to Geometry, the standard research on geometry defines a part as a segment of the whole, where both the segment and the whole are defined independently of recoverable causal history. That is, standardly, a part is defined *non-causally*. For example, the literature in computer vision is full of diagrams of an animal body which is separated into rigid closed shapes corresponding to arms, legs, torso, head, etc., all floating independently in space. Thus, there is no conception that an arm actually grew out of the body. That is, in the standard approach, the relation between a part and the whole does not embody any *understanding* of the part.

In contrast, as stated on the previous page, my New Foundations define a part as necessarily capturing understanding. With respect to the Process Grammar, which is a component of the New Foundations, we have this:

**DEFINITION OF PART IN THE PROCESS GRAMMAR**

**Recall that my New Foundations to Geometry define *parts* as *phases* of the recovered causal history, such that they *maximize transfer and recoverability* in the recovered causal history.**

**Therefore, in the Process Grammar, a part is defined as a recovered process and the recovered subsequent history of that process.**

Since my book *A Generative Theory of Shape* gives a theory of recoverable processes, and their recoverable subsequent histories, i.e. their transfer, the entire theory can be viewed as equivalent to a theory of parts. While *A Generative Theory of Shape* develops the general theory for recovering generative history from every property of the data set, the present book focusses on recovering generative history from curvature extrema.

Therefore, based on the role of maximization of transfer and recoverability, in the definition of part, given in the New Foundations , the reader should observe the following: Even though the title of the present chapter *specifically* refers to the concept of part, everything else in the book is necessarily about parts, because it is about recoverable and transferred processes; i.e., everything we have said so far in this book is about parts; and the entire remainder of the book is about parts.

As a result of this, let us now consider some basic concepts: Recall that our theory gives the following correspondence between the four kinds of extrema and the processes recoverable from them:

| | |
|---|---|
| $M^+$ | protrusion |
| $m^-$ | indentation |
| $m^+$ | squashing |
| $M^-$ | resistance |

Therefore, the four kinds of extrema must correspond to four kinds of part. Notice that, because the standard literature fails to understand compressive extrema $m^+$ and $M^-$, i.e., fails to understand the processes that produced them, it fails to understand them as parts. This is a fundamental reason why the standard literature is stupid, in accord with the technical definition of stupid given by the New Foundations.

According to our theory, the compressive extrema give fundamentally important parts to morphology, as will be shown in this chapter.

It is important now to observe this:

**THE PROCESS-GRAMMAR IS A THEORY OF PARTS.**

**Recalling that our definition of part in the Process Grammar (page 127) said that a part is a recovered process, and also the recovered subsequent history of that process,**

**and using the crucial fact that the subsequent history is given by the** *operations of the Process Grammar*, **we conclude this:**

**The Process Grammar gives a theory of** *part-development*.

Thus, we see that, the standard research literature in computer vision and perception *not only* fails to recognize the range of parts recognized by our theory, *but also* fails to recognize that the types of parts are related to each other. This is another consequence of the fact that the standard literature views parts as rigid segments. In contrast, in our system, the types of parts are related to each other by the transfer by which our theory defines causal explanation. Again, this causal relationship is provided by the Process Grammar, which gives a *developmental* relationship between the types of parts.

## 5.2 Two Fundamentally Important Parts

According to our theory, in major types of morphology, the two compressive extrema, $m^+$ and $M^-$, correspond, respectively, to two types of structural phenomena: *shields* and *bays*, that our theory regards as fundamentally important in morphology. We will be studying these in detail, in the following sections.

First, on the following page, we give a summary of a number of things that we will demonstrate:

## SHIELDS AND BAYS

**According to my theory, *shields* and *bays* should be regarded as extremely important to many disciplines such as biological morphology, automotive, aerospace and architectural design, computer vision, robotics, etc.**

**Nevertheless, these disciplines tend not to notice shields and bays; and on the rare occasions when the disciplines notice them, the disciplines describe them incorrectly.**

**In contrast, I argue that my geometry defines them correctly, and, in so doing, shows the fundamental importance of these structures.**

## EXAMPLES

**This chapter argues that examples of shields in biological morphology are the forehead, top and back of the human head, the chin and jaw, the chest, the tips of fingers, the knee-cap, the heel of the foot. Also, examples of shields in road vehicles are the front, top, back, and sides of cars, buses and vans. This unifies biological morphology and automotive design. For example, according to this chapter, the forehead of the human head is the same structure as the front of a van. In fact, both are what I call bi-directional shields in the sense that they are shields both in the vertical and horizontal direction.**

## GEOMETRIC DESCRIPTION

**Conventional geometry, such as the highly popular Medial Axis of Blum, completely fails to describe shields. Blum invented and published the Medial Axis as an analysis of biological morphology. Yet it completely fails to describe what I will demonstrate are fundamental and frequent in biological shape: i.e., shields. My argument is that the fundamental importance of shields in biological shape is because biology requires shields for *protection.* In contrast to Blum's Medial Axis, the symmetry axis I invented, the Process-Inferring Symmetry Analysis (PISA) not only recognizes the existence of shields but describes them correctly by revealing their protective structure. The same applies to shields in automotive, aerospace and architectural design.**

**I will also argue that bays are important structures in architectural design, such as the design of bridges, doorways, ceilings, windows, etc. Again, Blum's Medial Axis completely fails to describe these, and my PISA axis not only describes them but does so revealing their functional structure.**

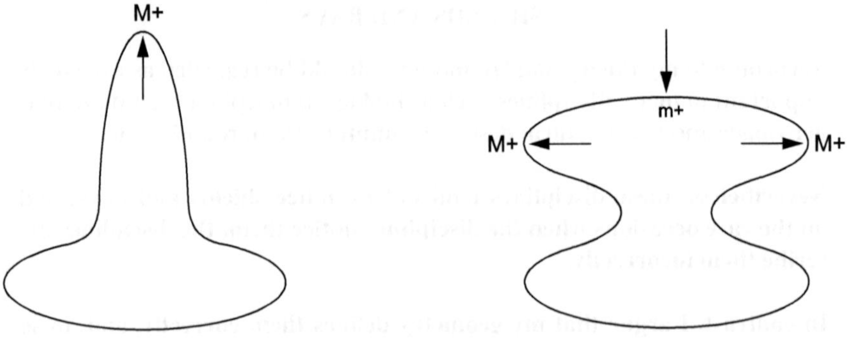


**Fig. 5.1.** Shield-formation.

## 5.3 Shield Structures

Recall the Process Grammar operation

$$BM^+ \;:\; M^+ \;\longrightarrow\; M^+m^+M^+$$

illustrated again in Fig 5.1. When this operation was introduced, in section 3.18, it was named *shield-formation*, because of the causal explanation, as follows: Under our process-inference rules, the domain extremum $M^+$ defines a protruding process, illustrated by the central upward arrow in the first shape in Fig 5.1. In the operation, this protrusion process undergoes a bifurcation into two copies of itself, shown as the left and right arrows in the second shape, and another process is introduced between these two copies: the downward squashing process shown at the top $m^+$ in the second shape. According to our theory of the *causal explanation* of the $B$ operator, the reason why the protrusion, of the first shape, splits into the two copies of itself in the second shape, is due to the *opposing* squashing process inferred at the $m^+$ extremum in the second shape.

The codomain triple $M^+m^+M^+$ has the shape of a *shield.* In fact, this triple is completely determined by the presence of the extremum $m^+$ because, for *any* occurrence of $m^+$, a singularity to the left or right of $m^+$ must necessarily be $M^+$, because the graph of the curvature function at a $m^+$ extremum can only go upwards to the left and right of $m^+$. In fact, I define a *shield* to be a neighborhood of a $m^+$ extremum on a curve, and I define a *complete shield* to be the triple $M^+m^+M^+$.

There is, in fact, an important reason why I named such a section of curve, a shield, as will now be discussed:

## THE ANTICIPATION ROLE

**This chapter will show that, in real-world objects, the central extremum of a part that acts as a shield is what our theory defines as the *compressive* extremum $m^+$, because a shield is designed to protect against an external force from the real world. In other words, a real-world shield is structured as a compressive extremum in *anticipation* of an external force.**

**Again, other morphology theories completely fail to represent this, because they fail to represent the force structure of compressive extrema.**

What our theory defines above, as the Anticipation Role, involves a profound concept to be described as follows: Our process-inference rules, including the Process Grammar, concern the recovery of process-history. In the case of a shield, the recovered process-history is the recovered sequence of *design* operations that created its shape; e.g., in biological evolution or industrial manufacturing. Most crucially, the design sequence must have contained a design operation that applied an external compressive process to the shape in order that the shape could fulfill the intended *function* of protecting against a future external force; i.e., the design force matches the anticipated force. Therefore, our theory has this profoundly important conclusion:

## INTENDED FUTURE FUNCTIONS AS RECOVERABLE PAST DESIGN OPERATIONS

**According to the New Foundations to Geometry, the intended functions of a shape can often be realized by generating the shape via recoverable design operations that have the same structure as those future functions.**

Thus, to return to our concept of a shield: We have defined a shield to be a neighborhood of a $m^+$ extremum, where $m^+$ corresponds to the inferred force that *causes* the compressive structure, and represents the anticipated function. We shall now see that shield structures occur with great frequency in biological morphology as well as manufacturing design.

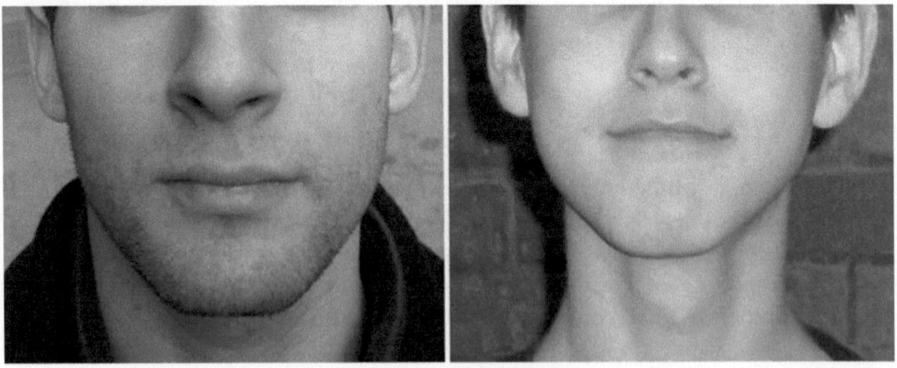

**Fig. 5.2.** Shield structures in the lower halves of faces.

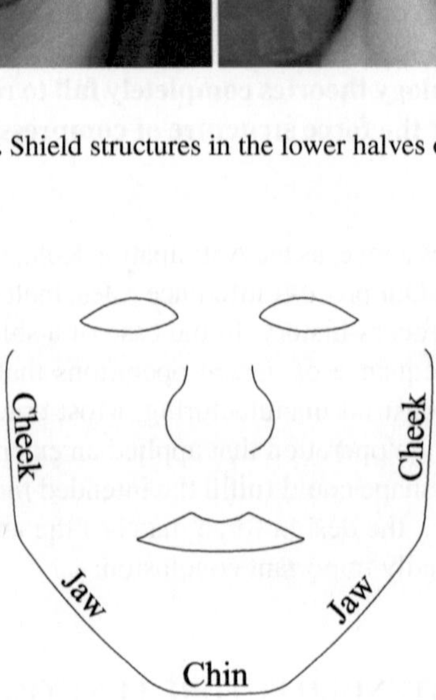


**Fig. 5.3.** Common terms correspond to the shield structures in the above faces.

## Shields In Biological Morphology

We begin by considering the human head. According to our theory, shields are very significant in the structure of heads. To see this, consider the photographs in Fig 5.2, which shows the lower halves of two human heads. As can be seen from these photographs, these lower halves of human heads are structured by what our theory calls *shields*. In fact, our claim, that these parts are significant, is corroborated by the fact that there are common terms used by people to label these parts. That is, as shown in Fig 5.3, the common terms used to label these parts are *chin*, *jaw*, and *cheek*. Observe that the central extremum for each of these parts is $m^+$. Furthermore, by our system of inference rules, one infers, from each $m^+$ extremum, an external compressive force. It is clear that

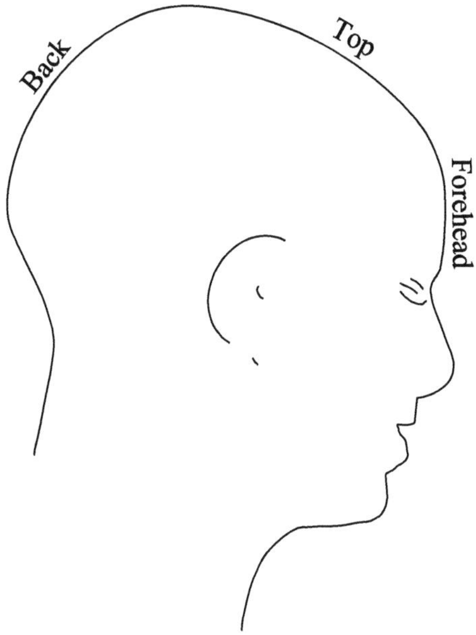


**Fig. 5.4.** Upper parts of the head defined by shield structures.

biology anticipates such an external force. That is, these shields are formed to protect the eating and talking functions of the mouth.

We have just seen that the *lower* half of the human head is structured as shields. In fact, the same is also true of the *upper* half of the human head. Fig 5.4 shows the outline of the head of the pop-singer Britney Spears after she shaved her head. We can clearly see that the *upper* half of the head is structured as shields: e.g., the *forehead*, the *top* of the head, and the *back* of the head, are shields.

Shields are therefore very significant to the structure of the head. The biological reason is clear: The head contains several of the body's most important survival systems, e.g., the brain, which must be protected, and the eating and communication functions.

The significant presence of shield structures in the head was noticed in the remarkable *Portrait of a Man* by the famous artist Hans Memling (c. 1440-1494), which consists almost entirely of shields, as shown by the outline of the portrait shown in Fig 5.5. In fact, to reinforce this, not only is the entire outline of the face constructed from shields, but so is the hair. Even the edge of the neck, as given by clothing just below the face, is drawn as the shape of a chin. The fist, at the bottom of the painting, indicates the strength, resolve, and determination involved in the concept of a shield. This is one of many paintings analyzed in my book *The Structure of Paintings* (Springer-Verlag).

**Fig. 5.5.** *Portrait of a Man* by Hans Memling (c. 1440-1494) emphasizes the shield structures of the human head.

Besides the head, shields are also significant in the remainder of the body. For example, the *chest* is a shield which has the role of protecting important organs such as the heart and lungs. The body moves forward, and therefore must be *shielded* from external opposing forces that could damage those internal organs. The reader can see that the top surface of the second shape in Fig 5.1 has the shape of a chest, and the external force inferred from the central $m^+$ extremum matches the external force that the chest can meet.

It is important now to observe the following:

> **Compressive parts, such as a chin, cheek, chest, etc., cannot be accounted for by the standard theory of parts in computer vision and perception. That is, compressive parts cannot be segmented as blobs (closed shapes) separated from the remainder of the body.**
>
> **In contrast, compressive parts are excellently accounted for by our process theory of parts. In fact, our theory gives a single framework that accounts for both compressive and penetrative parts: the theory defines them as processes recovered from extrema via the PISA axes.**

We will now see that, in biology, the functional strategy of flattening a part into a shield is adopted all over the body, in many parts, which according to our theory, were initially formed purely as protrusions. For example, observe first that, although, there are examples of fingers that, viewed from the front (or back) of the hand, have a penetrative extremum at the top, there are other fingers that have a compressive extremum at the top, i.e., in this case, the finger tip is, according to our theory, a *shield*. An example of such a finger tip is shown in Fig 5.6. Again, our theory, that relates the inferred process to the anticipated function, explains why this occurs as follows: Fig 5.7 shows the extrema structure of this finger tip; i.e., it is the triple $M^+m^+M^+$, which defines what our theory calls a complete shield. From the two $M^+$ extrema, our process-inference rules infer two protruding processes, given by the outward arrows shown in the figure; and from the central $m^+$ extremum, our process-inference rules infer a squashing process given by the inward-directed arrow shown in the figure. Furthermore, by the Process Grammar, the shield was formed as a result of the bifurcation operation:

$$BM^+ \ : \ M^+ \ \longrightarrow \ M^+m^+M^+$$

That is, prior to the shield, there was a single *penetrative* extremum $M^+$, given by an outward protruding process at the center of the top. Then, in the design of this finger, by biological evolution, the protruding process bifurcated, in order to introduce a $m^+$ extremum, in order to protect the finger with respect to an opposing force. We shall refer to such a developed finger as an *advanced* finger.

Another example of a part which, according to our morphological theory, is best defined as a protrusion that underwent the shield-formation operation $BM^+$ is a nose. As illustrated in Fig 5.8, the front outline of the tip of the nose is in fact a *shield*. Notice that, again, this is a complete shield $M^+m^+M^+$, as we saw in Fig 5.7 for a finger.

The above discussion has been illustrating our definition of part (section 5.1) which states that a part includes not only a recoverable process, but also the recoverable subsequent history of that process. This subsequent history is given by the operations of the Process Grammar. For example, our definition of a compressive finger tip includes both the initial penetrative process, and the subsequent bifurcation of that process under a compressive action.

In biological morphology, the Process Grammar operations also correspond to the history of the *intended* functions of the part. For example, the initial *intention* in the evolutionary development of a finger must have been an outward function, i.e., a function corresponding to the penetrative extremum; e.g., the function of digging. Later, because of the necessary survival of fingers, the top of the finger had to be protected with respect to an opposing force at the tip. Therefore, according to our theory, what had previously been a penetrative extremum, at the tip, underwent bifurcation in *anticipation* of an opposing force; i.e., it is for this reason that the initial penetrative extremum underwent the *shield-formation operation* of the Process Grammar. Thus, we can see that the Process Grammar, by recovering the succession of *causal* stages in the shape development of a part, also recovers the succession of *design intents* in the history of a part.

An important fact is that we have been applying the shield-formation operation to a penetrative extremum that is defined with respect to a line of curvature through the top point of a finger. This is a topic which will be defined in detail in section 5.8. That is, we argue that the Process Grammar operations for curves can be applied to the lines of curvature on a surface. Another important fact is this:

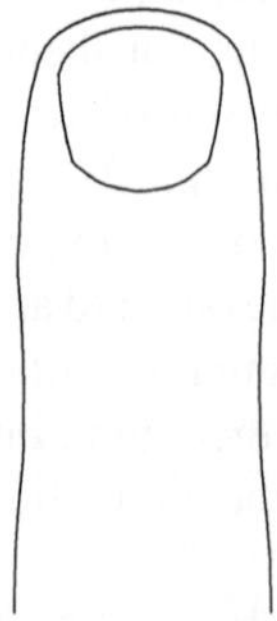

**Fig. 5.6.** The tip of this finger is a *shield*.

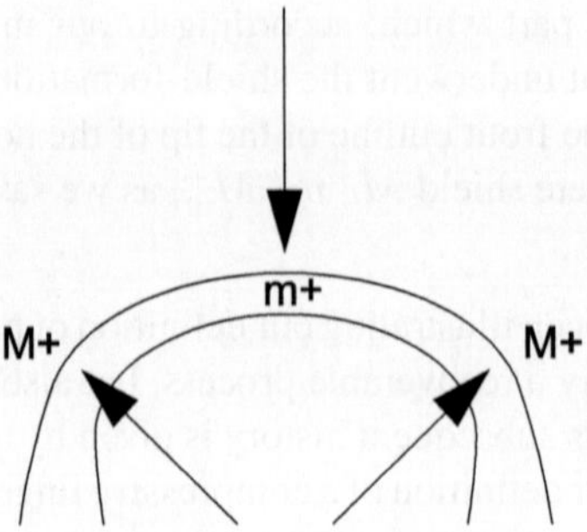


**Fig. 5.7.** The shield structure of the tip of this finger.

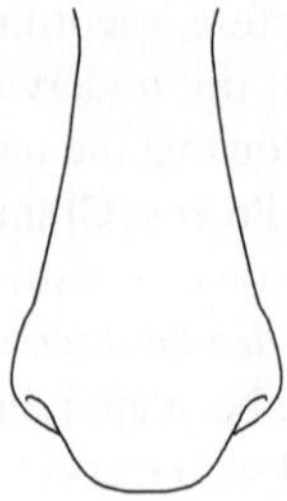

**Fig. 5.8.** The tip of this nose is a *shield*.

**The tip of what our theory calls an advanced finger illustrates the case, defined in our theory, where a point is a compressive extremum in one of its two lines of curvature, i.e., is a shield with respect to that line of curvature; and is a penetrative extremum in the orthogonal line of curvature through that point. According to our theory, this occurs because this type of finger tip has two functions, the protection which is realized by the compressive extremum and the digging which is realized by the penetrative extremum. This is a profound issue that will be discussed in section 5.8.**

We will also study cases where a surface point has a compressive extremum in both principal directions of curvature; i.e., is a shield with respect to both directions.

Before going onto that topic, we will further illustrate our definition of a part as involving recovered sequences of Process Grammar operations. To do so, let us return to the part called a *chin*. Occasionally, some people have a chin that is *pointed*; i.e., it is purely a protrusion $M^+$, as illustrated by the left-most shape in Fig 5.9. We can consider this outward protruding growth to be the initial morphological stage in evolution. In the subsequent development, the protrusion $M^+$ bifurcates under the Process Grammar operation

$$BM^+ \; : \; M^+ \longrightarrow M^+m^+M^+$$

to become a shield that is the typical chin; as illustrated by the transition from the left-most shape to the middle shape in Fig 5.9.

Furthermore, in some people, after that stage, an indentation can develop in the center of the chin. According to the Process Grammar, this is the result of the central compressive force, i.e., the central arrow in the middle shape in Fig 5.9, continuing till it indents, as shown in the third shape in Fig 5.9. This Process Grammar operation is coded as follows:

$$Cm^+ \; : \; m^+ \longrightarrow 0m^-0$$

and is called "squashing continues till it indents". This extra stage, where the compressive process of a shield can continue till it indents, can occur in a number of situations in biology, presumably due to a biological mistake in estimating the required magnitude of the central compressive process. For example, this extra transition can occur at the

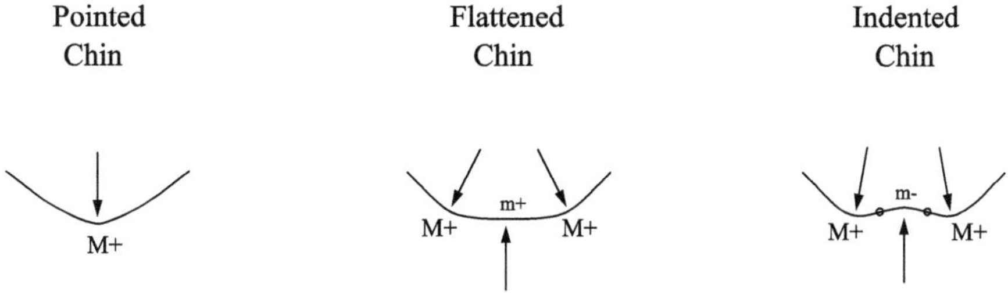


**Fig. 5.9.** The Process Grammar gives the successive relationship between types of chins.

top of the head; i.e., in older people, the compressive structure at the top of the head can indent producing a dip in the top of the head. Also, it can occur at the side of the jaw; i.e., in older people, the compressive structure at the side of the jaw can indent producing an inward bending in the jaw. Also, it can occur at the tip of the nose; i.e., the compressive structure at the tip of the nose can indent producing a dip into the nose-tip.

The sequence just described illustrates again what I meant by defining a part to be a recovered process, and also the recovered subsequent history of the process.

### Bi-Directional Shields

We saw in Fig 5.4 page 133 that a forehead is a shield. In fact, the diagram illustrated that the *vertical* central cross-section through the forehead is a shield. However, another important fact is that the *horizontal* central cross-section through the forehead is also a shield. Thus the forehead is an example of what my theory calls a bi-directional shield. In fact, I define a *bi-directional shield* to be a neighborhood of a point $p$ on a surface where $p$ is a $m^+$ extremum of each of the two principal lines of curvature through $p$.

Based on my theory, it is clear that, functionally, a bi-directional shield is used because it has a powerful protective role. In fact, I will show that bi-directional shields are extremely important protective structures in biological morphology and manufacturing design. Other biological examples of bi-directional shields are as follows:

Fig 5.10 shows the *heel of a foot*. In the first photograph in Fig 5.10, the heel is viewed from the back; and in the second photograph in this figure, the heel is viewed from the side. It is clear that, in both directions, the heel is a shield, i.e., is given by a $m^+$ extremum. Therefore it is a bi-directional shield. Notice also, in the second photograph, that the base of the toe is also a shield. In fact it is also a bi-directional shield.

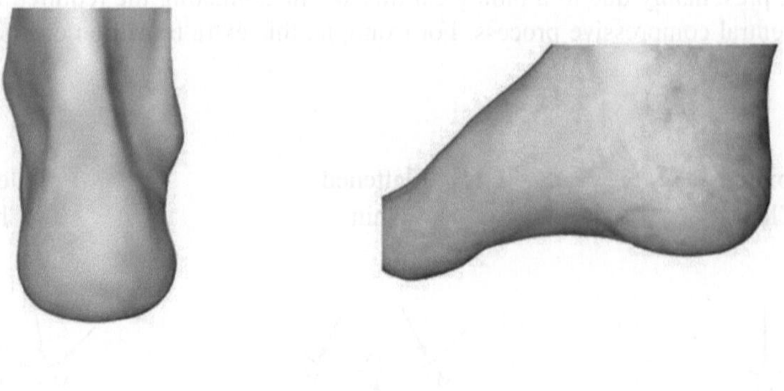

**Fig. 5.10.** The heel of a foot, viewed from the back and from the side, is a *shield.*

As another example, consider Fig 5.11 which shows the *cap of a knee*. It is viewed from the top in the first diagram, and from the side in the second diagram. It is clear that in both directions, the knee-cap is a shield, i.e., is given by a $m^+$ extremum. Therefore it is a bi-directional shield.

**Fig. 5.11.** The knee, viewed from the top and from the side, is a *shield.*

### Shields In Vehicles

Having shown the importance of shields in biological morphology, we will now show the importance of shields in vehicles. Fig 5.12 shows the front of a van. We see that it is almost entirely made of shields. A crucial fact is that each of the surfaces is a bi-directional shield, in order to powerfully realize the function of protection. For example, the roof is a bi-directional shield. This is in order to protect the passengers and the transported goods. Also, observe that the window is a bi-directional shield; to protect the passengers. Furthermore, observe that the hood (i.e., bonnet) is a bi-directional shield. Clearly this is to protect the engine. Also, observe that the front of the radiator is a bi-directional shield; again to protect the engine. Furthermore, observe that the bumper is a bi-directional shield in the horizontal and vertically diagonal direction, in order to protect the van from impacts that come both from the horizontal direction and upward from the road. Also, each side of the van constitutes what our theory calls a double-shield, which is a sequence of two shields along a cross-section curve, which will be discussed when we describe shields in architecture. In fact, each shield in the double-shield of a side of the van is also a bi-directional shield; i.e., a shield in both the horizontal and vertical directions.

Fig 5.13 then shows that the back of a van is also defined by shield structures. Once again, the surfaces are given by bi-directional shields.

Now, the fact that the surfaces of the van body are given by bi-directional shields is due to the crucial protective function of the surfaces. In contrast, consider any tire,

**Fig. 5.12.** The front of a van is defined by shield structures.

**Fig. 5.13.** The back of a van is defined by shield structures.

as can be seen in both Fig 5.12 and Fig 5.13. We can see that, in the lateral direction of the vehicle, the contact surface of a tire is given by a shield. Clearly the function of this is to deal with the compressive force of the road. In contrast, in the front-to-back direction along the contact surface of the tire, the surface is given by a circle, because the function in that direction is rotation. Thus the two principal directions on the tire's contact surface are a shield and a circle, because the two directions correspond to two different functions. This contrasts with the surfaces of the body of the van, which are bi-directional shields because the two directions correspond to the same function, that of protection.

Fig 5.14 shows that a city bus is also defined by shields. Note first that, since, in a *van*, the transported items are passengers in the *front* and goods *behind* them in the rest of the van, the shields which constitute the large main body of a van have the function of protecting passengers at the front and the goods behind them. In contrast, since, in a *bus*, the transported items are passengers located at positions along the entire length of the bus, the shields that constitute the entire body of a bus have the function of protecting passengers along the entire length.

Fig 5.15 shows the side of a city bus, and shows that, viewed from this direction, the body is also structured by shields. Notice the shield structure that defines the front which is designed also to protect the passengers as they enter. Observe also, in the same figure, the container of the mirror is also structured as a shield, and this is because the protection of the mirror is crucial in order to ensure the safety of navigation. Also, in the same photograph, observe the shield structure of the back of the van beyond the bus.

Fig 5.16 shows a school bus from the front. We can see the enormously powerful shield that defines the top of the bus. It has an obvious protective role with respect to the children.

Fig 5.17 shows a car viewed from a diagonal direction, in order to exhibit the enormous importance of shields in all directions. Note that each *city bus* shown in Figs 5.14 and 5.15 consists entirely of one compartment in order to contain the large set of passengers. In contrast, because a *car* transports only a small set of passengers, it consists of three small compartments: one in the center for the passengers; one at the front for the engine; and one at the back for the small set of goods belonging to the passengers. Nevertheless, each of these compartments is entirely defined by shields, because their contents have to be protected. In section 5.5, we will show that the Process Grammar defines the design of, and relationship between, the three compartments. The purpose of presenting the car photograph in the present section, i.e., Fig 5.17, is so that the reader can begin to see the considerable frequency of shields in a car. For example, observe from Fig 5.17 that the bumper in the front is a shield, again to fulfill its protective role for the car. Also observe that the roof is a shield to protect the passengers. Notice that even the edges of the head-lights are shields. Furthermore, the car-body around the wheel consists of shields to protect the wheel. Also, the edges of the mirror are shields. And of course, the hood is also a shield, for obvious reasons: It protects the engine.

**Fig. 5.14.** The front of a city bus is defined by shield structures.

**Fig. 5.15.** The side-view of the front of a bus is a shield structure.

**Fig. 5.16.** The front of a school bus is defined by shield structures.

**Fig. 5.17.** A car is defined by shield structures.

### Shields in Architecture

So far, in this section, we have looked at shield structures in biological morphology and in the design of vehicles. We will now show that shields are also important in architecture.

Fig 5.18 shows a building where all the balconies are defined by shield structures. A close-up view of one of the balconies can be seen in Fig 5.19. Again we can see that the structure is the result of the *anticipated function*: The central $m^+$ extremum of the balcony implies an external force, from which the balcony is trying to protect anyone who goes onto the balcony.

As another architectural example, consider Fig 5.20, which shows the entrance of a subway station in New York. The roof is actually an example of what our theory calls a *double-shield*. That is, the central extremum (at the top of this roof) is actually a protrusion $M^+$, with a squashing extremum $m^+$ on each side; i.e., as shown by the right diagram in Fig 5.21. The reader will recall, from section 3.21, that this structure is a consequence of the Process Grammar operation

$$Bm^+ \ : \ m^+ \ \longrightarrow \ m^+M^+m^+$$

which we call *breaking-through of a protrusion*. This operation is illustrated by the transition from the left shape to the right shape in Fig 5.21. That is, the left shape is a shield, given by the downward squashing process inferred from the central extremum $m^+$ by its PISA axis. In the transition to the right shape, this process undergoes a bifurcation due to an opposing force that breaks-though, shown as the bold upward arrow in the right shape. This breaking-through force has created a protrusion. The two resulting copies of the squashing process are on the sides of the central protruding process.

According to our theory, this Process Grammar operation defines important functional roles of the roof. First, in the left shape of Fig 5.21, the downward process at the central $m^+$ represents a downward external process (rain, snow, etc.) from which the subway roof is protecting the people who are entering. Then, the upward protruding process, which breaks-through, has the role of bifurcating the downward process, i.e., pushing that process (rain, snow, etc.) to each side so it can run down the roof – which is the function of the standard A-shape of the roofs of most houses. The bifurcation operation therefore bifurcates the protective function of the initial central shield into two side shields in order to *add* the function given by the protrusion. Therefore:

**Process Grammar tracks the functional hierarchy of the roof.**

**Fig. 5.18.** A building where all the balconies are defined by shields.

**Fig. 5.19.** A balcony defined by a shield structure.

**Fig. 5.20.** Entrance to a subway station defined by a double-shield structure.

**Fig. 5.21.** The $Bm^+$ operation from the Process Grammar.

A basic conclusion of this section is the following:

**FUNDAMENTAL IMPORTANCE OF SHIELDS**

**We have shown that what our theory defines as a shield structure, that is, a $m^+$ extremum with its crucial function given by its corresponding PISA axis, is fundamentally important in biological morphology and manufacturing design.**

**In contrast, the standard literature on shape image analysis completely fails to observe shields. The reason is that the standard literature uses the conventional definition of reflectional symmetry, e.g., the Medial Axis, which does not capture any meaningful information about compressive extrema.**

**It is important to understand that, in** *designs*, **the PISA axis gives both the** *recovered past history* **of compressive extrema, as well as the** *intended function* **of those extrema.**

Section 5.5, which gives a theory of car design, demonstrates how the shield structure is a primary factor in the design operations.

## 5.4 More Examples of Shields

When a person is walking, each foot makes a step forward in front of the rest of the body in the travel direction. Therefore each foot is threatened by a possible opposing environmental force.

In order to protect each foot, the front of a shoe tends to be structured as what our theory calls a shield, as shown in Fig 5.22. Note that the threatening opposing environmental force, which defines the purpose of the shield, is given by the PISA axis, whereas the other differential symmetry axes completely fail to define that force.

What our theory defines as a shield is used on the front of many shoes, as demonstrated by the many examples of shoes shown in Fig 5.23, which is the display of many shoes at a store in New York.

Next, Fig 5.24 and Fig 5.25 show significant shields in the head of a dog.

That is, Fig 5.24 shows the existence of a shield in both the lateral curve across the front of the nose, and the lateral curve across the top of the head. And Fig 5.25 shows the existence of a shield in both the vertical curve across the front of the nose, and the forward curve across the top of the head.

As a result of this, we can see that the front of a dog's nose and the top of a dog's head are examples of what our theory calls *bi-directional shields*.

The reason for the existence of these bi-directional shields is as follows: The front of a dog's nose is the major front of the dog's movement, and is therefore required to have a protective structure. In addition, the top of a dog's head is required to protect the dog's head from falling objects due to gravity.

We have seen that, according to our theory, the front and top of cars, vans, and buses, are bi-directional shields. Therefore, since we have also shown this for a dog's head, this is an example of how our theory unifies biological morphology and automotive design.

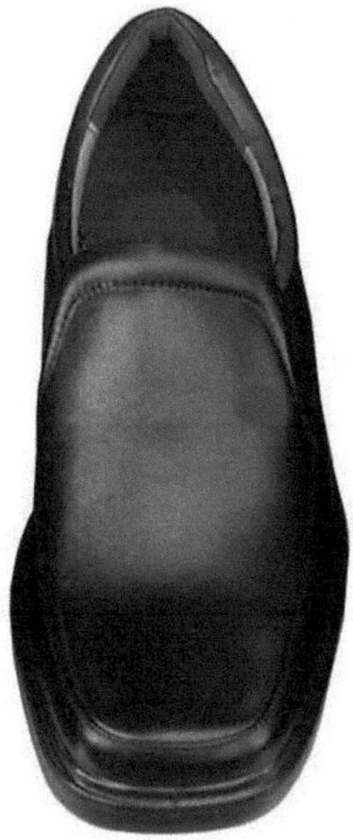

**Fig. 5.22.** The front of a shoe structured as a shield.

**Fig. 5.23.** Many shoes are examples of shields at the front.

**Fig. 5.24.** Shield in the lateral curve across the front of the nose, and the lateral curve across the top of the head.

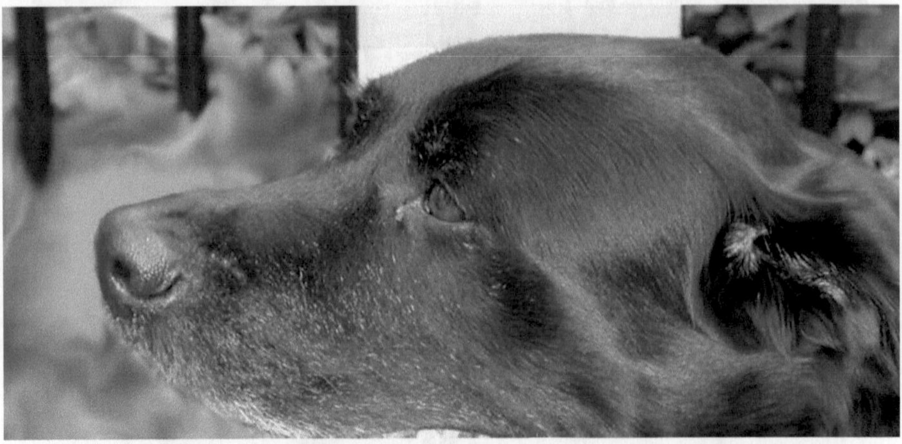

**Fig. 5.25.** Shield in the vertical curve across the front of the nose, and the forward curve across the top of the head.

## 5.5 Car Design

This section shows that the Process Grammar operations intelligently define both the shape generation of a car and also its *intended functions*.

In the design of a vehicle, the Process Grammar operations are applied to an ellipsoid, a 3D shape illustrated in Fig 5.26. The travel-direction is along the longest axis of the ellipsoid. For the application of the Process Grammar operations, a crucial fact is that this ellipsoid is symmetric about a vertical plane that is along the travel-direction, and the cross-section of the ellipsoid given by this vertical symmetry plane is an ellipse whose long axis is along the travel-direction. Furthermore, this ellipsoid is symmetric about a horizontal plane; and the cross-section of the ellipsoid given by this horizontal symmetry plane is an ellipse whose long axis is also along the travel-direction. Notice, we have defined two ellipses, one in the vertical symmetry plane and one in the horizontal symmetry plane. Both of these ellipses are shown in Fig 5.26.

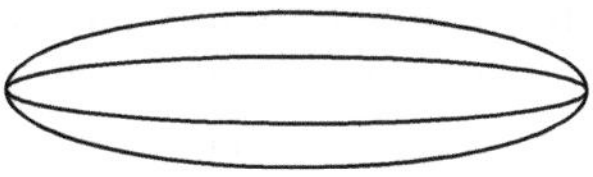

**Fig. 5.26.** The Process Grammar operations will be applied to this ellipsoid.

According to our theory, these two ellipses are fundamentally important in determining functions of the vehicle. To understand these functions, consider the structure of an ellipse. It is important to notice that an ellipse has four extrema of the same sign of curvature. Both the ellipses we are considering are of positive curvature. On an ellipse of positive curvature, the extrema are two $M^+$ and two $m^+$, as illustrated in Fig 5.27. By our process-inference rules, based on PISA, there are four processes: two outward *penetrative* processes and two inward *compressive* processes, as also illustrated in Fig 5.27. Therefore observe that PISA correctly gives the processes that formed the ellipse, whereas the conventional symmetry axes, such as the Medial Axis, do not.

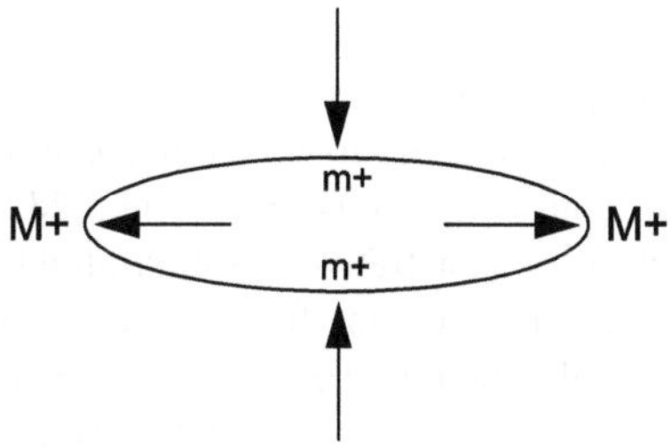


**Fig. 5.27.** An ellipse has four extrema, with four processes correctly defined by PISA.

Our claim is that these four extrema, in each of the two ellipses we are considering, have fundamentally important functions for the vehicle, as follows: To understand this, consider either one of these ellipses. Its two penetrative extrema are in the *direction* of travel (i.e., forward and backward) and they are required because, in the travel-direction, the vehicle has to *penetrate* the environment. Furthermore, the two compressive extrema, on the ellipse, act as *shields* for the vehicle, with respect to external forces. For example, in the ellipse on the vertical symmetry plane, the two compressive extrema give top and bottom shields to the vehicle, i.e., create a protective role for the roof and the floor. Furthermore, in the ellipse on the horizontal symmetry plane, the two compressive extrema give side shields to the vehicle, i.e., create a protective role for the left and right sides of the vehicle.

Our next claim is this: Consider one of these ellipses. Our claim is that there are reasons, in the design of certain vehicles, why one or both of its penetrative extrema have to undergo the Process Grammar operation $BM^+$, shield-formation; i.e., it is necessary for a penetrative extremum in the travel-direction to bifurcate into a complete shield, the triple $M^+m^+M^+$, with the compressive extremum $m^+$ along the travel-direction. In such a vehicle, it is usually necessary for this to happen to each penetrative extremum on each of the two ellipses. For example, we claim that, in the design of a *car*, each of the front and back penetrative extrema usually has to undergo the shield-formation operation $BM^+$ in order to anticipate a compressive force against which the car might crash. The possibility of crashing is very frequent with a car, due to the fact that it travels in a road with a high density of nearby cars, and is also faced with walls on the streets, etc. The crucial consequence is this:

> **Consider the ellipse of the vertical symmetry plane and the ellipse of the horizontal symmetry plane, as defined above, in the design procedure we are defining for a car. The result of applying the Process Grammar operation $BM^+$ to each of their penetrative extrema is that the front and back of a car are structured by what our theory calls *bi-directional shields*.**

Now, besides cars, there are certain types of non-road vehicles where the possibility of crashing is considerably less frequent than a car, and therefore the penetrative extrema are maintained. Furthermore, in these vehicles, the penetrative extrema are maintained for additional reasons. We will now give two examples of such vehicles:

The design of an airplane preserves the penetrative extrema along the travel-direction for the following reasons: An airplane has to penetrate the air faster than a car, and has to maintain its penetration through winds and clouds. Fig 5.28 illustrates, for an airplane, the considerable preservation of penetrative extrema along the travel-direction. This includes not only penetrative extrema at the front and back, but also along the edges of the wings. This preservation of penetrative extrema is also due to the fact that, unlike a car, its path is not filled with many other close vehicles using the same path. Notice also, as a result of this, that, unlike a car, an airplane does not have a bumper at the front and back. Recall that a bumper is structured vertically and horizontally by what our theory calls compressive extrema. That is, a bumper is an example of what our theory calls a bi-directional shield. Now, observe from Fig 5.28 that, despite the fact that the front and back of the airplane are penetrative, the airplane body in the perpendicular direction to the travel-direction is structured by compressive extrema $m^+$, i.e., shields.

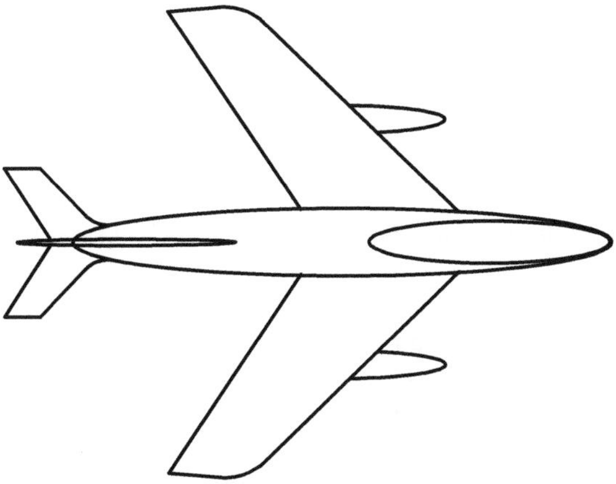

**Fig. 5.28.** In an airplane, several of the penetrative extrema along the travel-direction are preserved.

**Fig. 5.29.** In a submarine, several of the penetrative extrema along the travel-direction are preserved.

Similar properties apply to a submarine. A submarine has to penetrate a thicker fluid than a car. Therefore, the design of a submarine preserves its penetrative extrema along the travel-direction (both forward and backward). Fig 5.29 illustrates, for a submarine, the considerable preservation of penetrative extrema along the travel-direction. Observe that this is true not only for the main body, but also for all the attachments to the main body. Another reason why the design of a submarine preserves the penetrative extrema in the travel-direction is this: Unlike a car, which is restricted to a road, a submarine can vary its travel direction. Furthermore, unlike a car, its travel-direction is not filled with many other close vehicles using the same travel-direction. Notice also, as a result of this, that, unlike a car, a submarine does not have a bumper at the front and back. Again, recall that a bumper is structured vertically and horizontally by what our theory calls compressive extrema. That is, a bumper is an example of what our theory calls a bi-directional shield. Now, observe from Fig 5.29 that, despite the fact that the front and back of the submarine are penetrative, the submarine body in the perpendicular direction to the travel-direction is structured by compressive extrema $m^+$, i.e., shields. From Fig 5.29, we can see that this is also true of the attachments to the main body.

We are now going to return to the analysis of cars. An important fact is this:

**We will now show that the Process Grammar gives an extensive understanding of the generation of both the shape of a car and its *intended functions*.**

**A crucial consequence of this is that the Process Grammar gives the ability to recover the *design intents* for which the shape features were created.**

**This is important for solving major problems that exist in large-scale engineering systems and product lifecycle management. That is, it is internationally known that integration of the many design teams, involved in the manufacturing of a large-scale product, fails because the teams use different design softwares and thus cannot infer the design intents from each others' CAD models.**

**The Process Grammar, as well as the other components of the New Foundations to Geometry, solve this problem.**

According to our theory, the shape and intended functions of the car are generated by the application of the Process Grammar operations to the two ellipses we defined earlier, that on the vertical symmetry plane, and that on the horizontal symmetry plane. These two ellipses will be called, respectively, the *vertical basic ellipse* and the *horizontal basic ellipse*. In fact, our theory will show that most of the generation of the shape and functions of the car is necessary to accomplish on the vertical basic ellipse.

To help the reader understand our generative theory, we will examine the car-body shown in Fig 5.30 page 155.

It is first necessary to see the level and height of the vertical basic ellipse. This is shown in Fig 5.31. To help the reader see the vertical basic ellipse, the figure has placed it on the side of the car, although it is actually in the center of the car. The crucial fact, that is shown in this figure, is that, according to our theory, the vertical basic ellipse is on the level below the windows which will later emerge from the shape generation.

Notice that the arrows in Fig 5.31 show the processes, inferred by our theory, from the curvature extrema of this ellipse.

Observe that the arrows given by the penetrative extrema, which, as stated earlier, are in the travel-direction and define the movement function of the car, are shown as going beyond the front and back of the car. Thus, the Process Grammar shield-formation operation $BM^+$, that creates the front and back shields, will flatten these extrema back towards the car. This again accords with our definition of a *part* as a recovered process together with its recovered subsequent history.

Observe that the vertical arrows shown in Fig 5.31 are the processes, inferred by our theory, as corresponding to the compressive extrema on the ellipse. As stated earlier, since these inferred process-arrows are in the vertical direction, these compressive extrema correspond to the *shielding function* of the floor and roof. In fact, we will see how the

**Fig. 5.30.** The car that we are analyzing.

**Fig. 5.31.** According to our theory, the lower half of the car defines a long ellipse. Our inference rules imply two outward penetrative arrows giving the travel-direction, and two inward compressive processes defining a significant shield structure.

**Fig. 5.32.** The front and back door-lines have extrema which imply two arrows that emphasize the travel-direction.

**Fig. 5.33.** The arrows leading to the extrema of the window. Notice that the outward arrows are coincident with the reflection lines on the glass.

**Fig. 5.34.** Arrows leading to extrema on the top boundary.

**Fig. 5.35.** Some of the important processes in the creation of the car shape.

Process Grammar raises the upper central region of the ellipse to the roof-height needed to contain and protect the passengers. Furthermore, in doing so, the Process Grammar operations will also create the windows. We will refer to the passenger roof and the windows as the *upper compartment* of the car. Again, this accords with our definition of a *part* as a recovered process together with its recovered subsequent history. The part given by the compressive extremum on the top of the ellipse corresponds to the protective role of the roof; and the recovered subsequent application of the Process Grammar operations to this extremum will produce the upper compartment of the car.

Before giving the operations that produce the upper compartment from this compressive extremum on the vertical basic ellipse, we will show how the vertical basic ellipse is visually reinforced, very explicitly, by certain features on the car. For example, observe from the photograph, Fig 5.30, the long horizontal ridge, at the level of the door-handles. This ridge helps to emphasize the top side of the vertical basic ellipse. Observe also that the two outward penetrative arrows are on the same level as the thin lower rib that the designer explicitly put along the body of the car.

Next, observe that the back line of the back door and the front line of the front door have curvature extrema indicated by the two arrows shown in Fig 5.32. These two extrema are visually important, because they help to emphasize, to any viewer, the travel function of the car.

Notice that, above the car's vertical basic ellipse, there is a deformed ellipse defined by the window structure. Processes inferred from this shape, by our rules, are shown in Fig 5.33. Concerning this, observe the following: The two outward arrows correspond to the *reflection lines* of *light* on these windows. In industrial design, the enabling of certain particular reflection lines is one part of making the design object attractive. We can therefore see that our rules explain why the designers choose the particular reflection lines they create.

Next observe that the top line of the car is given by a set of curvature extrema, most of which are indicated by the process-arrows shown in Fig 5.34. Once again, the structure and function of these processes are generated by the Process-Grammar, as will be shown soon, when we define the generation of the upper compartment.

Observe that the full set of the arrows discussed so far is shown in Fig 5.35. In addition, a horizontal inward arrow has been placed on the left, corresponding to the penetrative extremum in the back light. This forms a *pinching* relationship with the extremum in the back door. Pinching is given by Process Grammar operations that will be defined later.

The components of the process-structure in Fig 5.35 are used throughout the car, on every level of scale, giving the car a powerful unity, as will now be illustrated:

To understand this, we will first introduce the reader to a new terminology that will help understand the Process Grammar operations applied to create a car. In this terminology, an ellipse will be referred to as a *2-sided oval*, or simply *2-oval*. What we call the *two sides* are the *two shields*. Thus, the sides are defined by the PISA axes, and therefore the terminology will be based on our process-inference rules.

As we will see, Process Grammar operations will generate, from the 2-sided oval, ovals of a successively greater number of sides, i.e., shields. Generally we will refer to these as *$n$-sided ovals*, or simply *$n$-ovals*; i.e., *ovals consisting of $n$ shields*.

So far we have seen the importance of a 2-sided oval as giving the largest main structure of the car. But we will now see that ovals are on every level of scale, bringing shield structures to all levels.

First, as shown in Fig 5.36, the oval is used as the shape of the front grill, the back trunk, the door handles, etc.

**Fig. 5.36.** Ovals as the front grill, back trunk, door handle.

Each of these is an example of a 2-sided oval; i.e., an oval consisting of 2 shields. Observe that this gives important shields to the car. Notice, on the front grill, the top and bottom compressive extrema protect the grill with respect to forces in that direction. Furthermore, these shields are extensions of the lateral shield structure on the front hood. We will also see later how the travel-direction shield on the hood is created; thus giving the bi-directional shield structure of the hood. Notice also, from Fig 5.36, that, on the back face of the lid of the back trunk, the top and bottom compressive extrema protect this face, e.g., in the slamming down of the lid. Furthermore, these shields are extensions of the lateral shield structure on the trunk. We will also see later how the travel-direction shield on the trunk is created; thus giving the bi-directional shield structure on the trunk. Also observe, from Fig 5.36, that the compressive extrema on the door handle are shields that protect both the handle and the hand that uses the door handle.

The next important thing to understand is that, according to our generative theory of a car, the 2-sided oval undergoes a transition to a 3-sided oval; i.e., a shape consisting of 3 shields. This transition is shown in Fig 5.37.

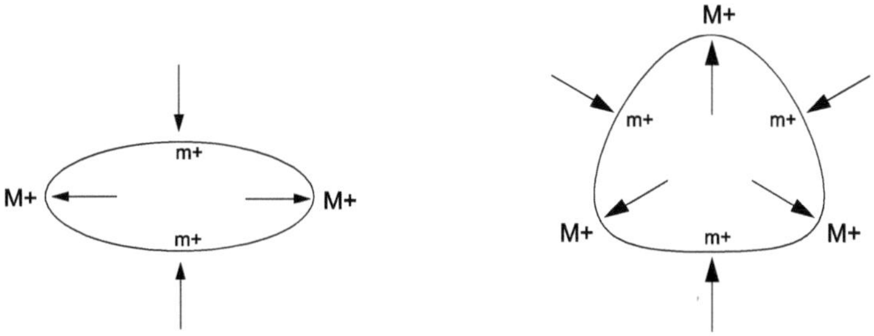


**Fig. 5.37.** The transition from the 2-oval to the 3-oval, given by the Process Grammar operation $Bm^+$.

**Fig. 5.38.** Some examples of $n$-sided ovals in the car: the back light, front light, and mirror holder.

The transition from the 2-sided oval to the 3-sided oval is given by the Process Grammar operation defined in section 3.21:

$$Bm^+ \ : \ m^+ \ \longrightarrow \ m^+M^+m^+$$

which we call *breaking through of a protrusion*; i.e., where one of the shields bifurcates into two shields as shown in Fig 5.37. Furthermore, the 3-sided oval can undergo a further such transition to a 4-sided oval, and so on. Generally, $n$-sided ovals are a significant shape in a car, as illustrated in Fig 5.38, which shows the back light, the front light, and the mirror holder, all structured by shields to protect their contents.

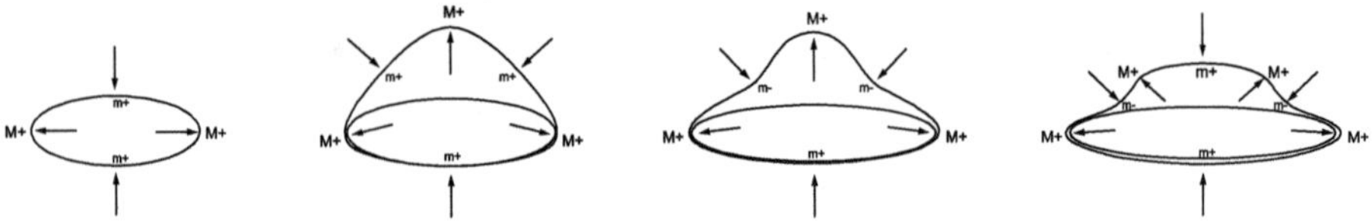


**Fig. 5.39.** The successive stages in the creation of the car's upper compartment from the car's vertical basic ellipse.

We now make a fundamental claim that the transition from the 2-oval to the 3-oval is a stage in the development of the *large-scale structure* of the car. What we will do now is show how the Process Grammar explains the creation of the car's upper compartment from its vertical basic ellipse. According to our theory, the successive stages in this creation are as shown Fig 5.39.

Before examining this important sequence, it is necessary to understand that there is one stage prior to the sequence. The reader will recall that, according to the Maximization of Recoverability principle, of the New Foundations to Geometry, the starting shape, in an *intelligent* shape generation, i.e., one that maximizes recoverability, must have had maximal symmetry. In the case of the ellipsoid we defined for a car, the starting shape must have therefore been a sphere. This means that the starting shape of the vertical basic ellipse must have been a vertical circle, and the starting shape of the horizontal basic ellipse must have been a horizontal circle.

Therefore, the initial transition, in the vertical symmetry plane of the car, goes from the vertical circle to the vertical basic ellipse which is the first shape shown in Fig 5.39.

We now examine the sequence of transitions shown Fig 5.39. After the prior stage from the vertical circle to the vertical basic ellipse, the next stage, i.e., the transition from the first to the second shape in Fig 5.39, is this: The vertical basic ellipse, i.e., *2-sided oval*, undergoes the Process Grammar operation $Bm^+$ to become a *3-sided oval*. That is, the $m^+$, at the top of the 2-sided oval, bifurcates into the two side $m^+$ extrema in the 3-sided oval. This is due to the breaking-through of the upward protrusion $M^+$ that we see in the 3-sided oval. Therefore, this breaking-through begins to produce the upper compartment of the car to accommodate the passengers.

In the next stage, i.e., the transition from the second to the third shape in Fig 5.39, the inward processes at the two side $m^+$ extrema continue till they indent; i.e., they each undergo the operation $Cm^+$ from the Process Grammar. This action creates the end-points $m^-$ of the upper compartment of the car. The effect is to *restrict* the zone of influence of the upward protrusion – a concept which will be fully analyzed later.

In the final stage in Fig 5.39, i.e., the transition from the third to the fourth shape, the upward protrusion itself undergoes a bifurcation; i.e., the application of the Process Grammar operation $BM^+$, *shield-formation*. That is, the $M^+$ at the top of the third shape bifurcates into the triple $M^+m^+M^+$ at the top of the fourth shape. This results in the fundamentally important shield, the roof of the upper compartment, shown in the fourth shape; i.e., the roof that protects the passengers.

### Triple-Shields in Cars

In many cars, there is an additional important stage in the creation of the upper compartment, as follows: Observe, in the fourth shape in Fig 5.39, only one shield has been created in the upper compartment: the shield defined by the roof. In the stage given by the fourth shape, the front and back of the upper compartment are each a spiral, going downwards from a $M^+$ extremum to a $m^-$. These spirals are the front and back windows. In some cars, these windows are kept as spirals. However, in many cars, these windows are developed further to become *shields*. An example is the New York taxi shown in Fig 5.40, where we can see that the front and back windows are structured as shields. There is an obvious function to this: The front and back windows have the important role of *protecting* the passengers. Exactly the same structure is seen also, for example, in police cars, where again, the protective role of the front and back windows is important.

**Fig. 5.40.** A New York taxi, showing the triple-shield structure of the upper compartment.

The crucial fact that can be seen from Fig 5.40 is that the upper compartment, in this type of car, is what I call a **triple-shield**. That is, the roof and the front and back windows constitute three adjacent shields. Observe that we also saw that the top of a person's head is a triple-shield, as was illustrated in Fig 5.4 (page 133). Furthermore, we also saw that the jaw and chin structure of a person's face is a triple-shield, as was illustrated in Fig 5.3 (page 132). In fact, we will see that the following claim is crucially important:

> **We claim that the structure defined in our theory as a *triple-shield* is a fundamental structure in many designs – from biological morphology to automotive design.**

Now, according to our theory, a car of the type in Fig 5.40 is the result of an *additional design stage* that converts the spiral front window and spiral back window of the fourth diagram in Fig 5.39 into the shield front window and shield back window in Fig 5.40. The operations that achieve this are pair-creation operations from the Process Grammar. That is, to convert the spiral back window into a shield, one applies the downward operation

$$\pi\emptyset^{\downarrow+} \; : \; \emptyset^{\downarrow+} \quad \longrightarrow \quad m^{+}M^{+}$$

to a positive curvature point on that spiral. Correspondingly, to convert the spiral front window into a shield, one applies the upward operation

$$\pi\emptyset^{\uparrow+} \; : \; \emptyset^{\uparrow+} \quad \longrightarrow \quad M^{+}m^{+}$$

to a positive curvature point on that spiral. Clearly, these two operations can be regarded as symmetrically paired on the car. Chapter 6 will discuss such symmetrical pairings.

We now claim that the car in Fig 5.40 exhibits another additional design stage. To understand this, observe that, in the fourth diagram of Fig 5.39, the top of the front hood is a spiral, and the top of the back trunk is a spiral. The additional design stage that we now define is the transition that converts those two spirals into shields, thus protecting the front hood and back trunk from above. We can see these resulting shields in Fig 5.40. Again this is achieved by pair-creation operations from the Process Grammar.

Furthermore, our theory gives an extra operation that is also applied to the front hood and back trunk. Recall that, as we said earlier, to create the *vertical* front shield of the front hood, and the *vertical* back shield of the back trunk, one applies the shield-formation operation $BM^{+}$, respectively, to the front penetrative extremum and to the back penetrative extremum of the vertical basic ellipse of the car.

A fundamental fact is that, as stated earlier, on the *horizontal basic ellipse*, the shield-formation operation $BM^{+}$ is applied also to the penetrative extrema at the front and back of that ellipse. The crucial fact is that this produces what our theory calls the *bi-directional shield* structure of the front face of the front hood and back face of the back trunk.

By the Maximization of Transfer principle of my New Foundations to Geometry, the operations we have listed for the shape generation of the car are transferred to corresponding points that exist in directions perpendicular to the bifurcation directions of the operations. The results of this include bi-directional shields created on all faces of the car. The mathematics of the transfer of the operations is defined in section 5.8.

The next page gives a fundamental fact about the system that has been described in this section.

## RECOVERY OF DESIGN INTENT

**We have shown that the important functions of a car are processes inferred from the curvature extrema in the successive stages defined by the Process Grammar operations.**

**This means that the Process Grammar recovers the** *layers* **of** *design intents* **in the car.**

**In fact, the recovery of design intent is one of the main achievements of the laws of recoverability given by the New Foundations to Geometry.**

**This is relevant for solving major problems that exist in the integration of large-scale engineering systems and product lifecycle management.**

**For example, the manufacturing of a major engineering product, such as an aircraft, involves supply chains that tend to use different design softwares that cannot exchange their model information. What would solve this interoperability problem is the recovery of design intent. Furthermore, over the product lifecycle, the recovery of design intent would also solve the reuse, adaptability, and maintenance problems.**

**A crucial fact is that the mathematics of recoverability, invented by my New Foundations to Geometry, solves these problems. This is corroborated by the fact that aspects of this mathematics have been applied by scientists and engineers in over 40 disciplines. Furthermore, certain members of the International Standardization Organization (ISO) are implementing my mathematics to establish international standards in engineering design.**

## 5.6 Grasping

So far, in our analysis of shields, we have seen that the PISA axis models the function of *protection*. Now we will see that the PISA axis models also the function of *grasping*.

Let us begin by considering a bowl as shown in Fig 5.41. According to our theory, the penetrative extrema $M+$, shown in the figure, correspond to the outward arrows which create the significant width of a bowl so that it fulfills the *function* of *containing*. However, the penetrative extrema make the grasping of the bowl unstable. Therefore, in order to enable the function of grasping, the bowl is compressed at the sides, as shown in Fig 5.42. This figure illustrates an object such as a bottle, a vase, a lamp-stand, etc. Such objects are structured so that they can be *grasped*.

A crucial fact is that, in Fig 5.42, the arrows are given by the PISA axes inferred from the compressive $m^+$ extrema. The first important thing to observe is that these PISA axes give the forces that are used to grasp the object. In contrast, the Medial Axis symmetry sets would give a totally incorrect representation of these forces. That is, the Medial Axis symmetry sets would have the following properties: (1) at the two $m^+$ extrema they would be on the opposite side of the curve, and far away from the curve; and (2) they would point away from the curve. These properties make the Medial Axis symmetry sets completely fail to describe the grasping forces. We thus conclude:

**PISA axis is totally correct in describing the forces of grasping.**

**Medial Axis symmetry set is totally wrong in describing the forces of grasping.**

Now, it is clear that, in the *design* of the object, the PISA axes give the compressive forces used to *deform* the shape into that required for the intended future function of grasping. This gives another example of the theory we proposed on page 131, which states that, according to the New Foundations to Geometry, the intended functions of a shape can often be realized by generating the shape via recoverable design operations that have the same structure as those future functions.

Therefore, once again, we see that the New Foundations to Geometry have the unique power of recovering the **design intent**.

Now, according to our theory, the transition from the bowl in Fig 5.41 to the bottle in Fig 5.42 is given by the shield-formation operation

$$BM^+ \ : \ M^+ \ \longrightarrow \ M^+m^+M^+$$

Observe that this not only creates the compressive extremum $m^+$ required for the function of grasping, but, in bifurcating the $M^+$ to the top and bottom, it preserves the function of containing. Thus, again we see the crucial fact that the Process Grammar, by recovering the succession of *causal* stages in the shape development of an object, also recovers the succession of *design intents* in the history of the object.

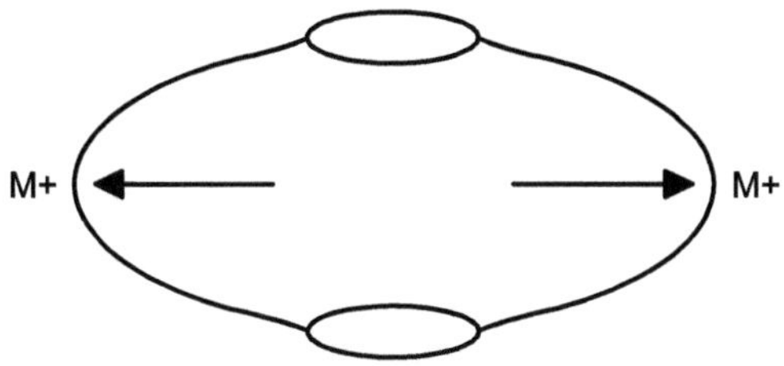


**Fig. 5.41.** A bowl with a penetrative structure in order to enable the function of containing.

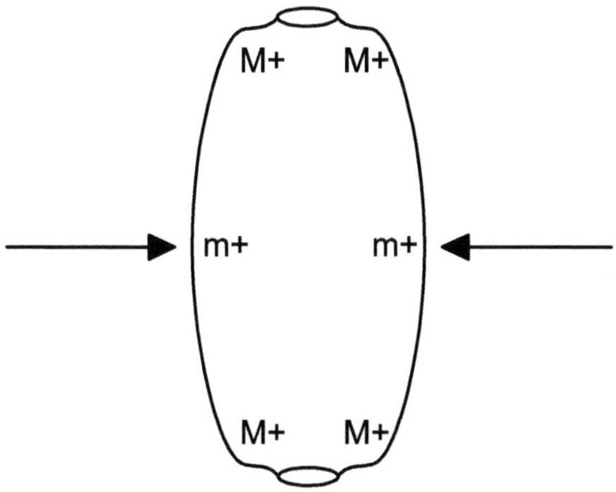


**Fig. 5.42.** In this figure, the bowl in Fig 5.41 has undergone the Process Grammar shield-formation operation in order to enable the function of grasping. The arrows in this figure are the PISA axes at the compressive extrema $m^+$, and these axes correctly describe the grasping function.

The next important thing to observe is this: In order to grasp the object, one needs to apply *two* forces *simultaneously*, as illustrated by the two arrows in Fig 5.42. Furthermore, they need to be *reflected* versions of each other. Without this, the object could not be held and lifted. For example, the PISA axis on one side can represent the pressure applied by a thumb, and the PISA axis on the other side can represent the opposing pressure applied simultaneously by a finger.

A further thing to observe is that, when the object is a bottle, a vase, or lamp-stand, its horizontal cross-section is typically circular. In conventional differential geometry, one would therefore describe the object as a surface of revolution, i.e., a profile curve that undergoes circular rotation.

However, to describe the grasping structure, our theory gives an extra level of transformation, as follows: Observe that the function of the circular structure of a bottle, vase, or lamp-stand, is to allow the person to approach and grasp the object from any direction. Since, as we have seen, the grasping requires the two grasping forces to be reflectional versions of each other, it is this *reflectional* structure that must be rotationally equivalent around the object.

To mathematically understand this, we use fundamental concepts from my New Foundations to Geometry. Recall that a basic principle of these foundations is Maximization of Transfer. Furthermore, recall that I argue that transfer is modeled as follows:

$$\textbf{Fiber Group} \ \circledR \ \textbf{Control Group}$$

where I define the fiber group as the set of actions that are transferred, and the control group as the set of actions that transfer the fiber group; and I argue that the wreath product Ⓦ models the relation *is transferred by*. That is, in my New Foundations, the above expression should be interpreted as meaning:

**Fiber Group** *is transferred by* **Control Group**

I now claim that transfer is basic to the structure of grasping. First consider one of the two compressive forces used in grasping; i.e., one of the two arrows shown in Fig 5.42. Let us model it by the action of the translation group $\mathbb{R}$, that is, corresponding to the trajectory given by one of the arrows. Now, we have said that the two forces must be reflectionally related to each other. In my geometry, this will mean that the translation group $\mathbb{R}$, modeling one of the two forces, is *transferred* by the reflection group $\mathbb{Z}_2$ in order to model the other force. That is, we have this transfer structure:

$$\mathbb{R} \ ⓦ \ \mathbb{Z}_2 \tag{5.1}$$

where the fiber group $\mathbb{R}$ represents one of the two compressive forces, and the control group $\mathbb{Z}_2$ is the reflection which *transfers* that compressive force to become the other compressive force.

Now recall that we noted that the circular structure, of such objects as bottles, vases, lamp-stands, etc., is to allow the person to approach and grasp the object from any direction. According to my geometric theory, this means that the reflectional structure can itself be *transferred* around the object. Therefore the reflectional structure given in expression (5.1) is transferred by the rotation group $SO(2)$ around the object. Thus, we take the structure in expression (5.1) as a fiber group and add the rotation group $SO(2)$ as a control group that transfers that structure. In doing so, we obtain this:

$$\mathbb{R} \,\textcircled{w}\, \mathbb{Z}_2 \,\textcircled{w}\, SO(2) \tag{5.2}$$

where the first two levels $\mathbb{R} \,\textcircled{w}\, \mathbb{Z}_2$ come from expression (5.1), and the third level is the added level of transfer. We see therefore that this new expression describes a *hierarchical* structure of transfer, i.e., the *transfer of transfer*. This is an example of the Maximization of Transfer principle in my New Foundations to Geometry. That is, in the current situation, the transfer structure given by the reflectional relation between the two compressive forces is itself transferred by the rotation group.

Now, also according to my geometry, there are additional layers of transfer, *below* expression (5.2), as follows. We saw that the shield-formation operation

$$BM^{+} \,:\, M^{+} \longrightarrow M^{+}m^{+}M^{+}$$

of the Process Grammar, describes the transition from the bowl in Fig 5.41 to the bottle in Fig 5.42.

In section 3.27, we showed that any bifurcation operation of the Process Grammar is given by a hierarchical structure of transfer, defined by a class of groups, called *unfolding groups*, invented in the New Foundations. That is, by the Symmetry-Curvature Duality Theorem, each of the extrema is given by a symmetry group. Furthermore, these symmetry groups are initially aligned, i.e., given by what my theory calls the *alignment kernel* of an unfolding group (the alignment kernel also containing the world-frame symmetry group). The bifurcation transfers these symmetries from their alignment, hierarchically in accord with my mathematical theory of *object-linked inheritance*.

Now, according to my theory, in the inheritance hierarchy, the arrow at the $m^{+}$ on the profile curve in Fig 5.42 represents the parent control group, with respect to which the bifurcating movement of the two $M^{+}$ are given by child control groups. That is, the movement of the two $M^{+}$ are defined relative to the movement of the $m^{+}$. Again, by my theory, the *design movement* of the $m^{+}$ corresponds to the *grasping force* given by $\mathbb{R}$ in expression (5.2). Therefore, the creation of the grasping structure on the object involves additional levels of transfer below the $\mathbb{R}$ level in expression (5.2). These additional levels are the alignment kernel and the relative movements of the $M^{+}$ extrema.

The crucial fact is that, because the bifurcation structure is a fiber below the $\mathbb{Z}_2 \,\textcircled{w}\, SO(2)$ structure given in expression (5.2), it is transferred by $\mathbb{Z}_2 \,\textcircled{w}\, SO(2)$ around the entire object. This expresses the important fact, in our theory, that the shield-formation operation $BM^{+}$ is transferred from the profile curve of the object to its reflectionally opposite curve, and this reflectional pair of the shield-formation operations is transferred around the entire object.

### DESIGN INTENT: GRASPING

**The New Foundations to Geometry, in defining grasping, have again the unique power of recovering the** *design intent*: **i.e., the compressive function of the $m^{+}$ extrema, its reflected structure in grasping, the rotational transfer of this reflected compressive structure in order to allow the person to approach and grasp the object from any direction.**

## 5.7 Bay Structures

We will now turn to our theory of the importance of bay structures in morphology.

Recall, from section 3.19, the Process Grammar operation called *bay-formation*

$$Bm^- \ : \ m^- \ \longrightarrow \ m^- M^- m^-$$

illustrated again in Fig 5.43. That is, the $m^-$ extremum, shown in the first shape, splits into two copies of itself shown in the second shape, with a new extremum $M^-$ introduced between those two copies. Recall, also, that the causal explanation of this operation is as follows: Under our process-inference rules, the domain extremum $m^-$ defines an indenting process, illustrated by the central upward arrow in the first shape in Fig 5.43. Then, according to our theory of the *causal explanation* of the $B$ operator, the splitting that this indenting process undergoes into two copies of itself is due to an *opposing* process inferred from the central extremum $M^-$ in the codomain, illustrated by the downward arrow in the second shape in Fig 5.43.

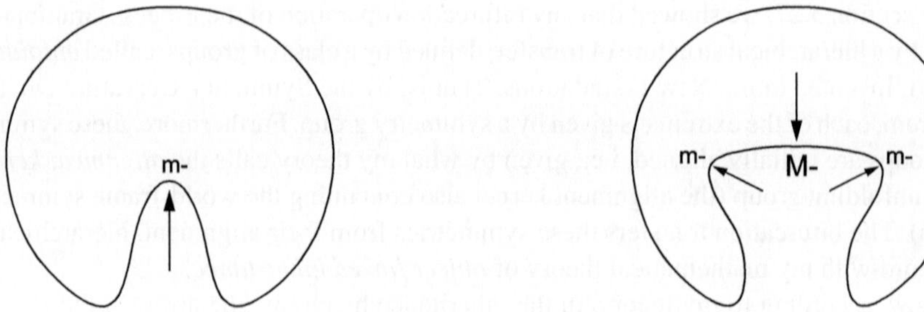


**Fig. 5.43.** The bay-formation operation of the Process Grammar.

The codomain triple $m^- M^- m^-$ has the shape of a *bay*. In fact, this shape is completely determined by the presence of the extremum $M^-$ because, for *any* occurrence of $M^-$, the singularity to the left and right of $M^-$ must necessarily be $m^-$, because the graph of the curvature function at a $M^-$ extremum can only go downwards to the left and right of $M^-$. In fact, I define a *bay* to be a neighborhood of a $M^-$ extremum on a curve, and I define a *complete bay* to be the triple $m^- M^- m^-$.

As stated previously, in our system, a *bay* is the *figure-ground reversal* of a *shield*. Therefore, whereas the shield bends outwards from the object under consideration, a bay bends inwards into the object.

When we introduced the concept of a bay in section 3.19, we illustrated it within the framework of geology. Indeed the Process Grammar has undergone implementation in geology, Larsen [9]. What we will do now is show that a bay is an important structure also in *architecture*; in particular, in the design of *arches*.

Our basic claim will be this: An important understanding of arches is given by our classification of curvature extrema and the corresponding processes given by PISA.

Now, recall that, since, according to our theory, each curvature extremum (on a shape of non-constant curvature) was created by the process inferred from the extremum, the starting shape must have been a shape of constant curvature. Our theory is corroborated in the history of arches: Arches were mainly invented by the Romans, and their most frequent arch was semi-circular, as illustrated in Fig 5.44.

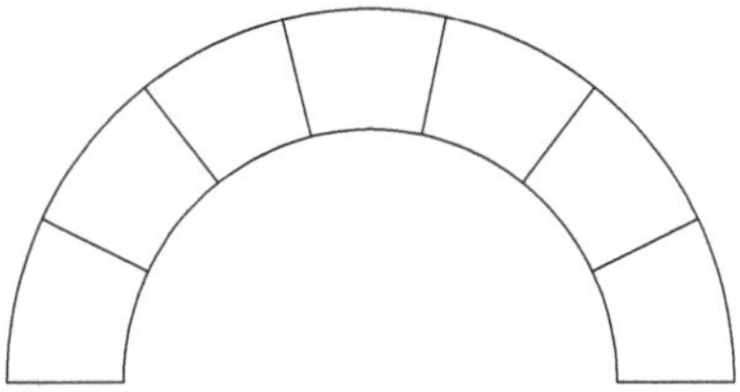

**Fig. 5.44.** The Roman arch.

As can be seen from Fig 5.44, this is built from an odd number of wedge-shaped bricks called *voussoirs*. The odd number is due to the fact that there is a central top stone called the *keystone*. The keystone receives the downward pressure from above, and this pressure is then relayed from the keystone to each of the voussoir wedge-shaped bricks successively down the two sides of the arch.

For obvious reasons, the Roman arch is the easiest to build. However, it is not the strongest type of arch because the sides of the arch tend to bulge outward. Therefore, alternative shaped arches were developed.

According to our theory, the most common property of the alternatives was that the keystone became a curvature extremum. Furthermore, using our classification of curvature extrema, this curvature extremum became either *penetrative* or *compressive*. As a result of this, we propose the following definition:

## OUR DEFINITION OF TWO NON-ROMAN TYPES OF ARCHES

**Our theory defines the following two types of arches, where these definitions are based on our Symmetry-Curvature Duality Theorem and the processes corresponding to their PISA axes:**

**Penetrative Arch**

**This is an arch whose top point is a penetrative extremum.**
**Therefore, from the PISA axis, one infers an *upward* process, on the convex side of the arch, acting to penetrate the space above the arch.**

**Compressive Arch**

**This is an arch whose top point is a compressive extremum.**
**Therefore, from the PISA axis, one infers a *downward* process, on the concave side of the arch, acting to compress the space below the arch.**

The above distinction between penetrative and compressive arches have never been given in architecture, because, as stated, the distinction is based on our PISA method of defining symmetry, which, as we have seen before, gives a fundamentally different view of symmetry than that which has been used in the previous thousands of years. That is, whereas the standard view of symmetry always places the symmetry axis on the convex side of the curve, the PISA definition places it on the convex side of a penetrative extremum and the concave side of a compressive extremum.

An example of a penetrative arch in architecture is Fig 5.45, which shows the bottom entrance to an escalator in a public building. Thus the photograph views the escalator, from the bottom, upwards. According to our process-inference rules, the inferred process that created the central extremum is an upward one between the two sides of the arch. It is clear that this process corresponds to a crucial *function* of the arch, which is to allow the upward movement of people without damaging their heads.

The central extremum is clearly the major extremum of the arch. However, notice that, at each of the two sides, there is also a compressive extremum. According to our theory, the inferred processes from these two compressive extrema are on the other side of the curve from the central penetrative process, and are directed inwards. It is clear that the function of these two compressive processes is to prevent the people from falling outward.

Notice that, because the empty space is below the curve of the arch, the central penetrative extremum is $m^-$. Furthermore, the two side compressive extrema are $M^-$. In particular, this means that the two compressive extrema define the sides of the arch as *bays*.

Now observe the following: Not only does the entrance to this elevator have the shape of a penetrative arch, but also so does the ceiling inside the escalator, as shown in Fig 5.46. This figure shows the same escalator viewed from the inside downwards. The penetrative arch, which we have just seen, is at the bottom; but the photograph shows that the ceiling of the escalator is also shaped as a penetrative arch. I will call such a ceiling a **penetrative-arched ceiling**.

To observe the powerful difference between a penetrative and compressive arch, the reader should compare the penetrative arch with the compressive arch shown in Fig 5.47. According to our process-inference rules, the inferred process that created the central compressive extremum in the latter arch is a downward one above the curve of the arch. It is clear that this inferred process corresponds to the downward pressure of the traffic on the top of the arch, and therefore defines the crucial *function* of the arch.

Again, according to our theory, the downward force that created the shape, in the *design* process, corresponds to the intended function, i.e., the downward force that the built arch is intended to deal with.

The central extremum is clearly the major extremum of the arch. However, notice that, at each of the two sides, there is also a penetrative extremum. According to our theory, the inferred processes from these two penetrative extrema are on the other side of the curve from the central compressive process, and are directed outwards. That is,

**Fig. 5.45.** A penetrative arch at the entrance of an escalator.

**Fig. 5.46.** The penetrative-arched ceiling over the same escalator.

**Fig. 5.47.** An example of what our theory calls *a compressive arch.*

these two side extrema imply two horizontal forces stretching the arch outward. It is clear that an intended function of these two penetrative processes is to widen the arch so that two lanes of traffic can go under the arch.

Notice that, because the empty space is below the curve of the arch, the central compressive extremum is $M^-$. Furthermore, the two side penetrative extrema are $m^-$. Thus, the extremum structure of the arch is the triple $m^-M^-m^-$ which defines what I call a *complete bay*.

The example we have just looked at is that of a compressive arch in a bridge. However, architects also use what we call compressive arches within walls. Fig 5.48 shows a compressive arch over a doorway; and Fig 5.49 shows a compressive arch over a window. Notice that, in both cases, there is a prominent block at the center of the arch. This is a typical device used by architects to emphasize the center of an arch. According to our theory, in the case of a compressive arch, the block emphasizes what our theory defines, using the PISA axis, as the downward force creating the central compressive extremum.

**Fig. 5.48.** A compressive arch over a doorway.

**Fig. 5.49.** A compressive arch over a window.

**Fig. 5.50.** A doorway compressive arch with a large ornament over the central extremum.

**Fig. 5.51.** A window compressive arch with the largest brick over the central extremum.

Other devices are also used in architecture to emphasize the central extremum of what our theory calls a compressive arch. Fig 5.50 shows the compressive arch of a doorway, where there is a large ornament over the central compressive extremum. Again this emphasizes what our theory defines, using the PISA axis, as the downward compressive force creating the extremum.

Consider also Fig 5.51. This shows a compressive arch over a window. Observe that the bricks over the arch successively increase in size from the sides of the arch to the central extremum. Again this emphasizes what our theory defines, using the PISA axis, as the downward compressive force creating the extremum.

Now, just as we previously saw that a penetrative arch can have a corresponding penetrative-arched ceiling, we will now see that a compressive arch can have a corresponding **compressive-arched ceiling**. Fig 5.52 shows a magnificent compressive-arched ceiling in a public building. The cross-section is a bay because it encloses empty space.

Similarly, Fig 5.53 shows a compressive-arched ceiling of a train. When viewed from outside the train, the roof is a shield, because one is viewing the *upper* surface of the roof, and the empty space one sees is above the surface. When viewed from inside the train, the roof is a bay, because one is viewing the *lower* surface of the roof, and the empty space one sees is below the surface.

Finally, note that compressive arches can occur in *biological morphology*. An example is the arch that defines the central region of the foot, as illustrated in Fig 5.10 (page 138). Again, this has the structure of what our theory calls a *bay*.

**Fig. 5.52.** A compressive-arched ceiling in a public building.

**Fig. 5.53.** A compressive-arched ceiling in a train.

## 5.8 Double Design Intent

We now come on to another important topic concerning compressive extrema.

**AN ADDITIONAL IMPORTANT PRESENCE AND ROLE
OF
COMPRESSIVE EXTREMA**

**In this chapter, we have proposed and demonstrated the enormous importance of shields and bays in biological morphology and manufacturing design, and seen how the PISA axis correctly defines them as** *compressive extrema,* **thus giving the correct causal explanation both as the past history of those extrema, as well as the intended function of those extrema. Also we have seen how the Process Grammar correctly defines their morphological developments in design sequences.**

**In contrast, we have seen that the conventional image-research literature completely fails to recognize compressive extrema and uses conventional symmetry definitions that completely fail to represent their past history, design intent, and morphological developments.**

**As another example of the crucially important presence and role of compressive extrema, we now define a further entirely new concept that has never been recognized in the conventional literature, and is, we claim, a fundamentally important phenomenon in biological morphology and manufacturing design. We will call this phenomenon: double design intent.**

In order to understand this additional important example, it is necessary to consider in detail the following issues concerning the way the Process Grammar handles the differential geometry of surfaces.

In this chapter, we have been seen that certain important phenomena, in the morphology of surfaces, are crucially defined by applying, to particular curves on surfaces, the Process Grammar operations for curves. That is, these operations are applied to *one-variable* functions, the curvature functions of particular curves on surfaces. However, the reader should note this: The causal techniques of the Process Grammar can also be used to define, on surfaces, operations that are defined individually with respect to *two-variable* functions on a surface. We will give those operations in another document. However, in the current book, the Process Grammar operations that we are applying to *surfaces* are the operations defined on curves, an important case being: the *lines of curvature*, and another case being the image contour of a surface. The reason we are concentrating on curves, in the present book, is that we are demonstrating that this captures enormously important *design intents* on *surfaces*.

Now, to define the additional important presence and role of compressive extrema, it is first necessary to define the following:

Since, at each point on a smooth surface, there are two lines of curvature going through the point orthogonally, our theory calls the application of the Process Grammar operations for curves, to lines of curvature on a surface, the **Bi-Directional Process Grammar**. That is, a Bi-Directional Process Grammar operation is applied, at a point, to a line of curvature through the point, and another Bi-Directional Process Grammar operation can be applied, at the same point, to the orthogonal line of curvature through that point.

The elements, $M^+, m^+, M^-, m^-, 0$, used to define a Bi-Directional Process Grammar operation, are singularities in the function of principal curvature along a line of curvature, where, at each point on the line of curvature, the principal curvature selected to define the function is that given by the principal direction that is tangential to the line of curvature. Thus the extrema $M^+, m^+, M^-, m^-$ are *ridge points*, i.e., extrema of that principal curvature along the line of curvature.

Another crucial fact is this: A Bi-Directional Process Grammar operation can be *transferred* to corresponding points on adjacent lines of curvature of the same principal foliation of the surface. Thus, supposing that a point is a ridge point, its corresponding points on adjacent lines of curvature are also ridge points, and therefore, together, these points form a ridge line. The consequence of the transfer of a Bi-Directional Process Grammar operation along a ridge line is that it creates a bifurcation of the ridge line. For example, a 3-fold process-bifurcation operation (e.g., $BM^+$) will bifurcate the ridge line into three ridge lines; and a process-continuation $C$ operation ($Cm^+$ or $CM^-$) will bifurcate the ridge line into a ridge line and two parabolic curves.

With respect to the phenomenon of transfer, consider a curve $\gamma$ of corresponding points. Suppose that the Process Grammar operation is transferred not along the entire curve, but along a segment $L$ of the curve. In the transfer, the width of bifurcation is successively decreased to zero at the two end-points of the segment $L$. Thus, each point along this segment $L$, except the end-points, undergoes that Process Grammar operation, and the end-points do not. The end-points adopt the transition state of the operation, and therefore, the transition, from a point on the curve $\gamma$ beyond an end-point, across the end-point into the changed segment $L$ of the curve $\gamma$, is also given by the same Process Grammar operation.

A crucial fact is that certain important morphological changes, which have been represented, in the standard research literature, as changes in two-variable functions on surfaces, are represented differently in our theory as the *transfer* of one-variable Process Grammar operations, i.e., the Process Grammar operations for curves. For example, in Leyton [27], I have shown that the creation of any parabolic loop on a surface is given by the transfer of a Process Grammar $C$ continuation operation ($Cm^+$ or $CM^-$) along a segment $L$ of a ridge line; and also the creation of any ridge loop is given by the transfer of a Process Grammar pair-creation operation (one of the four non-zero pair creation operations) along a segment $L$ of a curve that is not a ridge line.

The important fact about this is that it accords with a fundamental principle of my New Foundations to Geometry, the principle of Maximization of Transfer (within the generative representation). That is, whereas the creation of parabolic loops and ridge loops are conventionally represented as transitions in two-variable functions on surfaces, my theory represents them as the maximization of transfer of transitions given by one-variable Process Grammar operations.

Another important fact is that the one-variable Process Grammar operations, i.e., the operations on lines of curvature, also provide the important concept that we will define called *double design intent*, which becomes crucial in giving a deeply important issue concerning compressive extrema, as we shall see.

As will be explained, what we mean by double design intent is *two* design intents at a surface point. In order to understand this important concept, it is first necessary to give my Symmetry-Curvature Duality Theorem for surfaces, as follows:

**SYMMETRY-CURVATURE DUALITY THEOREM FOR SURFACES**

**Leyton**

**Given a ridge point on a surface, a PISA axis terminates at the ridge point. (A SLS axis also terminates at a ridge point.)**

**A Medial axis does not.**

**A surface point that is two ridge points, i.e., a ridge point on each of its two lines of curvature, has two PISA axes terminating at the point. (It also has two SLS axes terminating at the point.)**

In my theory, a surface point that is simultaneously two ridge points, one for each of its two lines of curvature, is called a **bi-directional ridge point**. Before we examine such points in more detail, it is important to understand that the distinction that my theory makes between penetrative and compressive extrema is relevant for ridge points. As a result of this, it is relevant for ridge lines, as follows:

## PENETRATIVE vs. COMPRESSIVE RIDGE LINES

**According to the Interaction Principle (p43), the process-history of a ridge line on a surface is given by the PISA axis sheet leading to that ridge line.**

**Using the PISA axes, the Bi-Directional Process Grammar classifies ridge points as penetrative or compressive extrema.**

**Furthermore, in the Bi-Directional Process Grammar, a ridge line of penetrative extrema is called a** *penetrative ridge line,* **and a ridge line of compressive extrema is called a** *compressive ridge line.*

**Because the Medial Axis symmetry set is directed away from what our theory calls a compressive ridge line, the Medial Axis symmetry set fails to correctly describe a compressive ridge line.**

**As a result of this, compressive ridge lines are totally ignored in the standard research literature.**

**According to our theory, this is a fundamental mistake in the standard literature, because, according to our theory, compressive ridge lines are fundamentally important to biological morphology and manufacturing design.**

**PISA gives the correct representation of both penetrative and compressive ridge lines.**

To illustrate penetrative and compressive ridge lines, consider a triaxial ellipsoid, an example of which is shown Fig 5.54. A triaxial ellipsoid is an ellipsoid of equation

$$\frac{x^2}{a^2} + \frac{y^2}{b^2} + \frac{z^2}{c^2} = 1$$

in which $a$, $b$, $c$, which are the radii respectively in the $x$, $y$, $z$ directions, are each of a different length. In the particular example illustrated in Fig 5.54, $a > b > c$.

The consequence of this, in Fig 5.54, is that the curve shown in the $z = 0$ plane is a penetrative ridge line, i.e., every line of curvature crossing this ridge line has a penetrative extremum at the ridge line. Furthermore, the curve shown in the $x = 0$ plane is a compressive ridge line, i.e., every line of curvature crossing this ridge line has a compressive extremum at the ridge line.

The crucial fact we will now consider is that a point where these two ridge lines intersect, as shown as the left intersection point on the figure, is a penetrative ridge point

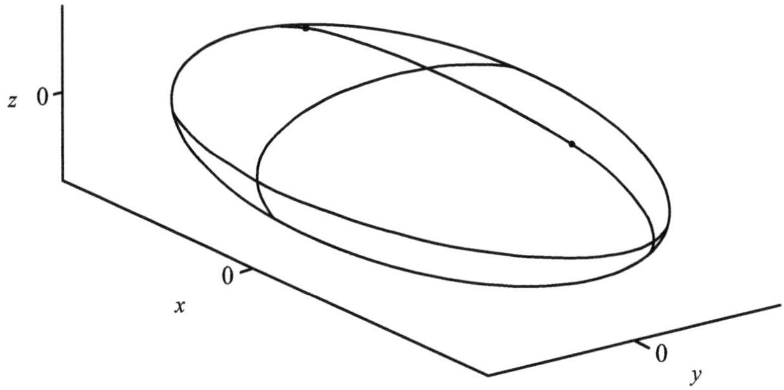


**Fig. 5.54.** Triaxial ellipsoid

(extremum) in one direction (i.e., in the line of curvature that vertically crosses this point) and is also a compressive ridge point (extremum) in the other direction (i.e., in the line of curvature that horizontally crosses this point). Our theory claims the following crucial consequences with respect to *biological morphology* and *manufacturing design*:

**BI-DIRECTIONAL PENETRATIVE-COMPRESSIVE RIDGE POINT**

**Consider a surface point that is a penetrative ridge point (extremum) on one of its two lines of curvature and is simultaneously a compressive ridge point (extremum) on the other of its two lines of curvature.**

**In our theory, such a point will be called a** *bi-directional penetrative-compressive ridge point.*

**As an example of such a point, let us consider the case where the curvature is the same sign in both principal directions at the point. A crucial fact that our theory gives concerning this point will be illustrated using the example of the left intersection point shown on the ellipsoid in Fig 5.54, as follows:**

**The crucial fact is that the PISA axis, corresponding to the** *penetrative* **extremum at this point, is** *inside* **the ellipsoid, leading to this point. In contrast, the PISA axis, corresponding to the** *compressive* **extremum at this point, is** *outside* **the ellipsoid, leading to this point.**

**That is, this point has, simultaneously, two PISA axes that lead to the point**

**in *opposite* directions, one from the inside, one from the outside.**

**According to our theory, this gives the point** *two design intents*, **and therefore** *two functional roles.*

**Our theory claims that this structure we are proposing:**
**the *opposite directed* PISA axes at a bi-directional penetrative-compressive ridge point, is a *fundamentally important* structure in biological morphology and engineering design.**

An example of this structure occurs at the tip of what, in section 5.3, we called an *advanced* finger. Recall that on page 135, we stated that the top point of such a finger is a compressive extremum in one of its two lines of curvature, i.e., is a shield with respect to that line of curvature; and is a penetrative extremum in the orthogonal line of curvature through that point. According to our theory, this type of finger tip occurs because it has two functions: the *protection* function which is realized by the compressive extremum at that point, and the *digging* function which is realized by the penetrative extremum at that point.

Prior to that statement (in bold-print) on page 135, we described the creation of the compressive extremum at the top point. Based on this, we now claim the following:

According to our theory, in the evolution of fingers, the first type of finger had the following type of top point: Its top point was a penetrative extremum in *both* lines of curvature through that point. The reason was that, with respect to both lines of curvature, the top point had the function of digging.

Then, according to our theory, in the development of an advanced finger, the penetrative extremum in one of these two lines of curvature next underwent the shield-formation operation $BM^+$ to realize the *additional* function of *protection*. This resulted in the top point of the finger being an example of what our theory calls a bi-directional penetrative-compressive ridge point. The crucial consequence is that, according to our theory, the top point of that finger realizes *two different design intents*.

## DOUBLE DESIGN INTENT AT A POINT

**A ridge point, being a curvature extremum on one of the two lines of curvature through the point, has a PISA axis terminating at it, and the PISA axis gives the design intent.**

**As a result of my Symmetry-Curvature Duality Theorem for surfaces, if a surface point is a ridge point on both lines of curvature through the point; i.e., is a bi-directional ridge point, then our theory says that it has** *two design intents.* **Our theory calls this the** *double design intent at the point.*

**If the two ridge points at the point are of different types (the types being $M^+$, $m^+$, $M^-$, $m^-$), then there is a combination of two different design intents at the point.**

**For example, our theory states that the top point of what we call an advanced finger has a** *penetrative* **extremum $M^+$ in one principal direction of curvature to realize the function of digging, and a** *compressive* **extremum $m^+$ in the other principal direction to realize the function of protection.**

**Furthermore, our theory gives an equivalent account for the human chin and the human nose. That is, both of these have an end-point that is a penetrative ridge point $M^+$ in one of its principal directions and a compressive ridge point $m^+$ in its other principal direction. The penetrative ridge point realizes the outward penetrative function of the chin and nose, and the compressive ridge point realizes the shielding of this biological part.**

**The important fact is that PISA is the only symmetry analysis that creates the symmetry axis for the compressive extremum on the** *opposite* **side of the same sign penetrative extremum, and thus represents the correct relationship between these two design intents.**

**Based on this, a further important fact is that the two opposite-directed PISA axes imply that the two design intents, the compressive and penetrative ridge points at the tip point, were created at two different times. The penetrative extremum was created first, and the compressive extremum was created later by the shield-formation operation applied to the penetrative extremum that was perpendicular to the penetrative extremum that remained at the tip point. The shield-formation operation, which pushed inwards at the tip point, must have transferred that remaining penetrative extremum inwards. Furthermore, it transferred that penetrative extremum along the bifurcation direction of the shield-formation operation. This illustrates the fact that the Double Design Intent accords with the fundamental principle of my New Foundations to Geometry, the representation of an object in such a way that** *maximizes transfer.*

## 5.9 Unification of Biological Morphology and Vehicle Design

**UNIFICATION**
**OF**
**BIOLOGICAL MORPHOLOGY AND VEHICLE DESIGN**

**According to the New Foundations to Geometry, the origin of *both* an animal *and* a vehicle is a sphere.**

**The reason is this:**

**According to the theory of intelligence, invented by the New Foundations to Geometry, understanding the origin of both an animal and a vehicle as a sphere is *intelligent* for the following reason:**

**By understanding the origin to be a sphere, the New Foundations account for important intended functions by characterizing the features of the final shape as *asymmetries* that have *causal* explanations that give the *design intents*.**

**Thus, according to the New Foundations, by describing the generation of a vehicle and animal as the generation of a hierarchy of asymmetries, one is then able to describe a vehicle and animal as a hierarchy of design intents.**

**Using this claim, we propose that, to establish the layers of design intent, one should understand that the first generative stage is that the sphere is stretched, in the travel direction, to become an ellipsoid. The penetrative extrema in this direction correspond to the design intent of movement.**

**Additionally, in ground vehicles and ground animals, these penetrative extrema undergo bifurcation in the Process-Grammar shield-formation operation $BM^+$, creating what our theory calls a *complete shield* $M^+m^+M^+$. This is to fulfill the extra design intent of protection along the movement direction.**

**Furthermore, on the outward growths, most penetrative extrema also undergo the shield-formation operation $BM^+$, thus putting shields on those growths to fulfill the design intent of protecting them along their direction of movement.**

**This chapter has shown that what our theory defines as shields are crucial in both biological morphology and vehicle design. We showed that examples of shields in biological morphology are the forehead, top and back of the human head, the chin and jaw, the chest, the tips of fingers, the knee-cap, the heel of a foot. We also showed that examples of shields in road vehicles are the front, top, back, and sides of cars, buses and vans.**

**Again, this unifies biological morphology and automotive design. For example, according to this chapter, the forehead of the human head is the same structure as the front of a van. In fact, both are what our theory calls bi-directional shields in the sense that they are shields both in the vertical and horizontal direction.**

**Since the travel-direction for four-legged animals and ground vehicles is horizontal, the initial sphere is stretched horizontally, and the penetrative extrema in that direction undergo bifurcation into shields; and therefore a bi-directional shield appears at both the front and back.**

**Correspondingly, because human beings are two-legged animals, an important travel-direction is vertical, because standing-up is a primary direction, and therefore, the initial sphere is stretched in the vertical direction, and therefore a bi-directional shield is formed at both the top of the head and the heal of the foot. Furthermore, because of the forward movement of the body, a bi-directional shield appears at both the front and back of the head. In addition, people put a shield on the front of a shoe.**

**A crucial fact that we gave for the type of car illustrated in Fig 5.40 is that its upper compartment is what our theory calls a triple-shield. That is, the roof and the front and back windows constitute three adjacent shields. Also we saw that the top of a person's head is a triple-shield, as was illustrated in Fig 5.4 (page 133). Furthermore, we also saw that the jaw and chin structure of a person's face is a triple-shield, as was illustrated in Fig 5.3 (page 132). Again, this unifies biological morphology and automotive design.**

**We have seen that the conventional definitions of symmetry, including the highly popular Medial Axis symmetry set, completely fail to represent the causal force, design intent, and morphological development of a shield in vehicle design and biological morphology. In contrast, the PISA axis correctly defines and describes the fundamental shields of vehicle design and biological morphology.**

## 5.10 Zone of Influence

In this section, I will propose a concept that will be useful for the analysis of penetrative parts. To introduce the concept, it is first necessary to recall a situation we considered earlier, illustrated again here in Fig 5.55. We will then add to this situation a new concept that will be useful.

So first, we recall the basic facts concerning this situation. The left-most shape in the figure is an ellipse. The processes inferred by our rules are shown by the arrows on the ellipse. Let us consider the downward compressive process that our theory gives at the top of the ellipse. Now let us apply the Process Grammar operation

$$Bm^+ : m^+ \longrightarrow m^+M^+m^+$$

to the top extremum of the ellipse; i.e., *breaking-through of a protrusion*. The result is the second shape in Fig 5.55, where the bold upward arrow shows the protrusion that broke-through the downward compressive process at the top of the ellipse. By our *causal* theory of the $B$ operator (section 3.24), this protruding process *caused* the bifurcation.

What we now need to observe, from the second shape, is that the range of influence of this upward protruding process has been the entire top half of the ellipse; i.e., the entire top half has moved upward under the action of this protruding process. We will therefore say that the top half is in the **zone-of-influence** of the protruding process.

The concept we will now introduce is that of *restricting* the zone-of-influence of the protruding process. To carry out this restriction, we *continue* the two side compressive $m^+$ processes on the second shape, which bifurcated from the top compressive process on the first shape, until they indent, as shown in the third shape. That is, we apply, to each of the two side $m^+$ extrema, in the second shape, the Process Grammar operation

$$Cm^+ : m^+ \longrightarrow 0m^-0$$

i.e., squashing continues till it indents.

An important thing to notice about the third shape is this: Below the two new indentation extrema $m^-$, the third shape is almost exactly the same as the first shape, thus illustrating the restricting role that these two indentations have on the zone-of-influence of the central protruding process.

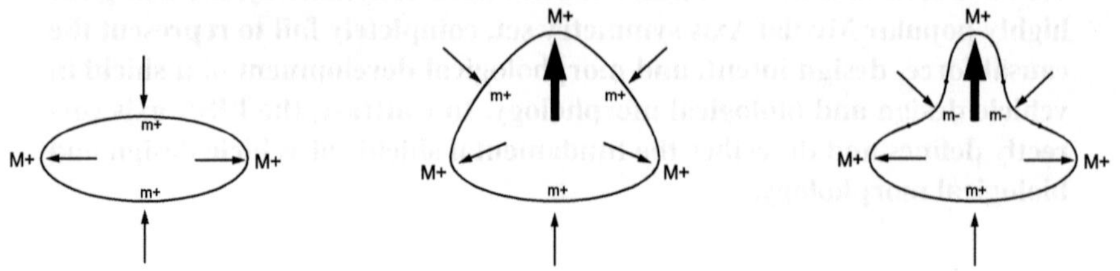


**Fig. 5.55.** Illustration of the zone-of-influence of the breaking-through of a protrusion, and subsequent restriction of that zone-of-influence.

Therefore, the transition from the second to the third shape in Fig 5.55 will be called **zone-of-influence restriction**.

We now observe extra properties concerning the zone-of-influence and its restriction. Begin by considering again the transition from the first to the second shape in Fig 5.55, that is, the application of the $Bm^+$ operation, breaking-through of a protrusion. We see that the result of applying the $Bm^+$ to the central extremum $m^+$ of the top *shield* of the first shape is that this top shield has bifurcated, in the second shape, into *two adjacent shields* which are the two sides of the protruding process that has broken through in the second shape. Recall that, as noted earlier in this chapter, this is a general fact concerning the $Bm^+$ operation; i.e., the protruding process, that breaks through the compressive process of a shield, *bifurcates the shield into two adjacent shields*. Also, recall that our theory calls two adjacent shields a *double-shield.*

We can now characterize the zone-of-influence and its restriction in terms of the double-shield. First recall that, as we said earlier in this section, one can see, from the second shape, that the zone-of-influence of the upward protruding process has been the entire top half of the ellipse; i.e., the entire top half has moved upward under the action of this protruding process.

Based on that, what we observe now is that this zone-of-influence on the ellipse has become the double-shield in the second shape, i.e., the double-shield gives the region that moved upward due to the protruding process.

Next consider the subsequent *restriction* of the zone-of-influence, i.e., the transition from the second shape to the third shape. In this transition, the squashing processes, on the left and right side of the double-shield, continued till they indented. We will call this the **pinching of the double-shield**.

Therefore we see the crucial fact that, by pinching the double-shield that was produced by the protruding process, the zone-of-influence of the protruding process has been restricted.

We summarize what we have been describing by stating three of the main facts thus:

**The Process-Grammar operation $Bm^+$, i.e., breaking-through of a protruding process, creates a double-shield.**

**The double-shield gives the zone-of-influence of that protruding process.**

**Pinching the double-shield causes a restriction of the zone-of-influence of the protruding process.**

Let us now consider the *dual* situation. That is, we will now consider the breaking-through of an *indentation* at the compressive extremum of a *bay*, and the subsequent zone-of-restriction of that indentation. This is illustrated in Fig 5.56.

Observe that the middle region of the left-most shape is a bay. Its compressive extremum is the $M^-$ shown in the center of the bay. At this extremum, our theory infers the process to be the upward compressive arrow shown at this extremum. Recall that our theory calls the compressive process at a $M^-$ extremum, a *resistance* process. For example, if the shape were an island, the process would be the resistance that prevents the water from entering further.

Now, to the central extremum of the bay, let us apply the Process Grammar operation

$$BM^- : M^- \longrightarrow M^- m^- M^-$$

i.e., ***breaking-through of a indentation.*** The result is the second shape in Fig 5.56, where the bold downward arrow shows the indentation that broke-through the upward compressive $M^-$ process at the center of the bay. Notice therefore that the upward compressive $M^-$ process, in the first shape, bifurcated into the two compressive $M^-$ processes shown in the second shape. By our *causal* theory of the $B$ operator (section 3.24), the downward indenting process, in the second shape, *caused* the bifurcation.

What we now need to observe, from the second shape, is that the range of influence of this downward indenting process has been the entire bay in the first shape; i.e., that entire region from the first shape has moved downward under the action of this indenting process. We will therefore say that this region is the **zone-of-influence** of the indenting process.

The next stage will *restrict* the zone-of-influence of the indenting process. To carry out this restriction, we *continue* the two compressive $M^-$ processes shown in the second shape, till they *protrude*, as shown in the third shape. That is, we apply, to each of the two $M^-$ extrema shown in the second shape, the Process Grammar operation

$$CM^- : M^- \longrightarrow 0M^+0$$

which our theory calls, *resistance continues till it protrudes.*

An important thing to notice about the third shape is this: Above the two new protrusion extrema $M^+$, the third shape is almost exactly the same as the first shape, thus illustrating the restricting role that these two protrusions have on the zone-of-influence of the central indenting process.

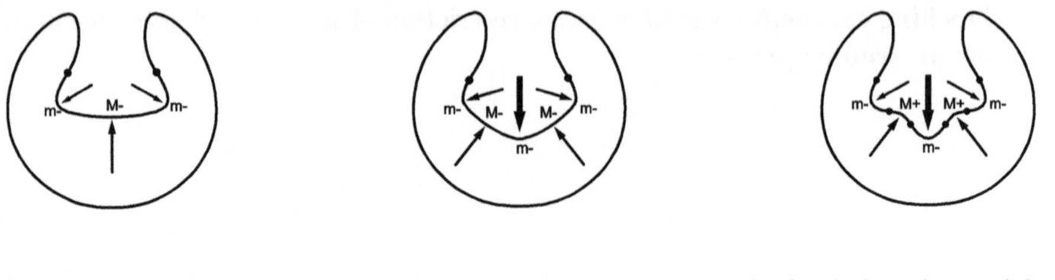


**Fig. 5.56.** Illustration of the zone-of-influence of the breaking-through of an indentation, and the subsequent restriction of that zone-of-influence.

Let us now observe the following properties concerning the zone-of-influence and its restriction. Begin by considering again the transition from the first to the second shape in Fig 5.56, that is, the application of the $BM^-$ operation, breaking-through of an indentation. We see that the result of applying the $BM^-$ to the central extremum $M^-$ of the *bay* in the first shape is that this bay has bifurcated, in the second shape, into *two adjacent bays* which are the two sides of the downward indenting process that has broken-through in the second shape. It is a general property of the Process Grammar operation $BM^-$ that a *bay bifurcates into two adjacent bays*. Our theory calls two adjacent bays a **double-bay**.

We can now characterize the zone-of-influence and its restriction in terms of the double-bay. First recall that, as we said earlier in this section, one can see, from the second shape, that the zone-of-influence of the downward indenting process has been the entire bay of the first shape; i.e., this entire region from the first shape has moved downward under the action of the indenting process.

Based on that, what we observe now is that this zone-of-influence has become the double-bay in the second shape, i.e., the double-bay gives the region that moved downward due to the indenting process.

Next consider the subsequent *restriction* of the zone-of-influence, i.e., the transition from the second shape to the third shape. In this transition, the resistance processes, on the left and right $M^-$ of the double-bay, continued till they protruded. We will call this the **pinching of the double-bay**.

Therefore we see the crucial fact that, by pinching the double-bay that was produced by the indenting process, the zone-of-influence of the indenting process has been restricted.

We summarize what we have been describing by stating three of the main facts thus:

**The Process-Grammar operation $BM^-$, i.e., breaking-through of an indenting process, creates a double-bay.**

**The double-bay gives the zone-of-influence of that indenting process.**

**Pinching the double-bay causes a restriction of the zone-of-influence of the indenting process.**

In this section, we have defined:

**Zone-of-influence restriction of a penetrative process that breaks-through a compressive process.**

In fact, we have described the effect of the scenario on both types of compressive process. In the case of a squashing process (i.e., at extremum $m^+$) the scenario was illustrated in Fig 5.55. In the case of a resistance process (i.e., at extremum $M^-$) the scenario was illustrated in Fig 5.56. In both figures, the scenario was given by a sequence of three shapes. This is because the scenario is a sequence of two transition stages:

**Stage 1** (from the first to the second shape): In this stage, a penetrative process breaks-through a compressive process, as a result of using the $B$ operator. Therefore, the compressive process bifurcates into two copies, one on each side of that penetrative process.

**Stage 2** (from the second to the third shape): This stage restricts the zone-of-influence of the penetrative process that has broken-through. It is achieved by pinching: That is, the two compressive processes that resulted from the breaking-through bifurcation are continued till they become penetrative. This is due to their compressive extrema each undergoing the $C$ operator, thus forcing those compressive extrema to become penetrative and change sign.

Now we will distinguish between two kinds of scenario:

**Two-Stage Scenario:** Here the pinching is applied *after* the breaking-through of the penetrative process. This scenario is exactly the morphological sequence that has been described in this section; i.e., as illustrated in Fig 5.55 and also in Fig 5.56.

**One-Stage Scenario:** Here the pinching is applied *simultaneously* to the breaking-through of the penetrative process. In this case, the shape goes directly from the first to the third shape, in Fig 5.55, and in Fig 5.56, without going through the intervening second shape. This means that the two pinching processes will bifurcate out, in their penetrative state, from the application-point of the breaking-through penetrative process. This scenario will be described in section 5.12.

Both scenarios can occur in the real world, and it is important to add new operations to the Process Grammar to fully realize them, as we will do in the following sections.

From the above discussion, we can see the following property: In each scenario, there are *three* penetrative processes, defined as follows:

**Definition 5.1.** *In each of the above scenarios, the penetrative process that breaks-through will be called the* **opposing penetrative process***, since it opposes the compressive process which it breaks-through.*

**Definition 5.2.** *The two penetrative processes, that restrict the zone-of-influence of the opposing penetrative process, will be called the* **restricting penetrative processes.**

## 5.11 Pinching Operator $CC$

We now formalize Stage 2 of the Two-Stage Scenario defined in the previous section, that is, the *pinching* stage, i.e., the *zone-of-influence restriction*. This will be captured by a derived operator in the Process Grammar. Since it is the *simultaneous application* of the continuation operator $C$ to each of two separate compressive extrema, we can use the notation in section 4.9 to denote this operator as $CC$.

The operator $CC$ will be called *2-fold* pinching because, as will be shown later, there can be $n$-fold pinching, $CC \ldots C$, where $C$ is applied simultaneously to $n$ separate points. Also, $CC$ will be called *pinching of type 1*, because, as will also be shown later, there is another type of pinching, based not on $C$ but on the pair-creation operator $\pi$.

In defining the $CC$ operations, below, the over-braces, $\overbrace{\quad}$, on the final string of each operation, indicate the effect of the two simultaneous uses of the $C$ operator.

**2-FOLD PINCHING OPERATOR (TYPE 1)** $CC$

**Given a penetrative extremum, with a compressive extremum of the same sign, on each side of that extremum, the Process Grammar operator** $CC$ **restricts the zone-of-influence, of the process inferred from the penetrative extremum, by the simultaneous application of the continuation operator** $C$ **to each of the two compressive extrema. The application of the** $CC$ **operator is realized by two operations:**

**Pinching a Double-Shield:**

$$CCm^+ : \ m^+M^+m^+ \quad \longrightarrow \quad \overbrace{0m^-0}\, M^+ \overbrace{0m^-0}$$

**Pinching a Double-Bay:**

$$CCM^- : \ M^-m^-M^- \quad \longrightarrow \quad \overbrace{0M^+0}\, m^- \overbrace{0M^+0}$$

**Generalization: In the above two operations, replace the central penetrative extremum, by a symmetric string** $X$ **terminating at each end by a penetrative extremum of the same sign:**

$$CCm^+ : \ m^+Xm^+ \quad \longrightarrow \quad \overbrace{0m^-0}\, X \overbrace{0m^-0}$$

$$CCM^- : \ M^-XM^- \quad \longrightarrow \quad \overbrace{0M^+0}\, X \overbrace{0M^+0}$$

**The length of the string** $X$ **will be called the** *order* **of the operator. In contrast, the length of the string** $CC$ **will be called the** *degree* **of the operator (which is 2 in the present case).**

## 5.12 The One-Stage Scenario

The two stages of the Two-Stage Scenario defined in sections 5.10 – 5.11 are an *extremum-bifurcation* and a *zero-bifurcation*. Whereas the Two-Stage Scenario goes through these two stages *successively*, the corresponding One-Stage Scenario can be viewed as going through these two stages *simultaneously*.

As an example of the One Stage-Scenario, consider the case where the initial compressive extremum is positive, $m^+$. Recall that, with this initial extremum, the Two-Stage scenario was illustrated as going through the three successive shapes shown in Fig 5.55 page 186. In contrast, given this same initial extremum $m^+$, the One-Stage Scenario goes directly from the first to the third shape, in that illustration, without going through the second shape. This direct transition is shown here in Fig 5.57.

The reader should understand that a crucial property of the One-Stage Scenario is this:

> **In the One-Stage Scenario, the two pinching processes will bifurcate out, in their penetrative state, from the application-point of the opposing penetrative process.**

To illustrate this concept, consider Fig 5.57. The upward bold arrow in the second shape represents the penetrative process that *opposed* the downward compressive force in the first shape. In the second shape, the two indenting penetrative processes, that restrict the zone-of-influence of the upward penetrative process, bifurcated out, in their indenting penetrative state, from the application-point of the upward penetrative process, at the moment that the upward process broke-through. This means that the upward protruding process *pushed* the two indenting processes away from itself as it appeared.

From the above discussion, we see that, in the One-Stage Scenario, the zone-of-influence of the opposing penetrative process is *always* restricted, from its beginning, rather than starting to be restricted *later* (as occurs in the Two-Stage Scenario). Furthermore, in the One-Stage Scenario, the restricting penetrative processes do not initially appear at a finite distance from the center (as occurs in the Two-Stage Scenario). Therefore, this type of part will be called a **radially-restricted part**; meaning that, in its

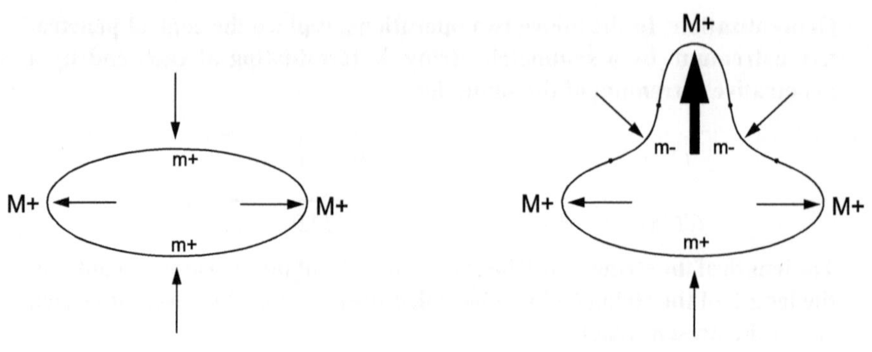


**Fig. 5.57.** The One-Stage Scenario.

formation, the zone-of-influence, and its boundary, radiate outwards from the central extremum.

To visualize this scenario, imagine that the zone-restricted protrusion, on the top of the second shape in Fig 5.57, first emerges as an infinitely small version itself.

Now observe that, because the two restricting penetrative processes are of opposite curvature signs from the rest of the shape, they are separated from the rest of the shape by four zeros, as illustrated in the second shape in Fig 5.57 (the four dots). Therefore, the creation and bifurcation of the restricting penetrative processes requires the creation and bifurcation also of those four zeros.

To fully understand this scenario, consider its *time-reversal*. Again, we will illustrate this using Fig 5.57. In running time backwards from the second shape, *both* the protrusion, and its boundary, *shrink* till they coalesce at the center. If they did not, i.e., if, backwards in time, the two indentations stopped at a finite distance from the central protrusion, no matter how small that distance was, this would mean that the scenario which created them would have to be the Two-Stage Scenario, i.e., their creation would be explained by the application of two $Cm^+$ operations a finite distance from the central extremum. Therefore, in the One-Stage Scenario, backward-time must mean that the two *negative* minima $m^-$ and the central *positive* maximum $M^+$ must eventually coalesce. Clearly, they can do this only at a point of *zero* curvature. Thus, the four zeros must coalesce with them at the same time. This is the structure of the backward-time *shrinking* of the zone-restricted protrusion till it disappears.

Thus to emphasize: For the One-Stage Scenario, the backward-time direction requires that the three extrema and four zeros must coalesce at a zero in the *transition-state*.

Therefore, going in the forward-time direction, the transition from the first to the second shape, in Fig 5.57, must be given qualitatively by the three successive curvature functions shown in Fig 5.58, where the middle function gives the transition-state. Thus we see that the three extrema and the four zeros in the last function must, backwards in time, coalesce at a common zero as seen in the middle function. This corresponds to the complete shrinkage of the zone-restricted protrusion. In the forward-time direction, this necessarily corresponds to the simultaneous bifurcation of the three extrema and the four zeros.

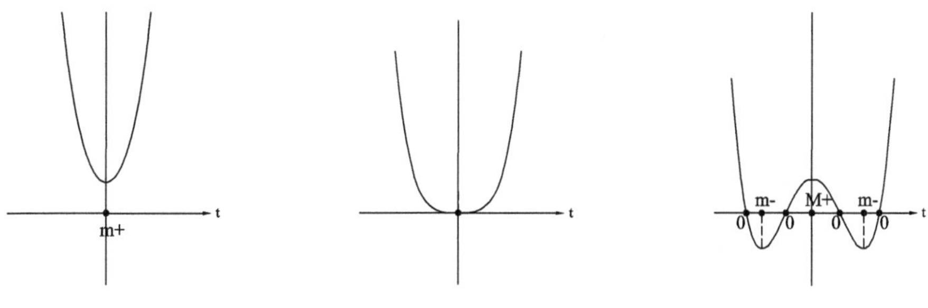


**Fig. 5.58.** The successive curvature functions in the One-Stage Scenario.

Thus we see that, via the zero minimum in the middle graph, the positive minimum $m^+$ in the first graph has undergone two *simultaneous* bifurcations in the final graph: a bifurcation into three extrema, and a bifurcation into four zeros. These two bifurcations are produced by two operators in the Process Grammar: The bifurcation into three extrema is produced by the 3-fold extremum-bifurcation operator $B(3)$, which we usually write as $B$, and which we have already seen examples of in the Process Grammar. The bifurcation into four zeros is produced by a new operator $C(4)$, which we will now add to the Process Grammar.

The reader will recall that the operator $C$, given in Chapter 3, expresses the *continuation* of a single process, as the bifurcation of two zeros, on the sides of the single process. It can be notated as $C(2)$. Also, we have seen that the operator $C(3)$, introduced in sections 4.5 and 4.6, can be regarded as expressing the continuation of two processes, as the bifurcation of three zeros, on the sides of the two processes. Furthermore, these two processes are going in opposite directions. Correspondingly, the new operator $C(4)$, can be regarded as expressing the continuation of three processes, as the bifurcation of four zeros, on the sides of the three processes. Furthermore, these three processes are in alternating opposite directions, as we see in Fig 5.57.

Based on what we have just said, the transition from the first graph to the third graph, in Fig 5.58, is given by the simultaneous application of the two operators, $C(4)$ and $B(3) = B$. This simultaneous application will be denoted by $[C(4)B(3)]$ or simply $[C(4)B]$, which is a new operator to be added to the Process Grammar. Recall, from Notation 4.1 (page 111) that the square brackets, $[\ ]$, mean that the two component operators, $C(4)$ and $B$, are applied *simultaneously to the same point.*

Thus, the new operator $[C(4)B]$ must describe the transition from the $m^+$, in the first graph in Fig 5.58, to the *seven* singularities $0m^-0M^+0m^-0$ in the third graph; i.e., the transition is a *7-fold bifurcation*. Therefore the application of the new operator $[C(4)B]$ to the $m^+$ in the first graph is given by the following new operation:

$$[C(4)B]m^+ \ : \ m^+ \quad \longrightarrow \quad 0m^-0M^+0m^-0 \tag{5.3}$$

It is easy to check that, when the operator $[C(4)B]$ is applied to the dual extremum $M^-$, it produces the dual of the configuration of the above seven singularities. That is, it is given by the following new operation:

$$[C(4)B]M^- \ : \ M^- \quad \longrightarrow \quad 0M^+0m^-0M^+0 \tag{5.4}$$

**Comment 5.3** The new operator $[C(4)B]$ accords with Definition 4.15 (page 122) of an *n-fold radial-undulation operator*. In particular, it is the *3-fold radial-undulation operator*, for the following reasons: The term *3-fold undulation* refers to the created configuration, which consists of 4 zeros and 3 extrema that alternate with the zeros. That is, because of this alternation, the 3 extrema must necessarily be *penetrative*, and in fact of opposite signs, $M^+$ and $m^-$; i.e., implying penetrative processes of alternating opposite directions; i.e., an *undulation*. The term *radial-undulation* means that, under the operator, all 7 singularities of the undulation (4 zeros and 3 extrema) *radiate* out simultaneously from a *single* central application-point.

The *causal* structure of this operator is given as follows:

**CAUSAL STRUCTURE**
**OF THE**
**3-FOLD RADIAL-UNDULATION OPERATOR $[C(4)B]$**

**The Process Grammar operator $[C(4)B]$ describes a process that opposes and breaks-through a compressive process, as a zone-restricted penetrative process. That is, the two penetrative processes, that restrict it, bifurcate out, in their restricting state, from the application-point of $[C(4)B]$ at the same time that the opposing penetrative process breaks-through.**

**Thus, the operator $[C(4)B]$ is the *simultaneous* application of two operators, the 4-fold zero-bifurcation operator $C(4)$ and the 3-fold extremum-bifurcation operator $B(3) = B$, to the *same application-point,* the initial compressive extremum. It is realized by the following two operations:**

**Zone-Restricted Protrusion Formation**

$$[C(4)B]m^+ \; : \; m^+ \quad \longrightarrow \quad 0m^-0M^+0m^-0$$

**In this case, a squashing force is opposed by a process that breaks-through it as a zone-restricted protrusion, where the two penetrative processes that restrict the zone-of-influence of the protrusion are indentations.**

**Zone-Restricted Indentation Formation**

$$[C(4)B]M^- \; : \; M^- \quad \longrightarrow \quad 0M^+0m^-0M^+0$$

**In this case, a resistance force is opposed by a process that breaks-through it as a zone-restricted indentation, where the two penetrative processes that restrict the zone-of-influence of the indentation are protrusions.**

Observe that this is the *second* example we have defined of an $n$-fold radial-undulation operator. The first was the 2-fold radial-undulation operator, which is the zero pair-creation operator $[C(3)\pi]$, defined on page 112. That operator $[C(3)\pi]$ is the simultaneous application of the 3-fold zero-bifurcation operator $C(3)$ and the 2-fold extremum-bifurcation operator $\pi$ to a shared application-point. Therefore, it is the particular example of $[C(n+1)B(n)]$ where $n = 2$. That is, it can be written as $[C(3)B(2)]$. Notice, in that example, $n$ is even; and, as stated in Definition 4.15 (page 122), this means that its application-point is a curvature zero that is regular, $0^{\uparrow}$ or $0^{\downarrow}$.

In contrast, in the case that is defined in the present section, $[C(n+1)B(n)] = [C(4)B(3)]$, we have $n = 3$, which is odd; and, as also stated in Definition 4.15 (page 122), this implies that its application-point is a compressive extremum, either $m^+$ or $M^-$. Notice that it is impossible to have an occurrence of the $C(4)$ operator without the simultaneous occurrence of the $B(3) = B$ operator.

Note also that the $C(4)$ operator, in this One-Stage Scenario, takes the place of the $CC$ operator in the Two-Stage Scenario. That is, both operators are responsible for a bifurcation into four zeros. However, whereas $CC$ bifurcates the four zeros as two pairs simultaneously from *two* separate points, in contrast, $C(4)$ bifurcates the four zeros simultaneously from *one* point. Thus we can understand the two separate application-points of the $CC$ operator as having coalesced into the one application-point of $C(4)$. Furthermore, this coalescence simultaneously pulls the other stage $B$ of the Two-Stage Scenario into the single stage of the One-Stage Scenario. As a result of this, we claim the following:

**$[C(4)B]$ AS THE *COALESCENCE* OF THE OPERATORS $CC$ and $B$**

**According to our theory, the operator $[C(4)B]$ is the *coalescence* of the two operators $CC$ and $B$ of the Two-Stage Scenario, in the following sense, given by our theory of the *causal* structure of the One-Stage Scenario:**

**This causal theory is again based on the fundamentally important causal structure that our theory gives of a compressive extremum, as well as the fundamentally important structure that our theory gives of the continuation of a compressive process.**

**That is, for the One-Stage Scenario, we claim the following:**

**The $[C(4)B]$ operator is the *coalescence* of the two operators $CC$ and $B$ in the sense that, whereas in the Two-Stage Scenario the initial compressive process undergoes continuation after it has bifurcated into two copies, in contrast, in the One-Stage Scenario, the continuation of the initial compressive process occurs at the same time as its bifurcation.**

**That is, understanding the $[C(4)B]$ operator as the *coalescence* of the two operators $CC$ and $B$ means that the continuation of the initial compressive process has coalesced with the bifurcation of that process.**

**Because standard theories of shape, e.g., the Medial axis, completely fail to represent the causal process of a compressive extremum, as well as the continuation of a compressive process, the standard theories completely fail to understand the causal structure we have described for a zone-restricted penetrative process breaking-through a compressive process.**

## **THE ORGANIZATION OF THE OPERATORS** $[C(4)B], CC, B$

**The organizational relationship between the operators $[C(4)B]$, $CC$, $B$, will be given in Chapter 12, during the elaboration of the singularity theory of shape, which we have invented, which we call** *Interactive Singularity Theory.*

**A crucial aspect of Interactive Singularity Theory is that it helps to precisely understand the organization of the relationships between operators in a way that matches the** *causal* **structure of the operators given by the correct process-inference rules, i.e., the PISA axes.**

**According to our theory, this causal structure is fundamental to understanding biological morphology and manufacturing design.**

**In contrast to Interactive Singularity Theory, the conventional singularity theories of shape fail to match the fundamentally important causal structure of biological morphology and manufacturing design.**

**PROCESS-GRAMMAR SO FAR**

**Process-Continuation**

$Cm^{+}: \; m^{+} \longrightarrow 0m^{-}0$ (squashing continues till it indents)
$CM^{-}: \; M^{-} \longrightarrow 0M^{+}0$ (resistance continues till it protrudes)

**3-Fold Process-Bifurcation**

Pure 3-fold Process-Bifurcation

$BM^{+}: \; M^{+} \longrightarrow M^{+}m^{+}M^{+}$ (shield-formation)
$Bm^{-}: \; m^{-} \longrightarrow m^{-}M^{-}m^{-}$ (bay-formation)
$Bm^{+}: \; m^{+} \longrightarrow m^{+}M^{+}m^{+}$ (breaking-through of a protrusion)
$BM^{-}: \; M^{-} \longrightarrow M^{-}m^{-}M^{-}$ (breaking-through of an indentation)

3-fold Radial-Undulation

$[C(4)B]m^{+}: \; m^{+} \longrightarrow 0m^{-}0M^{+}0m^{-}0$ (zone-restricted protrusion formation)
$[C(4)B]M^{-}: \; M^{-} \longrightarrow 0M^{+}0m^{-}0M^{+}0$ (zone-restricted indentation formation)

**Pair-Creation**

Pure 2-fold Process-Bifurcation

$\pi\emptyset^{\uparrow+}: \; \emptyset^{\uparrow+} \longrightarrow M^{+}m^{+}$ (application-point: upward positive)
$\pi\emptyset^{\uparrow-}: \; \emptyset^{\uparrow-} \longrightarrow M^{-}m^{-}$ (application-point: upward negative)
$\pi\emptyset^{\downarrow+}: \; \emptyset^{\downarrow+} \longrightarrow m^{+}M^{+}$ (application-point: downward positive)
$\pi\emptyset^{\downarrow-}: \; \emptyset^{\downarrow-} \longrightarrow m^{-}M^{-}$ (application-point: downward negative)

2-fold Radial-Undulation

$[C(3)\pi]0^{\uparrow}: \; 0^{\uparrow} \longrightarrow 0M^{+}0m^{-}0$ (application-point: upward zero)
$[C(3)\pi]0^{\downarrow}: \; 0^{\downarrow} \longrightarrow 0m^{-}0M^{+}0$ (application-point: downward zero)

**2-Fold Pinching (Type 1)**

$CCm^{+}: \; m^{+}\,M^{+}\,m^{+} \longrightarrow \overbrace{0m^{-}0}\;M^{+}\;\overbrace{0m^{-}0}$ (pinching a double-shield)

$CCM^{-}: \; M^{-}\,m^{-}\,M^{-} \longrightarrow \overbrace{0M^{+}0}\;m^{-}\;\overbrace{0M^{+}0}$ (pinching a double-bay)

# 6

# Process-Theory of Parts (2)

## 6.1 Part-Formation as Bifurcation

We now need to observe the following:

**PART-FORMATION AS BIFURCATION**

> **Recall that, within the Process Grammar, a part is defined as a recovered process and the recovered subsequent history of that process. (Note that this can include operations on the side of the process.)**
> **Therefore,** ***process-bifurcation*** **is a fundamental aspect of many part-formations.**
> **A crucial distinction between several types of parts is the number $n$ of processes that have created the part.**
> **By our process-inference rules, these correspond to the final number $n$ of extrema in the part.**
> **The resulting part will be called an** ***$n$-fold part,*** **and will be said to result from** ***$n$-fold part-formation.***

For example, several of the parts in Chapter 5 were created by 3-fold process-bifurcation. Therefore, we will say that these parts are *3-fold parts*, and were created by *3-fold part-formation.*

3-fold parts are of great significance in the world. However, we shall now show that there are also 5-fold parts, and that these are also very significant.

In order to begin to understand crucial issues, it is necessary to compare Fig 6.1 and Fig 6.2 which respectively show an example of a 3-fold part and a 5-fold part. Observe first that, to fully contrast the two kinds of parts, the figures show both as starting on the same shape, an ellipse of positive curvature. Furthermore, both are formed with respect to the same starting extremum, the $m^+$ at the bottom of the ellipse.

Next observe the number of extrema that result from the formation. In going from the left to the right shape in Fig 6.1, the bottom $m^+$ of the ellipse has undergone a transition to three extrema. In contrast, in going from the left to the right shape in Fig 6.2, the

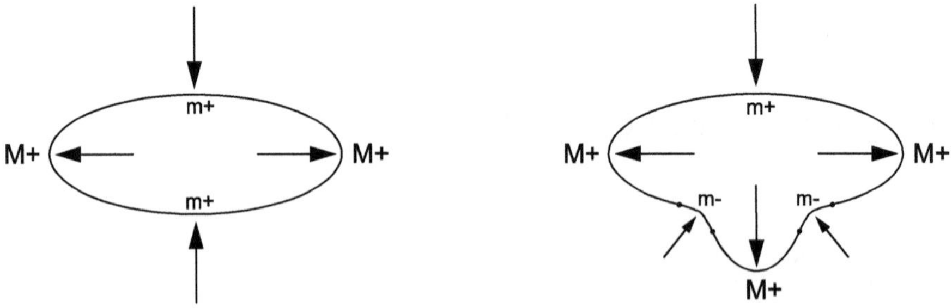


**Fig. 6.1.** Formation of a 3-fold part.

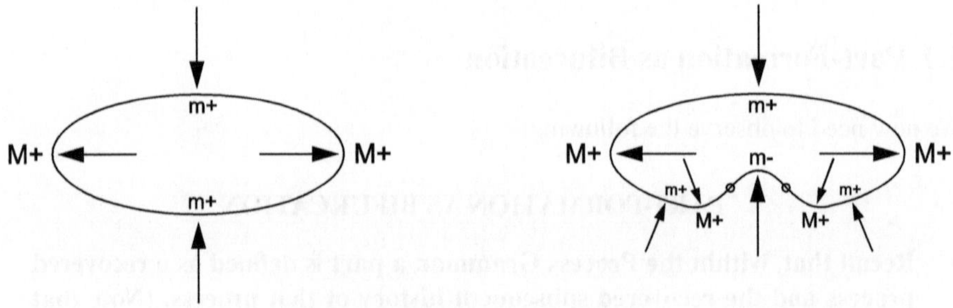


**Fig. 6.2.** Formation of a 5-fold part.

bottom $m^+$ of the ellipse has undergone a transition to five extrema. That is why we will call the respective parts, 3-fold and 5-fold; and the respective histories, 3-fold and 5-fold part-formation.

One crucial difference between the two situations is seen by the direction of the central process of the final part: In the right shape of Fig 6.1, the central process is in the *opposite* direction from the compressive process at the corresponding point in the left shape. In contrast, in the right shape of Fig 6.2, the central process is in the *same* direction as the compressive process at the corresponding point in the left shape.

It will now be argued that the 5-fold example covers a large number of real-world situations. For example, consider biology. One tends to think of *growth* as an *outward* process. However, there are many embryological situations in which the growing body creates a part by *indentation* rather than protrusion. Furthermore, for such an indentation, protrusions are often created to restrict the zone-of-influence of the indentation. This fact is captured by the right shape in the 5-fold example, Fig 6.2. The indentation is given by the central $m^-$ process, and the protrusions that restrict the zone-of-influence of the indentation are given by the $M^+$ on each side of the central process.

It is now important to understand that the zone-of-influence restriction given in Fig 6.2 is not an example of the zone-of-influence scenarios given in sections 5.10 - 5.12. That is, the zone-of-influence restriction in Fig 6.2 is not the figure-ground reversal of the zone-of-influence restriction shown in Fig 6.1. The reason is this: Observe that, in the right shape in Fig 6.1, no additional extrema occur between the restricting $m^-$ extrema and the $M^+$ extrema at the left and right ends of that shape. In contrast, observe that, in the right shape in Fig 6.2, two *additional* extrema occur between the restricting $M^+$ extrema and the $M^+$ extrema at the left and right ends of that shape. These additional extrema are the two diagonal $m^+$ shown at the bottom of that shape.

Thus, whereas the "pinching", which was created by the two $m^-$ extrema at the bottom of the right shape in Fig 6.1, did not require extra extrema to the side of those extrema, the "pinching" created by the two $M^+$ extrema at the bottom of the right shape in Fig 6.2 did require two extra extrema to the side of those extrema. The pinching in the two cases are therefore not figure-ground reversals of each other. They are indeed structurally different.

To understand this even further, consider Fig 6.2 in more detail, as follows: Observe that since, in going from the left to right shape in Fig 6.2, the upward central process can be regarded as merely continuing, one can regard this movement to be the result of the $Cm^+$ operation of the Process Grammar; i.e., squashing continues till it indents.

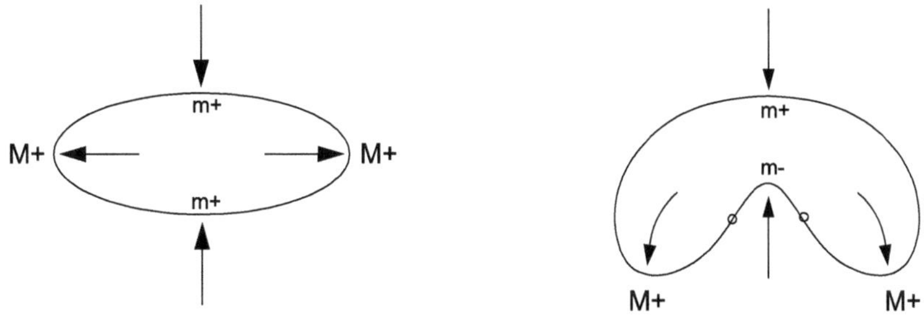


**Fig. 6.3.** Continuation of a compressive process, without zone-of-influence restriction.

If the zone-of-influence of that upward continued process were not restricted, i.e., it did not undergo pinching, the transition from the left to right shape would be as shown in Fig 6.3. In fact, Fig 6.3 can be understood as the first stage in a two-stage sequence that produced the transition shown Fig 6.2.

Thus, in going from the right shape in Fig 6.3 to the right shape in Fig 6.2, we see that the pinching (i.e., restriction of zone-of-influence) was carried out by creating the pair $m^+M^+$ to the left of the zero which is left of $m^-$, and creating the pair $M^+m^+$ to the right of the zero which is right of $m^-$. These pairs are therefore produced, respectively, by our positive downward pair-creation operation

$$\pi\emptyset^{\downarrow+} \ : \ \emptyset^{\downarrow+} \quad \longrightarrow \quad m^+M^+$$

and positive upward pair-creation operation

$$\pi\emptyset^{\uparrow+} \ : \ \emptyset^{\uparrow+} \quad \longrightarrow \quad M^+m^+$$

The restriction effect of applying these two operations can be seen by observing that, in the transition from the left to right shape in Fig 6.2, most of the ellipse has been preserved in the right shape.

Thus, if we view both Fig 6.1 and Fig 6.2 to be the result of a zone-of-restriction applied *after* the emergence of the central penetrative extremum, then the restriction in Fig 6.1 was carried out by the double continuation operator $CC$, and the restriction in Fig 6.2 was carried out by the pairing of $\pi^{\downarrow}$ with $\pi^{\uparrow}$. Therefore, the case in Fig 6.2 gives us a new zone-of-restriction operator. Often, we will use the term *pinching* to mean the restriction of the zone-of-influence of a process by the application of symmetrical-paired processes around that process. Whereas we called the type of pinching created by the $CC$ operator *Type 1*, the type of pinching we have just described will be called *Type 2*.

The symmetrical pairing of two pair-creation operators yields another set of operators to be added to the Process Grammar. These will be defined in the following bold-print. For convenience, we will assume that each operator is applied to the sides of a symmetric configuration of processes; and each operation will be written as a case where this symmetrical configuration contains only one extremum, the *anchor-extremum*. The generalizations to symmetrical configurations of multiple extrema are obvious.

### SYMMETRICALLY-PAIRED PAIR-CREATION OPERATORS

**When the upward pair-creation operator $\pi^{\uparrow}$ and downward pair-creation operator $\pi^{\downarrow}$ are applied simultaneously to symmetrically-corresponding points on opposite sides of a process, or on opposite sides of a bifurcation of a process, then this combination of the two operators will be called a** *symmetrically-paired pair-creation operator.*

**When the two side application-points have non-zero curvature, the operator will be called a** *non-zero* **symmetrically-paired pair-creation operator. There are two such operators: an** *upward-downward* **operator $\pi^{\uparrow}\pi^{\downarrow}$, and a** *downward-upward* **operator $\pi^{\downarrow}\pi^{\uparrow}$.**

**When the two side application-points have zero curvature, the operator will be called a** *zero* **symmetrically-paired pair-creation operator. There are two such operators: an** *upward-downward* **operator $[C(3)\pi^{\uparrow}][C(3)\pi^{\downarrow}]$, and a** *downward-upward* **operator $[C(3)\pi^{\downarrow}][C(3)\pi^{\uparrow}]$.**

**The operations that realize the above operators are as follows:**

### UPWARD-DOWNWARD OPERATIONS

**The upward-downward operators, $\pi^{\uparrow}\pi^{\downarrow}$ and $[C(3)\pi^{\uparrow}][C(3)\pi^{\downarrow}]$, are necessarily applied to the sides of maxima $M^{\pm}$. These operators are realized by a total of** *three* **operations. They are written here for a single central anchor-extremum.**

**Negative Upward-Downward Operation: Its two applications are (1) to the sides of a resistance process, or (2) to the sides of a protruding process in its negative tails:**

$$\pi^{\uparrow}\pi^{\downarrow}\emptyset^{-} \;:\; \emptyset^{-}\ X\ \emptyset^{-} \;\longrightarrow\; \overbrace{M^{-}m^{-}}\ X\ \overbrace{m^{-}M^{-}}$$

**where $X$ is $M^{-}$ or $0M^{+}0$.**

**Positive Upward-Downward Operation: Its application is to the sides of a protruding process in its same-sign (i.e., positive) components:**

$$\pi^{\uparrow}\pi^{\downarrow}\emptyset^{+} \;:\; \emptyset^{+}\ M^{+}\ \emptyset^{+} \;\longrightarrow\; \overbrace{M^{+}m^{+}}\ M^{+}\ \overbrace{m^{+}M^{+}}$$

**Zero Upward-Downward Operation: Its application is to the sides of a protruding process, at the zero-points:**

$$[C(3)\pi^{\uparrow}][C(3)\pi^{\downarrow}]0 \;:\; 0^{\uparrow}\ M^{+}\ 0^{\downarrow} \;\longrightarrow\; \overbrace{0M^{+}0m^{-}0}\ M^{+}\ \overbrace{0m^{-}0M^{+}0}$$

## DOWNWARD-UPWARD OPERATIONS

**The downward-upward operators, $\pi^{\downarrow}\pi^{\uparrow}$ and $[C(3)\pi^{\downarrow}][C(3)\pi^{\uparrow}]$, are necessarily applied to the sides of minima $m^{\pm}$. These operators are realized by a total of *three* operations. They are written here for a single central anchor-extremum.**

**Positive Downward-Upward Operation: Its two applications are (1) to the sides of a squashing process, or (2) to the sides of an indenting process in its positive tails:**

$$\pi^{\downarrow}\pi^{\uparrow}\emptyset^{+} \;:\; \emptyset^{+}\ X\ \emptyset^{+} \;\longrightarrow\; \overbrace{m^{+}M^{+}}\ X\ \overbrace{M^{+}m^{+}}$$

**where $X$ is $m^{+}$ or $0m^{-}0$.**

**Negative Downward-Upward Operation: Its application is to the sides of an indenting process in its same-sign (i.e., negative) components:**

$$\pi^{\downarrow}\pi^{\uparrow}\emptyset^{-} \;:\; \emptyset^{-}\ m^{-}\ \emptyset^{-} \;\longrightarrow\; \overbrace{m^{-}M^{-}}\ m^{-}\ \overbrace{M^{-}m^{-}}$$

**Zero Downward-Upward Operation: Its application is to the sides of an indenting process, at the zero-points:**

$$[C(3)\pi^{\downarrow}][C(3)\pi^{\uparrow}]0 \;:\; 0^{\downarrow}\ m^{-}\ 0^{\uparrow} \;\longrightarrow\; \overbrace{0m^{-}0M^{+}0}\ m^{-}\ \overbrace{0M^{+}0m^{-}0}$$

## GENERALIZATIONS

**The cases listed on this and the previous page can obviously be generalized to situations in which the symmetrically-paired pair-creation operations are applied to the two sides of a configuration which satisfy the corresponding upward and downward conditions.**

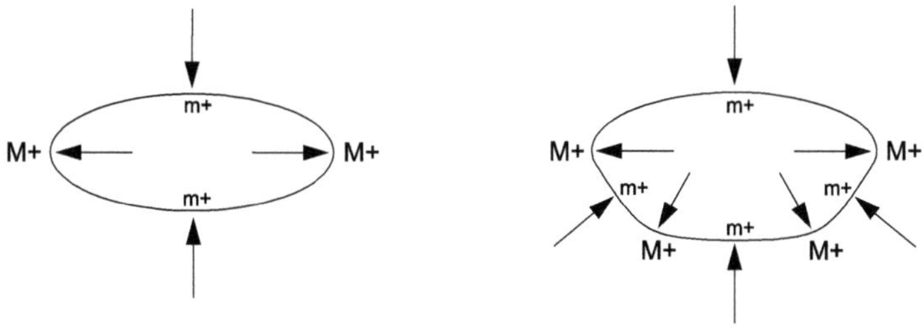


**Fig. 6.4.** The operation $\pi^{\downarrow}\pi^{\uparrow}\emptyset^{+}$ applied to the sides of a squashing process produces a *triple-shield*.

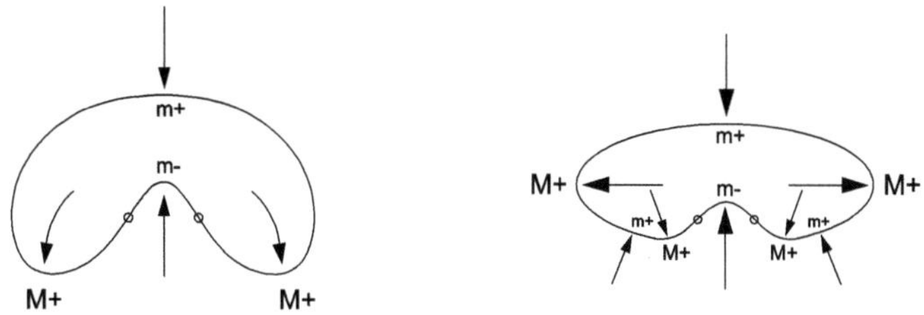


**Fig. 6.5.** The operation $\pi^{\downarrow}\pi^{\uparrow}\emptyset^{+}$ applied to the positive tails of an indenting process *pinches* that process.

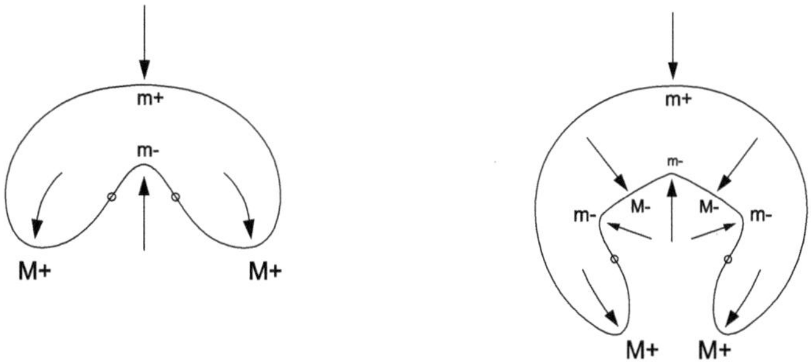


**Fig. 6.6.** The operation $\pi^{\downarrow}\pi^{\uparrow}\emptyset^{-}$ applied to the same-sign (negative) components of an indenting process produces a *complete double-bay*.

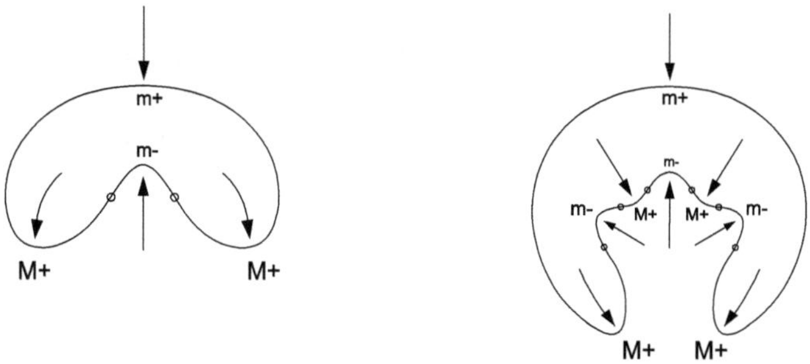


**Fig. 6.7.** The operation $[C(3)\pi^{\downarrow}][C(3)\pi^{\uparrow}]0$ applied to the zero-points on the sides of an indenting process produces a *pinched complete double-bay*.

The reader will recall, from Table 4.1 (page 105), that each of the *non-zero* pair-creation operations has three applications, given by the anchor-extrema involved. Therefore, each operator $\pi^{\downarrow}$ or $\pi^{\uparrow}$ has *six* non-zero applications, as can be seen from that table. That is, $\pi^{\uparrow}$ has a total of three applications to left sides in the collection of maxima, and a total of three applications to right sides in the collection of minima. Similarly, $\pi^{\downarrow}$ has a total of three applications to right sides in the collection of maxima, and a total of three applications to left sides in the collection of minima.

However, using those facts, the upward-downward operator $\pi^{\uparrow}\pi^{\downarrow}$ can be applied only to the sides of a maximum, because, *within* the operator $\pi^{\uparrow}\pi^{\downarrow}$, the operator $\pi^{\uparrow}$ can be applied only to the *left* side of a maximum, and the operator $\pi^{\downarrow}$ can be applied only to the *right* side of a maximum. Thus, of the six non-zero applications of $\pi^{\uparrow}$ in Table 4.1 (page 105), only three are valid when $\pi^{\uparrow}$ is used within the combined operator $\pi^{\uparrow}\pi^{\downarrow}$; that is, the three that are applied to the left side of a maximum.

Thus, we have just shown that the upward-downward operator $\pi^{\uparrow}\pi^{\downarrow}$ has three applications: Two are the two applications of the negative upward-downward operation $\pi^{\uparrow}\pi^{\downarrow}\emptyset^{-}$, and the third is the single application of the positive upward-downward operation $\pi^{\uparrow}\pi^{\downarrow}\emptyset^{+}$. These three applications were listed in the definition of the upward-downward operations on page 203.

The corresponding arguments for the downward-upward operator $\pi^{\downarrow}\pi^{\uparrow}$ are as follows: This operator can be applied only to the sides of a minimum, because, *within* the operator $\pi^{\downarrow}\pi^{\uparrow}$, the operator $\pi^{\downarrow}$ can be applied only to the *left* side of a minimum, and the operator $\pi^{\uparrow}$ can be applied only to the *right* side of a minimum. Thus, of the six applications of $\pi^{\downarrow}$ in Table 4.1 (page 105), only three are valid when $\pi^{\downarrow}$ is used within the combined operator $\pi^{\downarrow}\pi^{\uparrow}$; that is, the three that are applied to the left side of a minimum.

Thus, we have just seen that the downward-upward operator $\pi^{\downarrow}\pi^{\uparrow}$ has three applications: Two are the two applications of the positive downward-upward operation $\pi^{\downarrow}\pi^{\uparrow}\emptyset^{+}$, and the third is the single application of the negative downward-upward operation $\pi^{\downarrow}\pi^{\uparrow}\emptyset^{-}$. These three applications were listed in the definition of the downward-upward operations on page 204.

Now recall, from section 4.5, that each of the *zero* pair-creation operations $[C(3)\pi]0^{\uparrow}$ and $[C(3)\pi]0^{\downarrow}$ has *two* applications, corresponding to two different anchor-extrema, as follows: The upward zero pair-creation operation $[C(3)\pi]0^{\uparrow}$ can be applied to the zero on the left side of $M^{+}$ and the zero on the right side of $m^{-}$. Conversely, the downward zero pair-creation operation $[C(3)\pi]0^{\downarrow}$ can be applied to the zero on the right side of $M^{+}$ and the zero on left side of $m^{-}$.

However, it is clear that when $[C(3)\pi]0^{\uparrow}$ and $[C(3)\pi]0^{\downarrow}$ are symmetrically-paired, they have only one application. That is, the zero upward-downward operator $[C(3)\pi^{\uparrow}][C(3)\pi^{\downarrow}]$ can be applied only to the sides of a maximum, and the zero downward-upward operator $[C(3)\pi^{\downarrow}][C(3)\pi^{\uparrow}]$ can be applied only to the sides of a minimum.

To gain a fuller understanding of the symmetrically-paired pair-creation operators, we take the *downward-upward* operations listed on page 204, and illustrate them in the four figures shown the next page, page 205. Thus, for convenience, the reader will be able to go forwards and backwards between the two adjacent pages, and see how the operations on page 204 are illustrated by the four figures on page 205. The following facts should be observed:

There are three downward-upward operations, as listed on page 204. The first two are the non-zero operations (i.e., applied to points of non-zero curvature), and the third is the zero operation (i.e., applied to points of zero curvature).

Now, recall that the above discussion showed that the non-zero operations have a total of three applications. These are distributed as follows: The first non-zero operation, listed on page 204, is the positive one, $\pi^{\downarrow}\pi^{\uparrow}\emptyset^{+}$. It has two applications: To the sides of a squashing process, and to the opposite-sign components of an indenting process. These two applications are illustrated by the first and second figures on page 205. The first application creates a *triple-shield.* The second application *pinches* an indenting process. Notice, in the latter case, that the operator is applied *outside* the two dots representing the zeros.

The second operation, listed on page 204, is the negative one, $\pi^{\downarrow}\pi^{\uparrow}\emptyset^{-}$. This is illustrated in the third figure on p205. Notice that the operation is applied *between* the two dots representing the zeros. The result is what our theory calls a *complete double-bay.* Our theory defines a double-bay as *complete*, when it is based on the definition we gave of a *complete* bay, $m^{-}M^{-}m^{-}$; that is, a complete double-bay consists of five extrema $m^{-}M^{-}m^{-}M^{-}m^{-}$; that is, it contains two complete bays, which share a boundary point at the center.

Finally, the third operation, listed on page 204, is the zero one, $[C(3)\pi^{\downarrow}][C(3)\pi^{\uparrow}]0$. It is illustrated in the fourth figure on p205. Notice that the operation is applied to the two dots themselves. The result is what our theory calls a *pinched complete double-bay.* Notice that, when the final codomain of the Two-Stage Scenario and One-Stage Scenario of sections 5.10 - 5.12 was a *pinched double-bay*, that codomain had only three extrema, i.e., was a 3-fold part; that is, the codomain was not a pinched *complete* double-bay. In contrast, the operation $[C(3)\pi^{\downarrow}][C(3)\pi^{\uparrow}]0$, which we defined on page 204, has a codomain that contains five extrema, i.e., is a pinched *complete* double-bay. That is, this operation creates a 5-fold part. Observe that, in this case, the 5-fold part is a 5-fold undulation, i.e., it consists of five penetrative extrema of alternating opposite signs.

It will be crucial for the reader to clearly understand that the three downward-upward operations, which are listed on page 204, correspond to the four figures on p205 (the first operation having two applications that are illustrated by the first two figures).

With respect to the issue of two applications for the first operation, let us return to Fig 6.2 (p200). We will consider two alternative routes in going from the left shape to the right shape. These two shapes are shown again on the left and right of Fig 6.8 where the two middle shapes show the two alternative intervening stages between left and right.

The route via the upper shape is the application first of the positive downward-upward operation

$$\pi^{\downarrow}\pi^{\uparrow}\emptyset^{+} \;:\; \emptyset^{+}\ X\ \emptyset^{+} \;\longrightarrow\; \overbrace{m^{+}M^{+}}\ X\ \overbrace{M^{+}m^{+}}$$

where $X = m^{+}$, followed by the application of the continuation operation $Cm^{+}$. The route via the lower shape is the application first of the continuation operation $Cm^{+}$, followed by the same downward-upward operation, but where $X = 0m^{-}0$.

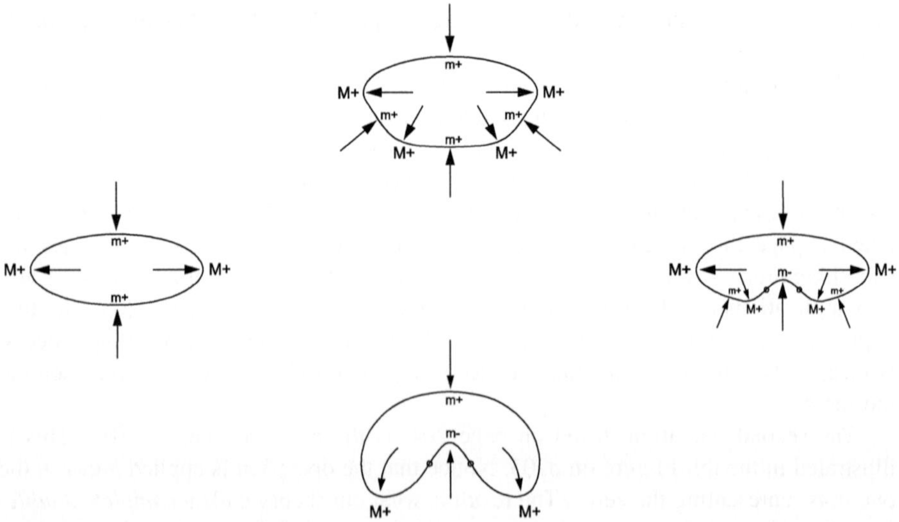


**Fig. 6.8.** Five-fold part creation.

That is, the figure illustrates the *sequential order in which one chooses to restrict the zone-of-influence*. The route from the left to right shape via the upper middle shape can be understood as setting up the restriction that will limit the zone-of-influence of the subsequent indentation. In contrast, the route via the lower middle shape can be understood as creating the restriction *after* the indentation was introduced.

## 6.2 The Three Scenarios

Each of the four figures on p205 is an example of 5-fold part-formation, i.e., each is a transition from one extremum to five. The scenario used to explain these transitions was given by symmetrically-paired pair-creation operators. Such operators will be referred to as *Scenario 1*.

However, the same transition can be created by two alternative scenarios, which we will refer to as *Scenarios* 2 and 3. In both of these, the two side-pairs are not created by bifurcation at points away from the central extremum. Instead, they bifurcate out from the central extremum. To illustrate, let us consider the transition given by the top figure on p205, which is shown again here, as Fig 6.9.

Before we begin, we remind the reader of our notation that an $n$-fold process-bifurcation operator $B(n)$ can be notated as $B_n$ when the operator is applied to a minimum, and can be notated as $B_{-n}$ when the operator is applied to a maximum (Notation 4.6 page 121).

Now, in what will be called Scenario 2, the bifurcation occurs in two stages, both of which are the application of the 3-fold process-bifurcation operator $B$. Thus, the

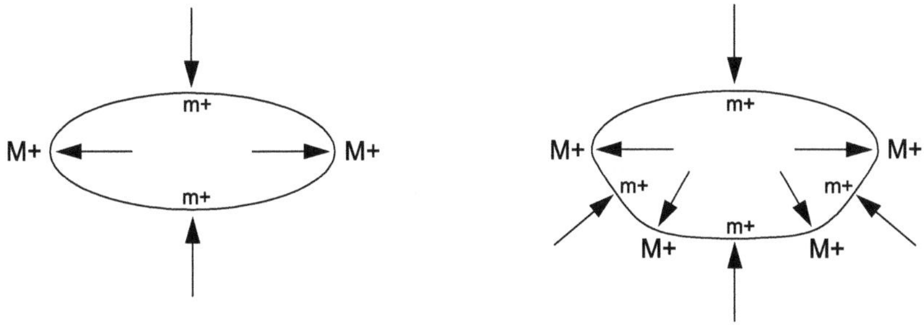


**Fig. 6.9.** Creation of a 5-fold part, a triple-shield.

transition from the left to the right shape in Fig 6.9 is decomposed into the two stages, shown in Fig 6.10. That is, since the starting central extremum is $m^+$, one first applies its corresponding $B$ operator thus:

$$B_3 m^+ \; : \; m^+ \; \longrightarrow \; m^+ M^+ m^+$$

which is the Process Grammar operation called *breaking-through of a protrusion.* It gives the transition from the first shape to the second in Fig 6.10, and indeed we see that a protrusion has broken-through at the bottom of the second shape. Next, to achieve the transition from the second shape to the third, we will apply the $B$ operator to the central downward protrusion of the second shape, thus:

$$B_{-3} M^+ \; : \; M^+ \; \longrightarrow \; M^+ m^+ M^+$$

which is the *shield-formation* operation from the Process Grammar. We see that this has produced an additional shield at the bottom of the third shape.

In this scenario, it is fundamentally important to understand that, even though the five resulting extrema $m^+M^+m^+M^+m^+$, which define the resulting lower region of the final shape, were created in successive stages, *all five* extrema bifurcated out from the initial central extremum.

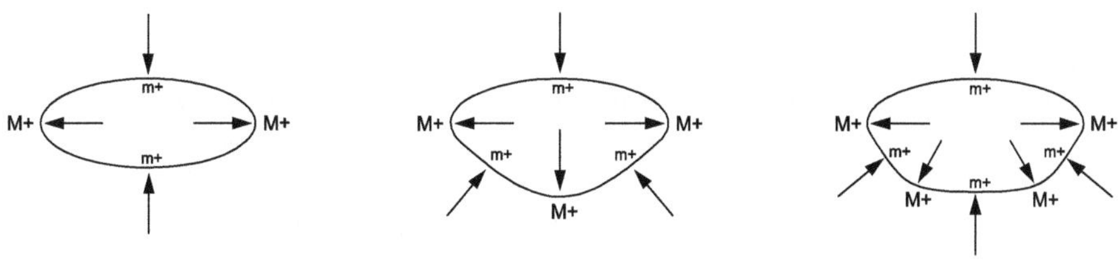


**Fig. 6.10.** Two successive 3-fold bifurcations.

Now, in the next method, Scenario 3, these five extrema will not only bifurcate out from the central extremum, but will do so in *one* stage. That is, the right-hand shape in Fig 6.9, will be produced by a *single simultaneous* 5-fold bifurcation out from the central extremum in the left-hand shape. This will be given by a new operator $B(5)$, which will be called the **5-fold process-bifurcation** operator. Since, in the present example, Fig 6.9, the operator is applied to a minimum, it will be denoted also by $B_5$. Therefore, in this example, it will be realized as the following new operation:

$$B_5 m^+ \ : \ m^+ \ \longrightarrow \ m^+ M^+ m^+ M^+ m^+ \qquad (6.1)$$

As with all our operators, there is a single intervening state, the transition-state, which is a curvature function that contains a *coalescence* of singularities that appear in the codomain. In the operation in expression (6.1), the transition-state contains one singularity that is a coalescence of *all* the five extrema that appear in the codomain. It can easily be seen that this transition-state is entirely different from the transition-states in the other two scenarios.

We have therefore seen that there are three alternative scenarios for the transition shown in Fig 6.9. They are defined as follows:

**THE 3 SCENARIOS**

**(1) Symmetrically-paired pair-creation: The simultaneous use of the pair-creation operators $\pi^{\downarrow}$ and $\pi^{\uparrow}$ around the central extremum.**

**(2) Two successive 3-fold process-bifurcations from the central extremum.**

**(3) One simultaneous 5-fold process-bifurcation from the central extremum.**

We can see that all three scenarios have a *symmetrical* structure with respect to the central extremum. The first occurs within the two sides of that extremum; and the others are symmetrical bifurcations from the central extremum and are therefore of odd-order.

Recall that our theory calls the 5-fold part in the right shape of Fig 6.9 a **triple-shield**. Also recall that, in the previous chapter, we showed that triple-shields are a fundamental structure in many designs – from biological design to automotive design.

Now, we have been examining the formation of a 5-fold part in which the central extremum is *positive compressive*, i.e., a shield.

Let us next examine a 5-fold part in which the central extremum is *negative penetrative*, i.e., an indentation. An example of such a 5-fold part is the *complete double-bay* illustrated in the right shape in Fig 6.11. It is crucial to understand that a complete double-bay is a 5-fold part of entirely negative curvature, whose central extremum is penetrative.

Again, since a complete double-bay is a 5-fold part, it is formed by any one of the three scenarios listed above. Scenario 1 is the application of a symmetrically-paired pair-creation operator, and, in the case of the formation of a complete double-bay, this was already defined as the downward-upward operation $\pi^{\downarrow}\pi^{\uparrow}\emptyset^{-}$, on p204. It was illustrated

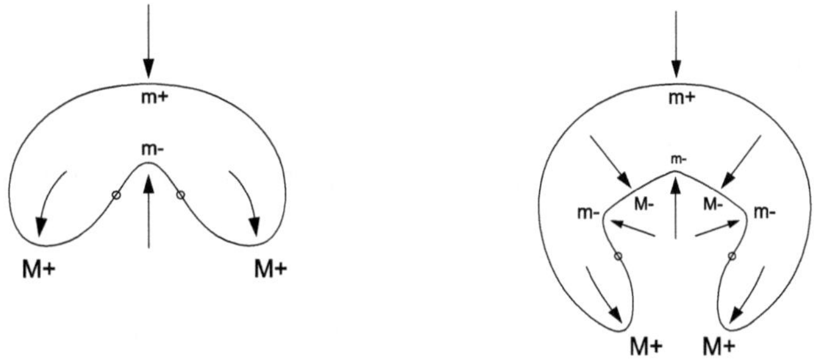


**Fig. 6.11.** Creation of a 5-fold part, a complete double-bay.

by the same diagram as given here, that shown on the third level on page 205. Observe, as stated earlier, the operation is applied between the two zeros of the indenting process.

Now let us look at Scenario 2, the two-stage formation of such a part. This will again be the successive application of two 3-fold bifurcations. Therefore, since the starting central extremum is $m^-$, one first applies its corresponding $B$ operator thus:

$$B_3m^- \; : \; m^- \;\longrightarrow\; m^-M^-m^-$$

i.e., the *bay-formation operation*. It gives the transition from the first shape to the second in Fig 6.12, where the indentation in the center of the first shape has bifurcated into the *complete bay* in the second shape. Next, to achieve the transition from the second shape to the third, we will apply the $B$ operator to the central extremum $M^-$ of the complete bay in the second shape, thus:

$$B_{-3}M^- \; : \; M^- \;\longrightarrow\; M^-m^-M^-$$

This is the Process Grammar operation called *breaking-through of an indentation*. Indeed, we see that, at the $M^-$ extremum of the second shape, an indentation has broken-through in the third shape.

What is important to understand is that, in this scenario, all five extrema of the resulting complete double-bay were created out from the central extremum in the first shape.

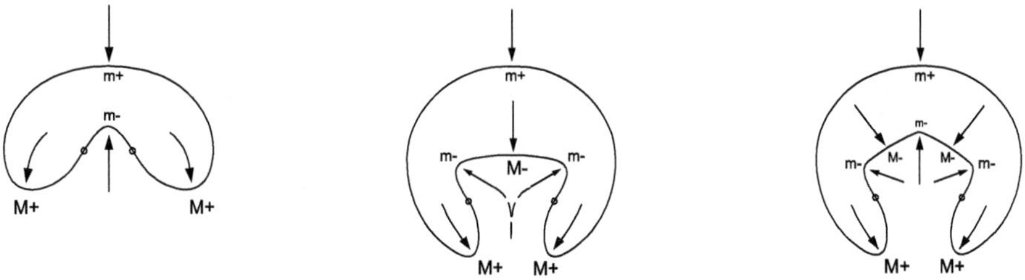


**Fig. 6.12.** The first 3-fold bifurcation at $m^-$ followed by 3-fold bifurcation at $M^-$.

It is worthwhile for the reader to contrast the *triple-shield*, in the third shape of Fig 6.10, with the *double-bay*, in the third shape of Fig 6.12, as follows: The central process of the triple-shield is *compressive*, i.e., *crushes* its opposing space; whereas the central process of the double-bay is *penetrative*, i.e., *invades* its opposing space.

Nevertheless, it is important to understand that the scenarios in the two cases were given by the same sequence of *operators*:

$$B_{-3} \circ B_3$$

However, a crucial difference is that this operator-sequence is applied to a compressive vs. penetrative process in the two respective cases. This emphasizes the fact that the *Process Grammar*, as a *grammar of processes*, requires the full specification of the *operations* that realize the *operators*.

Now let us turn to Scenario 3, where the transition from the first to the third shape in Fig 6.12 is direct (i.e., not via the second shape); i.e., as shown in Fig 6.11. In this scenario, the complete double-bay is created by *simultaneous* 5-fold bifurcation, outward from the initial central extremum. This is given by the same 5-fold bifurcation operator $B_5$, as in our previous example of Scenario 3, but is realized as the following new operation:

$$B_5 m^- \ : \ m^- \ \longrightarrow \ m^- M^- m^- M^- m^-$$

because the application-point is the *penetrative* extremum $m^-$, instead of the *compressive* extremum $m^+$ of the previous case.

So far, in this section, we have looked at 5-fold part-formation originating at each of the two curvature *minima*, $m^+$ and $m^-$. Now let us look at 5-fold part-formation originating at each of the two curvature *maxima*, $M^-$ and $M^+$.

Consider first the case of $M^-$. The transition from the left to the right shape in Fig 6.13 illustrates 5-fold part-formation at $M^-$. The single extremum $M^-$ in the left shape has been replaced by the string of 5 extrema $M^- m^- M^- m^- M^-$ in the right shape. Thus the structure goes from the single bay in the left shape to what our theory calls the **triple-bay** in the right shape. Notice that the *triple-bay* is, of course, the figure-ground reversal of the *triple-shield*.

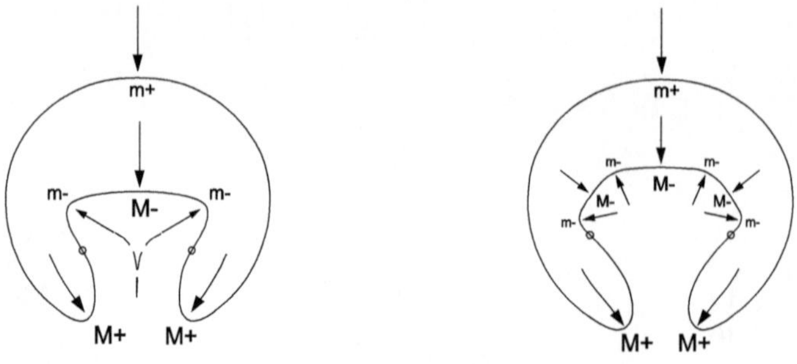


**Fig. 6.13.** At extremum $M^-$, the creation of a 5-fold part, a *triple-bay*.

Again, such a part is formed by any one of the three scenarios listed above. In Scenario 1, the relevant operation was defined on p203 as

$$\pi^{\uparrow}\pi^{\downarrow}\emptyset^{-} \ : \ \emptyset^{-} \ X \ \emptyset^{-} \ \longrightarrow \ \overbrace{M^{-}m^{-}} \ X \ \overbrace{m^{-}M^{-}}$$

where, in the present case, $X$ is the extremum $M^{-}$. This is the first time we are giving an illustration of an *upward-downward* operation. All illustrations so far have been of *downward-upward* operations. By careful examination of Fig 6.13, the reader can see that pair-creation has been applied to the spiral on each side of the central $M^{-}$.

Now let us look at Scenario 2, the two-stage formation of such a part. This will be a succession of two 3-fold bifurcations. Fig 6.14 exhibits this scenario. It is the succession:

**Bay** $\longrightarrow$ **Double-Bay** $\longrightarrow$ **Triple-Bay**

The two successive 3-fold bifurcations that generate this are in the reverse order of the previous cases in this section, thus: $B_3 \circ B_{-3}$. That is, since the starting central extremum is $M^{-}$, one first applies the $B$ operator corresponding to a maximum, thus:

$$B_{-3}M^{-} \ : \ M^{-} \ \longrightarrow \ M^{-}m^{-}M^{-}$$

This is the Process Grammar operation called *breaking-through of an indentation.* It gives the transition from the first shape to the second in Fig 6.14, and indeed we see that an indentation has broken-through at the center of the second shape, thus creating a double-bay in the second shape. Next, to achieve the transition from the second shape to the third, we will apply the $B$ operator to that central upward indentation of the second shape, thus:

$$B_3m^{-} \ : \ m^{-} \ \longrightarrow \ m^{-}M^{-}m^{-}$$

which is the *bay-formation* operation from the Process Grammar. We see that this has produced, in the third shape, an additional bay between the two side bays in the second shape.

In this scenario, it is fundamentally important to understand that, even though the five extrema $M^{-}m^{-}M^{-}m^{-}M^{-}$, which constitute the resulting part, were created in successive stages, *all five* extrema bifurcated out from the initial central one.

Finally, consider Scenario 3. Here the five extrema will not only bifurcate out from the central extremum, but will do so in *one* stage; i.e., as illustrated in Fig 6.13. This

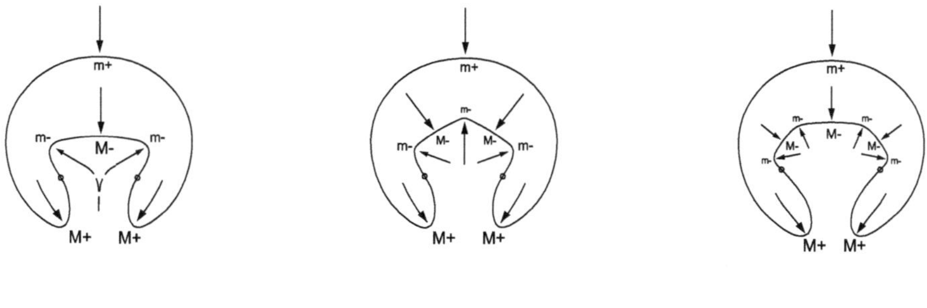


**Fig. 6.14.** Two successive 3-fold bifurcations producing a triple-bay.

therefore uses the operator $B(5)$. Since, in the present example, the operator is applied to a maximum, it will be denoted also by $B_{-5}$. Therefore, in this example, the operator will be realized as the following new operation:

$$B_{-5}M^- \;:\; M^- \;\longrightarrow\; M^-m^-M^-m^-M^-$$

The $B_{-5}$ operator can be viewed as a *coalescence* of the $B_{-3}$ and $B_3$ operators.

In the above example, we have examined the three scenarios for 5-fold part-formation where the starting extremum is a *compressive maximum* $M^-$. By the logic of 5-fold part-formation, the central extremum of the resulting structure is also a compressive maximum $M^-$. Let us now turn to the case of a *penetrative maximum*; i.e., where the starting extremum, and the central extremum of the resulting 5-fold part, are both $M^+$.

The transition from the left to the right shape in Fig 6.15 illustrates 5-fold part-formation at $M^+$. The single extremum $M^+$, at the bottom of the left shape, has been replaced by the string of 5 extrema $M^+m^+M^+m^+M^+$, at the bottom of the right shape. In the Process Grammar, this resulting 5-fold structure is called a *complete double-shield*. It is, of course, the figure-ground reversal of a complete double-bay.

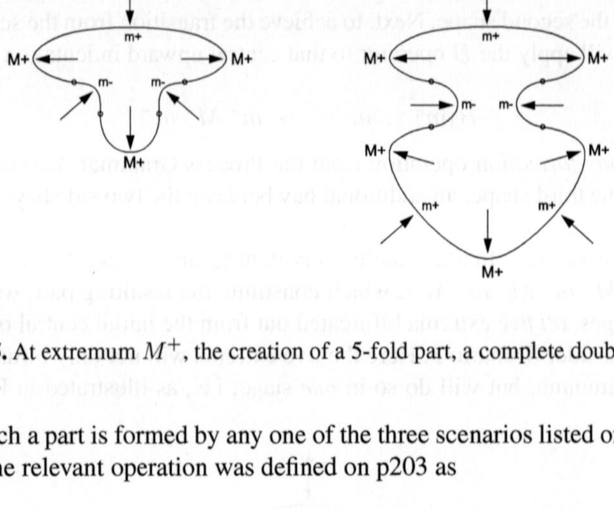


**Fig. 6.15.** At extremum $M^+$, the creation of a 5-fold part, a complete double-shield.

Again, such a part is formed by any one of the three scenarios listed on page 210. In Scenario 1, the relevant operation was defined on p203 as

$$\pi^\uparrow\pi^\downarrow\emptyset^+ \;:\; \emptyset^+\; M^+\; \emptyset^+ \;\longrightarrow\; \overbrace{M^+m^+}\; M^+\; \overbrace{m^+M^+}$$

By careful examination of Fig 6.15, the reader can see that this operation has been applied to the region *between the two zeros* on each side of the central $M^+$ in the first shape.

Now let us look at Scenario 2, the two-stage formation of such a structure. This will be the succession of two 3-fold bifurcations exhibited in Fig 6.16. Therefore, since the starting central extremum is $M^+$, one first applies its corresponding $B$ operator thus:

$$B_{-3}M^+ \ : \ M^+ \longrightarrow M^+m^+M^+$$

i.e., the *shield-formation operation.* It gives the transition from the first shape to the second in Fig 6.16, where the protrusion at the bottom of the first shape has bifurcated into the *complete shield* at the bottom of the second shape. Next, to achieve the transition from the second shape to the third, we will apply the $B$ operator to the central extremum $m^+$ of the complete shield at bottom of the second shape, thus:

$$B_3m^+ \ : \ m^+ \longrightarrow m^+M^+m^+$$

This is the Process Grammar operation called *breaking-through of a protrusion.* Indeed, we see that, in the complete shield at the bottom of the second shape, a protrusion has broken-through in the third shape, resulting in a complete double-shield.

In this scenario, it is fundamentally important to understand that, even though the five extrema $M^+m^+M^+m^+M^+$, which constitute the resulting part, were created in successive stages, *all five* extrema bifurcated out from the initial central one.

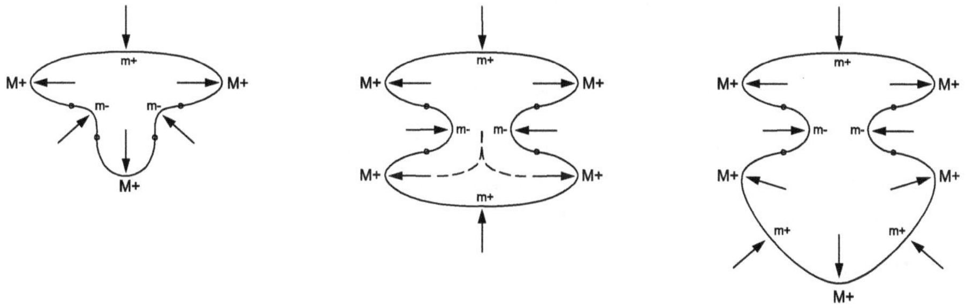


**Fig. 6.16.** Two successive 3-fold bifurcations producing a complete double-shield.

It is worthwhile for the reader to contrast the *triple-bay*, in the third shape of Fig 6.14, with the *double-shield*, in the third shape of Fig 6.16. The central process of the triple-bay is *compressive*, i.e., *resists* its opposing space; whereas the central process of the double-shield is *penetrative*, i.e., *invades* its opposing space.

Nevertheless, it is important to understand that the scenarios in the two cases are given by the same sequence of *operators*:

$$B_3 \circ B_{-3}$$

However, a crucial difference is that this operator-sequence is applied to a compressive vs. penetrative process in the two respective cases. Again, this emphasizes the fact that the *Process Grammar*, as a *grammar of processes*, requires the full specification of the *operations* that realize the *operators*.

Now let us turn to Scenario 3, where the transition from the first to the third shape in Fig 6.16 is direct (i.e., not via the second shape); i.e., as shown in Fig 6.15. In this scenario, the complete double-shield is created by *simultaneous* 5-fold bifurcation, outward from the initial central extremum. This is given by the same 5-fold bifurcation operator $B_{-5}$, as in our previous example of Scenario 3, but is realized as the following new operation:

$$B_{-5}M^{+} \;:\; M^{+} \;\longrightarrow\; M^{+}m^{+}M^{+}m^{+}M^{+}$$

because the application-point is the *penetrative* extremum $M^{+}$, instead of the *compressive* extremum $M^{-}$ of the previous case.

Above, we have defined four classes of 5-fold part-formation, i.e., one class for each of the four extrema $m^{+}, m^{-}, M^{-}, M^{+}$. Each class consisted of three scenarios. Scenario 1 is symmetrically-paired pair-creation, defined on p203. Scenario 2 is two successive uses of the 3-fold bifurcation operator, that was already defined in an earlier chapter. Scenario 3 is the 5-fold bifurcation introduced in the present section. Since this third case bifurcates the five extrema *simultaneously* from a single extremum, we will say that it generates a 5-fold part *radially*. The four operations for Scenario 3 are summarized in the following bold print. Instead of using the above notation, $B_5m^{+}$, $B_5m^{-}$, $B_{-5}M^{-}$, $B_{-5}M^{+}$, for these operations, we will use the alternative notation $B(5)m^{+}$, $B(5)m^{-}$, $B(5)M^{-}$, $B(5)M^{+}$, to emphasize the use of the single master operator $B(5)$:

**PURE 5-FOLD PROCESS-BIFURCATION OPERATOR $B(5)$**

**Under the Process Grammar operator $B(5)$, a single extremum bifurcates simultaneously into five extrema of the same sign as the original extremum. This is realized by four operations:**

**Radial Formation of a Triple-Shield**

$$B(5)m^{+} \;:\; m^{+} \;\longrightarrow\; m^{+}M^{+}m^{+}M^{+}m^{+}$$

**Radial Formation of a Complete Double-Bay**

$$B(5)m^{-} \;:\; m^{-} \;\longrightarrow\; m^{-}M^{-}m^{-}M^{-}m^{-}$$

**Radial Formation of a Triple-Bay**

$$B(5)M^{-} \;:\; M^{-} \;\longrightarrow\; M^{-}m^{-}M^{-}m^{-}M^{-}$$

**Radial Formation of a Complete Double-Shield**

$$B(5)M^{+} \;:\; M^{+} \;\longrightarrow\; M^{+}m^{+}M^{+}m^{+}M^{+}$$

## 6.3 Applying Continuation Operators

Let us now return to the creation of a triple-shield from a compressive extremum, as illustrated again in Fig 6.17. In the resulting part, there are three compressive extrema, $m^+$, corresponding to three squashing processes.

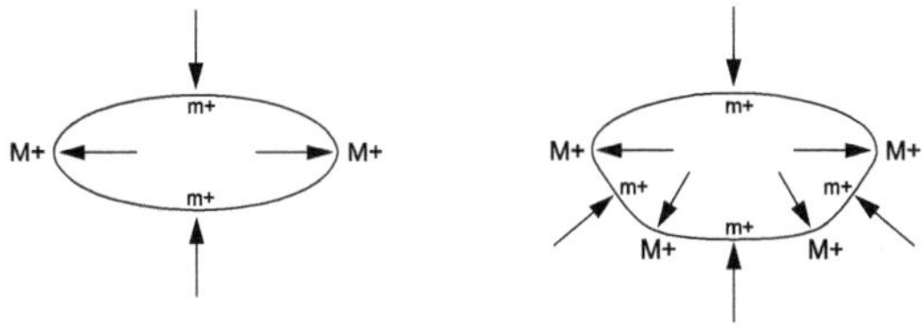


**Fig. 6.17.** Creation of a 5-fold part, a triple-shield.

These three squashing processes can continue till they indent. Since, in each of the three alternative scenarios that created the triple-shield, the three squashing processes were created *symmetrically* with respect to the starting extremum in the first shape, it is likely that the continuations of the squashing processes would be symmetrically organized as follows:

First, the central squashing process of the triple-shield can continue till it indents, as shown in the transition from the first to the second shape in Fig 6.18. This is a use of the Process Grammar operation

$$Cm^+ \; : \; m^+ \longrightarrow 0m^-0$$

Next, the two side squashing processes, which are symmetric with respect to the central extremum, can continue simultaneously till they indent. This is an example of the Type 1 *pinching* operator $CC$. Thus, as illustrated in Fig 6.18, what was a triple-shield, in the first shape, is pinched at its two side $m^+$ extrema, in the second shape, till they indent, in the third shape.

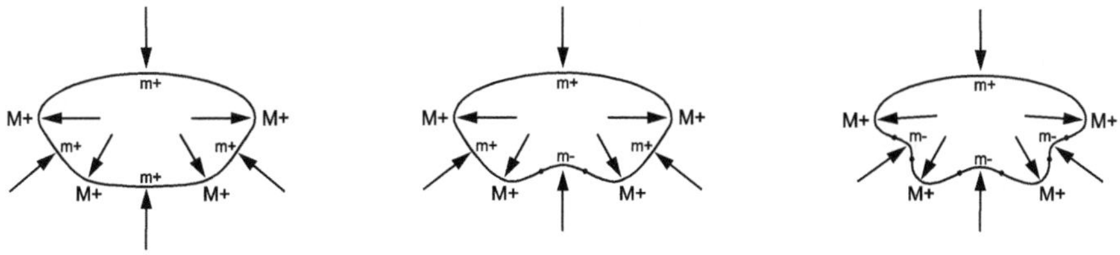


**Fig. 6.18.** Continuation $C$ followed by 2-fold pinching $CC$.

Notice that, in this second transition, we are applying the generalization of $CC$ that was defined on page 191 as

$$CCm^+ : \; m^+ X \, m^+ \quad \longrightarrow \quad \overbrace{0m^-0} \; X \; \overbrace{0m^-0}$$

where $X$ is a symmetrical sequence terminating at each end with a $M^+$ extremum. In the present case, $X$ is the sequence between the two side $m^+$ extrema in the second shape in Fig 6.18.

Now, keeping the symmetrical arrangement of the continuation operators, with respect to the central extremum, we could of course apply the sequence we have just given, $C$ followed by $CC$, in the reverse order. That is, we could first apply the pinching operator $CC$, as illustrated in the transition from the first shape to the second in Fig 6.19. The reader can see that the two side squashing processes in the triple-shield, in the first shape, are continued, creating the pinching, in the second shape. Then, we can apply the continuation operator $C$ to the central squashing, at the bottom of the second shape, till it indents as shown at the bottom of the third shape.

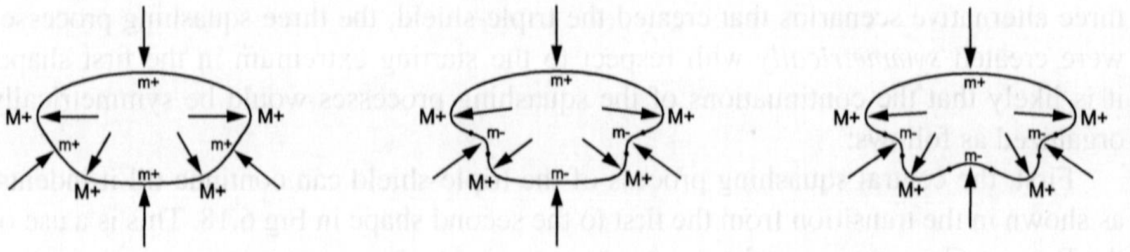


**Fig. 6.19.** 2-fold pinching $CC$ followed by continuation $C$.

What is important to emphasize is that in both scenarios, $C$ followed by $CC$, and $CC$ followed by $C$, all three squashing processes in the triple-shield were continued till they indented, and that this was done in a symmetrical way.

There is one more symmetrical case that remains: In this, *all three* squashing processes continue *simultaneously* till they indent, as illustrated in the transition from the left to the right shape in Fig 6.20.

I will call this action *3-fold pinching*. It is as if there are three fingers placed on the three minima of the left shape, and they press inwards together.

The situation will be considered as defining a new compound operator for the Process Grammar, which will be denoted by $CCC$. In the present example, this operator is realized by the following operation

$$CCCm^+ \; : \; m^+M^+m^+M^+m^+ \quad \longrightarrow \quad \overbrace{0m^-0} \; M^+ \; \overbrace{0m^-0} \; M^+ \; \overbrace{0m^-0}$$

where the three over-braces, in the codomain, indicate the three simultaneous uses of the continuation operator $C$. Thus, the new operator can be defined as follows:

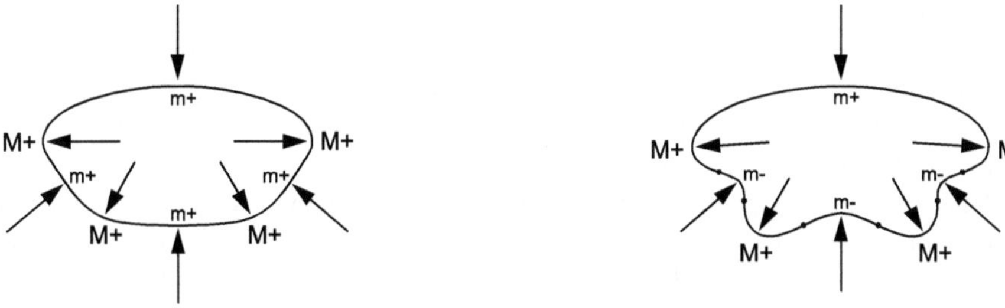


**Fig. 6.20.** 3-fold pinching.

**3-FOLD PINCHING OPERATOR (TYPE 1)** $CCC$

**Applying the Process Grammar operator $CCC$ to a 5-fold part, which contains three compressive extrema (necessarily of the same sign), is the triple-simultaneous application of the continuation operator $C$ to those three compressive extrema. The operator is realized by two operations:**

**3-Fold Pinching of a Triple-Shield**

$$CCCm^+ \;:\; m^+M^+m^+M^+m^+ \;\longrightarrow\; \overbrace{0m^-0}\;M^+\;\overbrace{0m^-0}\;M^+\;\overbrace{0m^-0}$$

**3-Fold Pinching of a Triple-Bay**

$$CCCM^- \;:\; M^-m^-M^-m^-M^- \;\longrightarrow\; \overbrace{0M^+0}\;m^-\;\overbrace{0M^+0}\;m^-\;\overbrace{0M^+0}$$

Generalizations of this, where there are strings between the three compressive domain extrema, are also valid.

## 6.4 5-fold Radial-Undulation Operator $[C(6)B(5)]$

In the previous section, we started by considering 5-fold part-formation in which the bottom compressive extremum $m^+$ of the ellipse in Fig 6.17 underwent a transition to a triple-shield. Next we considered the set of alternative continuation operations that can be applied at the three compressive extrema of the triple-shield to continue their squashing processes till they indented.

Observe now that, when all three squashing processes were continued, the triple-shield was converted into what we call a **5-fold undulation**. This is a configuration of 6 zeros alternating with 5 penetrative extrema of alternating opposite signs, $M^+$ and $m^-$; i.e., implying penetrative processes of alternating opposite directions. Indeed, the reader can check that a 5-fold undulation was the final configuration in each of the continuation sequences considered in the previous section; i.e., in each of the figures *after* the first figure (Fig 6.17) in that section.

Thus, the entire shape development, considered in the previous section, started with the bottom compressive extremum $m^+$ of the ellipse in Fig 6.17, and ended with the 5-fold undulation in each of the subsequent figures. This overall transition is therefore shown in Fig 6.21.

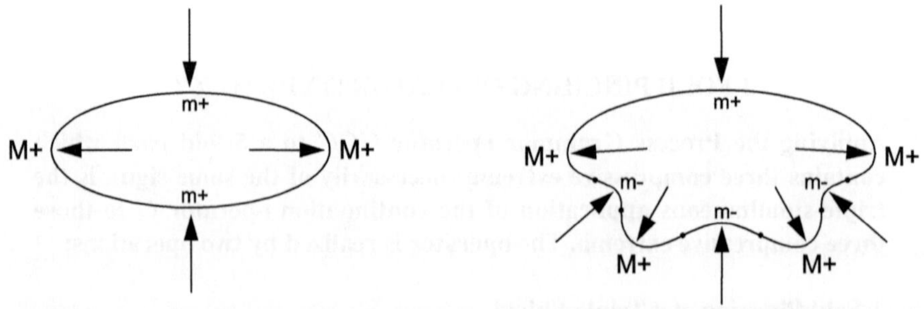


**Fig. 6.21.** Transition from a compressive extremum $m^+$ to a 5-fold undulation.

What we want to define now is the operator that goes *directly* from the starting $m^+$ extremum to the final 5-fold undulation, without going through the intervening stages; i.e., exactly as shown in Fig 6.21. In this direct transition, it must be necessarily the case that there is a single transition-state which contains a singularity that is the coalescence of the six zeros and five extrema of the final created undulation. Observe that this transition singularity is necessarily a zero minimum.

Thus we deduce that, via this zero minimum, the starting $m^+$ undergoes two *simultaneous* bifurcations: a bifurcation into five extrema, and a bifurcation into six zeros.

We will understand these two bifurcations as produced by two operators in the Process Grammar: The bifurcation into five extrema is produced by the 5-fold process-bifurcation operator $B(5)$. The bifurcation into six zeros is produced by a new operator $C(6)$, which we will now add to the Process Grammar.

The reader will recall that the operator $C$, given in Chapter 3, expresses the *continuation* of a single process, as the bifurcation of two zeros, on the sides of the single process. It can be notated as $C(2)$. Also, we have seen that the operator $C(3)$, introduced in sections 4.5 and 4.6, can be regarded as expressing the continuation of two processes, as the bifurcation of three zeros, on the sides of the two processes. Furthermore, these two processes are going in opposite directions. Also, we have seen that the operator $C(4)$, introduced in section 5.12, can be regarded as expressing the continuation of three processes, as the bifurcation of four zeros, on the sides of the three processes. Furthermore, these three processes are in alternating opposite directions. Correspondingly, the new operator $C(6)$ can be regarded as expressing the continuation of five processes, as the bifurcation of six zeros, on the sides of the five processes. Furthermore, these five processes are in alternating opposite directions, as we see in Fig 6.21.

Therefore, based on our analysis, the transition from the left shape to the right shape, in Fig 6.21, is given by the simultaneous application of the two operators, $C(6)$ and $B(5)$. This simultaneous application will be denoted by $[C(6)B(5)]$, which is a new operator to be also added to the Process Grammar. Recall, from Notation 4.1 (page 111) that the square brackets, [ ], mean that the two component operators are applied *simultaneously to the same point.*

Thus, the new operator $[C(6)B(5)]$ must describe the transition from the bottom $m^+$, in the first shape in Fig 6.21, to the configuration $0m^-0M^+0m^-0M^+0m^-0$ of *eleven* singularities in the second shape; i.e., the transition is an *11-fold bifurcation.* Therefore the application of the new operator $[C(6)B(5)]$ to the $m^+$ in the first shape is given by the following new operation:

$$[C(6)B(5)]m^+ \; : \; m^+ \quad \longrightarrow \quad 0m^-0M^+0m^-0M^+0m^-0 \tag{6.2}$$

It is easy to check that, when the operator $[C(6)B(5)]$ is applied to the dual extremum $M^-$, it gives the dual of the configuration of the above eleven singularities. That is, it is given by the following new operation:

$$[C(6)B(5)]M^- \; : \; M^- \quad \longrightarrow \quad 0M^+0m^-0M^+0m^-0M^+0 \tag{6.3}$$

Observe that the new operator $[C(6)B(5)]$ conforms to Definition 4.15 (page 122) of an *$n$-fold radial-undulation operator*. In particular, it is the *5-fold radial-undulation operator*. The term *radial-undulation* means that, under the operator, all 11 singularities of the undulation (6 zeros and 5 opposite penetrative extrema) *radiate* out simultaneously from a *single* central application-point, the initial compressive extremum.

**5-FOLD RADIAL-UNDULATION OPERATOR $[C(6)B(5)]$**

**The operator $[C(6)B(5)]$ is the *simultaneous* application of two operators, the 6-fold zero-bifurcation operator $C(6)$ and the 5-fold extremum-bifurcation operator $B(5)$, to the *same application-point*, a compressive extremum. It is realized by the following two operations:**

**5-fold Radial Undulation from a Squashing Process**

$$[C(6)B(5)]m^+ \; : \; m^+ \quad \longrightarrow \quad 0m^-0M^+0m^-0M^+0m^-0$$

**5-fold Radial Undulation from a Resistance Process**

$$[C(6)B(5)]M^- \; : \; M^- \quad \longrightarrow \quad 0M^+0m^-0M^+0m^-0M^+0$$

## 6.5 Double-Continuation of Complete Double-Compressive Parts

When we studied the continuation operators, $C$, $CC$, $CCC$, in section 6.3, we applied them to a 5-fold part consisting of *three* compressive extrema and *two* penetrative extrema. In the case of positive curvature, such a part is a triple-shield; and in the case of negative curvature, such a part is a triple-bay. We can call such a structure, a **triple-compressive part**.

However, the operator $CC$ can be applied also to a 5-fold part consisting of *two* compressive extrema and *three* penetrative extrema. In the case of positive curvature, such a part is a complete double-shield; and in the case of negative curvature, such a part is a complete double-bay. We can call such a structure, a **complete double-compressive part**.

It is important to understand the issues involved. To do so, recall first that, in section 6.3, our consideration of the continuation operators, $C$, $CC$, $CCC$, began by the formation of a 5-fold part from a single *compressive* extremum, a *positive minimum* $m^+$, as illustrated in Fig 6.17 (page 217).

In the present case, our consideration of the operator, $CC$, will begin by the formation of a 5-fold part from a single *penetrative* extremum, a *negative minimum* $m^-$. This is illustrated in Fig 6.22. The left shape shows the indentation process associated with the initial $m^-$ extremum at the top of the shape. The transition from the left shape to the right shape is given by applying any one of the three scenarios, $\pi^{\downarrow}\pi^{\uparrow}\emptyset^-$, $B_{-3} \circ B_3 m^-$, or $B_5 m^-$. The result is a *complete double-bay*.

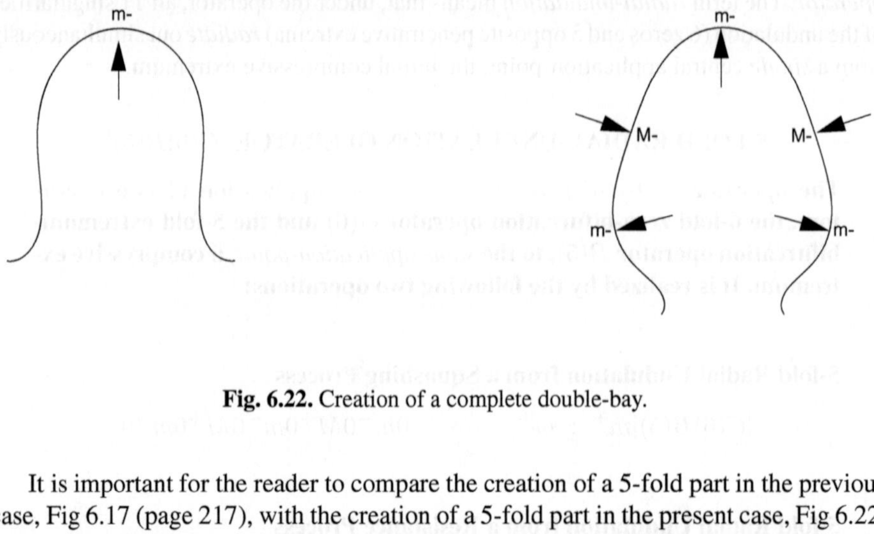


**Fig. 6.22.** Creation of a complete double-bay.

It is important for the reader to compare the creation of a 5-fold part in the previous case, Fig 6.17 (page 217), with the creation of a 5-fold part in the present case, Fig 6.22. Both start with a single *minimum*. In the previous case, the minimum is *compressive*. In the present case, the minimum is *penetrative*.

The crucial consequence is this: In the previous case, the resulting 5-fold part consists of *three* compressive extrema and *two* penetrative extrema. In the present case, the resulting 5-fold part consists of *two* compressive extrema and *three* penetrative extrema.

Thus, an important fact, is that, in the previous case, each of the three operators $C$, $CC$, $CCC$, has an individual application that is symmetric with respect to the extrema (the case of $C$ being to the central extremum). However, in the present case, only the operator $CC$ has an individual symmetric application. The latter application is illustrated in Fig 6.23. The first shape in this figure is the 5-fold part that was produced in the above transition, Fig 6.22; that is, the complete double-bay. We now apply the 2-fold pinching operator $CC$ to the two side $M^-$ extrema in the double-bay; thus making their two arrows continue till they indent, as shown in the second shape in Fig 6.23. That is, the transition is given by the operation $CCM^-$. The result is what our theory calls the *pinched complete double-bay*.

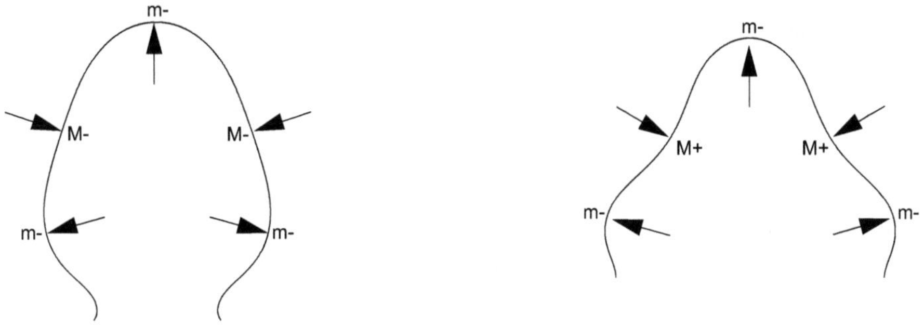


**Fig. 6.23.** Application of the 2-fold pinching operation $CCM^-$.

It is worth seeing this transition in relation to the figures shown on page 205. The transition from the single indentation to the complete double-bay was shown in the third level on that page. Then the transition from the complete double-bay to the pinched complete double-bay can be understood as going from the complete double-bay in this third level, downward to the pinched complete double-bay in the fourth level; i.e., directly downward from the right shape in the third level to the right shape in the fourth level.

Also, we observe that the dual situation is the application of $CCm^+$ to a complete double-shield, thus producing a pinched complete double-shield.

## 6.6 The $n$-fold Process-Bifurcation Operators $B(n)$ with the Odd Numbers $n$

Several of the Process Grammar operations are given by operators of the form $B(n)$, where $n$ is an odd number. For example, several of the Process Grammar operations we have described so far are given by operators of that form, i.e., the operations based on the $B(3)$ and $B(5)$ operators. Furthermore, the Process Grammar contains process-bifurcation operators $B(n)$ with all odd numbers $n$. We are now going to see important *general* properties of those operators.

First consider the case of applying an operator $B(n)$, with odd number $n$, to a protrusion extremum $M^+$. It has this effect:

For odd number $n$:

$$B(n)M^+ \quad : \quad M^+ \quad \longrightarrow \quad \overbrace{M^+m^+ \ldots\ldots\ldots\ldots\ldots\ldots}^{(n-1)/2 \text{ copies of } M^+m^+} M^+ \tag{6.4}$$

That is, the codomain is a singularity-configuration consisting of $n$ extrema, and this configuration is $(n-1)/2$ successive copies of $M^+m^+$ followed by $M^+$.

This means that the codomain is a set of $(n-1)/2$ adjacent shields along the curve. Our theory calls this structure, an $[(n-1)/2]$-fold shield. Furthermore, since the codomain string begins and ends with a $M^+$ extremum, our theory calls that structure a *complete* $[(n-1)/2]$-fold shield.

Now, to illustrate, we give examples of expression (6.4), for the values $n = 3, n = 5, n = 7, n = 9$:

$$B(3)M^+ \quad : \quad M^+ \quad \longrightarrow \quad M^+m^+M^+$$

That is, $B(3)M^+$ gives a complete shield.

$$B(5)M^+ \quad : \quad M^+ \quad \longrightarrow \quad M^+m^+M^+m^+M^+$$

That is, $B(5)M^+$ gives a complete double-shield.

$$B(7)M^+ \quad : \quad M^+ \quad \longrightarrow \quad M^+m^+M^+m^+M^+m^+M^+$$

That is, $B(7)M^+$ gives a complete triple-shield.

$$B(9)M^+ \quad : \quad M^+ \quad \longrightarrow \quad M^+m^+M^+m^+M^+m^+M^+m^+M^+$$

That is, $B(9)M^+$ gives a complete 4-fold shield.

Now observe that applying an operator $B(n)$, with odd number $n$, to a protrusion extremum $M^+$, gives the same result as applying the $B(3)$ operator $(n-1)/2$ times, starting with the same extremum.

For example, consider the transitions we have just listed:

$$\begin{array}{lcccc} B(3)M^+ & : & M^+ & \longrightarrow & M^+m^+M^+ \\ B(5)M^+ & : & M^+ & \longrightarrow & M^+m^+M^+m^+M^+ \\ B(7)M^+ & : & M^+ & \longrightarrow & M^+m^+M^+m^+M^+m^+M^+ \\ B(9)M^+ & : & M^+ & \longrightarrow & M^+m^+M^+m^+M^+m^+M^+m^+M^+ \end{array}$$

Observe that we can go from any one of these codomains to the next by applying the $B(3)$ operator, to any extremum in that codomain. That is, we can apply either the $Bm^+$ operation or the $BM^+$ operation, to get from any one of these codomains to the next. Note, however, that the operation that lead to the first codomain was $BM^+$.

Now let us consider the case of applying an operator $B(n)$, with odd number $n$, to the shield extremum $m^+$. It has this effect:

For odd number $n$:

$$B(n)m^+ \quad : \quad m^+ \quad \longrightarrow \quad \overbrace{m^+M^+ \ldots\ldots\ldots\ldots}^{(n-1)/2 \; copies \; of \; m^+M^+} m^+ \tag{6.5}$$

That is, the codomain is a singularity-configuration consisting of $n$ extrema, and this configuration is $(n-1)/2$ successive copies of $m^+M^+$ followed by $m^+$.

This means that the codomain is a set of $1 + (n-1)/2$ adjacent shields along the curve. Our theory calls this structure, a $[1 + (n-1)/2]$-fold shield. However, since the codomain string begins and ends with a $m^+$ extremum, it is a $[1+(n-1)/2]$-fold shield without the two completing $M^+$ end-points. However, since the application-point is $m^+$, an existing singularity to the left and also to the right of the application-point must both be $M^+$. So, if there is a singularity to the left and right of the application-point, the shape already has a complete shield in that region, and the result of applying the operation $B(n)m^+$, to its central extremum, will be a complete $[1 + (n-1)/2]$-fold shield in the shape, even though the two $M^+$ end-points of that complete $[1+(n-1)/2]$-fold shield are not within the codomain of $B(n)m^+$, but are adjacent to the codomain because they were adjacent to the application point.

Based on the discussion so far in this section, we have seen the following remarkable fact: Given an odd number $n$, when $B(n)$ is applied to a protrusion extremum $M^+$, it results in an $[(n-1)/2]$-fold shield; but, when the same operator $B(n)$ (i.e., with the same odd number $n$) is applied to the shield extremum $m^+$, it results in an extra shield, that is, a $[1+(n-1)/2]$-fold shield.

Now, to illustrate, we give examples of expression (6.5), for the values $n=3$, $n=5$, $n=7$, $n=9$:

$$B(3)m^+ \quad : \quad m^+ \quad \longrightarrow \quad m^+M^+m^+$$

That is, $B(3)m^+$ gives a double-shield, whereas $B(3)M^+$ gives a single complete shield.

$$B(5)m^+ \quad : \quad m^+ \quad \longrightarrow \quad m^+M^+m^+M^+m^+$$

That is, $B(5)m^+$ gives a triple-shield, whereas $B(5)M^+$ gives a complete double-shield.

$$B(7)m^+ \quad : \quad m^+ \quad \longrightarrow \quad m^+M^+m^+M^+m^+M^+m^+$$

That is, $B(7)m^+$ gives a 4-fold shield, whereas $B(7)M^+$ gives a complete triple-shield.

$$B(9)m^+ \quad : \quad m^+ \quad \longrightarrow \quad m^+M^+m^+M^+m^+M^+m^+M^+m^+$$

That is, $B(9)m^+$ gives a 5-fold shield, whereas $B(7)M^+$ gives a complete 4-fold shield.

Now observe that applying an operator $B(n)$, with odd number $n$, to a shield extremum $m^+$, gives the same result as applying the $B(3)$ operator $(n-1)/2$ times, starting with the same extremum. That is, even though the same number $(n-1)/2$ of applications of $B(3)$ to a protrusion extremum $M^+$ gives the same result as the $B(n)M^+$ operation, i.e., a complete $[(n-1)/2]$-fold shield, this same number $(n-1)/2$ of applications of $B(3)$ to a shield extremum $m^+$ gives the same result as the $B(n)m^+$ operation, that is, an extra shield, i.e., a $[1+(n-1)/2]$-fold shield.

Given the discussion so far in this section, it is easy to deduce the equivalent properties of the application of the operators $B(n)$, with odd numbers $n$, to the dual extrema $m^-$ and $M^-$.

## THE $n$-FOLD PROCESS-BIFURCATION OPERATORS $B(n)$ WITH ODD NUMBERS $n$

**Applying operator $B(n)$, with odd number $n$, to a protrusion extremum $M^+$ is this**

$$B(n)M^+ \quad : \quad M^+ \quad \longrightarrow \quad \overbrace{M^+m^+ \ldots\ldots\ldots\ldots\ldots\ldots\ldots M^+}^{(n-1)/2 \text{ copies of } M^+m^+}$$

**That is, the codomain is a complete $[(n-1)/2]$-fold shield.**

**Applying operator $B(n)$, with odd number $n$, to a shield extremum $m^+$ is this**

$$B(n)m^+ \quad : \quad m^+ \quad \longrightarrow \quad \overbrace{m^+M^+ \ldots\ldots\ldots\ldots\ldots\ldots\ldots m^+}^{(n-1)/2 \text{ copies of } m^+M^+}$$

**That is, the codomain is a $[1+(n-1)/2]$-fold shield, without the two completing $M^+$ end-points. However, if there is a singularity to the left and right of the application-point $m^+$, the result of applying the operation $B(n)m^+$ will be a complete $[1+(n-1)/2]$-fold shield in the shape.**

**Applying operator $B(n)$, with odd number $n$, to an indentation extremum $m^-$ is this**

$$B(n)m^- \quad : \quad m^- \quad \longrightarrow \quad \overbrace{m^-M^- \ldots\ldots\ldots\ldots\ldots\ldots\ldots m^-}^{(n-1)/2 \text{ copies of } m^-M^-}$$

**That is, the codomain is a complete $[(n-1)/2]$-fold bay.**

**Applying operator $B(n)$, with odd number $n$, to a bay extremum $M^-$ is this**

$$B(n)M^- \quad : \quad M^- \quad \longrightarrow \quad \overbrace{M^-m^- \ldots\ldots\ldots\ldots\ldots\ldots\ldots M^-}^{(n-1)/2 \text{ copies of } M^-m^-}$$

**That is, the codomain is a $[1+(n-1)/2]$-fold bay, without the two completing $m^-$ end-points. However, if there is a singularity to the left and right of the application-point $M^-$, the result of applying the operation $B(n)M^-$ will be a complete $[1+(n-1)/2]$-fold bay in the shape.**

**All cases of $B(n)$, with odd number $n$, give the same result as applying $B(3)$ a total of $[(n-1)/2]$ times, starting with the same extremum.**

## 6.7 Additional Process Grammar Operators

This book has shown several of the Process Grammar operators. Based on this, one can understand the nature of other Process Grammar operators. For example, there are $B(n)$ operators, with any odd number $n$. We saw their general properties in section 6.6. Also, there are $B(n)$ operators, with any even number $n$. For example, we have seen the nature of the pair-creation operator $B(2)$; that is, the fact that it is applied to points on spirals. There is, of course, a 4-fold process-bifurcation operator $B(4)$; and, based on our description of $B(2)$, the reader can understand that $B(4)$ is also applied to points on spirals. Furthermore, there are continuation operators $C(n)$, with any number $n$, and therefore radial undulation operators $C(n)B(n-1)$, with any number $n$. What this book has shown is the nature of all the above types of operators, and the crucial fact that they give the important causal structure, including the design intent, in biological morphology and manufacturing design.

The remainder of the book is going to show how these operators are structurally related to each other via the singularity theory of shape I invented called Interactive Singularity Theory. A crucial aspect of Interactive Singularity Theory is that the relationship it gives between operators matches the causal structure of the operators given by the PISA axes. According to our theory, this causal structure is fundamental to understanding biological morphology and manufacturing design. In contrast to Interactive Singularity Theory, other singularity theories of shape fail to match the fundamentally important causal structure of biological morphology and manufacturing design.

## 6.8 Important New CAD Operations

This book has described, in detail, 30 Process Grammar operations. These operations are listed on pages 230–231. In addition, sections 6.6–6.7 have shown how certain properties of these operations can be generalized to define additional operations. Therefore, this book defines over 30 Process Grammar operations.

There is the following important consequence:

## IMPORTANT NEW CAD OPERATIONS

**We have shown that the Process Grammar operations capture important *intended functions*, in both biological morphology and manufacturing design. Consequently, the Process Grammar operations define new CAD operations that achieve fundamentally important *design intents*. Therefore, since this book has defined over 30 Process Grammar operations, it has defined over 30 important new CAD operations.**

**For example, the process-bifurcation operators $B(n)$, with odd numbers $n$, achieve the shield structures, which our theory has shown are fundamentally important to biological morphology and vehicle design.**

**In a CAD program, these operations would work in the following way:**

**(1) Given a stage in the design, the designer would request the CAD program to show the curvature extrema on the shape at that stage.**

**(2) The designer would then select one of the curvature extrema, and from the CAD program menu, the designer would select, to apply at that point, the particular $B(n)$ operator that would give the shield structure that achieves the design intent, as well as the required penetrative structure.**

**(3) The CAD program would also show, at each of the extrema, the PISA vectors that give the design intents, i.e., the crucial functions of the extrema.**

**A similar design process would exist for the application of the continuation operators; i.e., the designer would request the CAD program to show the curvature extrema, and then the designer would select particular curvature extrema, e.g., two compressive extrema at which pinching is needed, and would select the particular continuation operator, e.g., the CC operator, to apply to those curvature extrema. Again the program would exhibit the PISA vectors that give the design intents.**

**In the case of the process-bifurcation operators $B(n)$, with even numbers $n$, again the designer would request the CAD program to show the curvature extrema, and then the designer would select particular points on spirals, e.g., two points on a symmetrical pair of spirals, and would select the particular operator, e.g., a symmetrically-paired pair-creation operator, to apply to those spiral points. Again the program would exhibit the PISA vectors that give the design intents.**

## PROCESS-GRAMMAR SO FAR

### Process-Continuation

| | | | |
|---|---|---|---|
| $Cm^+$: | $m^+ \longrightarrow 0m^-0$ | (squashing continues till it indents) |
| $CM^-$: | $M^- \longrightarrow 0M^+0$ | (resistance continues till it protrudes) |

### 3-Fold Process-Bifurcation

Pure 3-fold Process-Bifurcation

| | | |
|---|---|---|
| $BM^+$: | $M^+ \longrightarrow M^+m^+M^+$ | (shield-formation) |
| $Bm^-$: | $m^- \longrightarrow m^-M^-m^-$ | (bay-formation) |
| $Bm^+$: | $m^+ \longrightarrow m^+M^+m^+$ | (breaking-through of a protrusion) |
| $BM^-$: | $M^- \longrightarrow M^-m^-M^-$ | (breaking-through of an indentation) |

3-fold Radial-Undulation

| | | |
|---|---|---|
| $[C(4)B]m^+$: | $m^+ \longrightarrow 0m^-0M^+0m^-0$ | (zone-restricted protrusion formation) |
| $[C(4)B]M^-$: | $M^- \longrightarrow 0M^+0m^-0M^+0$ | (zone-restricted indentation formation) |

### Pair-Creation

Pure 2-fold Process-Bifurcation

| | | |
|---|---|---|
| $\pi\emptyset^{\uparrow+}$: | $\emptyset^{\uparrow+} \longrightarrow M^+m^+$ | (application-point: upward positive) |
| $\pi\emptyset^{\uparrow-}$: | $\emptyset^{\uparrow-} \longrightarrow M^-m^-$ | (application-point: upward negative) |
| $\pi\emptyset^{\downarrow+}$: | $\emptyset^{\downarrow+} \longrightarrow m^+M^+$ | (application-point: downward positive) |
| $\pi\emptyset^{\downarrow-}$: | $\emptyset^{\downarrow-} \longrightarrow m^-M^-$ | (application-point: downward negative) |

2-fold Radial-Undulation

| | | |
|---|---|---|
| $[C(3)\pi]0^{\uparrow}$: | $0^{\uparrow} \longrightarrow 0M^+0m^-0$ | (application-point: upward zero) |
| $[C(3)\pi]0^{\downarrow}$: | $0^{\downarrow} \longrightarrow 0m^-0M^+0$ | (application-point: downward zero) |

### 2-Fold Pinching (Type 1)

| | | |
|---|---|---|
| $CCm^+$: | $m^+\,M^+\,m^+ \longrightarrow \overbrace{0m^-0}\;M^+\;\overbrace{0m^-0}$ | (pinching a double-shield) |
| $CCM^-$: | $M^-\,m^-\,M^- \longrightarrow \underbrace{0M^+0}\;m^-\;\underbrace{0M^+0}$ | (pinching a double-bay) |

**Symmetrically-Paired Pair-Creation**

Upward-Downward
Negative, Positive and Zero Operations:
($X$ is $M^-$ or $0M^+0$)

$$\pi^\uparrow\pi^\downarrow\emptyset^- : \quad \emptyset^- \; X \; \emptyset^- \quad \longrightarrow \quad \overbrace{M^-m^-} \; X \; \overbrace{m^-M^-}$$

$$\pi^\uparrow\pi^\downarrow\emptyset^+ : \quad \emptyset^+ \; M^+ \; \emptyset^+ \quad \longrightarrow \quad \overbrace{M^+m^+} \; M^+ \; \overbrace{m^+M^+}$$

$$[C(3)\pi^\uparrow][C(3)\pi^\downarrow]0 \; : \quad 0^\uparrow \; M^+ \; 0^\downarrow \quad \longrightarrow \quad \overbrace{0M^+0m^-0} \; M^+ \; \overbrace{0m^-0M^+0}$$

Downward-Upward
Positive, Negative and Zero Operations:
($X$ is $m^+$ or $0m^-0$)

$$\pi^\downarrow\pi^\uparrow\emptyset^+ : \quad \emptyset^+ \; X \; \emptyset^+ \quad \longrightarrow \quad \overbrace{m^+M^+} \; X \; \overbrace{M^+m^+}$$

$$\pi^\downarrow\pi^\uparrow\emptyset^- : \quad \emptyset^- \; m^- \; \emptyset^- \quad \longrightarrow \quad \overbrace{m^-M^-} \; m^- \; \overbrace{M^-m^-}$$

$$[C(3)\pi^\downarrow][C(3)\pi^\uparrow]0 \; : \quad 0^\downarrow \; m^- \; 0^\uparrow \quad \longrightarrow \quad \overbrace{0m^-0M^+0} \; m^- \; \overbrace{0M^+0m^-0}$$

**5-Fold Process-Bifurcation**

Pure 5-fold Process-Bifurcation

$B(5)m^+$: $m^+ \longrightarrow m^+M^+m^+M^+m^+$ (radial-formation of a triple-shield)
$B(5)m^-$: $m^- \longrightarrow m^-M^-m^-M^-m^-$ (radial-formation of a complete double-bay)
$B(5)M^-$: $M^- \longrightarrow M^-m^-M^-m^-M^-$ (radial-formation of a triple-bay)
$B(5)M^+$: $M^+ \longrightarrow M^+m^+M^+m^+M^+$ (radial-formation of a complete double-shield)

5-fold Radial-Undulation

$[C(6)B(5)]m^+$: $m^+ \longrightarrow 0m^-0M^+0m^-0M^+0m^-0$ (from a squashing process)
$[C(6)B(5)]M^-$: $M^- \longrightarrow 0M^+0m^-0M^+0m^-0M^+0$ (from a resistance process)

**3-Fold Pinching (Type 1)**

$CCCm^+$: $m^+M^+m^+M^+m^+ \longrightarrow \overbrace{0m^-0} \; M^+ \; \overbrace{0m^-0} \; M^+ \; \overbrace{0m^-0}$ (of a triple-shield)

$CCCM^-$: $M^-m^-M^-m^-M^- \longrightarrow \overbrace{0M^+0} \; m^- \; \overbrace{0M^+0} \; m^- \; \overbrace{0M^+0}$ (of a triple-bay)

**In addition, Sections 6.6–6.7 have shown how certain properties of the Process Grammar operations on pages 230–231 can be generalized to define more Process Grammar operations.**

# 7

# The Inadequacy of Catastrophe Theory for Morphology

## 7.1 Introduction

As I have said, the standard singularity theories of shape fail to give an understanding of biological morphology and manufacturing design. As an example of this, we will show how René Thom's Catastrophe Theory fails. However, I will first state this:

> **I regard René Thom's Catastrophe Theory as one of the most important branches of mathematics. It lead to a wonderful development in mathematics, as well as wonderful applications in physics.**
>
> **The only problem is that it failed to achieve one of Thom's main goals, to model spatial morphology, and in particular biological morphology.**

A basic way of understanding the reasons why Thom's theory failed to model morphology is that Thom did not understand morphology within the framework of my New Foundations to Geometry:

**Within my New Foundations to Geometry:**
**Morphology is an example of**
**my Mathematical Theory of Memory Storage.**

In particular, this means that, according to my New Foundations to Geometry, the laws of morphology must include my laws of *recoverability*. These laws lead to the Symmetry-Curvature Duality Theorem (i.e., as a partitioning rule between asymmetries and symmetries), the Interaction Principle, the PISA method of defining symmetry axes, the use of iso-regular groups, etc. It is the absence of these laws from Thom's theory that prevented it from reaching his goal of providing a theory of spatial morphology.

There is also another related and important way of understanding why Thom's theory failed to reach his goal. According to my New Foundations to Geometry, shape is equivalent to causal explanation. A basic fact that we will see about Thom's mathematical formulation of spatial morphology is that it is not compatible with plausible causal explanation. That is, it fails to fulfill the following condition:

**CAUSAL COMPATIBILITY**

**We will say that a technique of shape-analysis is** *causally compatible* **if it is compatible with the plausible causal explanation that defines the shape.**

In order to create a singularity theory that fulfills this condition, I invented what I call *Interactive Singularity Theory*. This will be elaborated in the following chapters of the book. I argue that this is the appropriate singularity theory for spatial morphology.

It is important to understand that conventional singularity theories of shape do not accord with the real causal structure that explains the shape. Most importantly, they fail to model compressive extrema which our theory gives as fundamental to morphology. Consequently, they fail to understand the crucial role that our theory gives of the causal structure of process-bifurcation, which is dependent on the oppositional role of compressive extrema; and they fail to understand the crucial role that our theory gives of the causal structure of process-continuation, which is also dependent on the role of compressive extrema.

An important aspect of Interactive Singularity Theory is that it accords with the fundamental role of compressive extrema in process-bifurcation as well as process-continuation.

Despite the causal incompatibility of Thom's Catastrophe Theory, it created methods and results that, when used within Interactive Singularity Theory, are profoundly valuable. For example, the space of functions in the unfolding of an interactive singularity contains simultaneously *two* of the separatrix structures created by Thom. However, in doing so, the central issue becomes the *interaction* between the two structures in organizing the space – an issue that Thom did not consider. Furthermore, the space, and morphological movements through it, become organized in a much more complex way than Thom envisioned. In fact, this particular form of complexity is crucial for ensuring the causal compatibility of that singularity structure.

The purpose of the present chapter is to examine the relation between the Process Grammar and Catastrophe Theory. This will lead to showing why Catastrophe Theory cannot account for the crucial spatial morphologies: biological morphology and manufacturing design. After that, we will devote chapters 8 – 13 to elaborating the structure of Interactive Singularity Theory.

## 7.2 Process-Bifurcations and Continuations

The operations of the Process Grammar are of two types: *process-bifurcations* and *process-continuations*. The first type is coded by the bifurcation of curvature extrema, and the second type is coded by the bifurcation of curvature zeros.

The present chapter will look at the valuable things that Catastrophe Theory says generally about the bifurcation of extrema of functions. However, we shall eventually find the following serious problem that Catastrophe Theory has with respect to spatial morphology: Because Catastrophe Theory does not classify extrema in accord with our 4-fold classification, $M^+, m^-, m^+, M^-$, it does not structure the system of extremum-bifurcations in a way that ensures their compatibility with plausible causal explanation. In fact:

> **Catastrophe Theory does not recognize the following basic fact that is recognized in the Process Grammar:**
>
> **To ensure compatibility with plausible causal explanation, the system of extremum-bifurcations must be structured by a system of zero-bifurcations, thus revealing the powerful relation between process-bifurcations and process-continuations, and the fundamental role of compressive extrema in both of these.**

## 7.3 Canonical forms in Catastrophe Theory

Before beginning our argument, this section gives a overview of canonical forms in Catastrophe Theory, together with the basic concepts on which they are based. Readers who are familiar with these can go directly to section 7.4 page 250 to begin our argument. We will try to make parts of the present section as intuitive as possible, to help readers who are unfamiliar with these concepts. Furthermore, this description will concentrate on *one-variable functions*.

Consider a smooth one-variable function

$$f : \begin{cases} \mathbb{R} \longrightarrow \mathbb{R} \\ t \longmapsto f(t) \end{cases}$$

A point, $t$, in the domain of the function, is called a *critical point* of $f$ if the first derivative $Df|_t$ at $t$ is zero; i.e., the graph of $f$ has a horizontal tangent at $t$. For a 1-variable function such as $f$, the critical points occur at local maxima, minima, and inflection points, which are each illustrated in Fig 7.1. It is important to emphasize that the critical points are points on the *domain* axis $t$; for example, the points marked $t_1$, $t_2$, and $t_3$, in Fig 7.1. In contrast, the *critical value* of a critical point $t$ is the value $f(t)$ of the function at the critical point; i.e., the graph value on the *codomain* axis.

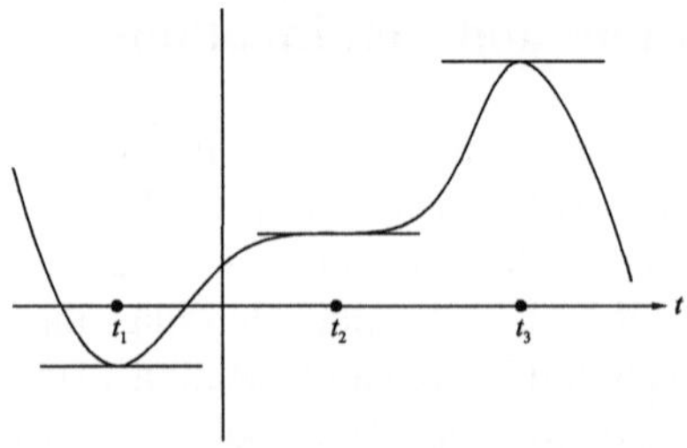


**Fig. 7.1.** Critical points of a 1-variable function.

A critical point $t$ is an individual point that is either the position of a *single* critical point or the position of *multiply coincident* critical points. This is illustrated as follows: Fig 7.2 shows, on the left, the graph of the function $t^2$, and shows, on the right, the graph of the function $t^4$. They both have a critical point (horizontal tangent) at the origin. Now the left function $t^2$ has only a single critical point at the origin. We can see this because its first derivative $2t$ has only one solution at the origin. In contrast, the right function $t^4$ has three coincident critical points at the origin. We can see this because its first derivative $4t^3$ has three solutions at the origin.

In the first case, that of a non-coincidence, the critical point is called *non-degenerate* or *Morse*. In the second case, that of a coincidence, the critical point is called *degenerate* or *non-Morse*. A critical point is called *$n$-fold degenerate* if its multiplicity as a critical point is $n$ (i.e., it is the position of $n$ coincident critical points). Thus returning to Fig 7.2, the critical point of the function $t^2$ is non-degenerate; whereas the critical point of the function $t^4$ is 3-fold degenerate. A smooth function whose critical points are all non-degenerate is called a *Morse function*; and a smooth function with at least one degenerate critical point is called a *non-Morse function*. Therefore, the function $t^2$ is Morse, whereas the function $t^4$ is non-Morse.

A simple test distinguishes between a non-degenerate and degenerate critical point: At a non-degenerate critical point $t$, the second derivative $D^2f|_t$ is non-zero; whereas at a degenerate critical point $t$, the second derivative $D^2f|_t$ is zero.

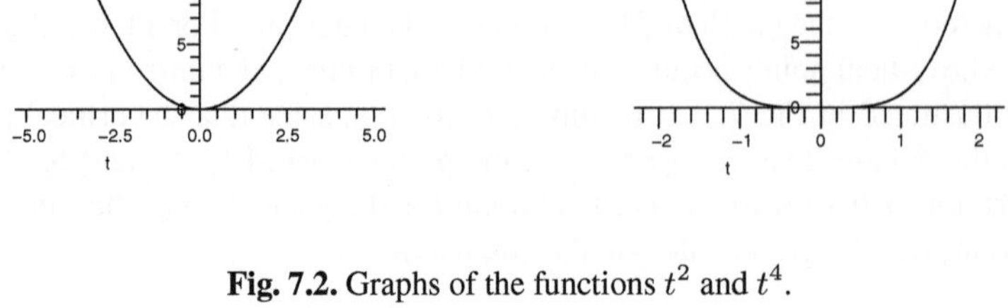


**Fig. 7.2.** Graphs of the functions $t^2$ and $t^4$.

The points in the domain of a smooth function $f : \mathbb{R} \longrightarrow \mathbb{R}$ can be classified into three types:

(1) **Non-Critical Points.**
(2) **Non-Degenerate Critical Points.**
(3) **Degenerate Critical Points.**

In a neighborhood of each of these three types of point, the function $f$ can be put in a *canonical form* that is *qualitatively equivalent* to $f$ in that neighborhood. The type of qualitative equivalence, used in Catastrophe Theory, is called *right-equivalence*. Right-equivalence preserves the *qualitative type* of a point, as classified in the above list. One defines right-equivalence as follows: A smooth function $f : \mathbb{R} \longrightarrow \mathbb{R}$ is called *right-equivalent*, at a point $t_1$ in the domain $\mathbb{R}$, to a smooth function $g : \mathbb{R} \longrightarrow \mathbb{R}$ at a point $t_2$ in the domain $\mathbb{R}$, if there exist, in the domain, an open neighborhood $U_1$ of $t_1$ and an open neighborhood $U_2$ of $t_2$, and a smooth reversible local change of coordinates $y : U_1 \longrightarrow U_2$ which sends $t_1$ to $t_2$, (i.e., $y$ is a local diffeomorphism), and a constant $z$, such that for all $t \in U_1$,

$$f(t) \;=\; g(y(t)) \;+\; z \tag{7.1}$$

The points $t_1$ and $t_2$ can be called the *base-points* respectively of $f$ and $g$, with respect to the equivalence. Thus, equation (7.1) says that, in their respective neighborhoods, $f$ is exactly the same as $g$, after the local change of domain coordinates and the constant shift $z$ along the co-domain axis, where obviously $z$ must simply be the difference $f(t_1) - g(t_2)$ between the value of the two functions at their respective base-points.

Based on this particular concept of *qualitative* equivalence, i.e. *right-equivalence*, Catastrophe Theory therefore gives the following as the three canonical forms that a smooth one-variable function $f : \mathbb{R} \longrightarrow \mathbb{R}$ can take around the three types of point listed above, respectively:

(1) **Non-Critical Point:** A smooth function $f : \mathbb{R} \longrightarrow \mathbb{R}$, at a non-critical point, is right-equivalent to the linear function $\pm t$, at the point 0.
(2) **Non-Degenerate Critical Point:** A smooth function $f : \mathbb{R} \longrightarrow \mathbb{R}$, at a non-degenerate critical point, is right-equivalent to the function $\pm t^2$, at the point 0.
(3) **Degenerate Critical Point:** For $k \geq 3$, a smooth function $f : \mathbb{R} \longrightarrow \mathbb{R}$, at a $k-1$ degenerate critical point, is right-equivalent to the function $\pm t^k$, at the point 0.

If one moves the base-point of $f$ to the origin, and shifts its codomain axis along itself so that $f(0) = 0$, then one can understand the above classification in terms of the following concept: The above classification follows from the fact that, around the base-point 0, there is a neighborhood in which $f$ is right-equivalent to the first non-zero term of its Taylor series expansion. In other words, the higher terms in the Taylor series expansion can be removed by a smooth change of coordinates around the base-point. (For this, one is of course assuming that $f$ has a non-zero Taylor series expansion).

Notice what the above classification (1)-(3) says about the two functions $t^2$ and $t^4$ shown in Fig 7.2. First it says that there is no smooth reversible local change of coordinates that can make the two functions the same as each other around their minima. This means that the minima of these two functions are *qualitatively different*. This reflects the fact which was already given: the minimum of $t^2$ is non-degenerate, whereas the minimum of $t^4$ is 3-fold degenerate.

In contrast, consider the remaining points of both functions. These points are all non-critical; i.e., they are all in category (1) of the above classification (1)-(3). Therefore, the two functions *are* qualitatively equivalent around all these other points. Now for the following comment:

**Throughout this book, the term equivalence, between functions, will mean qualitative equivalence, unless otherwise stated.**

**In Catastrophe Theory, the assumption is that, in morphology, the type of equivalence that gives qualitative equivalence is right-equivalence.**

**However, we will later show that this type of equivalence is not sufficient to model spatial morphology.**

Let us now consider the concept of structural stability in Catastrophe Theory. A smooth function $f : \mathbb{R} \longrightarrow \mathbb{R}$ is regarded as *locally structurally stable* at $t$ if, for *all* sufficiently small smooth functions $p : \mathbb{R} \longrightarrow \mathbb{R}$, the function $f$ and its perturbed version $f + p$ are right-equivalent to each other around $t$. That is, a re-parametrization of the domain will make the function and its perturbed version the same around $t$, up to the *possible addition of a constant*. Most significantly, this will mean that $f$ and $f + p$ are *qualitatively* the same around $t$, in the particular view of qualitative equivalence in Catastrophe Theory. A function that is locally structurally stable at all points $t$ in its domain is *globally structurally stable*. Usually, the context of discussion shows whether one is referring to local or global structural stability, and therefore the adjectives local and global are often omitted.

Concerning the classification of points $t$ into three types, the following results hold:

**(1) Given a non-critical point, its property of non-criticality is structurally stable.** That is, any sufficiently small perturbation of a function around a non-critical point will preserve the non-criticality. *However, as we shall see later, according to our theory, other singularity properties of that point could be non-structurally stable, and this is crucial to morphology.*

**(2) Given a non-degenerate critical point, its property of non-degenerate criticality is structurally stable.** That is, any sufficiently small perturbation of a function around a non-degenerate critical point will preserve its non-degenerate criticality. *However, as we shall see later, according to our theory, other singularity properties of that point could be non-structurally stable, and this is crucial to morphology.*

**(3) Given a degenerate critical point, its property of degenerate criticality is not structurally stable.** That is, there exist perturbations, no matter how small, of a function around an $n$-fold degenerate critical point, that will destroy its $n$-fold degeneracy.

To illustrate the above structural stability classifications given above by Catastrophe Theory, let us return to the function $t^4$ shown on the right in Fig 7.2. As we have seen, it has a *degenerate* critical point at the origin. Now add, to this function, an arbitrarily small perturbation of the form $\epsilon t^2$, thus giving the function:

$$t^4 + \epsilon t^2$$

Then, this new function still has a critical point at the origin, because its first derivative $4t^3 + 2\epsilon t$ is zero at the origin. However, the critical point has now become a *non-degenerate* critical point, as can easily be seen by the fact that the second derivative is $12t^2 + 2\epsilon$, which is non-zero at the origin. These facts hold true for the perturbed function, no matter how small the perturbation $\epsilon t^2$ contained in it. Therefore, we see that the degenerate critical point in $t^4$ is unstable, because the moment we add the perturbation $\epsilon t^2$, no matter how small, the degenerate critical point becomes a non-degenerate critical point. Thus we have just illustrated fact (3) in the above list: Degenerate critical points are unstable.

In contrast, consider the non-degenerate critical point at the origin of the perturbed version $t^4 + \epsilon t^2$. Any arbitrarily small perturbation of this function, will not alter the non-degeneracy of the critical point. For example, any sufficiently small change in $\epsilon$ will not make the critical point degenerate, because the second derivative $12t^2 + 2\epsilon$ will remain non-zero at the origin. This illustrates fact (2) above, that non-degenerate critical points are stable (with respect to their property of non-degenerate criticality).

Notice, by what was said earlier, the perturbed version $t^4 + \epsilon t^2$ is actually right-equivalent, in a neighborhood of the origin, to the function $\pm t^2$. In other words, understanding the expression $t^4 + \epsilon t^2$ as the Taylor series expansion of this perturbed function (around the origin), a local smooth change of coordinates will remove the higher-order term $t^4$ in this expansion, leaving only the lowest term $\epsilon t^2$, which is right-equivalent to

$t^2$. Thus, in a neighborhood of the origin, the perturbed version of the right function $t^4$ in Fig 7.2, will actually be right-equivalent to the left function $t^2$ (with a possible change of sign), but will not be right-equivalent to the right function.

Let us now turn to the concept of *transversality*. Consider two manifolds $M_1$ and $M_2$ in $R^n$; and suppose that they intersect at a point $p$. One says that they *intersect transversally at* $p$, if their two tangent spaces, at $p$, *together* generate the tangent space $R^n$ at $p$. Based on this, one then defines the concept of transverse intersection without reference to a particular point: One says that the two manifolds *intersect transversally* if either of the following two conditions are fulfilled: (1) They intersect transversally at all their points of intersection; or (2) they do not intersect at all. The definition is illustrated in Fig 7.3, where, in the first and third case, the curve intersects the horizontal axis transversally, whereas, in the second case, the intersection is non transversal.

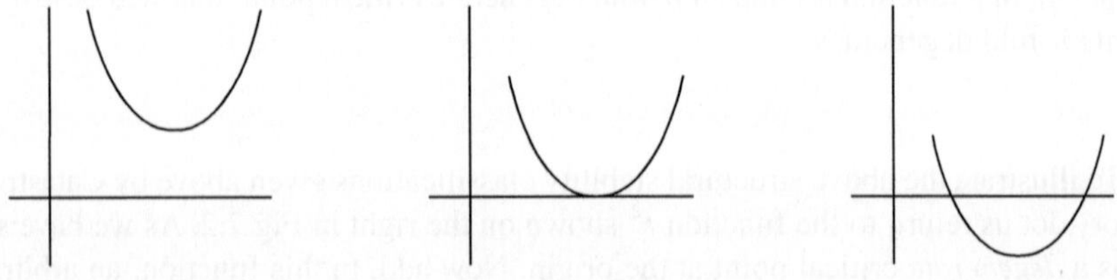


**Fig. 7.3.** Illustration of the definition of transverse intersection.

It is important to observe that transversality is a structurally stable property, as can be illustrated again in Fig 7.3. An arbitrarily small perturbation of the first and third curves will not alter the fact that their intersection with the horizontal axis is transversal. However, there are perturbations, no matter how small, of the second curve, that will change its intersection from being non-transversal to transversal.

We note also that the definition of the transversal intersection of two manifolds can be appropriately modified for the transversal intersection of a function with a manifold, and the transversal intersection of a function with another function.

Now we note that, according to Catastrophe Theory, the concept of right-equivalence and structural stability can be extended from individual functions, to families of functions. Define a family of functions as $F : \mathbb{R} \times \mathbb{R}^k \longrightarrow \mathbb{R}$, where $\mathbb{R}^k$ is the space that parameterizes the family, and each function in the family is of the form $F_c : \mathbb{R} \longrightarrow \mathbb{R}$ for each point $c$ in the parameter space $\mathbb{R}^k$. Then, two families

$$F, G : \mathbb{R} \times \mathbb{R}^k \longrightarrow \mathbb{R}$$

are *right-equivalent* if one can enforce a relationship between them that simply extends the definition given earlier, in expression (7.1), for right-equivalence between two individual functions. That is, one must ensure the following three types of changes of coordinates:

**Map $y$: Coordinate-Change of the domain $\mathbb{R}$ for each individual function.**
For each $c \in \mathbb{R}^k$, one must be able to have a smooth reversible change of coordinates $y_c$ on the domain of the individual function $F_c$ located at $c$, and one must be able to tie these individual coordinate-changes (for all $c$) smoothly together across the family by putting them all into a single overall smooth map $y : \mathbb{R} \times \mathbb{R}^k \longrightarrow \mathbb{R}$.

**Map $\phi$: Coordinate-change of the parameter space.**
This is a diffeomorphism $\phi$ of the parameter space $\mathbb{R}^k$ itself.

**Map $z$: Coordinate-shift of the codomain $\mathbb{R}$ for each individual function.**
For each $c \in \mathbb{R}^k$, one must be able to shift the co-domain of the individual function $F_c$ by a constant $z(c)$, and one must be able to tie these individual shifts in co-domain (for all $c$) smoothly together across the family by putting them all into a single overall smooth map $z : \mathbb{R}^k \longrightarrow \mathbb{R}$.

Then the two families of functions $F$ and $G$ are right-equivalent, if there exist three such functions $y$, $\phi$, $z$, defined in a neighborhood of 0 (for convenience, take the base-points to be 0), such that for all $(t, c) \in \mathbb{R} \times \mathbb{R}^k$ in the neighborhood,

$$F(t, c) \quad = \quad G(\, y_c(t),\ \phi(c)\,) \quad + \quad z(c) \tag{7.2}$$

The reader can see the analogy between this expression for families $F$ and $G$, and expression (7.1) for individual functions $f$ and $g$.

A crucial concept is that, just as right-equivalence between individual functions $f$ and $g$ ensures that they are what Catastrophe theory regards as *qualitatively* equivalent, the corresponding situation holds here: i.e., right-equivalence between the families of functions $F$ and $G$ ensures that they are what Catastrophe theory regards as *qualitatively* equivalent.

Now recall that the concept of qualitative equivalence of individual functions $f$ and $g$ allowed one to define the structural stability of an individual function $f$. That is, an individual smooth function $f : \mathbb{R} \longrightarrow \mathbb{R}$ is structurally stable if, for *all* sufficiently small smooth functions $p : \mathbb{R} \longrightarrow \mathbb{R}$, the function $f$ and its perturbed version $f + p$ are equivalent to each other. In other words, the perturbations of $f$ do not loose the qualitative structure of $f$.

Similarly, the concept of qualitative equivalence of families of functions $F$ and $G$ allows one to define the structural stability of a family $F$. That is, a smooth family of functions $F : \mathbb{R} \times \mathbb{R}^k \longrightarrow \mathbb{R}$ is structurally stable if, for *all* sufficiently small smooth families $P : \mathbb{R} \times \mathbb{R}^k \longrightarrow \mathbb{R}$, the family $F$ and its perturbed version $F+P$ are equivalent to each other. In other words, the perturbations of $F$ do not loose the qualitative structure of $F$.

A fundamental concern of Catastrophe Theory is the structure of qualitative changes that can happen to a degenerate critical point under all possible perturbations. This is captured by a family of functions, $F : \mathbb{R} \times \mathbb{R}^k \longrightarrow \mathbb{R}$, in which the particular function $f_0 : \mathbb{R} \longrightarrow \mathbb{R}$ at the origin 0 of the parameter space $\mathbb{R}^k$ is the canonical form of the degenerate critical point, and in which movement to other members of the family reveal qualitative changes that can happen to the degenerate critical point, in all possible perturbations. Such a family $F$ is called a *versal unfolding*, and a versal unfolding is called a *universal unfolding* if the dimension $k$ of the parameter space $\mathbb{R}^k$ is the lowest dimension needed to reveal all the qualitative changes that can happen to the degenerate critical point. Frequently, the parameter space $\mathbb{R}^k$ is called the *control space*, and the $k$ parameters are called the *control parameters*.

In the family, the particular member $f_0 : \mathbb{R} \longrightarrow \mathbb{R}$, that is the canonical form of the degenerate critical point, is sometimes called the *germ* of the unfolding. (More technically, the germ is the *class* of functions that are coincident with $f_0$ in some open neighborhood around its degenerate critical point.)

Consider an $n$-fold degenerate critical point. As stated earlier, the canonical form of this critical point is the function $\pm t^{n+1}$. In this section, we will omit mentioning the negative version $-t^{n+1}$, simply to cut down the amount of notation. Thus, the function $t^{n+1}$ becomes the germ in the universal unfolding of that critical point. Thom's theorem in Catastrophe Theory [39] establishes that a universal unfolding of the germ $t^{n+1}$ requires the dimension $k$ of the control space to be $n-1$, and the universal unfolding can be given as

$$F(t; a_1, \ldots a_{n-1}) \quad = \quad \overbrace{t^{n+1}}^{germ} + \overbrace{a_1 t + a_2 t^2 + \ldots + a_{n-1} t^{n-1}}^{perturbation} \tag{7.3}$$

In this expression, the word "perturbation" is put over the set of terms that unfold the germ; i.e., they perturb the germ in a way that reveals the qualitative changes that all possible perturbations can produce.

Notice that the perturbation in expression (7.3) is simply the addition of successively higher powers of $t$. Furthermore, the control parameters appear in the perturbation as the coefficients $a_1, \ldots, a_{n-1}$ of those powers; the $i$-th control parameter $a_i$ being the coefficient of the $i$-th power of $t$. Given any specific set of values, $a_1, \ldots, a_{n-1}$, for the control parameters, expression (7.3) can be viewed as the Taylor series expansion of a function, whose highest order term is the germ. As is well known, the lowest terms of a Taylor expansion determine the qualitative structure of a function. Here, the control variables determine the weight of the low-order terms. When the values of the control parameters are moved to zero, thus erasing the low-order terms, then the lowest term in the expansion becomes the highest term, the germ $t^{n+1}$, which is $n$-fold degenerate. Therefore, the maximal degeneracy that can occur in the space of functions (7.3) must be $n$-fold degeneracy, which occurs only on reaching the germ. This means that, as we move the values of the control parameters away from zero, the $n$-fold degeneracy must necessarily diminish. This can happen in only two ways, some of the $n$ coincident critical points in the germ can simply disappear, or some can bifurcate away from each other; i.e., loosing their coincidence. Most crucially, the unfolding (7.3) gives all the qualitative changes that can happen to the $n$ coincident critical points, and does so with the minimal required control dimension. This is why the unfolding (7.3) is called *universal*. Any other perturbation of an $n$-fold degenerate critical point will not exhibit qualitative changes that are not exhibited by the unfolding (7.3). Thus the perturbation in (7.3) can be understood as the *canonical* way of representing the perturbation of an $n$-fold degenerate critical point; and the control variables $a_1, \ldots, a_{n-1}$ can be viewed as the canonical parameters that carry out that perturbation. The unfolding (7.3) is called a *cuspoid catastrophe* $A_n$. To emphasize the numbers involved:

---

**Numbers associated with a Cuspoid Catastrophe $A_n$**

$n$-fold degenerate critical point

| | | |
|---|---|---|
| $t^{n+1}$ | = | germ |
| $n-1$ | = | dimension of control space |

---

Now for each $n \geq 2$ there is a cuspoid catastrophe. The lowest $n$ is obviously 2, because this is the lowest degeneracy possible for a critical point; i.e., the coincidence of 2 critical points. Table 7.1 lists the first five (positive) cuspoid catastrophes $A_2$-$A_6$.

To illustrate a universal unfolding, let us consider the universal unfolding given for the 3-fold degenerate critical point. The canonical form of this degenerate critical point is the quartic $f(t) = t^4$. Observe that the function $t^4$ is the germ in the second line of Table 7.1. Thus the table says that the dimension $k$ of the control space must be 2, and the universal unfolding is the germ $t^4$ plus the perturbation $a_1 t + a_2 t^2$, that is:

$$F(t; a_1, a_2) = t^4 + a_1 t + a_2 t^2 \tag{7.4}$$

The two control parameters appear in the perturbation as the coefficients $a_1$ and $a_2$. The unfolding (7.4) is called the *cusp catastrophe* $A_3$.

The 2-dimensional control space is given by the plane in Fig 7.4, with the axes marked $a_1$ and $a_2$. At each point $(a_1, a_2)$ in the control space, there is a particular function, which is the function obtained by substituting the coordinates $(a_1, a_2)$ for that point into the unfolding given in expression (7.4) above. In this way, the control space in Fig 7.4 represents the 2-dimensional space of functions defined by the unfolding.

Fig 7.4 also shows examples of functions at various points in the control space. First let us locate, in this space, each of the functions that have a *degenerate* critical point.

| | | Control Dimension | Germ | Perturbation |
|---|---|---|---|---|
| Fold | $A_2$ | 1 | $t^3$ | $a_1t$ |
| Cusp | $A_3$ | 2 | $t^4$ | $a_1t + a_2t^2$ |
| Swallowtail | $A_4$ | 3 | $t^5$ | $a_1t + a_2t^2 + a_3t^3$ |
| Butterfly | $A_5$ | 4 | $t^6$ | $a_1t + a_2t^2 + a_3t^3 + a_4t^4$ |
| Wigwam | $A_6$ | 5 | $t^7$ | $a_1t + a_2t^2 + a_3t^3 + a_4t^4 + a_5t^5$ |

**Table 7.1.** The first five (positive) cuspoid catastrophes.

The most obvious such function is the germ $t^4$, which we know is located at the origin $(a_1, a_2) = (0, 0)$ of that space – illustrated by the graph placed at the origin (the center of Fig 7.4). In the entire control space, the only other functions that have degenerate critical points are the functions located at the points along the two curves which are shown as going downward from the origin. For example, consider the left curve. At any point on the left curve, the graph of the function located at that point is illustrated by the graph, in Fig 7.4, shown at an arbitrary point on this curve. As shown in this graph, such a function has two critical points: (1) a horizontal inflection point, and (2) a minimum. The inflection point is 2-fold degenerate, and the minimum is non-degenerate.

Note that, at any point on the right curve, the graph of the function located at that point must be the reflected version of the graph of the function located at the point on the same level on the left curve, as illustrated in Fig 7.4.

To summarize: In the entire 2D control space, the only functions that have degenerate critical points, are (1) the function at the origin, i.e., the germ, and (2) the functions along the two downward curves from the origin.

The origin and the two curves are called the *separatrix*. The reason is that they *separate* the entire control space into two regions in which all the functions are Morse, i.e., have no degenerate critical points. These two regions are labeled Region 1 and 3 in Fig 7.4.

Region 1 is that surrounding the separatrix. Every function in this region has only 1 critical point, a minimum which is non-degenerate; i.e., Morse. This fact is illustrated by the top graph and the left-most and right-most graphs in Fig 7.4. In contrast, Region 3 is the region between the separatrix curves. Every function in this region has 3 critical points – minimum, maximum, minimum – all of which are non-degenerate; i.e., Morse. An example of such a function is illustrated by the lowest graph shown in Fig 7.4.

In the control space of any catastrophe, the separatrix is defined as the set of functions that have degenerate critical points, i.e., the set of non-Morse functions. The separatrix divides the control space into a set of open regions of Morse functions. Such a region will be called a *Morse region* or simply a *region*. The functions in a region all have the

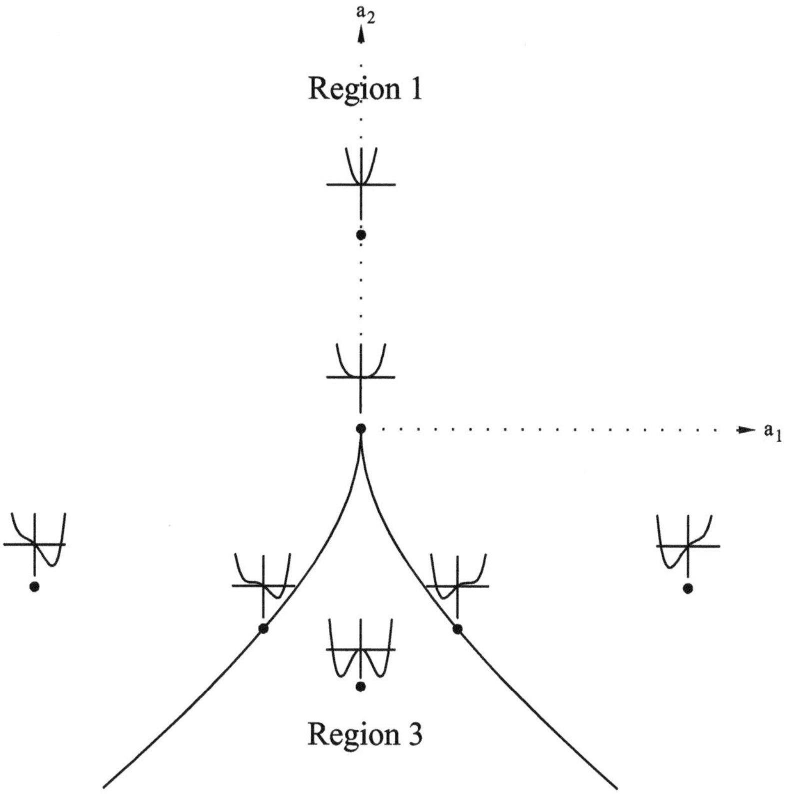


**Fig. 7.4.** Representative functions at points in the control space of the cusp catastrophe.

same critical-point structure, which we will call the critical-point structure *belonging to* the region. The region is connected. Furthermore, the separatrix itself is partitioned into connected components, where all the functions in a component have the same critical-point structure. We will call such a connected component, a *component* of the separatrix. The critical-point structure of the functions, in the component, will be called the critical-point structure *belonging to* the component. Therefore to emphasize: the critical-point structure belonging to a region will have no degenerate critical points, because a region consists of only Morse functions; whereas the critical-point structure belonging to a component will have degenerate critical points, because a component consists of only non-Morse functions.

As an example of the partitioning of a separatrix into components, consider again the control space of the cusp catastrophe, as shown in Fig 7.4. We observe that the separatrix has three components: (1) the origin which will be called the *cusp-point*; and (2 & 3) the two branches which will be called the *separatrix curves*. It is important to emphasize that the term "separatrix curve" will mean a branch *without* the cusp-point, because the cusp-point has a different critical-point structure from the points on each of the branches.

Generally, in the control space of a catastrophe, the union of the *regions* forms a dense open subset, and the separatrix is thin; i.e., of measure zero. In topology, one says that a property belonging to a dense open subset is *generic*, meaning *typical*. Thus we see that, in the control space of a catastrophe, the property of being a Morse function is generic; i.e., it is a property of the dense open subset which is the union of the regions. Observe that a consequence of this is that any non-Morse function in the space is arbitrarily close to a Morse function.

Fig 7.4, for the cusp catastrophe, illustrates the fact that one can set up a correspondence between *genericity* and *structurally stability*. Observe that any Morse function is structurally stable in that any arbitrarily small perturbation leads to another Morse function of the same type. Conversely, any non-Morse function (i.e., on the separatrix) is not structurally stable in that there are arbitrarily small perturbations which will immediately make the function Morse (i.e., move it off the separatrix).

Now let us consider the issue of bifurcation in Catastrophe Theory. To illustrate, return to the control space of the cusp catastrophe $A_3$, as shown in Fig 7.4 (p245). Consider a path from any point in Region 1 to any point in Region 3, such that the path crosses the separatrix only at the origin. The point in Region 1 corresponds to a function with a single critical point, a non-degenerate minimum; and the point in Region 3 corresponds to a function with 3 critical points, a non-degenerate minimum, non-degenerate maximum, non-degenerate minimum. By making the path between them *cross the origin*, one ensures that, along the path, the single non-degenerate minimum of the function in Region 1 will become the 3-fold degenerate minimum of the germ at the origin, and this 3-fold degenerate minimum will then undergo a 3-fold bifurcation into the non-degenerate minimum, non-degenerate maximum, non-degenerate minimum of the function in Region 3. At first, the reader might think that this bifurcation represents the 3-fold bifurcation operator $B$ of the Process Grammar. We will see later that this is not completely correct because Catastrophe Theory fails to define qualitative equivalence correctly for morphology, and correspondingly fails to have the full transversality structure required for morphology.

Let us return now to the control space. It is important to observe that the 3-fold bifurcation, described in the previous paragraph, depended crucially on the fact that the path crossed the separatrix at the origin, rather than another point on the separatrix, as follows: Consider a path between exactly the same two points in the control space (i.e., in Regions 1 and 3 respectively), but make the path cross the separatrix at a point on one of the two separatrix curves rather than the origin. Because the path goes between the same two points, it must be a bifurcation from 1 to 3 critical points. However, the type of bifurcation that achieves this is now 2-fold rather than 3-fold. The reason is that the path goes via a point on a separatrix curve, and we saw in Fig 7.4 that a function at any point on a separatrix curve contains a 2-fold degenerate critical point (given by a horizontal inflection in the graph) and a non-degenerate minimum. The non-degenerate minimum in the separatrix function was previously the non-degenerate minimum in the starting function in Region 1. The 2-fold degenerate critical point in the separatrix function was previously a non-critical point in the starting function in Region 1. Subsequently, the non-degenerate minimum in the separatrix function remains a non-degenerate minimum in the function in Region 3; that is, this non-degenerate minimum, which was the non-degenerate minimum in the starting function, has remained a non-degenerate minimum in the entire path. Furthermore, the 2-fold degenerate critical point in the separatrix function bifurcates to become the two other critical points in the function in Region 3.

It is important to understand that the set of critical points in the cusp catastrophe forms a manifold, as follows: To illustrate how to construct this, return again to the control space of the cusp catastrophe $A_3$, as illustrated in Fig 7.4. At each point in the control space, place the $t$-axis (domain axis) of its function perpendicularly to the control space at that point, and mark on its $t$-axis the critical points of that function. The entire set of critical points, given by the entire set of $t$-axes on the control space, will form a manifold. This is illustrated in Fig 7.5. The horizontal plane at the bottom of the figure is the control space, and the manifold shown above it is the set of critical points from all the functions

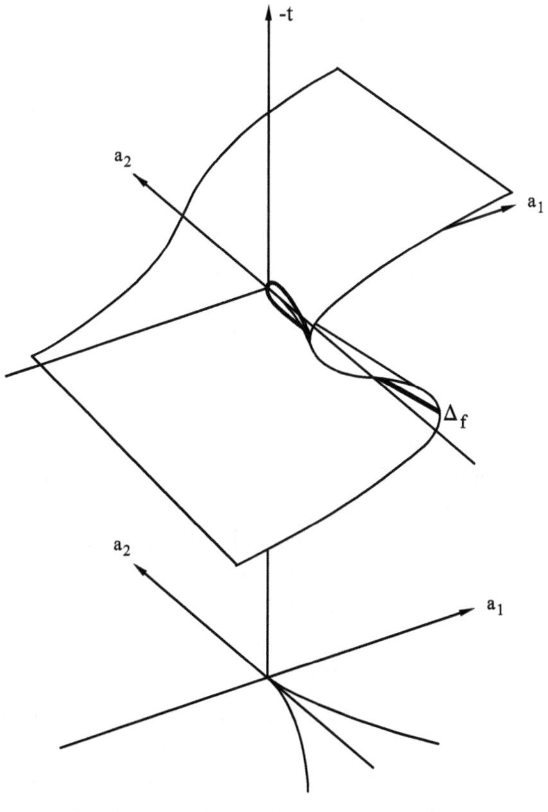


**Fig. 7.5.** The critical manifold over the control space of the cusp catastrophe $A_3$.

in the control space. That is, select any point $(a_1, a_2)$ in the horizontal control plane, and consider the vertical line through that point. This vertical line is the $t$-axis of the function located at that point in the control plane. The vertical line will cut through the manifold at a number of points. These points, on this $t$-axis, are the critical points of the function located at that point $(a_1, a_2)$ in the control plane.

Notice, in the manifold, the bold curve marked $\Delta_f$. This is the set of *degenerate* critical points that occur in the functions. Obviously, the curve projects down onto the separatrix in the control space; i.e., the separatrix is the set of functions that have degenerate critical points, and the bold curve $\Delta_f$ consists of those degenerate critical points. Thus, all the points in the manifold, except the points on the curve $\Delta_f$, are the *non-degenerate* critical points. In fact, the curve $\Delta_f$ divides the manifold into two connected components: (1) the sheet surrounding the curve $\Delta_f$ is the set of all non-degenerate *minima* in the functions, and (2) the middle sheet of the manifold is the set of all non-degenerate *maxima* in the functions.

We will refer to the manifold as the **critical-point manifold**.

Observe that the central point on the bold curve $\Delta_f$, i.e., the origin $(0, 0, 0)$, is the 3-fold degenerate minimum, i.e. the minimum of $t^4$. Furthermore, every other point on the bold curve $\Delta_f$ is a 2-fold degenerate critical point. This can be viewed as the *coalescence* of two non-degenerate critical points, a maximum and minimum, because it is a point where the maximum sheet and the minimum sheet meet each other.

Something we should also note is this: We saw that the space of functions illustrated by representative functions in Fig 7.4 (p245) is the space of functions of the control space of the cusp catastrophe $A_3$. If we choose to invert each of the functions of this space, relative to the function's own codomain axis, then the space of functions would be that of the *dual* cusp catastrophe $A_{-3}$. Thus, observe the following: We saw in Fig 7.4 (p245) that each of the functions in Region 1 of the control space of the cusp catastrophe has a single critical point which is a *minimum*. In contrast, each of the functions in Region 1 of the control space of the dual cusp catastrophe has a single critical point which is a *maximum*. Correspondingly, it is easy to see how the other functions have changed.

Now consider another example of a catastrophe: the first item listed in Table 7.1 (p244), called the fold catastrophe $A_2$. Whereas the cusp catastrophe is the universal unfolding of the 3-fold degenerate critical point, the fold catastrophe is the universal unfolding of the 2-fold degenerate critical point. The canonical form of the 2-fold degenerate critical point is $t^3$. Therefore, $t^3$ is the germ of the fold catastrophe, and, as stated in Table 7.1 (p244), its universal unfolding is given as:

$$F(t; a_1) \quad = \quad t^3 + a_1 t \tag{7.5}$$

In this expression, observe therefore that the control space is 1-dimensional given by the control variable $a_1$. That is, the control space is simply a single line, the $a_1$-axis. There are two Morse regions in the control space. The positive $a_1$-axis of the control space is a Morse region consisting of functions that have no critical points. Therefore this region will be called Region 0 of the control space. An example of a function in this region is the left-most graph shown in Fig 7.6. The negative $a_1$-axis of the control space is a Morse region consisting of functions that have 2 critical points, a non-degenerate minimum and non-degenerate maximum. Therefore this region will be called Region 2 of the control space. An example of a function in this region is the right-most graph shown in Fig 7.6. The germ $t^3$ is, of course, located at the origin $a_1 = 0$ of this control space, and is shown as the middle graph in Fig 7.6. Notice that this graph has a horizontal inflection-point at $t = 0$, and this point is 2-fold degenerate as a critical point.

Thus the sequence of three graphs in Fig 7.6 illustrates what happens as one moves along the $a_1$-axis that defines the control space, from a point in Region 0 to a point in Region 2. A non-critical point in the starting function (the left-most graph) becomes the 2-fold degenerate critical point in the germ (the middle graph) which bifurcates into a non-degenerate minimum and non-degenerate maximum in the final function (the right-most graph).

A crucial fact is this: Because, in the control space $a_1$ of the fold catastrophe, the only function that has a degenerate critical point is the germ $t^3$ at the origin $a_1 = 0$, the *entire separatrix* of the fold catastrophe is only a *single point* in the control space $a_1$, the origin point $a_1 = 0$.

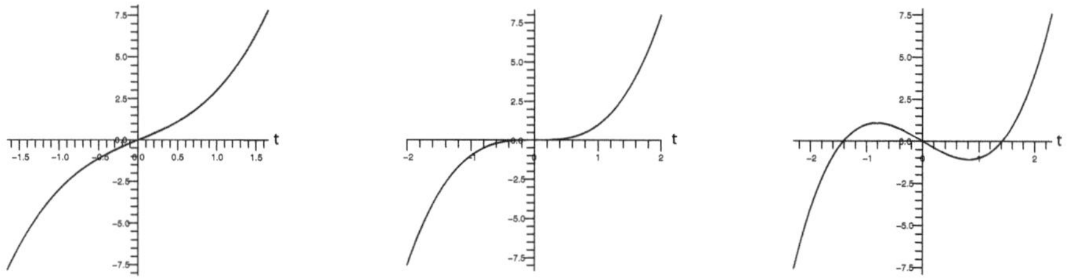

**Fig. 7.6.** The functions at points $a_1 = 2, 0, -2$ in the 1-dimensional control space $a_1$ of the fold catastrophe $A_2$.

Note that, according to Catastrophe Theory, the fold catastrophe is the canonical form of 2-fold critical-point bifurcation, and therefore is the canonical form of crossing a separatrix curve of the cusp catastrophe. In particular, the 2-fold degenerate critical point (horizontal inflection) in any function in the separatrix curve of the cusp catastrophe is right-equivalent to $t^3$, the germ of the fold catastrophe.

From this, we see that the cusp catastrophe $A_3$ actually contains two trajectories of fold catastrophes $A_2$ along its two separatrix curves. This is a standard situation in Catastrophe Theory: Given any cuspoid catastrophe $A_n$, the germ of $A_n$ is at the origin of its control space, and emanating from the origin, there are two curves of catastrophes of type $A_{n-1}$. This will of course mean that, for $n \geq 4$, there will be surfaces of catastrophes $A_{n-2}$ along those curves of $A_{n-1}$ (and so on).

## 7.4 Causal Incompatibility of Catastrophe Theory

We will now begin our argument that shows that the reason why Catastrophe Theory fails to achieve Thom's program for modeling spatial morphology, and, in particular, biological morphology, is that it is *incompatible* with *plausible causal explanations*.

First, we argue that a basic reason why Catastrophe Theory fails to model spatial morphology is that, given any critical point, Catastrophe Theory fails to qualitatively distinguish the case where the critical point has a positive value, from the case where the critical point has value zero, from the case where the critical point has a negative value. According to our theory, this is a failure to understand that this distinction is fundamental to the *causal structure of morphology*.

For example, consider the case of distinguishing an extremum having positive value from that extremum having negative value:

**Because Catastrophe Theory defines an extremum of positive value as right-equivalent to that extremum having negative value, it defines them as** *qualitatively equivalent.*

**However, according to our theory, a positive and negative version of an extremum are fundamentally different in morphology. That is, one of them defines a** *compressive* **action in the morphology and the other defines a** *penetrative* **action in the morphology.**

**Therefore, they are** *qualitatively* **different with respect to the morphology.**

**Therefore, Catastrophe Theory's view of qualitative equivalence is fundamentally wrong for morphology.**

The fact that Catastrophe Theory also fails to distinguish the positive and negative cases from the case where the extremum has value zero is also a fundamental mistake, as we will now demonstrate.

To illustrate, return to Fig 7.4 (page 245), which shows the control space of the cusp catastrophe. Recall that, in this figure, the term "Region 1" labels the set of Morse functions with only one critical point, a non-degenerate minimum; i.e., this region of the control space is the region around the cusp-shaped separatrix.

Now, according to our theory, a crucial mistake of Catastrophe Theory is that it classifies all the functions in Region 1 as having the *same* critical-point structure. That is, as we have seen, Catastrophe Theory characterizes the critical-point structure of each function in Region 1 purely as a non-degenerate minimum rather than additionally considering the crucial *value* properties of these minima.

This means that Catastrophe Theory does not distinguish between the fact that the minimum of any function located in the positive $a_2$-axis (of the control space) has value zero, and the fact that the minimum of every other function in Region 1 has a negative value. According to our theory, this is a fundamental mistake.

That is, according to our theory, the zero minimum of any function in the positive $a_2$-axis is itself a crucial degeneracy: It is the 2-fold zero-degeneracy that occurs in the transition state of the *process-continuation* operator $C$ of the Process Grammar, and, according to our theory, this operator defines an enormously important stage in morphological development, the continuation of a *compressive* force till it penetrates.

Furthermore, notice that the 2-fold zero-degeneracy of the minimum of any function in the positive $a_2$-axis of the cusp-catastrophe implies, according to our theory, that this minimum is *not structurally stable*. However, Catastrophe Theory classifies the minimum in each of the functions of Region 1, including the positive $a_2$-axis, as structurally stable.

**An additional reason why Thom's Catastrophe Theory fails to model morphology is this: Consider the positive $a_2$-axis of the control space of the cusp catastrophe, and the remainder of Region 1 (the region of Morse functions with only one critical point).**

**Whereas, according to Catastrophe Theory, the functions in the positive $a_2$-axis are *qualitatively equivalent* to the functions in the remainder of Region 1, our theory demonstrates that this is profoundly wrong for the following reason.**

**Catastrophe Theory regards functions on the positive $a_2$-axis as structurally stable.**

**This fails to model a crucial developmental stage defined by our theory of morphology: the continuation of a *compressive* process till it penetrates.**

**Therefore, according to our theory, the positive $a_2$-axis must be part of a separatrix, i.e., a set of functions with degenerate singularities, which are transition states in morphology.**

**Thus, according to our theory, for the singularities of the functions in the cusp catastrophe, the correct separatrix not only includes the cusp-shaped separatrix, but *also* the positive $a_2$-axis (and therefore more). This also means that the correct control space of functions is much larger. The full space, and its much more complex separatrix, will be explained in detail in the next chapter.**

Now, according to our theory, another reason why Thom's cusp catastrophe fails to model crucial aspects of morphology is this:

We will now consider the crossing of a *branch* of the separatrix given by Thom's cusp catastrophe. As stated on page 245, the two branches of the separatrix will also be called the *separatrix curves*. And again, it is important to emphasize that, in using the terms *branches* and *separatrix curves*, we will understand them as not including the cusp-point. Recall that a function at any point on a separatrix curve of the cusp catastrophe has a 2-fold degenerate critical point. This is a *horizontal inflection* in the graph of the function, as illustrated in the graph shown at a point on the left separatrix curve in Fig 7.4 (p245), and illustrated also in the graph shown at a point on the right separatrix curve in that figure.

The crucial fact is that, in Catastrophe Theory, this critical point in a function is regarded as qualitatively equivalent to the critical point in the function $t^3$, which is also a 2-fold degenerate critical point; and $t^3$ is regarded as the canonical form of this critical point, and is labeled $A_2$ as given in Table 7.1 page 244.

According to our theory of morphology, Catastrophe Theory's claim of this qualitative equivalence to $t^3$ is a fundamental mistake. The reason is this:

**Another reason why Thom's Catastrophe Theory fails to model morphology is this.**

**Thom's Catastrophe Theory regards the 2-fold degenerate critical point of any function in a branch of the separatrix of the cusp catastrophe as qualitatively equivalent to the 2-fold degenerate critical point of the canonical form $t^3$ that is the separatrix of the fold catastrophe.**

**However, according to our theory of morphology, this is a fundamental mistake for the following reason:**

**A 2-fold degenerate critical point in the cusp catastrophe has a** *non-zero* **value, whereas the 2-fold degenerate critical point in the fold catastrophe has value** *zero*.

**According to our theory, this difference defines a fundamental morphological difference:**

**In the control space of the cusp catastrophe, a path that crosses a branch of the separatrix, from Region 1 to Region 3, i.e., that involves a transition through a 2-fold degenerate critical point, is given by the** *positive* **pair-creation operator of the Process-Grammar.**
**This results in a** *compressive* **and** *penetrative* **extremum.**

**In contrast, in the control space of the fold catastrophe, a path that crosses the separatrix, from Region 0 to Region 2, i.e., that involves a transition through its 2-fold degenerate critical point, is given by the** *zero* **pair-creation operator of the Process-Grammar.**
**This results in** *two penetrative* **extrema.**

**Therefore, the morphologies are fundamentally different.**

Furthermore, an additional reason why Thom's cusp catastrophe fails to model crucial aspects of morphology is as follows:

It is important now to consider a particular type of path that crosses a separatrix curve (a branch of the separatrix) of the cusp catastrophe. Consider a path that starts with the left-most graph in Fig 7.4 (p245) and goes to the lowest central graph shown in the figure, via the graph shown between them on the separatrix. Observe that, on the left-most graph, there is a single critical point, the non-degenerate minimum. Note that this minimum is *preserved* along the entire path, and becomes the right minimum in the final graph. As noted in section 7.3, this single minimum remains *non-degenerate* along the entire path. Also we noted that, in the first graph, there is a non-critical point (point of *non-horizontal* tangency) which becomes the point of *horizontal* inflection in the graph on the separatrix.

However, now we observe what our theory regards as a crucial fact: This horizontal inflection, in the graph on the separatrix, has *positive value*. From this, we infer that it bifurcates into a positive minimum and positive maximum, after one crosses the separatrix curve.

Then observe what our theory regards as the next crucial fact: In the lowest central graph shown in Fig 7.4 (p245), the positive minimum which resulted from that bifurcation has become *negative*. Therefore, we see this:

**A further reason why Thom's Catastrophe Theory fails to model morphology is this.**

**Crossing a separatrix curve (a branch of the separatrix) in the cusp catastrophe, from Region 1 to Region 3, the 2-fold bifurcation results in the creation of a minimum that is** *positive.*

**However, as one then moves** *within* **Region 3 to a central point, this minimum becomes** *negative.*

**According to our theory, this means that the minimum, created by crossing the branch of the separatrix, has changed from a** *compressive* **extremum to a** *penetrative* **extremum.**

**Therefore, according to our theory, this minimum underwent a crucial morphological change.**

**Thus, according to our theory,** *inside* **Region 3, there is an important morphological transition, and the cusp catastrophe completely fails to notice this.**

**Notice furthermore that, since, according to our theory, this morphological transition,** *inside* **Region 3, is a transition from a compressive extremum to a penetrative extremum, it is the application of the process-continuation operation $Cm^+$ of the Process Grammar.**

**Therefore, according to our theory, Region 3 within the cusp catastrophe contains a separatrix that is essential to morphology, and has not been given by the cusp catastrophe. This separatrix will be given in the next chapter.**

**Furthermore, whereas, according to Catastrophe Theory, the functions in Region 3 are *qualitatively equivalent* to each other, this is profoundly wrong, according to our theory. That is, according to our theory, the functions in Region 3 are *not qualitatively equivalent* to each other.**

**Understanding this is essential for understanding morphology.**

# 8

# Interactive Singularity Theory

**This chapter and the following chapters describe the singularity theory of shape invented in the Process Grammar. It is called** *Interactive Singularity Theory.*

**A crucial fact is that the functions, whose singularities are considered in Interactive Singularity Theory, are curvature functions.**

**In contrast, conventional singularity theories of shape are based on functions that are not curvature functions.**

**Furthermore, Interactive Singularity Theory's definition and organization of singularities match the** *causal* **structures which the Process Grammar has proved are fundamentally important to biological morphology and manufacturing design. That is, Interactive Singularity Theory accords with my New Foundations to Geometry.**

**Conventional singularity theories of shape have completely failed to match the fundamentally important causal structures of biological morphology and manufacturing design.**

## 8.1 The Transversality Issue

In the previous chapter, we gave important reasons why Catastrophe Theory fails to model spatial morphology. It is now necessary to give an additional very important reason. It is based on the property of *transversality*. This property, of course, relates to the properties considered in the previous chapter, but it is important now to consider it directly.

To begin to understand this, let us consider the operator $B$ from the Process Grammar. In section 7.3, it was noted that, because $B$ is a 3-fold critical-point bifurcation, one might, at first, believe it to be modeled as the crossing from Region 1 to Region 3, via the origin, in the control space of the cusp catastrophe. However, we will now see that, using the unfolding function given in Catastrophe Theory, this crossing violates crucial mathematical conditions given by the causal structure defined by the Process Grammar.

To show this, let us choose three functions that are like the three functions shown down the central axis, $a_2$, of the control space, illustrated Fig 7.4 (p245). That is, the first function $f_1$ will be in the positive $a_2$-axis; the second function $f_2$ will be the germ (i.e., at the origin); and the third function $f_3$ will be in the negative $a_2$-axis. According to Catastrophe Theory, the first and third functions would be qualitatively equivalent to the domain and co-domain of the Process Grammar operator $B$, and the second function would be qualitatively equivalent to the transition-function of the operator.

However, in order to show the incorrectness of that view, let us choose a specific illustration: We will take the first and third functions to be respectively at the points 3 and $-3$ on the $a_2$-axis. Therefore, in the 2-dimensional control space, the coordinates $(a_1, a_2)$ for the three functions are, respectively,

$$\begin{aligned} f_1 &= (0,3) \\ f_2 &= (0,0) \\ f_3 &= (0,-3) \end{aligned}$$

Substituting these coordinates into the unfolding function, $F(t;a_1,a_2) = t^4 + a_1 t + a_2 t^2$ for the cusp catastrophe, we see that the functions located at these three points are therefore, respectively:

$$\begin{aligned} &f_1(t) = t^4 + 3t^2 \\ &f_2(t) = t^4 \\ &f_3(t) = t^4 - 3t^2 \end{aligned}$$

The graphs of these three functions are shown successively down the left side of Fig 8.1.

Now, a basic concept in Thom's Catastrophe Theory is that a smooth function $f$ is generic/structurally stable if and only if it is Morse, i.e., has no degenerate critical points. This is the case if and only if the graph of the first derivative $Df$ intersects the graph of the zero function (i.e., the $t$-axis) *transversally*. The concept is illustrated in Fig 8.1, where the right side of the figure shows the first derivatives of the three functions we placed on the left. Concerning the functions on the left, we know already that the first function (which is in Region 1) and the third function (which is in Region 3) are both Morse (non-degenerate). In contrast, the second function, which is the germ, is non-Morse (degenerate). Now, examining the first derivatives of the three functions, we see that, for the first and third functions, the graphs of the first derivatives are transversal

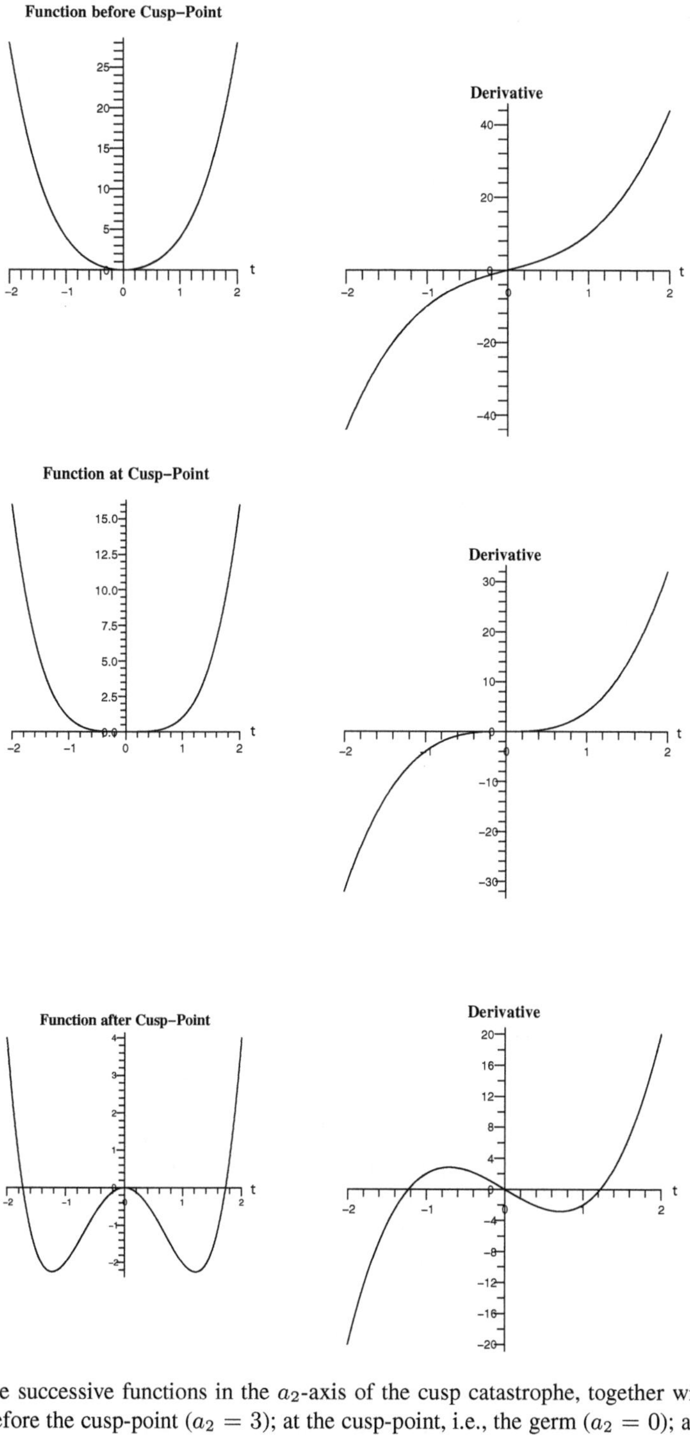


**Fig. 8.1.** Three successive functions in the $a_2$-axis of the cusp catastrophe, together with their derivatives: before the cusp-point ($a_2 = 3$); at the cusp-point, i.e., the germ ($a_2 = 0$); and after the cusp-point ($a_2 = -3$). The derivative of the first and third functions are transversal to the zero-function, whereas the derivative of the germ is not.

to the graph of the zero function (the $t$-axis). In contrast, for the second function, the graph of the first derivative is non-transversal to the $t$-axis.

This fact is used by Thom, in the following way: By the summary of his ideas given in section 7.3, transversality is a structurally stable property. Based on this, Thom argues that a Morse function $f$ is structurally stable because the graph of its first derivative $Df$ intersects the graph of the zero function transversally, and transversality is structurally stable.

However, I will argue that there is a serious problem with Thom's approach: It ignores the transversal intersection condition between the graph of the function $f$ *itself* and the graph of the zero function. This problem is serious because it prevents Thom's theory from fulfilling his goal of modeling spatial morphology such as biological development.

To illustrate this, let us return to Fig 8.1. We have seen that, in Catastrophe Theory, the first function $f_1$ is considered as being structurally stable because the graph of its first derivative $Df_1$ is transversal to the graph of the zero function. However, observe that the graph of the function $f_1$ itself is not transversal to that of the zero function. Because non-transversality is structurally *unstable*, this means that there are perturbations, no matter how small, that will change its relationship to the $t$-axis from non-transversal to transversal. For example, any upward movement, no matter how small, of this graph, will cause it to instantly loose its non-transversal contact with the $t$-axis and have no intersection at all with the $t$-axis; and non-intersection is an example of transversal intersection (as noted in the summary in section 7.3). Similarly, any downward movement, no matter how small, will make this graph cross the $t$-axis such that, at each of its two crossing points, its tangent and the tangent to the $t$-axis will together span 2D space – thus satisfying the definition of transversal intersection at each of the two crossing points. To summarize: Any vertical movement, no matter how small, of the graph of $f_1$ will change it from non-transversal to transversal.

Thus the contact condition of the first function $f_1$ in Fig 8.1 is non-stable. Furthermore, from what we have just seen, this contact condition is non-generic (non-typical), since the function is arbitrarily close to functions that are transversal to the $t$-axis.

Thus, the non-transversality of this function insures that it is non-stable and non-generic, and therefore inappropriate as a model. Yet, in Catastrophe Theory, this function would be regarded as appropriate for a model, because it is in the region outside the separatrix in the control space.

The situation is even worse when we observe that the non-transversality of the first function is a property that it *shares* with the middle function $f_2$ in Fig 8.1, i.e., the germ itself. In Catastrophe Theory, the germ is regarded as the ultimately inappropriate model (non-stable and non-generic) and the first function is regarded as an appropriate model (stable and generic). However, I will argue that the fact that the graph of the first function is non-transversal to the zero function gives it a commonality with the germ that puts the entire system on dangerous ground.

Genericity and stability relate to the structure of subsets of the control space. Generic functions form a dense open subset; and stable functions must preserve qualitative properties under perturbations.

Therefore, with respect to this, consider how functions are grouped into subsets in the space. Based on the above discussion, and the summary in section 7.3, we can see that the condition that the first derivative is transversal allows one to put the functions that are in the positive $a_2$-axis into the same class as the other functions in Region 1, the class of Morse functions, while putting the germ into a different class, the non-Morse functions. In contrast, the condition that a function itself is transversal results in a different grouping: The functions in the positive $a_2$-axis are put into the same class as the germ, since both these cases are all non-transversal to the zero function. Furthermore, this non-transversality puts the functions that are in the positive $a_2$-axis into a different class from the other functions in Region 1, because the other functions in Region 1 are transversal to the zero function.

We see therefore that the condition that a function's first derivative is transversal (the Morse condition) provides a different grouping from the condition that the function itself is transversal. Catastrophe Theory organizes the control space using the first condition. The singularity theory I am proposing organizes the control space with respect to *both* conditions – a requirement I will call the **double-transversality condition**. The result is that the control space must be extended, and extra separatrix surfaces will appear in the new space.

My argument is that this extension is fundamentally important for ensuring that the system models the causal structure of biological morphology and manufacturing design. Indeed I propose that the absence of this extension is one of the main reasons why Catastrophe Theory did not fulfill Thom's goal of giving a morphological theory of biology.

The compatibility with causal explanation, which does not exist in Catastrophe Theory, requires the continuation operator of the Process Grammar, for a crucial reason that we will now describe, and is basic to Interactive Singularity Theory.

## 8.2 Fundamental Transversality Role of the Continuation Operator

We will now see that the continuation operator of the Process Grammar has a fundamental transversality role that ensures compatibility with causal explanation – a crucial fact that is not understood by Catastrophe Theory.

It is necessary here to describe the reasoning I used, in 1985, to create the continuation operator: A Process Grammar operation must be a minimal singularity-increasing operation that realizes a transition between two generic singularity-configurations. By our theory, genericity requires that the singularity-configuration (of the domain and co-domain) must embody transversality not only in the first derivative but also in the function itself. Thus, consider the graph of a function *with only one extremum*, which will be assumed to be Morse (non-degenerate). To guarantee that this graph has a transversal relation to the $t$-axis, and has the lowest number of singularities, it should point *towards* the $t$-axis, as shown on the left in Fig 8.2. Now let us create a minimal singularity-increasing movement of the graph that sends it from this transversal position to another transversal position. Clearly, the only movement that fulfills this is to move the graph through the $t$-axis, as shown on the right in Fig 8.2. We observe that, the moment the extremum crosses the $t$-axis, two 0's appear in the function.

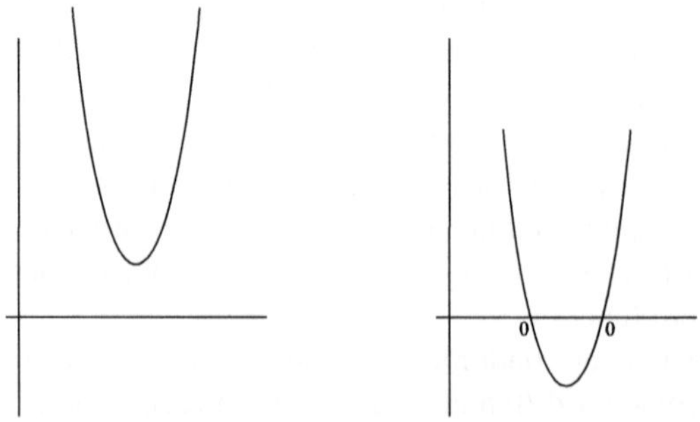


**Fig. 8.2.** The two transversality states of an extremum neighborhood in a function.

In the case of Fig 8.2, we chose the extremum to be a minimum. Thus, in the case of a minimum, the only possible singularity-configuration, that fulfills our starting transversality condition (pointing *towards* the $t$-axis), is the configuration $m^+$.

Therefore, the transition from the left to the right *transversality-states* in Fig 8.2 is given by the following transition of their respective singularity-configurations:

$$Cm^+ : m^+ \longrightarrow 0m^-0$$

We of course recognize this as a continuation operation of the Process Grammar. But, at this stage of the discussion, let us understand that our argument is considering only the issue of transversality. With respect to this, something else that should be observed is the following: In the transition between these two *transversality-states*, there must have been an intervening state in which the graph of the extremum was merely tangential to the $t$-axis; i.e., the extremum became a zero, and the *contact* between the graph and the $t$-axis was *non-transversal*. This is, of course, the function we have called the transition function. Therefore, the transition between the two *transversality-states*, shown in Fig 8.2, must have involved a 2-fold zero-bifurcation.

In the case where the starting extremum is a maximum, the only possible singularity-configuration, that fulfills our *starting transversality condition* (pointing *towards* the $t$-axis), is the configuration $M^-$. The corresponding operation in this case is

$$CM^- : M^- \longrightarrow 0M^+0$$

We of course recognize again that this is a continuation operation of the Process Grammar. But, observe again that we only used a transversality argument in producing this.

We now come onto a crucial correspondence:

### Transversality and the Causal Structure

We will now see that the two operations $Cm^+$ and $CM^-$, produced above purely by *transversality arguments*, fulfill our fundamental requirement of *compatibility* with *plausible causal explanation* as deduced by our process-inference rules. It is sufficient to demonstrate this for $Cm^+$, since the other operation $CM^-$ is simply its dual.

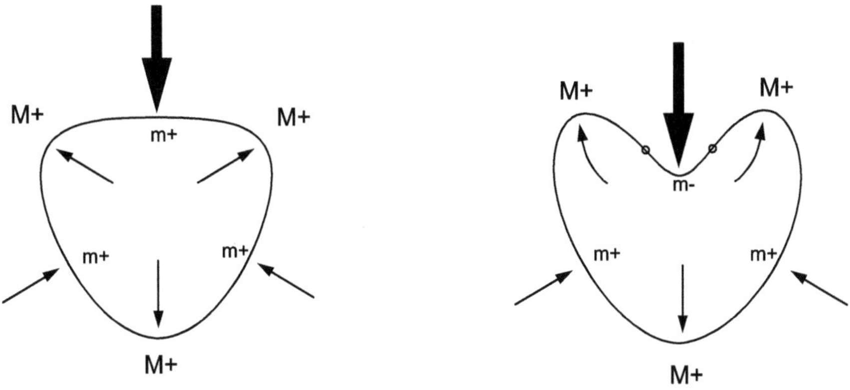


**Fig. 8.3.** Continuation at $m^+$.

We proceed as follows: In this paragraph, although we will be partially recapitulating an argument given in section 3.25, the starting point of the argument is different – we are starting only from the transversality basis given above. Now, the transversality operation $Cm^+$, defined above, is illustrated by the shape-transition shown in Fig 8.3. That is, a non-degenerate curvature minimum changes from positive, at the top of the left shape, into negative, in the right shape, with two zeros on its sides. Now apply our process-inference rules, one by one, to check *causal compatibility* of the transversality structure. By the Symmetry-Curvature Duality Theorem, the minimum must, throughout the transition, have a symmetry axis leading to and terminating at it. Furthermore, as demonstrated in section 3.25, the PISA definition of symmetry gives this axis consistently on the same side of the minimum, unlike the Medial Axis. Next, by our Interaction Principle, this axis is interpreted as a process whose application-point is the minimum. Thus, our rules infer that the process creating the $m^+$, at the top of the left shape, continued till it created the indentation, in the right shape. Furthermore, as the minimum changed from positive to negative, it must, for a moment, have become a zero, which then bifurcated into 2 zeros because these were *required* as the *boundary* between the central negative region, created by the emerging $m^-$, and the surrounding positive region.

From this, we can see that the operation $Cm^+$, understood purely as an operation between transversality-states of the curvature function itself (as opposed to only its first derivative), is compatible with the causal explanation produced by our process-inference rules.

One of the crucial results of the above discussion is this:

## PROCESS GRAMMAR OPERATOR $C$ AND TRANSVERSALITY

**Recall that two manifolds intersect transversally if either:**

**(T1) they do not intersect at all, or**
**(T2) they intersect transversally at all points of intersection.**

**FUNDAMENTAL FACT:**

**The continuation operator $C$ of the Process Grammar maps from transversality condition (T1) to transversality condition (T2).**

For example: The left and right functions in Fig 8.2 illustrate, respectively, the domain and co-domain functions of the Process Grammar operation $Cm^+$. It can be seen that, in relation to the horizontal axis, the graphs of these two functions fulfill the transversality conditions (T1) and (T2) respectively; i.e., the operation $Cm^+$ goes from the transversality condition (T1) to the transversality condition (T2).

Observe that, in contrast, the graph of the transition function, in which the minimum is zero, is not transversal to the $t$-axis. Furthermore, the transversality of the domain and co-domain functions is stable, i.e., any sufficiently small perturbation will not alter their transversality condition. In contrast, the non-transversality of the transition function is not stable; i.e., there exist perturbations, no matter how small, that will make the function loose its non-transversality.

Together with the transversality issue is the issue of genericity: The transversality property of the domain and codomain function is generic in the set of functions; i.e., the transversal functions form an open dense subset. In contrast the non-transversality property of the intervening function is non-generic. Non-transversal functions form a thin (measure zero) subset.

Now, an important difference between our view and Catastrophe Theory is in the classification of extrema: Because Catastrophe Theory classifies extrema only with respect to transversality of the first derivative, its classification of *structurally stable extrema* is into only 2 types: a maximum or a minimum. In contrast, because our theory classifies extrema with respect to *two* transversality criteria – that of the first derivative *and* that of the function itself – our classification of *structurally stable extrema* is into 4 types: $M^+, m^-, m^+, M^-$.

We have observed that the 2-type classification of structurally stable extrema does not account for the plausible extrema bifurcations. For example, it does not recognize that bay-formation $Bm^- : m^- \longrightarrow m^-M^-m^-$ and breaking-through of a protrusion $Bm^+ : m^+ \longrightarrow m^+M^+m^+$ are fundamentally different morphologies. To raise and lower the unfolding function of Catastrophe Theory is not enough because it does not deal with the following profound issues that are handled by Interactive Singularity Theory:

## INTERACTIVE SINGULARITY THEORY

**Interactive Singularity Theory ensures compatibility with plausible causal explanation by dealing with the following issues that are not handled in Catastrophe Theory:**

**1. In the transition between a positive and negative version of an extremum, there is a change in the transversality-state of the function.**

**2. This transition involves an additional unfolding that goes through a degeneracy of zeros, i.e., involves zero-bifurcation.**

**3. This extra unfolding must relate the unfoldings involved in the extremum-bifurcation operations. That is: the system of extremum-bifurcations must be structured by a system of zero-bifurcations.**

**4. The extra unfolding requires the control space to have additional separatrices through which the zero-bifurcations occur.**

**5. The crossing of these additional separatrices matches the fundamentally important *causal* structure of process-continuations, which are correctly defined by the PISA axes of the shape, and are completely failed to be defined by the Medial axes and MA symmetry sets.**

## 8.3 Control Space of an Interactive Unfolding

To gain the extra structure needed to ensure the double-transversality condition of Interactive Singularity Theory, we begin by including an extra control variable $a_0$ in the unfolding of the germ. This variable raises and lowers the unfolding function of a catastrophe. Thus, for example, given the germ $t^{n+1}$, the unfolding function of this germ in the cuspoid catastrophe $A_n$ is:

$$F(t; a_1, \ldots a_{n-1}) = \overbrace{t^{n+1}}^{germ} + \overbrace{a_1 t + a_2 t^2 + \ldots + a_{n-1} t^{n-1}}^{perturbation} \quad (8.1)$$

In contrast, given the same germ $t^{n+1}$, the unfolding function of this germ in Interactive Singularity Theory is:

$$F(t; a_0, a_1, \ldots a_{n-1}) = \overbrace{t^{n+1}}^{germ} + \overbrace{a_0 + a_1 t + a_2 t^2 + \ldots + a_{n-1} t^{n-1}}^{perturbation} \quad (8.2)$$

Consequently, given the germ $t^{n+1}$, its $n-1$ dimensional control space in Catastrophe Theory is increased to an $n$ dimensional control space in Interactive Singularity Theory.

However, if one only increased the dimension of the control space, this would deal only with the raising and lowering of the catastrophe unfolding function. It would not deal with the second transversality condition. For this, we need additional separatrix structures in the control space, as will be illustrated in the following sections.

The basis of this is that there is a fundamental difference in the analysis of the unfolding. In Catastrophe Theory, the singularity of $t^{n+1}$ is viewed only as a degenerate critical point; and $t^{n+1}$ is unfolded to reveal only the perturbations of this degenerate critical point. In contrast, in Interactive Singularity Theory, the singularity of $t^{n+1}$ is viewed *both* as a degenerate critical point and a degenerate zero; and $t^{n+1}$ is unfolded to reveal the perturbations of both its degenerate critical-point property and its degenerate zero property, and also to reveal the *interaction* between the resulting critical-point structure and zero structure in the unfolding.

A consequence of this is that there is a fundamental difference between the *meaning* of the label of the singularity in Catastrophe Theory and the *meaning* of the label of the singularity in Interactive Singularity Theory. Given the germ $t^{n+1}$, its singularity is labeled $A_n$ in Catastrophe Theory, where the notation $A_n$ refers only to the critical-point degeneracy of the singularity. In contrast, in Interactive Singularity Theory, the singularity of the same germ $t^{n+1}$ is labeled $EA_n$, where this notation $EA_n$ refers not only to the critical-point degeneracy of the singularity but also its zero degeneracy. Furthermore, $EA_n$ is called an **interactive singularity** because it is viewed as within the *interaction* of the critical point structure and zero structure that results in the unfolding.

The letter $E$ in $EA_n$ can be understood as an abbreviation of the word *Expanded*.

The unfolding given in expression in (8.2), together with the expanded set of separatrices, together with the interactive structure of the separatrices, as defined by Interactive Singularity Theory, will be called the **Interactive Unfolding of $EA_n$**.

## 8.4 Separatrices in an Interactive Unfolding

The separatrices in an Interactive Unfolding will consist of two types corresponding to the violation, respectively, of the two transversality conditions. The two types will be given the following names:

**(1) Critical-Point Separatrix:** This is the set of functions that contain a degenerate critical point.

**(2) Zero-Separatrix:** This is the set of functions that contain a critical point that has zero value, i.e., the graph of the function itself is in non-transversal contact with the $t$-axis.

To illustrate: We know that the cusp catastrophe $A_3$ has a 2-dimensional control space $(a_1, a_2)$; and therefore, according to Interactive Singularity Theory, the Interactive Unfolding of $EA_3$ has a 3-dimensional control space $(a_0, a_1, a_2)$. The first of the above separatrices can be easily given: It is merely the cusp-shaped separatrix of the cusp catastrophe $A_3$, swept along the entire third dimension $a_0$ of the control space of the Interactive Unfolding of $EA_3$, as illustrated in Fig 8.4.

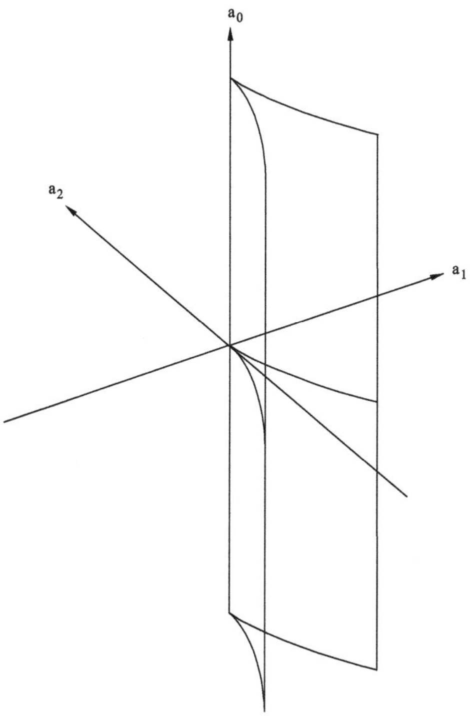


**Fig. 8.4.** The critical-point separatrix of the Interactive Unfolding of $EA_3$.

Two alternative methods will now be proposed to create the other separatrix in this control space, i.e., the zero-separatrix. Mathematically they are easy to understand. However, since they lead to complex results in the structure of the separatrix, it is important to examine them carefully, so that one is able to comprehend the properties of the different points on the separatrix.

## 8.5 Method 1 for Constructing the Zero-Separatrix

To provide the basis of the first method of constructing the zero-separatrix, it will be useful to examine some of the properties of functions in the 2-dimensional control space of the cusp catastrophe $A_3$. What we need to consider are the *values* of the functions at their critical points. Fig 8.5 will be used to illustrate our discussion.

First observe that any function on the positive $a_2$-axis (e.g., the top function in Fig 8.5), and the germ, have a single minimum with value 0. When we go from either of these functions leftward in the control space, the minimum moves downward and to the right within the function, as can be seen in the left-most function in Fig 8.5.

Next consider the row of functions along the middle of Fig 8.5, starting with this left-most function. As the function moves rightward along the row, we can see that the minimum moves upward within the function. Observe that this minimum disappears when the right separatrix curve is encountered.

Now, starting again from the top function in Fig 8.5, when we go rightward in the control space, the minimum moves downward and to the left within the function, as can be seen in the right-most function in Fig 8.5.

Consider again the row of functions along the middle of Fig 8.5, but this time start with this right-most function. As the function moves leftward along the row, we can see that the minimum moves upward within the function. Observe that this minimum disappears when the left separatrix curve is encountered.

At the center of the row, we see that the two minima – which have been moving *upward* respectively in the rightward and leftward direction in the row – have now reached the same height.

From this we conclude the following: As we move leftward *from* the central function, towards the separatrix curve, the left minimum will move up and the right minimum will move down within the function. That is, a function between the central function and the left separatrix curve must have two minima with the right minimum being lower than the left minimum.

Correspondingly, as we move rightward from the central function, towards the separatrix curve, the right minimum will move up and the left minimum will move down within the function. That is, a function between the central function and the right separatrix curve must have two minima with the left minimum being lower than the right minimum.

Also, as noted in section 7.4, the *pair*, minimum and maximum, *created* by crossing a separatrix curve, both initially have positive value; and then, in moving toward the center of Region 3, the minimum becomes negative and the maximum becomes zero.

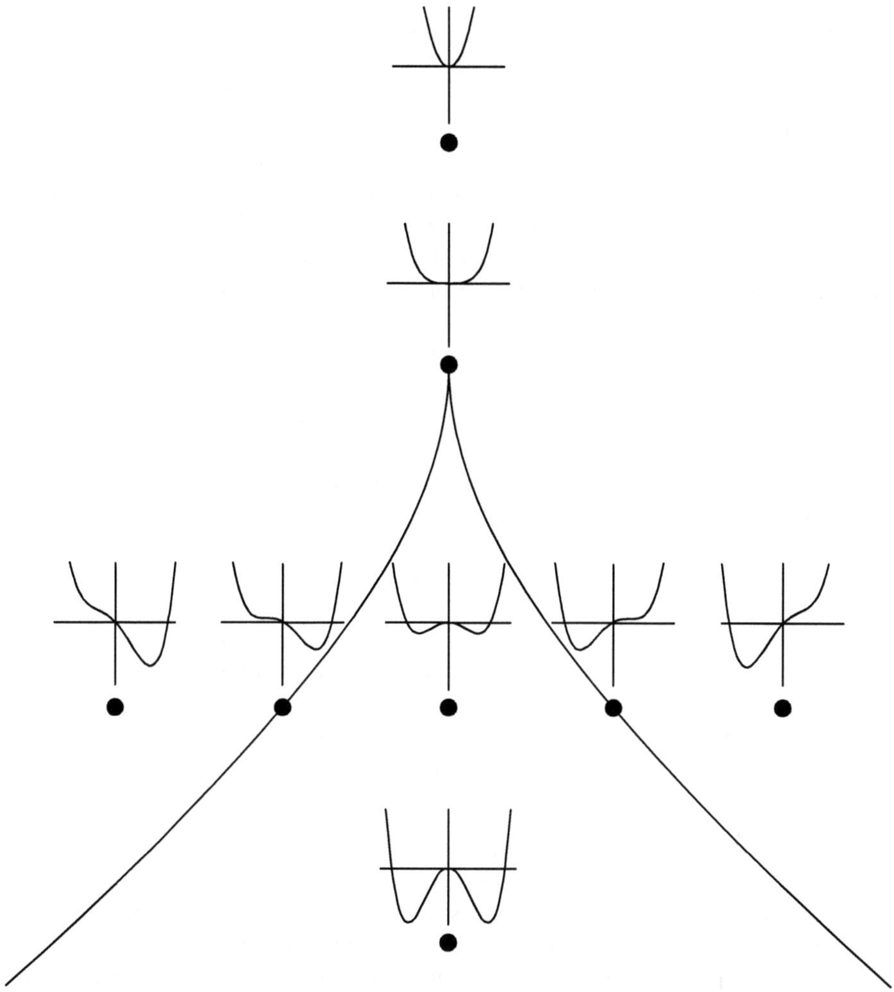

**Fig. 8.5.** Illustration of some important properties of the values of the critical points in the cusp catastrophe $A_3$.

Finally return to the central function of the row in Fig 8.5. As we move from that function downward in the control space to the lowest function shown, we see that the two minima become lower within the function.

From the above information, we will now proceed with the following construction. Note first the labeling of the pair of axes for any function shown: The horizontal axis of the function is the variable $t$, and the vertical axis of the function will be called the $F$-axis, since it corresponds to the *values* (levels) of the function $F(t; a_1, a_2)$.

Now the construction procedure, to be used, begins as follows: Given a position $(a_1, a_2)$ in the control space, locate, in its function $F(t; a_1, a_2)$, the critical points of that function, and mark the *values* of the critical points on the $F$-axis of that function.

When one puts together all the marked $F$-axes, for all the positions $(a_1, a_2)$ in the control space, one obtains the surface shown in Fig 8.6. That is, the surface is the collection of all the *marks*. To explain fully: Observe that the vertical axis in Fig 8.6 is the $F$-axis. Also notice that the bottom plane in Fig 8.6 represents the control space of the cusp catastrophe. Pick any position $(a_1, a_2)$ in this control space. The vertical line, through the position, is the $F$-axis of the function located at this position in the control space. The points where this axis intersects the surface are the *values* of the critical points of this function; i.e., the marks established earlier.

Since this surface is the collection of the *values* of the critical points of all the functions in the control space, it will be called the *critical-value surface* of the cusp catastrophe. This contrasts with the surface shown in Fig 7.5 (p247), which shows the surface of critical-point positions on the $t$-axis of each function in the control space. To emphasize: In both figures, the horizontal plane is the control space. However, in Fig 7.5, the vertical axis is the $t$-axis of the functions, whereas in Fig 8.6, the vertical axis is the $F$-axis of the functions. This is why the surface in Fig 7.5 is the set of critical points, i.e., locations on the $t$-axis of functions, whereas the surface in Fig 8.6 is the set of critical values, i.e., locations on the $F$-axis of functions.

Let us now observe how the critical-value surface Fig 8.6 embodies the properties we observed of the functions in Fig 8.5. Before beginning, note that the upper drawn version of the $a_1$-axis and $a_2$-axis in Fig 8.6 defines the plane at value 0 of the $F$-axis. The lower drawn version of the $a_1$-axis and $a_2$-axis is shown only to help the reader more easily visualize the cusp-shaped separatrix of the cusp catastrophe. However, this plane *should* be understood as coincident with the $F = 0$ plane at the middle height of the figure.

First, concerning Fig 8.5, we noted that any function on the positive $a_2$-axis, and the germ, have a single minimum with value 0. This is the reason why, in Fig 8.6, the positive $a_2$-axis is *coincident* with the critical-value surface, and any vertical line through this axis intersects the surface only once.

We also noted, concerning Fig 8.5, that when one goes to either side of a function on the positive $a_2$-axis, the minimum goes lower within the function. This is the reason why, in Fig 8.6, the surface moves downward on both sides of the positive $a_2$-axis.

Next we considered, in Fig 8.5, the row of functions along the middle of the diagram, starting with the left-most function. We saw that, as the function moves rightward along

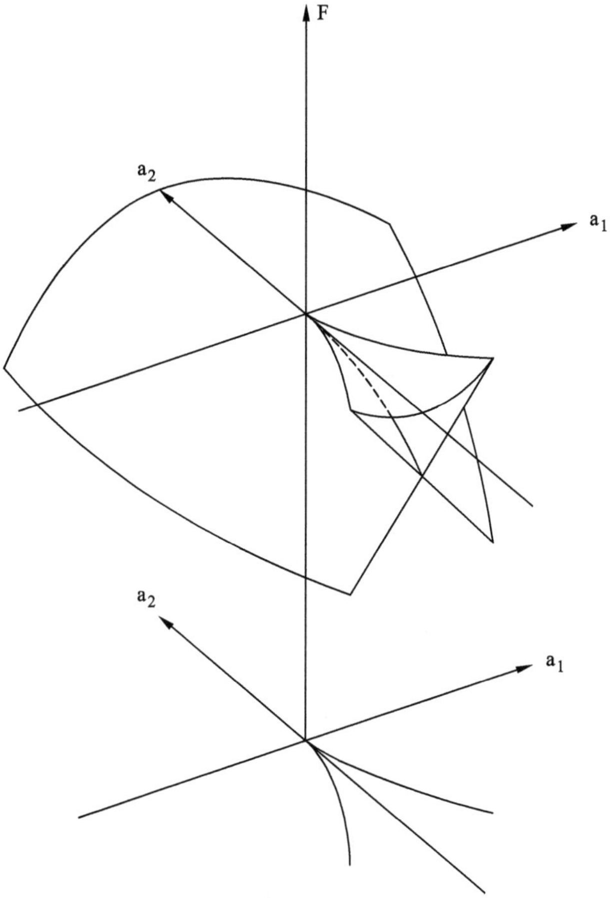


**Fig. 8.6.** The critical-value surface for the cusp catastrophe.

the row, the minimum moves upward within the function, and disappears when the right separatrix curve is encountered. This is the reason why, in Fig 8.6, the lower left sheet of the surface in the front half ($a_2 < 0$), rises and eventually terminates over the right separatrix curve.

Then we considered again, in Fig 8.5, the row of functions along the middle of the diagram, starting with the right-most function. We saw that, as the function moves leftward along the row, the minimum moves upward within the function, and disappears when the left separatrix curve is encountered. This is the reason why, in Fig 8.6, the lower right sheet of the surface in the front half ($a_2 < 0$), rises and eventually terminates over the left separatrix curve.

Then, at the center of the row in Fig 8.5, we saw that the two minima (which have been moving upwards respectively in the rightward and leftward direction in the row) have now reached the same height. The fact that the two minima are now on the same level corresponds, in Fig 8.6, to the central curve where the two rising sheets, that we have just discussed, *cross each other*; i.e., reach the same height on the $F$-axis.

Then we recalled, from section 7.4, the fact that the *pair*, minimum and maximum, *created* by crossing a separatrix curve, both initially have positive value; and then, in moving toward the center of Region 3, the minimum becomes negative and the maximum becomes zero. This is seen in Fig 8.6 as follows: Above a separatrix curve of the cusp catastrophe, the critical-value surface has a cuspidal edge. The cuspidal edge is above the $F = 0$ plane, and is therefore of positive $F$ value. From the cuspidal edge, there is the branching of two sheets: the higher one is a sheet of maxima and the lower one is a sheet of minima. Observe that, whereas the maxima sheet goes downward to become *tangential* to the $F = 0$ plane, the minima sheet goes downward *through* the $F = 0$ plane. This shows that the maxima do not get negative values, but the created minima go from positive to negative values.

Finally we noted, in Fig 8.5, that, as the central function of the row moves in the negative $a_2$ direction (i.e., towards the bottom of that figure), its minima become lower. This is why, in Fig 8.6, the central curve, where the two minima sheets cross each other, *bends downwards* towards the front.

Also observe this: All the sheets of the surface in Fig 8.6 correspond to minima, except the top sheet (between the two cuspidal edges) which corresponds to maxima, and the cuspidal edges which correspond to 2-fold degenerate critical points. Furthermore, the fact that the maxima of all the functions, along the negative $a_2$-axis, have zero value, explains why the negative $a_2$-axis, in Fig 8.6, is *coincident* with the maxima sheet.

Now let us return to the problem with which we are currently concerned: establishing the *zero-separatrix*, for the *Interactive Unfolding of* $EA_3$; i.e., establishing, in the 3D control space $(a_0, a_1, a_2)$ of that unfolding, those functions from which zeros will bifurcate as a result of perturbations of those functions.

Our proposed solution to this problem can now be stated: Recall that the first control variable $a_0$ moves any function $F(t; a_1, a_2)$, from the cusp catastrophe control space, up and down its $F$-axis. Now suppose that a critical point of a function $F(t; a_1, a_2)$ has value $c$ on its $F$-axis. Then, in order to raise (or lower) the function so that this critical point has value zero on its $F$-axis, it is necessary merely to add $-c$ to the function. That is, we must make its $a_0$ coordinate, in the Interactive Unfolding, equal to $-c$.

This means the following:

**In Method 1, the zero-separatrix for the Interactive Unfolding of $EA_3$ is obtained by vertically inverting the critical-value surface of the cusp catastrophe $A_3$, and replacing the inverted $F$-axis over the 2D control space of $A_3$ by the $a_0$-axis in the 3D control space of $EA_3$.**

This is illustrated by going from Fig 8.6 to Fig 8.7. Therefore, the surface in Fig 8.7 is the zero-separatrix for the Interactive Unfolding of $EA_3$.

While Fig 8.7 was derived from Fig 8.6 by inversion, it is important to understand that these two figures define two very different mathematical objects. To see this, we successively contrast them as follows:

Fig 8.6: The control space is only *two*-dimensional, i.e., it is only a subspace of the entire 3D space shown in this figure.

Fig 8.7: The control space is *three*-dimensional, i.e., it is the entire 3D space shown in this figure.

Fig 8.6: Given a point $(a_1, a_2)$ in the horizontal plane, the entire vertical line through that point corresponds to a *single* function $F(t; a_1, a_2)$.

Fig 8.7: Given a point $(a_1, a_2)$ in the horizontal plane, the vertical line through that point corresponds to an *entire family* of functions; i.e., all the raised and lowered versions of the function $F(t; a_1, a_2)$.

Fig 8.6: A single point in the vertical line through $(a_1, a_2)$ represents only one possible value of the function $F(t; a_1, a_2)$.

Fig 8.7: A single point in the vertical line through $(a_1, a_2)$ represents an entire function; i.e., a raised or lowered version of $F(t; a_1, a_2)$.

Now let us return to the important issue discussed earlier concerning the functions in Fig 8.1 p257. Recall that, using the criterion of transversality of the graph of a function *itself* with respect to the $t$-axis, the top two functions in that figure have a crucial similarity: they are both non-transversal, and therefore *both non-generic*. In contrast, in Catastrophe Theory, the top function, being in the positive $a_2$-axis, is regarded as *generic* but the middle function, being the germ (at the origin), is regarded as *non-generic*.

In Fig 8.7, we can now see that these two functions are *both* within the zero-separatrix surface; i.e., both the positive $a_2$-axis and the origin point are within the surface. Thus the diagram captures the fact that *both functions* are *non-generic*, and will therefore be *transitions in morphological operations* (in fact different ones), as we will illustrate.

The other transversality condition (that of the graph of the first derivative with respect to the $t$-axis) is still preserved because, given the critical-point separatrix in the *same* 3D control space, i.e., the separatrix shown in Fig 8.4, the function in the positive $a_2$-axis is outside that separatrix, but the other function, the germ, is within it.

Thus, by having *both separatrices* in the 3D control space, we will capture crucial relationships between different types of morphological operations.

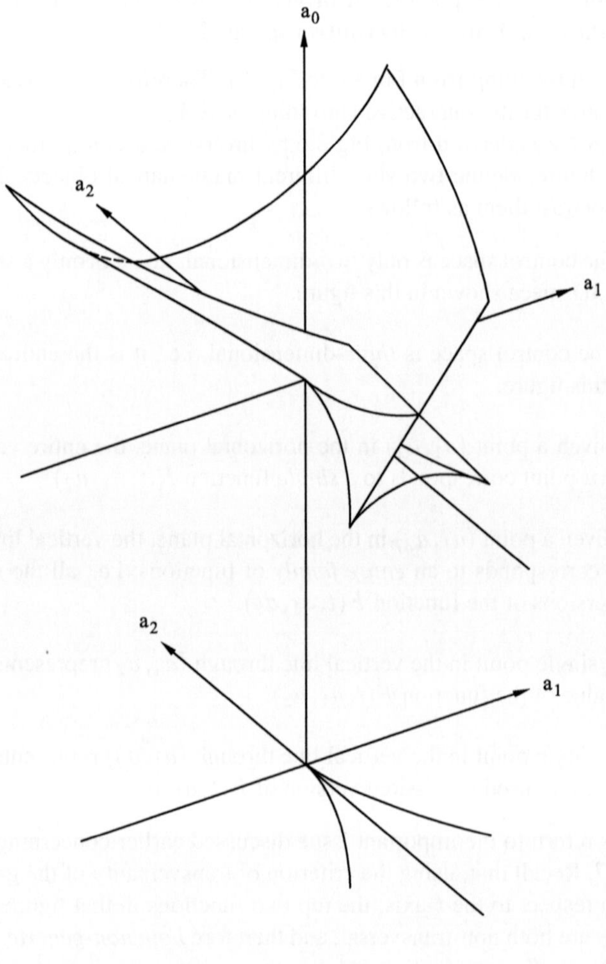


**Fig. 8.7.** The zero-separatrix for the Interactive Unfolding of $EA_3$.

Let us now return to the issue for which Fig 8.7 was constructed – that of defining the separatrix involved in the Process Grammar operation

$$Cm^+ : m^+ \longrightarrow 0m^-0$$

By our inversion method of producing the surface in Fig 8.7, we can see that all components of the surface correspond to minima that are non-degenerate as critical points, except the lowest sheet, i.e., the upward arch sheet (between the two cuspidal edges), which corresponds to maxima, and the cuspidal edges and origin, which correspond to degenerate critical points. The entire surface, without the upward arch sheet and cuspidal edges and origin, will be called the **minima zero-separatrix**, since it separates the positive minima $m^+$ (that are non-degenerate as critical points) above the surface, from their negative versions $m^-$ below the surface. The minima zero-separatrix is the set of functions that contain a minimum that is non-degenerate as a critical point, but has value 0, and is therefore 2-fold degenerate as a 0, and is therefore *non-generic*.

The operation $Cm^+$ is applicable to any positive minimum $m^+$ that is nondegenate as a critical point, in any function (in the entire 3D control space) that contains such a minimum. All such functions lie above *at least one component* of the minima zero-separatrix. The operation $Cm^+$ pulls the function downwards through the surface sheet corresponding to that minimum; i.e., lowering the value of that minimum to 0 when the function reaches that sheet, and then lowering the minimum further (making it negative), while bifurcating the 0, as the function goes below that sheet.

Except on the $a_0$-axis (i.e., vertically) through the origin, *all* functions in the control space have either *one* or *two* minima that are non-degenerate as critical points. The application of $Cm^+$ to positive minima in these two respective cases, i.e., *one* or *two* minima, will be illustrated in Fig 8.8 as follows: Consider the left vertical line, i.e., that through the point labeled $p$ on the positive $a_2$-axis. We will call it the $p$-line. Any point on the $p$-line represents a function with a single minimum. Consider, on this line, a point above the zero-separatrix. It represents a function whose graph is illustrated at the top of the $p$-line; i.e., the minimum is positive. Moving down the $p$-line is a process of simply lowering the function on its $F$-axis. Thus, on moving down, as one crosses the zero-separatrix, the minimum becomes 0 at the separatrix, and becomes negative below the separatrix, simultaneous to a 2-fold bifurcation of the 0. Thus, moving down the $p$-line realizes *one* application of the $Cm^+$ operation.

In contrast, consider the right vertical line in Fig 8.8, that through the point labeled $q$ in the $(a_1, a_2)$ control plane, between the two separatrix curves, but near one of them. We see that the $q$-line crosses two of the minima zero-separatrix sheets, i.e., at the points labeled 1 and 2. Now, any point on the $q$-line represents a function with two minimum, *one lower than the other*. Consider, a point that is on the $q$-line above *all* the separatrix sheets. It represents a function whose graph is illustrated at the top of the $q$-line; i.e., both minima are positive. Moving down the $q$-line will simply lower the function on its $F$-axis. Thus, on moving down, as we cross the upper separatrix sheet, the lower minimum of the function will reach 0 at that sheet and will bifurcate to $0m^-0$ below that sheet. This will be the first use of operation $Cm^+$ along the $q$-line. Then, moving further down the $q$-line, crossing the second separatrix sheet, the upper minimum of the function will reach 0 at that sheet and will bifurcate to $0m^-0$ below that sheet. This will be the second use of operation $Cm^+$ along the $q$-line.

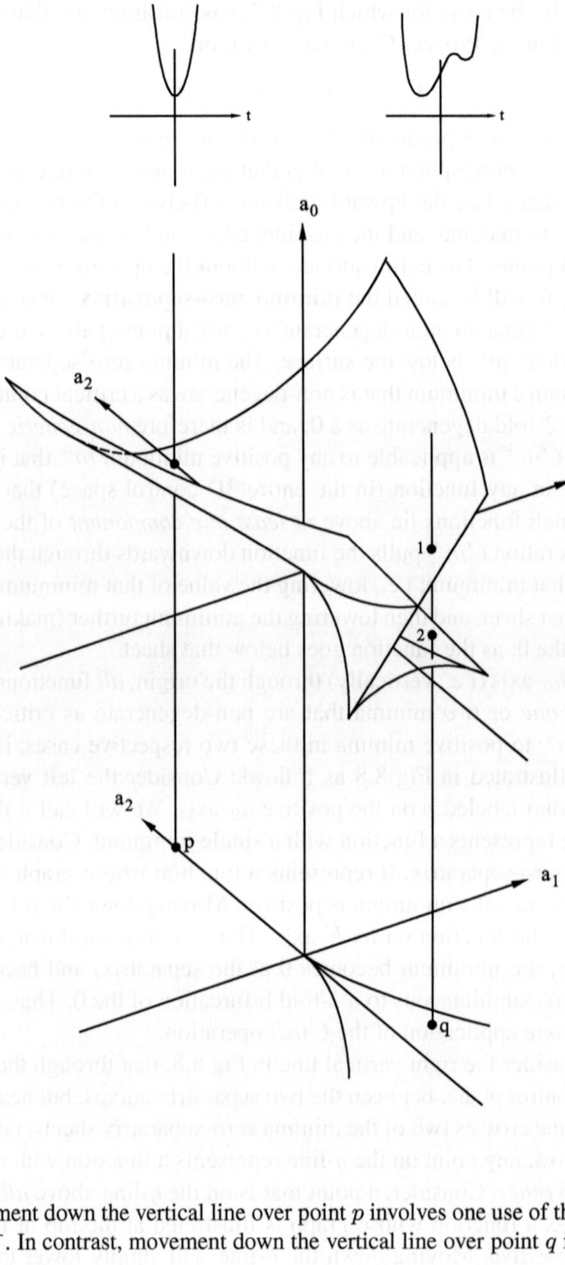


**Fig. 8.8.** Movement down the vertical line over point $p$ involves one use of the Process Grammar operation $Cm^{+}$. In contrast, movement down the vertical line over point $q$ involves two uses of $Cm^{+}$.

## 8.6 Quantitative Example of the $Cm^+$ Operation

We shall now give a *quantitative* example of the fact that the downward movement through the minima zero-separatrix realizes the operation

$$Cm^+ : m^+ \longrightarrow 0m^-0$$

of the grammar, and is *causally compatible* with our process-inference rules.

The three figures on the next three pages, i.e., Fig 8.9, Fig 8.10, Fig 8.11, show the curvature functions on three successive points down the line $(a_1, a_2) = (0, 3)$ in the control space. This is an example of the $p$-line in Fig 8.8. The three successive points are: (1) above the zero-separatrix surface, (2) on that surface, and (3) below that surface.

In our theory, the three points represent *curvature* functions of curves. In the three figures, Fig 8.9, Fig 8.10, Fig 8.11, we will make them curvature functions of planar curves. Thus, together with the three functions, the three figures will also give their resulting unit-speed parameterized curves, that are planar. In fact, throughout the book, the calculated parameterized curves, that we give in *quantitative* examples of the Process Grammar operations, will always be planar, to help the reader visualize them.

Throughout the book, the procedure of calculating and drawing the parameterized curve, from a curvature function, is as follows: Let the curvature function be $\kappa(t)$, and the unit-speed parameterized curve be given by $z(t)$ in the $(x, y)$ plane. Place the curve point $z(0)$ at the origin of the $(x, y)$-plane, and such that the curve tangent vector $z'(0)$ goes from the origin to the point $(1, 0)$ in the plane. The curve is then constructed in the usual way: Define $\theta(t)$ to be the angle between any tangent vector $z'(t)$ and $z'(0)$ such that the curvature condition $\theta'(t) = \kappa(t)$ is satisfied, and enforce the unit-speed conditions $x'(t) = cos(\theta(t))$ and $y'(t) = sin(\theta(t))$, to obtain the curve's parametric equations $z(t) = (x(t), y(t))$.

We take three pages to show the three figures because it is remarkable that, even people, who are familiar with differential geometry, find it difficult to use our 4-fold classification of extrema to understand the relation between curvature functions and parameterized curves.

**The text now continues in the captions of the three figures.**
**Therefore, the reader is recommended to read the captions of the three figures.**

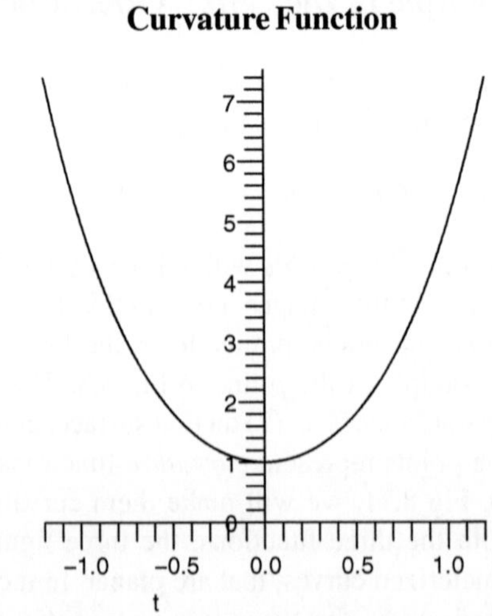


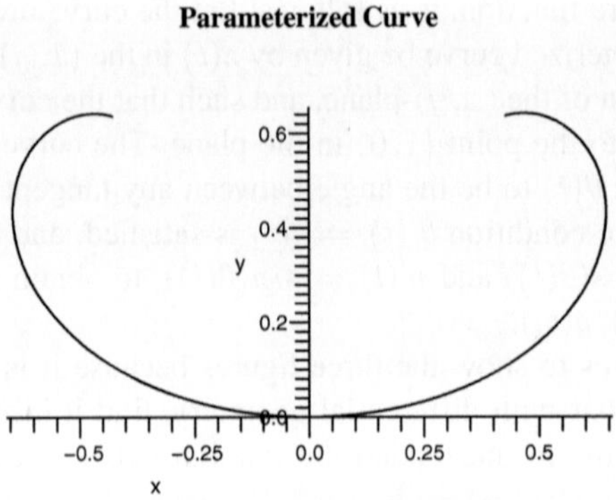


**Fig. 8.9.** Prior to zero-bifurcation: The upper figure shows the *curvature function* $F(t) = t^4 + 1 + 3t^2$ at the point $(a_0, a_1, a_2) = (1, 0, 3)$ which is *above* the zero-separatrix in the 3D control space of the Interactive Unfolding of $EA_3$. Therefore, as we can see, the minimum in the curvature function is positive, $m^+$; in fact, its value is $a_0 = 1$ as can be seen from both the equation and graph of the function. The lower figure shows the unit-speed *parameterized curve* produced by that curvature function. The positive minimum at the center of the curvature function occurs directly below it at the center of the parameterized curve.

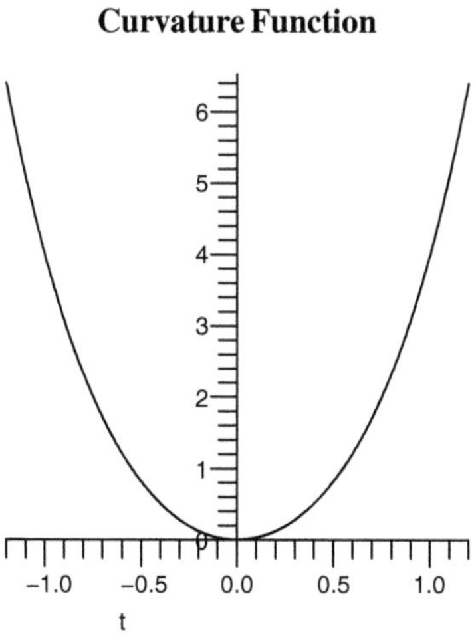


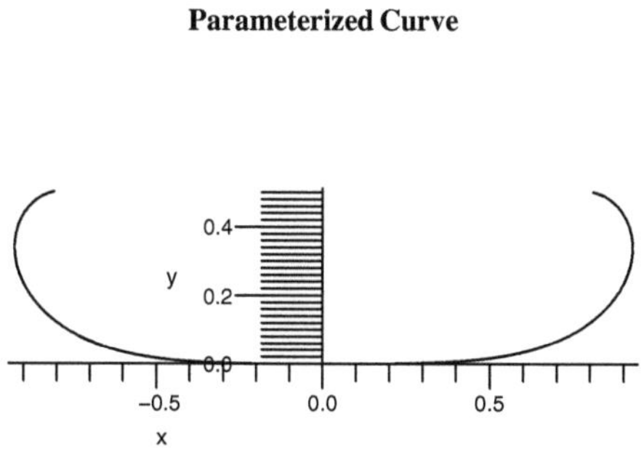


**Fig. 8.10.** On the zero-separatrix: The upper figure shows the *curvature function* $F(t) = t^4 + 3t^2$ at the point $(a_0, a_1, a_2) = (0, 0, 3)$ which is *on* the zero-separatrix in the 3D control space of the Interactive Unfolding of $EA_3$. Therefore, as we can see, the minimum in the curvature function is now zero; i.e., the extremum in the graph of the function is now in *non-transversal contact* with the $t$-axis. The lower figure shows the unit-speed *parameterized curve* produced by that curvature function. The zero minimum at the center of the curvature function occurs directly below it at the center of the parameterized curve. This is the first appearance of the curvature zero in the evolution.

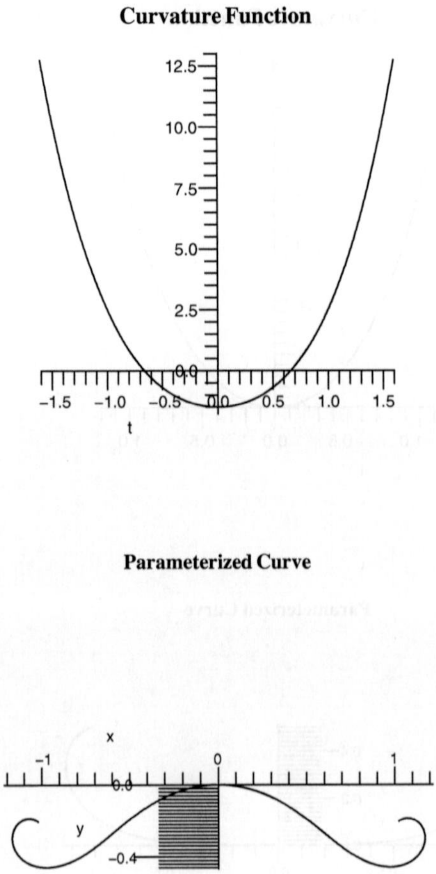


**Fig. 8.11.** After zero-bifurcation: The upper figure shows the *curvature function* $F(t) = t^4 - 1.5 + 3t^2$ at the point $(a_0, a_1, a_2) = (-1.5, 0, 3)$ which is *below* the zero-separatrix in the 3D control space of the Interactive Unfolding of $EA_3$. Therefore, as we can see, the minimum in the curvature function is now negative, $m^-$, and the zero of the previous function has bifurcated into two zeros at which the graph of the function intersects the $t$-axis *transversally*. The lower figure shows the unit-speed *parameterized curve* produced by that curvature function. The negative minimum $m^-$ is located at the center of the parameterized curve; and the two zeros (flat points) can be seen in the two sides of that extremum. The singularity-configuration is now $0m^-0$, which is the structure resulting from the application of $Cm^+$.

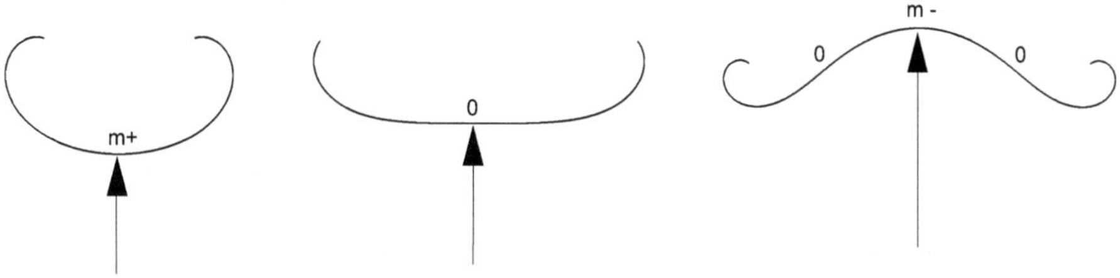


**Fig. 8.12.** The Process Grammar operation $Cm^+$ realized in the Interactive Unfolding of $EA_3$ is compatible with the force structure provided our process-inference rules.

Fig 8.12 shows the sequence of three curves from the previous three figures. The solid is on the upper side of the curve in each case. We can clearly see that the evolution is compatible with the force structure inferred by our process-inference rules, as shown by the arrows. That is, in the left-most figure, the arrow is a *squashing* force corresponding, in our theory, to the $m^+$ extremum at the bottom of this first curve. When this force continues pushing on the minimum, the curvature of the minimum decreases, first becoming zero, i.e., flat, as shown in the second shape. This is the transition state of the $Cm^+$ operation. Then, as the force continues pushing on the minimum, the curvature of the minimum becomes negative, as shown in the third shape, with the zero undergoing a 2-fold bifurcation that acts as the boundary between the emerging negative region and the surrounding positive region. By our process-inference rules, the negative minimum in this third shape corresponds to an *indentation*. Thus the sequence of shapes accords with our process-description of $Cm^+$ as *squashing continues till it indents*.

The figure therefore illustrates the important fact that the Interactive Unfolding of $EA_3$ is compatible with the causal structure defined by the $Cm^+$ operation.

## 8.7 Method 2 for Constructing the Zero-Separatrix

The first method of zero-separatrix construction (inversion of the critical-value surface, section 8.5) had the advantage that the *quantitative* properties of the separatrix surface were instantly understandable. For example, recall that, in the critical-value surface (Fig 8.6), the surface goes downward on each side of the positive $a_2$-axis because the minima move downward in the respective functions, as exhibited in Fig 8.5. This explains why the sides go upward in the zero-separatrix in Fig 8.7: The amount added to each function, in order to make the minimum 0, must be successively greater on each side.

We now give a second method (for constructing the zero-separatrix) which has a different advantage: It allows us to quickly understand the amount of degeneracy at the various types of points on the separatrix.

To describe this method, we start with the unfolding function defined earlier for the Interactive Unfolding of $EA_3$:

$$F(t; a_0, a_1, a_2) = t^4 + a_0 + a_1 t + a_2 t^2 \tag{8.3}$$

In the control space of this unfolding, the zero-separatrix is the set of functions $f$ where the domain of $f$ contains a point $t_0$ at which two conditions are satisfied simultaneously: both the function and its first derivative are zero. Obviously, this is the same as requiring that, at $t_0$, the *integral* of $f$ has its first derivative and second derivative equal to zero. But this is the same as saying that, for the integral of $f$, $t_0$ is a critical point that is degenerate, i.e., non-Morse. The conclusion is therefore this: The zero-separatrix of the unfolding (8.3) consists of those functions in the control space whose integrals are non-Morse.

Let us put the integral of the unfolding (8.3) in this form:

$$\tilde{F}(t; b_1, b_2, b_3) = t^5 + b_1 t + b_2 t^2 + b_3 t^3 \tag{8.4}$$

where the constant of integration has been set to zero and the coefficients have individually undergone the obvious scaling. The zero-separatrix of the unfolding (8.3) must have the same structure as the critical-point separatrix of (8.4) viewed as an unfolding on its control variables.

The usefulness of this conclusion lies in the crucial fact that (8.4) is the unfolding function for the next higher-order catastrophe $A_4$, called the *swallowtail* (listed in Table 7.1 p244). Most importantly, the critical-point separatrix of $A_4$ is exactly the same as the zero-separatrix shown in Fig 8.7 for the unfolding (8.3). This has therefore supplied us with the second method of producing that zero-separatrix. The above argument has been illustrating the following:

## THE BASIS OF METHOD 2

**The local structure of *zeros* , i.e., *contact* structure, in perturbations of the germ $t^{n+1}$, has universal unfolding**

$$F(t; a_0, a_1, \ldots a_{n-1}) = t^{n+1} + a_0 + a_1 t + \ldots + a_{n-1} t^{n-1} \tag{8.5}$$

**with control-space dimension $n$.**
**The local structure of *critical points*, in perturbations of the next higher-power germ $t^{n+2}$, has universal unfolding**

$$\tilde{F}(t; b_1, \ldots b_n) = t^{n+2} + b_1 t + \ldots + b_n t^n \tag{8.6}$$

**with the *same* control-space dimension $n$ as (8.5).**
**Furthermore, the zero-separatrix of the unfolding (8.5) has the same structure as the critical-point separatrix of the unfolding (8.6).**

**Method 2 uses this fact to study certain properties of the zero-separatrix of (8.5) as properties of the critical-point separatrix of (8.6).**

**We have seen that the zero-separatrix of the Interactive Unfolding of $EA_n$ has the same structure as the critical-point separatrix of the next higher order catastrophe $A_{n+1}$.**

There are several benefits to the above understanding. One of them is that, since, in our system, the unfolding (8.5) represents the curvature along a curve, its integral, which will correspond to the unfolding (8.6) represents the angle of rotation along the same curve. The above therefore means that one can handle bifurcations in the zero-structure of curvature, and bifurcations in the critical-point structure of angle, simultaneously. The benefits of this will be illustrated now with the Interactive Unfolding of $EA_3$.

To begin, it is useful to emphasize the following notational feature: In expression (8.3) we have given the 3D control space of $EA_3$ as $(a_0, a_1, a_2)$, and in expression (8.4) we have given the 3D control space of $A_4$ as $(b_1, b_2, b_3)$, where in both cases, the subscript $i$ on a control variable indicates that the variable controls the term $t^i$ in its unfolding function.

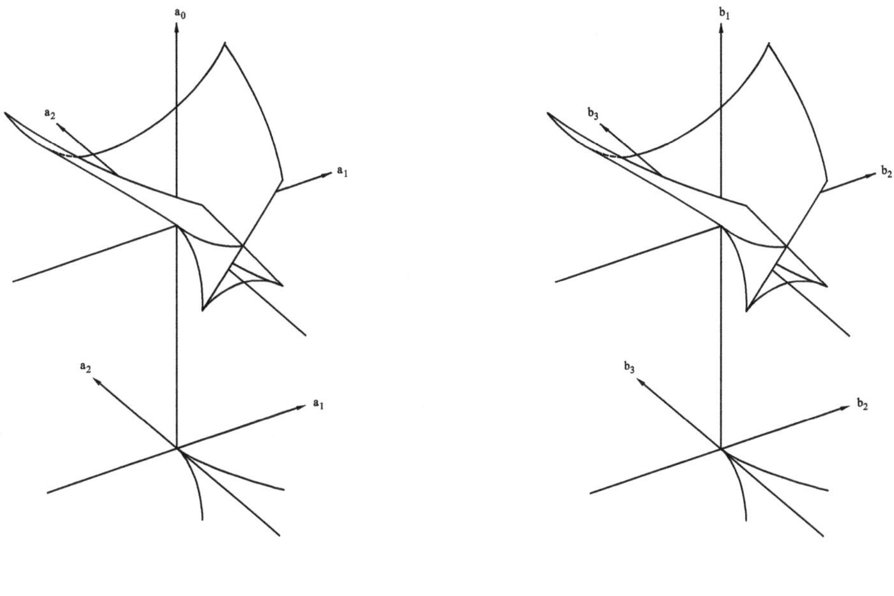


Zero-separatrix of $EA_3$ Critical-point separatrix of $A_4$

**Fig. 8.13.** The zero-separatrix of $EA_3$ and the critical-point separatrix of $A_4$.

Keeping this in mind, the left diagram in Fig 8.13 shows the zero-separatrix for $EA_3$; and the right diagram shows the critical-point separatrix for $A_4$. The two separatrices have been oriented in the way that illustrates the fact that they have the same structure, and this shared orientation corresponds of course to the relationship established by integration. Notice therefore that, for $EA_3$, the vertical axis $a_0$, which simply raises and lowers a function with respect to its $F$-axis, corresponds, in $A_4$, to the vertical axis $b_1$, which controls the *linear* term in the unfolding; i.e., it rotates the tangent to the graph of the function. Similarly, the axis $a_1$ which controls the linear term in the unfolding for

$EA_3$ is oriented in the same way as the axis $b_2$ which controls the quadratic term in the unfolding for $A_4$. Finally, the axis $a_2$ which controls the quadratic term in the unfolding for $EA_3$ is oriented in the same way as the axis $b_3$ which controls the cubic term in the unfolding for $A_4$. These facts, which obviously come from the integration relation between $EA_3$ and $A_4$, are worth pointing out in order to gain a qualitative understanding of the functions and bifurcations in the following discussion.

Now let us turn to the issue of degeneracy. Any function in the zero-separatrix for $EA_3$ has a point in its domain where the value of the function is 0 in a degenerate way; i.e., the graph of the function is in tangential contact with its $t$-axis. This type of degeneracy will be called **zero-degeneracy**, and the degree $n$ of this degeneracy will be called **$n$-fold zero-degeneracy**. From the above discussion, we can see this:

> **The degree of zero-degeneracy at positions in the zero-separatrix of the Interactive Unfolding of $EA_n$ is the degree of critical-point degeneracy at the corresponding positions in the critical-point separatrix of the next higher-order catastrophe $A_{n+1}$.**

We will see, throughout this book, that, to understand certain important operations in the Process Grammar, it will be valuable to specify the degree of degeneracy for the zero-degenerate points of functions in each of the different components of the zero-separatrix of the Interactive Unfolding of $EA_3$. By the above statement, these zero-degeneracies are the same as the critical-point degeneracies of the corresponding points in the critical-point separatrix of the swallowtail catastrophe $A_4$. We shall now elaborate these degeneracies, as follows:

First observe that, in the two spaces, that for $EA_3$ and that for $A_4$, shown respectively in the left and right diagrams in Fig 8.13, the germ for each space is located at the origin. In the unfolding (8.3) for $EA_3$, the germ is $t^4$. An important fact that we will exploit later is that this function contains a *4-fold degenerate zero*. Correspondingly, in the unfolding (8.4) for $A_4$, the germ is $t^5$. Clearly, this function contains a point that is also 4-fold degenerate, but as a *critical point*.

The two germs just described, i.e., for $EA_3$ and $A_4$, are shown in Fig 8.14. Their 4-fold degeneracy will become crucial later when we examine the **breaking-through of a zone-restricted penetrative force** in the Process Grammar.

Now let us consider the *minima zero-separatrix* of $EA_3$. In section 8.6, we saw that descending through any sheet of the minima zero-separatrix corresponds to the use of the Process Grammar operation $Cm^+$. The degeneracy fact that makes this operation

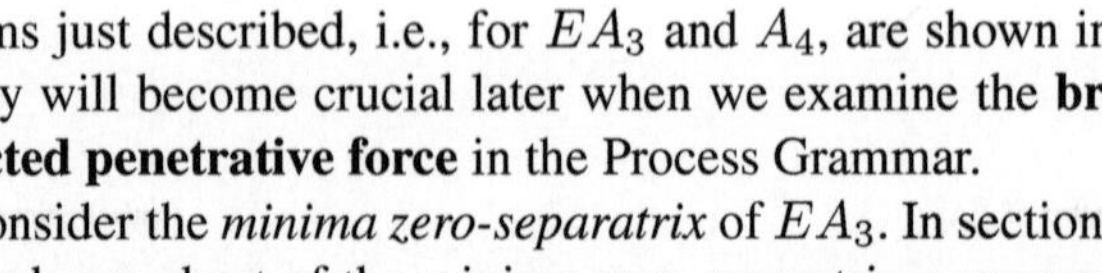


**Fig. 8.14.** Left: The germ for $EA_3$. Right: The germ for $A_4$.

possible is that the function at any point in the minima zero-separatrix has a 2-fold degenerate zero, and the downward movement through that point results in a 2-fold bifurcation of that zero. In particular, in section 8.6, we gave a quantitative illustration of the operation $Cm^+$ as a downward movement through the positive $a_2$-axis. To help the reader, we will currently regard this as illustrating a downward movement through the sheet region corresponding to the positive $a_2$-axis; i.e., the "curved back-sheet" of the zero-separatrix of $EA_3$, in the orientation illustrated in Fig 8.13. Later we will consider $Cm^+$ also as the downward movement through either of the front diagonal sheets of the minima zero-separatrix. We now see that the curved back-sheet of the zero-separatrix of $EA_3$ corresponds to the curved back-sheet of the critical-point separatrix of $A_4$, that is, the sheet region corresponding to the positive $b_3$-axis in the control space of $A_4$.

In the quantitative example, that we gave in section 8.6, of the $Cm^+$ operation, it will now be useful to exhibit the relations between the functions in the $EA_3$ space, i.e., the *curvature* functions, and the corresponding functions in the $A_4$ space, i.e., the *angle* functions involved.

This will be done in the three figures on the next three pages; i.e., Fig 8.15, Fig 8.16, Fig 8.17, which show the same quantitative example of $Cm^+$ as in section 8.6, but add the angle functions from $A_4$. Recall that, in the example of section 8.6, we moved down the particular vertical line $(a_1, a_2) = (0, 3)$ in the 3D control space of $EA_3$, through the zero-separatrix.

**The text now continues in the captions of these three figures.**

**It is necessary for the reader to read the three captions, because the information in the captions is subsequently used to describe the relation between various singularities that give insights concerning the operation $Cm^+$.**

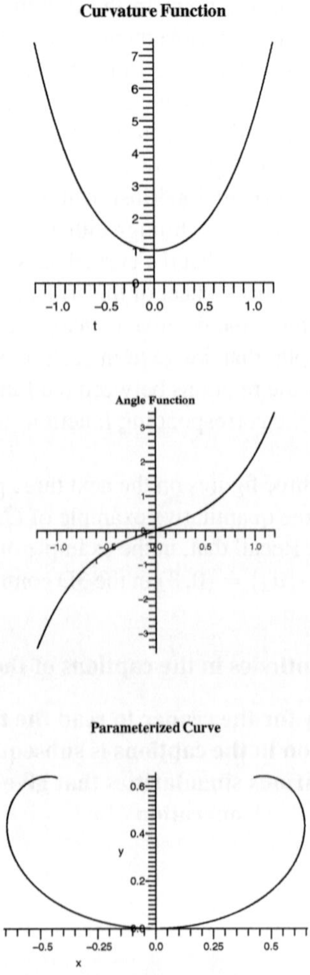


**Fig. 8.15.** Prior to zero-bifurcation: The figure shows the curvature function, angle function, and parameterized curve at the point $(a_0, a_1, a_2) = (1, 0, 3)$ in the control space of the Interactive Unfolding of $EA_3$. Observe from left to right along the $t$-axis that, whereas the curvature function decreases and then increases, the angle function purely increases. The effect of this can be seen in the parameterized curve where the anti-clockwise rotation of the tangent vector only increases as one moves along the curve. The successively decreasing value along the left half of the curvature function corresponds to the successively decreasing gradient along the left half of the angle function, which corresponds to the successive flattening along the left half of the parameterized curve. The successively increasing value along the right half of the curvature function corresponds to the successively increasing gradient along the right half of the angle function, which corresponds to the successive sharpening along the right half of the parameterized curve. The minimum in the curvature function corresponds to the point on the angle function at which the successively decreasing gradient changes to a successively increasing gradient, which corresponds on the parameterized curve to where the successive flattening along the curve changes to successive sharpening. This is the point to which the force is applied in our Process Grammar.

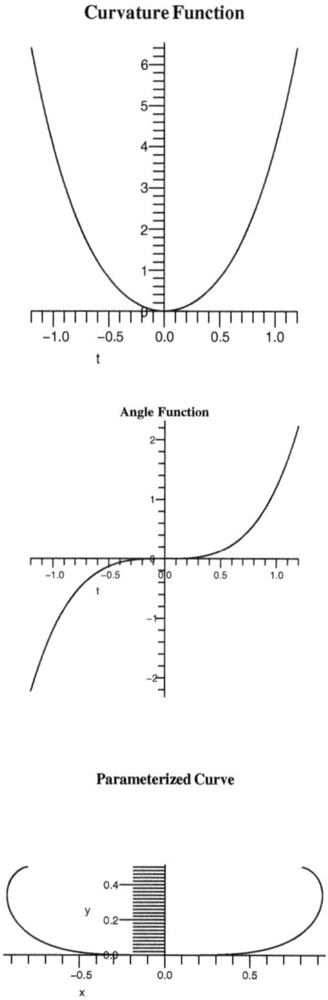


**Fig. 8.16.** On the zero-separatrix: The figure shows the curvature function, angle function, and parameterized curve at the point $(a_0, a_1, a_2) = (0, 0, 3)$ in the control space of the Interactive Unfolding of $EA_3$. All the decreasing/increasing features described in the caption of the previous figure are present in this figure. However, there is one extra feature: The critical-point minimum of the curvature function has now reached value zero. This corresponds to the fact that the angle function now has zero gradient at that point, i.e., the point is now a critical point *also* of the angle function. In fact, it must be a degenerate critical point of the angle function, because the function's first derivative (the curvature function) and its second derivative are both zero at this point. Thus, the angle function has now become non-Morse, and therefore lies on the critical-point separatrix of $A_4$. That is, the point we are considering here, a point on the zero-separatrix of $EA_3$, corresponds to a point on the critical-point separatrix of $A_4$.

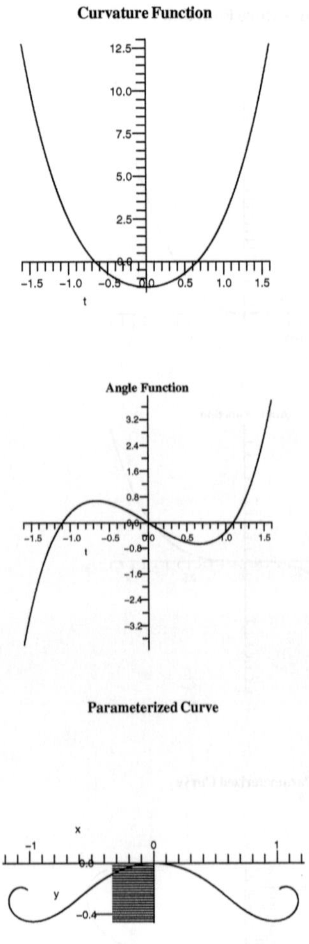


**Fig. 8.17.** After zero-bifurcation: The figure shows the curvature function, angle function, and parameterized curve at the point $(a_0, a_1, a_2) = (-1.5, 0, 3)$ in the control space of the Interactive Unfolding of $EA_3$. The left half of the curvature function (which is still purely decreasing) now has a zero point. This means that, in the left half of the angle function, although the gradient is always decreasing, there is a point where the angle value, which is initially increasing, starts to decrease. This corresponds, on the parameterized curve below, to the fact that, as one moves along the left half of the curve, the tangent vector starts by rotating anti-clockwise but then changes to rotating clockwise. This change occurs approximately half-way along the left half of the curve. The change-point corresponds to a zero in the curvature function and a critical point (maximum) in the angle function. The reflectionally-opposite situation occurs in the right half of the parameterized curve, due to a zero in the right half of the curvature function, and therefore a critical point (minimum) in the right half of the angle function. The two zeros in the curvature function bifurcated from the degenerate zero at the origin of the curvature function in the previous figure. Correspondingly, the two critical points in the angle function (maximum and minimum) bifurcated from the degenerate critical inflection point at the origin of the angle function in the previous figure.

It is important now to observe the following: In the three successive figures just given, the three *angle* functions are functions that characterize the *fold* catastrophe $A_2$. It will be valuable to quickly recall, in this paragraph, the following facts concerning the fold catastrophe: This catastrophe is the following unfolding of the cubic germ $t^3$:

$$F(t; a_1) \quad = \quad \overbrace{t^3}^{\text{germ}} \quad + \quad \overbrace{a_1 t}^{\text{perturbation}} \tag{8.7}$$

The control space is 1-dimensional given by the control variable $a_1$ shown in this equation. The germ $t^3$ is of course located at the origin $a_1 = 0$ of this control space, and is shown as the middle function in Fig 8.18. Notice that this function has an inflection point at $t = 0$, and this point is 2-fold degenerate as a critical point. Now, as the control variable $a_1$ is made negative, this degenerate critical point instantly undergoes a 2-fold bifurcation, giving functions that each have a non-degenerate maximum and non-degenerate minimum, an example of which is the right-most function in Fig 8.18. Conversely, as the control variable $a_1$ is made positive, the critical point of the germ disappears, giving functions that have no critical point, an example of which is the left-most function in Fig 8.18. Notice therefore that the separatrix in the 1D control space of $A_2$ is only a single point, the germ $t^3$. It separates two regions of Morse functions: The positive $a_1$-axis and the negative $a_1$-axis.

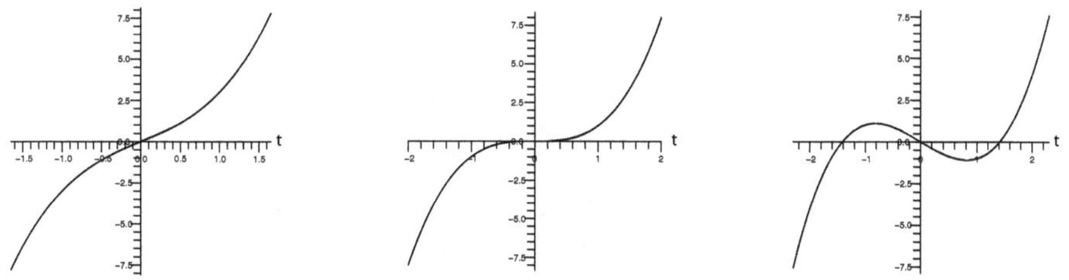

**Fig. 8.18.** The functions at points $a_1 = 2, 0, -2$ in the 1-dimensional control space $a_1$ of the fold catastrophe $A_2$.

Thus, the three successive functions from left to right in Fig 8.18, which illustrate the unfolding in the *fold* catastrophe $A_2$, exhibit the 2-fold critical-point bifurcation that we also see in the three successive *angle* functions on pages 284 – 286 that occur in the unfolding of the *swallowtail* catastrophe $A_4$.

The reason is this: Even though the three angle functions are taken from the control space of the swallowtail catastrophe $A_4$, the separatrix point that they cross has a critical-point that is 2-fold degenerate. In fact, while the functions in the swallowtail catastrophe $A_4$ are the unfolding (8.4) of the quintic $t^5$, Catastrophe Theory shows that crossing a point in the above separatrix-component of the swallowtail catastrophe $A_4$ is right-equivalent to crossing the single separatrix-point in the universal unfolding given by the fold catastrophe $A_2$.

Based on this, it is important now to observe the following:

## CONTINUATION OPERATION $Cm^+$

**The singularities of the Process Grammar operation $Cm^+$, which describes the continuation of a squashing process till it indents, are given by a downward movement of the curvature function through a point in the minima zero-separatrix of the Interactive Unfolding of $EA_3$. This movement corresponds to a downward movement of the angle function through the corresponding point in the critical-point separatrix of the swallowtail catastrophe $A_4$, this being right-equivalent to movement through the single-point separatrix of the fold catastrophe $A_2$.**

**However, observe the following crucial facts:**
**Catastrophe Theory defines each function in the positive $a_1$-axis of the fold catastrophe $A_2$ as having *no* singularity considered in Catastrophe Theory, and defines the same for each function in the corresponding trajectory in the swallowtail catastrophe $A_4$.**

**In contrast, Interactive Singularity Theory defines each function in the corresponding trajectory in the Interactive Unfolding of $EA_3$ as having a minimum, and, by the Symmetry-Curvature Duality Theorem, this minimum corresponds to a PISA axis which *causally explains* the minimum and also *causally explains* the transition through the separatrix.**

**Therefore, Catastrophe Theory fails to give the causal structure of the important morphology defined by $Cm^+$, whereas Interactive Singularity Theory gives the causal structure.**

Although the example, on pages 284 – 286, illustrated these statements by going downward through the "back curved sheet" of the zero-separatrix in Fig 8.13, these statements are true for a downward movement through *any* of the sheets in the minima zero-separatrix of the Interactive Unfolding of $EA_3$, e.g., through either one of the diagonal sheets at the "front" part of that separatrix. That is, a downward movement through either of the diagonal sheets of the Interactive Unfolding of $EA_3$ also induces a 2-fold zero-bifurcation of the type that defines $Cm^+$.

Note that, correspondingly, any point in either of the diagonal sheets in the critical-point separatrix of the swallowtail catastrophe $A_4$ corresponds to the critical-point separatrix in the fold catastrophe $A_2$. This is shown in Fig 8.19, which is a vertical planar slice through the front part of the critical-point separatrix of the swallowtail catastrophe $A_4$. In this diagram, the label $A_2$ applies to any point on the two diagonal sheets except the cusp-points at the bottom of those sheets.

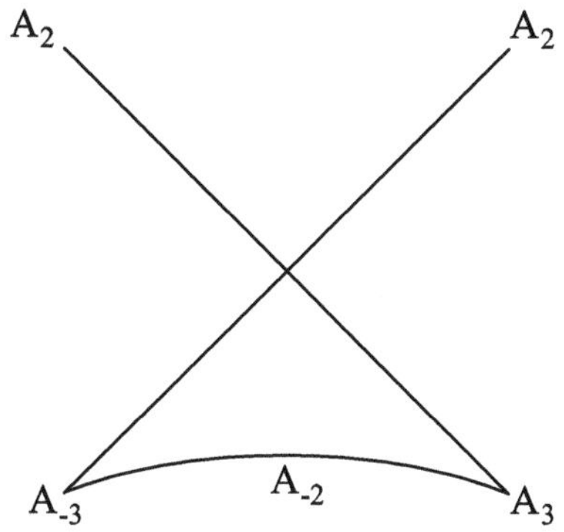


**Fig. 8.19.** The right-equivalence of points in the critical-point separatrix of the swallowtail catastrophe $A_4$ with points in critical-point separatrices of lower order catastrophes

To understand the relevance of the $A_2$ labels in Fig 8.19, and the other critical-point labels $A_k$ in this figure, consider Fig 8.20, which shows representative functions for that entire slice of the swallowtail catastrophe $A_4$. In particular, observe that the functions on the separatrix are the only functions in the space that have a degenerate critical point; and the degree of degeneracy of that point in a function has been indicated by the number we have put there on the function.

Thus, first observe that, in Fig 8.20, the functions shown along the two diagonal lines each have a 2-fold degenerate critical point (marked 2) and that locally to this critical point the function looks exactly like the germ $t^3$ of $A_2$ looks locally to its 2-fold degenerate critical point.

Next observe in Fig 8.19 that the bottom upward curve has the label $A_{-2}$. To understand this, note first that the bottom upward curve is a slice through the bottom upward arch sheet of the critical-point separatrix of the swallowtail catastrophe $A_4$ (shown in Fig 8.13). The label $A_{-2}$ given in Fig 8.19 marks the fact that any point in the bottom upward arch sheet of the critical-point separatrix of the swallowtail catastrophe $A_4$ corresponds to the single-point separatrix of the *dual* fold catastrophe $A_{-2}$. The functions in the dual fold catastrophe $A_{-2}$ are simply inversions of the functions in the fold catastrophe $A_2$. Thus, in particular, representative functions of the dual fold catastrophe $A_{-2}$ are inversions of those shown in Fig 8.18 for the fold catastrophe $A_2$.

Corresponding to this, observe in Fig 8.20 that the representative function shown on the bottom upward arch sheet contains a 2-fold degenerate critical point, within the graph of that function. Furthermore, observe that, *within that graph*, the graph goes downward from left to right in a neighborhood of its 2-fold degenerate critical point, whereas, in the case of an $A_2$ critical point, the graph goes upward from left to right in a neighborhood of its 2-fold degenerate critical point. This is why the degenerate critical point of any function on the bottom upward arch sheet of the critical-point separatrix of the swallowtail catastrophe $A_4$ is labeled with the singularity of the *dual* fold catastrophe $A_{-2}$.

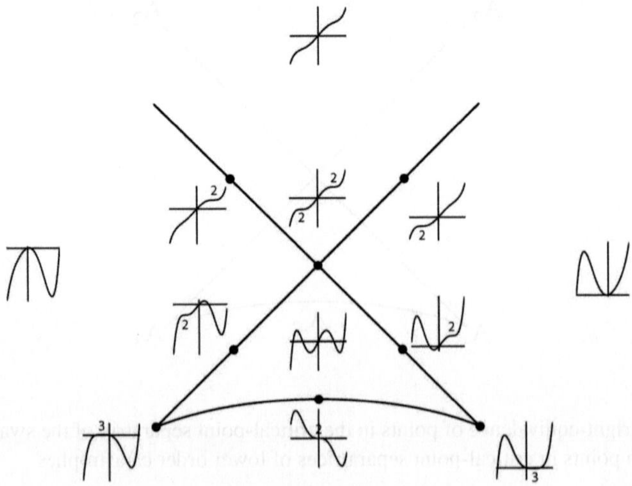

**Fig. 8.20.** Representative functions in the control space of the swallowtail catastrophe $A_4$.

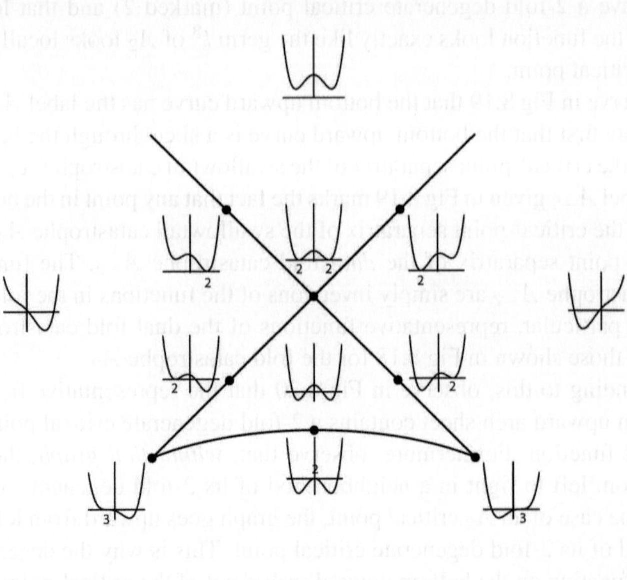

**Fig. 8.21.** Corresponding functions in the control space of the Interactive Unfolding of $EA_3$.

The next thing to observe in Fig 8.20 is this: Consider the *upward crossing* of the bottom upward arch sheet, given by the three shown successive representative functions, i.e., starting with the bottom representative function shown in the figure, followed by the representative function shown *on* the bottom upward arch sheet itself, and then followed by the representative function shown *above* the bottom upward arch sheet. Observe that this succession of three graphs exhibit, in a neighborhood of the *origin* of those three graphs, the 2-fold critical-point bifurcation that is simply the inversion of the 2-fold critical-point bifurcation given in the fold catastrophe $A_2$.

Now consider the remaining singularity labels shown in Fig 8.19. In this figure, we see that the *right* cusp-point, in the slice of the critical-point separatrix of the swallowtail catastrophe $A_4$, is labeled $A_3$. This is because it corresponds to the cusp-point in the critical-point separatrix of the cusp catastrophe $A_3$. This means that the function at this point contains a *3-fold degenerate minimum*. This fact is illustrated in the function shown at the right cusp-point in Fig 8.20, which is shown as containing a 3-fold degenerate minimum.

Also in Fig 8.19, we see that the *left* cusp-point, in the slice of the critical-point separatrix of the swallowtail catastrophe $A_4$, is labeled $A_{-3}$. This is because it corresponds to the cusp-point in the critical-point separatrix of the *dual* cusp catastrophe $A_{-3}$. This means that the function at this point contains a *3-fold degenerate maximum*. This fact is illustrated in the function shown at the left cusp-point in Fig 8.20, which is shown as containing a 3-fold degenerate maximum.

With respect to the slice shown in Fig 8.20, let us now consider the issues raised in the bold print on page 288. To do so, consider Fig 8.21, which shows the corresponding slice through the *zero*-separatrix of the Interactive Unfolding of $EA_3$. Fig 8.21 shows the functions in this space that correspond to the functions exhibited in the swallowtail catastrophe $A_4$ slice shown in Fig 8.20. That is, these are the *curvature functions* given by the Interactive Unfolding of $EA_3$ corresponding to the *angle functions* shown for the swallowtail catastrophe $A_4$.

By comparing the functions in these two figures, we can see illustrated the fact that an $n$-fold degenerate critical point within a function in the control space of the swallowtail catastrophe $A_4$ corresponds to an $n$-fold degenerate zero within a function in the control space of the Interactive Unfolding of $EA_3$. That is, the numbers put on the graphs in Fig 8.20 are the same as the numbers put on the graphs in Fig 8.21, where the former numbers are the degeneracies of critical points, and the latter numbers are the degeneracies of zeros.

Now we use the functions shown in these two figures to illustrate additional facts concerning the crucial issues given in the bold print on page 288.

First observe the following important fact: The top function shown in Fig 8.20 is a representative function of all the functions above the critical-point separatrix of the swallowtail catastrophe $A_4$. According to our theory, a crucial morphological mistake made by Catastrophe Theory is that it characterizes all these functions as being empty of the singularities considered by Catastrophe Theory. That is, it considers these functions as having no singularities of importance to morphology.

In contrast, consider the top function shown in Fig 8.21. This is a representative function of all the functions above the *front* part of the zero-separatrix of the Interactive Unfolding of $EA_3$. Observe that it has a positive maximum $M^+$ and two positive minima $m^+$. By the Symmetry-Curvature Duality Theorem, each of these extrema corresponds to a PISA axis that *causally explains* the extremum, which our theory claims is fundamental to the morphology, and which Catastrophe Theory completely fails to notice.

Furthermore, recall that a crucial fact is that, above the *back curved sheet* of the zero-separatrix of the Interactive Unfolding of $EA_3$, the functions all have *one* extremum, a positive minimum $m^+$. Thus, according to Interactive Singularity Theory, there is a crucial additional separatrix between the front and back regions above the zero-separatrix of the Interactive Unfolding of $EA_3$, and it gives the important morphological transition defined by the Process Grammar operation $Bm^+$. This additional separatrix and its relation to the operation will be explained in detail later.

Currently, we are considering the continuation operation $Cm^+$. Thus consider again the top function shown in Fig 8.21, that is, the representative function given of all the functions above the front part of the zero-separatrix of the Interactive Unfolding of $EA_3$. Observe in Fig 8.21 that, in crossing the left diagonal sheet, the right minimum of this function becomes zero in the function at this sheet and becomes negative after crossing the sheet. This corresponds to the continuation operation $Cm^+$.

Similarly, observe that, in going from the top function in Fig 8.21 across the right diagonal sheet, the left minimum of the function becomes zero in the function at this sheet and becomes negative after crossing that sheet. Again, this corresponds to the continuation operation $Cm^+$.

Again, because Catastrophe Theory regards the top function as not containing a singularity that it considers, it fails to give the causal structure of the transition across the left diagonal sheet and the causal structure of the transition across the right diagonal sheet. In contrast, because Interactive Singularity Theory gives the full singularities, the critical points and the zeros, it gives the full causal structure; i.e., *continuation of a compressive process till it penetrates.*

Recall that another reason why Catastrophe Theory fails to give the essential causal structure that defines the morphology given by these transitions is that Catastrophe Theory regards crossing the left diagonal separatrix sheet and the right diagonal separatrix sheet as both equivalent to crossing the single-point separatrix of the fold catastrophe $A_2$ and Catastrophe Theory regards the functions on the $a_1$-axis of the fold catastrophe $A_2$ is being empty of the singularities it considers.

A corresponding situation occurs at the bottom of the control space of the swallowtail catastrophe $A_4$. Consider in Fig 8.20 the three successive functions starting with the bottom function, then going to the function above it on the bottom upward arch sheet of the separatrix, and then going to the function above this. Observe that, at the domain point $t = 0$ in the bottom function, there is no singularity according to Catastrophe Theory, whereas, in the corresponding bottom function shown in Fig 8.21 for the Interactive Unfolding of $EA_3$, the point $t = 0$ has a negative maximum $M^-$. In fact, observe that, concerning the three successive functions we have just listed in Fig 8.20 in the swallowtail catastrophe $A_4$, the corresponding functions shown in Fig 8.21 in the Interactive Unfolding of $EA_3$ have their maximum rising from negative through zero to positive. That is, in the Interactive Unfolding of $EA_3$, this transition is given by the Process-Grammar operation $CM^-$. This will be analyzed in detail in Chapter 10. However, at the moment, we will observe the following important facts:

**CONTINUATION OPERATION $CM^-$**

**In the Interactive Unfolding of $EA_3$, the Process Grammar operation $CM^-$, which describes the continuation of a resistance process till it protrudes, is given by an upward movement of the curvature function through a point in the maxima zero-separatrix (the bottom upward arch sheet). This movement corresponds to an upward movement of the angle function through the corresponding point in the critical-point separatrix of the swallowtail catastrophe $A_4$, this being right-equivalent to movement through the single-point separatrix of the dual fold catastrophe $A_{-2}$.**

**However, observe the following crucial facts:**
**Catastrophe Theory defines each function in the positive $a_1$-axis of the dual fold catastrophe $A_{-2}$ as having *no* singularity considered in Catastrophe Theory, and defines the same for the corresponding domain region within each function that undergoes the same transition in the swallowtail catastrophe $A_4$.**

**In contrast, Interactive Singularity Theory defines each function in the corresponding trajectory in the Interactive Unfolding of $EA_3$ as having, in the same domain region, a maximum, and, by the Symmetry-Curvature Duality Theorem, this maximum corresponds to a PISA axis which *causally explains* the maximum and also *causally explains* the transition through the separatrix.**

**Therefore, Catastrophe Theory fails to give the causal structure of the important morphology defined by $CM^-$, whereas Interactive Singularity Theory gives the causal structure.**

## 8.8 Additional Inadequacy of Catastrophe Theory in Modeling Spatial Morphology

It is crucial now to recognize the following:

**Although there is a correspondence between the zero-separatrix of the Interactive Unfolding of $EA_3$ and the critical-point separatrix of the swallowtail catastrophe $A_4$, it is important to understand that this correspondence is not enough to realize the operations of the Process Grammar and achieve the goal of modeling spatial morphology.**

**The Interactive Unfolding of $EA_3$ contains much more than the swallowtail catastrophe $A_4$, and it is this that gives the Interactive Unfolding of $EA_3$ the power to model spatial morphology.**

To understand the above statement, first observe the following: We have said that Fig 8.20 gives the set of representative functions in the control space of the swallowtail catastrophe $A_4$. With respect to this, notice the following: In this control space, each of the volumes of functions, created by the separating effect of the separatrix, is represented by one of the functions shown in Fig 8.20, in the sense that the function exhibits the qualitative structure of all the functions in the volume of which it is a member.

Now, we have said that Fig 8.21 gives the set of functions in the control space of the Interactive Unfolding of $EA_3$ that correspond to the representative functions in the control space of the swallowtail catastrophe $A_4$. However, this set within the Interactive Unfolding of $EA_3$ is only a partial subset of the full set of representative functions for the Interactive Unfolding of $EA_3$. In fact, according to our theory, it is the limited nature of this subset that causes a major failure of the swallowtail catastrophe $A_4$ to model morphology.

To see this, let us consider the process-continuation operation $Cm^+$. We have seen that $Cm^+$ is given by a downward crossing of the minima zero-separatrix of the Interactive Unfolding of $EA_3$. With respect to this, consider the progression from the left-most function, in Fig 8.21, to the function on the lower part of the left diagonal sheet (not at the cusp-point), to the function inside the central enclosed volume. We will call these three functions respectively $g_1$, $g_2$, and $g_3$. Observe that, within the function $g_2$, the *left* minimum is labeled "2", indicating that this minimum is 2-fold degenerate as a zero. It is this left minimum that will move downwards through its $t$-axis to become the *left negative* minimum in the central function $g_3$. In other words, it is this minimum that is realizing the operation $Cm^+$. However, there is a problem: The function $g_1$ does not contain a *left* minimum. In particular, it does not contain the domain $m^+$ of the operation $Cm^+$.

The profound reason is this: While Fig 8.21 exhibits the zero-separatrix of the Interactive Unfolding of $EA_3$, it does not exhibit the critical-point separatrix of the Interactive Unfolding of $EA_3$. On the left side of the slice shown in Fig 8.21, the critical-point separatrix would be a vertical straight line that goes through the left cusp-point. This means that, in crossing this left line, the function $g_1$ would undergo a critical-point pair-creation, producing the pair $m^+M^+$ within the left side of the function. Thus, it is this crossing that creates the $m^+$ needed as the domain of the $Cm^+$ operation.

This function with the added pair $m^+M^+$ must therefore be an extra representative function corresponding to the extra volume that has been created *between* the critical-point separatrix and the zero-separatrix.

Thus, a crucial fact is that, by adding the critical-point separatrix to the control space of the Interactive Unfolding of $EA_3$, the volumes produced by the zero-separatrix are decomposed even further, and therefore the set of representative functions becomes larger than the set of representative functions for the swallowtail catastrophe $A_4$. Most crucially, it is this larger set of representative functions in the Interactive Unfolding of $EA_3$ that models the spatial morphology, whereas the smaller set induced by the swallowtail catastrophe $A_4$ does not. This means that, to capture the spatial morphology, we need the following concept:

## 8.9 Full Separatrix of the Interactive Unfolding of $EA_3$

I will define the **full separatrix** of an Interactive Unfolding to be the union of its critical-point separatrix and its zero-separatrix. The full separatrix of the Interactive Unfolding of $EA_3$ is illustrated in Fig 8.22. It consists of the critical-point separatrix, as illustrated in Fig 8.4, together with the zero-separatrix as illustrated in Fig 8.7.[1]

We have seen that one of the important features of the full separatrix is that it decomposes the control space into a sufficient number of volumes to capture the Process Grammar operations, and therefore to model spatial morphology. Another consequence of this is as follows: Because of the greater partitioning induced by adding the zero-separatrix to the critical-point separatrix that was derived by sweeping the critical-point separatrix from the cusp catastrophe $A_3$, it is the case that the *stable/generic functions* of the Interactive Unfolding of $EA_3$ no longer include certain functions that Catastrophe Theory classifies as stable/generic functions in the cusp catastrophe $A_3$, as follows: For example, in the cusp catastrophe, *all* the functions along the $a_2$-axis, except for the single point at the origin, are classified by Catastrophe Theory as stable/generic. However, in our Interactive Unfolding, they are not. This fact is illustrated in Fig 8.22, where we see that the $a_2$-axis is contained within the full separatrix. This is due to the fact that any function along the axis violates the transversality condition that our theory requires of the function itself, not just its first derivative.

Constructing the full separatrix allows us to deduce other important properties: Since the Process Grammar operations cross between the volumes produced by the full separatrix, the structural arrangement of these volumes allows us to clearly visualize the network of relationships that exist between the Process Grammar operations themselves. The importance of this will emerge in the following chapters.

[1] So that the reader can, in Fig 8.22, more easily distinguish the zero-separatrix from the critical-point separatrix, the diagram cuts off the zero-separatrix, horizontally, at a lower level than the critical-point separatrix. The upper diagonal sheets of the zero-separatrix should, of course, continue to extend upwards. Nevertheless, the correct number of volumes are present.

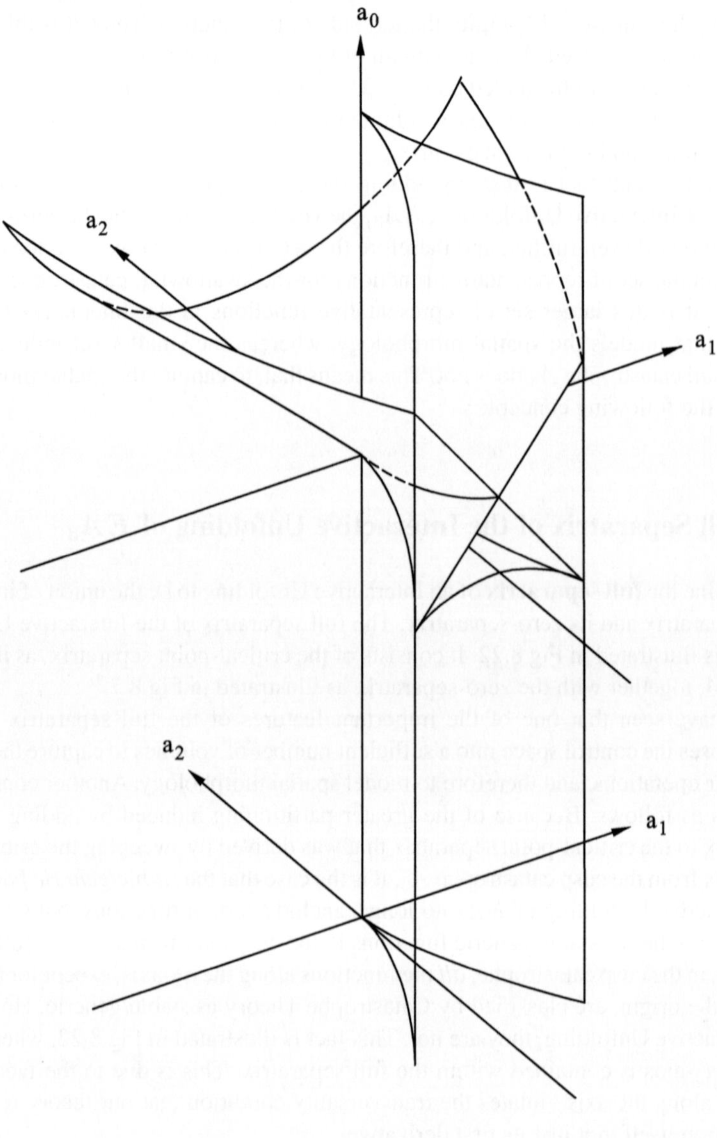


**Fig. 8.22.** The full separatrix of the Interactive Unfolding of $EA_3$.

# 9

# The Structure of Interaction

## 9.1 Interactive Structure

We have seen that a fundamental difference between the Process Grammar analysis of shape morphology and the Catastrophe Theory analysis of shape morphology is this:

**INTERACTIVE STRUCTURE**

**In the Process Grammar analysis of shape morphology, there is a fundamental morphological distinction between critical-point bifurcations whose transition-states are positive vs. zero vs. negative versions of a degenerate critical point. Therefore critical-point bifurcations are organized by zero-bifurcations. Thus critical-point bifurcations interact with zero-bifurcations. This will be called the** ***Interactive Structure*** **of the Process Grammar.**

**It is the Interactive Structure of the Process Grammar that makes it successful at modeling spatial morphology.**

**In contrast, Catastrophe Theory does not make a distinction between critical-point bifurcations whose transition-states are positive vs. zero vs. negative versions of a degenerate critical point. Therefore, critical-point bifurcations do not interact with zero-bifurcations. That is, there is no Interactive Structure.**

**It is the absence of Interactive Structure in Catastrophe Theory that makes it unsuccessful in modeling spatial morphology.**

**The singularity basis of the Interactive Structure in the Process Grammar is given by the organizational relationships between the critical-point structure and zero-structure in Interactive Unfoldings.**

**This will be called the Interactive Structure of Interactive Singularity Theory. It includes the interactive relationships between the critical-point manifold and zero-manifold within an Interactive Unfolding, the interactive relationships between the subsets of degenerate points of these two manifolds, and the interactive relationships between the critical-point separatrix and zero-separatrix.**

**One of the consequences of the Interactive Structure is that it leads to a different classification of singularities from that found in conventional singularity theory.**

## 9.2 Contact Structure

Our analysis of the Interactive Structure in Interactive Singularity Theory requires the use of the concept of *contact*. In fact, throughout our chapters on Interactive Singularity Theory, the term *contact* will always refer to the contact relation of the graph of a function with its domain axis $t$.

In this section, we briefly illustrate this type of contact, and its *labeling* in conventional singularity theory, for readers who are not familiar with it. Other readers can go directly to the next section on p300.

Consider for example, the function $f(t) = t^n$. Let us define the graph of this function as a *curve* parameterized by $t$ in the natural way; i.e., the points on the graph are of the form $(t, f(t)) \in \mathbb{R}^2$. Then, one says that, at the origin, the graph has *contact of order* $n$ with the $t$-axis. This means that the graph touches the $t$-axis at $n$ coincident points. The reason is simple: The function $f(t) = t^n$ has $n$ coincident zeros (roots) at the origin.

The reader who is not familiar with the concept of *order of contact* is recommended to go through the examples illustrated in Fig 9.1 and read each of the captions in the figure.

Now, in conventional singularity theory, the origin-point of each of these functions $f(t) = t^n$ is labeled as the singularity $A_{n-1}$, because each of these functions is $(n-1)$-fold degenerate as a critical-point at the origin. Again, Fig 9.1 illustrates this by writing under the graph of each function $f(t) = t^n$ the singularity $A_{n-1}$ that it has at the origin.

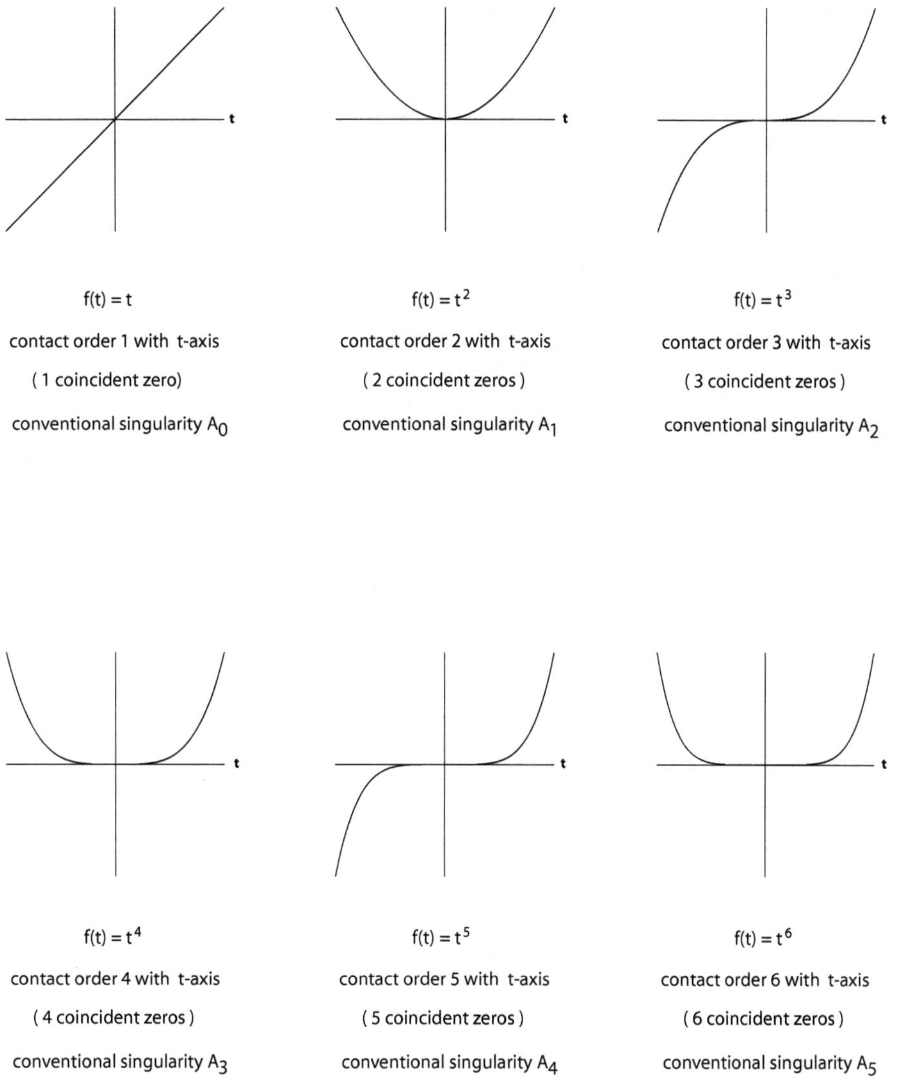


**Fig. 9.1.** The graphs of functions $f(t) = t^n$, for $n = 1 \ldots 6$, giving the order of contact of their graphs with the $t$-axis, and the conventional singularity label $A_{n-1}$ at the point of contact.

## 9.3 The Detailed Basis of Method 2

Before we describe the *Interactive* Structure of Interactive Singularity Theory, it is necessary to understand more details about the basis of Method 2 for constructing the zero-separatrix.

An important aspect of this will be to describe a crucial difference between specifying *right-equivalence* and specifying *contact-equivalence*, and to understand how this relates to the particular control parameters in the corresponding universal unfolding.

Another important issue is this: In our initial description of Method 2, we went directly to the separatrices, without looking at the solution manifolds of which the separatrices are a projection within the total domain of the universal unfoldings. It will also be valuable to examine these manifolds.

In order to do this, we will need to explore in more detail the relationship we saw in section 8.7 between the universal unfolding of $t^{n+2}$ with respect to critical-point structure and the universal unfolding of $t^{n+1}$ with respect to zero-structure. Based on what we said in section 8.7, this will at first appear obvious, but the details we now give are needed to understand the specification of the different types of *equivalence* involved and how they relate to the control parameters. Furthermore, the details will become essential for understanding the structure of the manifolds involved.

Thus, to begin, we first return to the type of unfolding that is considered in Catastrophe Theory. This type of unfolding reveals the local structure of *critical points* in the perturbations of a singularity, with no concern for the zero-structure, i.e., contact-structure. Thus, we have seen that, in Catastrophe Theory, the local structure of critical points in the perturbations of the singularity $A_{n+1}$ of the cuspoid $t^{n+2}$ is given by the canonical universal unfolding $\tilde{F} : \mathbb{R} \times \mathbb{R}^n \longrightarrow \mathbb{R}$ defined by

$$\tilde{F}(t; b_1, \ldots b_n) \quad = \quad t^{n+2} + b_1 t + \ldots + b_n t^n \tag{9.1}$$

A critical point in this unfolding is a point where $\partial \tilde{F}/\partial t = 0$. Based on this fact, let us now consider the derivative $\partial \tilde{F}/\partial t$ as an unfolding itself. That is, we can write this unfolding as $\partial \tilde{F}/\partial t(t; b_1, \ldots b_n)$ which has the same number of control variables as $\tilde{F}$. Obviously, the polynomial which gives this unfolding is $(n+2)t^{n+1} + b_1 + \ldots + n b_n t^{n-1}$, which is simply the derivative (with respect to $t$) of the polynomial in (9.1).

It is clear that a *critical point* of the unfolding $\tilde{F}$ must be a *zero* of the unfolding $\partial \tilde{F}/\partial t$. Therefore, a crucial result is that the critical-point structure of $\tilde{F}$ must be the same as the zero-structure (i.e., the contact-structure) of $\partial \tilde{F}/\partial t$.

Notice that, since we are considering the zeros of the derivative unfolding $\partial \tilde{F}/\partial t$, its zero-structure would not qualitatively alter if we divide this derivative unfolding by $n+2$ and write it in the form:

$$F(t; a_0, a_1, \ldots a_{n-1}) \quad = \quad t^{n+1} + a_0 + a_1 t + \ldots + a_{n-1} t^{n-1} \tag{9.2}$$

where, again, we have the same number of control variables, and the coefficients $ib_i$ of the powers $t^{i-1}$ in the exact derivative have simply been replaced by the coefficients $a_{i-1}$ of the same powers $t^{i-1}$ in this re-scaled version.

Observe that the germ in the unfolding (9.2) is the cuspoid $t^{n+1}$. Concerning this, we come to the following crucial fact: The unfolding (9.2) is the canonical universal unfolding that gives the local structure of zeros, i.e., the *contact*-structure, in the perturbations of the singularity $A_n$ of the cuspoid $t^{n+1}$. This singularity $A_n$ and cuspoid $t^{n+1}$ are one step down from the singularity $A_{n+1}$ and cuspoid $t^{n+2}$ in the first unfolding (9.1). Obviously this is the case because the second unfolding (9.2) is essentially the derivative of the first unfolding (9.1).

Therefore, observe this: We have seen that the local *critical-point* structure in the perturbations of the singularity $A_{n+1}$ of the cuspoid $t^{n+2}$ is given by the unfolding in (9.1); and we have seen that the local *zero*-structure (*contact*-structure) in the perturbations of the singularity $A_n$ of the cuspoid $t^{n+1}$ (one step down) is given by the unfolding in (9.2). Furthermore, because the second unfolding has qualitatively the same zero-structure as the derivative of the first unfolding, we have the fundamental result that the local critical-point structure of the first unfolding is the same as the local zero-structure of the second unfolding. That is, we have the following conclusion: The local *critical-point* structure of the universal unfolding of the critical point given by the singularity $A_{n+1}$ of the cuspoid $t^{n+2}$ is the same as the local *zero*-structure (*contact*-structure) of the universal unfolding of the zero given by the singularity $A_n$ of the cuspoid $t^{n+1}$ (one step down).

The two unfoldings will be called respectively the universal *critical-point unfolding* and the universal *zero-unfolding*. It is important to understand that these two universal unfoldings are defined relative to two different forms of **qualitative equivalence**, as follows:

The reader will recall, from section 7.3, that the qualitative equivalence used by the universal *critical-point* unfolding, i.e., the qualitative equivalence that *preserves* its critical-point structure, is *right-equivalence*. We saw that right-equivalence between two families of functions means that we can find three types of coordinate changes $y$, $\phi$, $z$, described on p241, where map $y$ gives a coordinate-change of the *domain* $\mathbb{R}$ for each individual function; map $\phi$ gives a coordinate-change of the control space; map $z$ gives a coordinate-shift of the *codomain* $\mathbb{R}$ along itself for each individual function; and the three coordinate changes must satisfy equation (7.2) which is also on p241.

The crucial issue we wish to focus on now is this: The reason why *right-equivalence* allows the third map $z$, which gives a coordinate-shift of the codomain $\mathbb{R}$ along itself for each individual function, is that right-equivalence is concerned with the preservation of only the critical-point structure. Clearly a critical point does not qualitatively change if the codomain $\mathbb{R}$ of a function is shifted along itself; i.e., the graph of the function is raised or lowered along the codomain axis.

Based on this, we can see why the universal critical-point unfolding $\tilde{F}$, given in expression (9.1) on p300, does *not* have a control parameter $b_0$; that is, a control parameter which is not a coefficient of a power of $t$, and which would therefore shift the height of the graph of a function. That is, the reason why the unfolding does not have such a control parameter is because the unfolding is concerned only with elaborating critical-point structure, and is therefore not concerned with the height of the graph of a function, since changing the height of a graph does not change the critical-point structure.

In contrast, the universal *zero*-unfolding $F$, given in expression (9.2) on p300, does have the control parameter $a_0$ which shifts the height of the graph of a function. This is required because, when a graph is moved to different heights, it can have different zero-structures, and the universal zero-unfolding is concerned with elaborating the different zero-structures.

As a result of this, the type of qualitative equivalence used by the universal zero-unfolding $F$, i.e., the qualitative equivalence that *preserves* its zero-structure, cannot be right-equivalence, because the map $z$ in right-equivalence allows the shift of a graph along the codomain axis, thus changing the zero-structure. Therefore, the qualitative equivalence required for the universal zero-unfolding $F$ must be *contact-equivalence* which does not have the map $z$.

> **The above logic therefore demonstrates this: When a particular type of qualitative equivalence does allow shift along the codomain axis, the corresponding universal unfolding does not. Conversely, when a particular type of qualitative equivalence does not allow shift along the codomain axis, the corresponding universal unfolding does.**

Now, despite the fact that the universal *critical-point unfolding* $\tilde{F}$ in expression (9.1) on p300, and the universal *zero-unfolding* $F$ in expression (9.2) on p300, use different types of equivalence, the two unfoldings give similar structures because the latter is essentially the derivative of the former and therefore the former's critical-point structure is the same as the latter's zero-structure.

We now need to explore more deeply the similarity between these structures. The first thing to observe is that the dimension of the control space is the same in both cases. That is, in the case of $\tilde{F}$ there are $n$ control variables $b_1, \ldots, b_n$; and in the case of $F$ there are also $n$ control variables $a_0, \ldots, a_{n-1}$. The control space dimension of a *universal* unfolding is crucial: It is the *minimal* dimension that reveals the full qualitative structure that can be locally unfolded from a singularity. Again, we have seen that this minimal dimension, in both cases, $\tilde{F}$ and $F$, is the same, despite the fact that the singularity in the first case is $A_{n+1}$ and the singularity in the second case is $A_n$ (one step down).

It will be useful to note the following simple fact concerning the numbers involved: In the case of the universal *critical-point* unfolding of the singularity $A_{n+1}$ of the cuspoid $t^{n+2}$, the dimension of the control space $n$ is *one number lower* than the index $n+1$ on that singularity $A_{n+1}$. In the case of the universal *zero*-unfolding of the singularity $A_n$ of the cuspoid $t^{n+1}$, the dimension of the control space $n$ is the *same* as the index $n$ on that singularity $A_n$.

Next, recall that $\tilde{F}$ is a map of the form $\mathbb{R} \times \mathbb{R}^n \longrightarrow \mathbb{R}$. We will call the full domain $\mathbb{R} \times \mathbb{R}^n$ of this map, the **total space**. Now, we know from Catastrophe Theory that, in the total space, the set of points $(t, b_1, \ldots b_n)$ that are critical points form an $n$-dimensional manifold. We will call this the **critical-point manifold** of $\tilde{F}$.

Recall that $F$ is also a map of the form $\mathbb{R} \times \mathbb{R}^n \longrightarrow \mathbb{R}$. Again, we will call the full domain $\mathbb{R} \times \mathbb{R}^n$ of this map, the **total space**. Now, we saw earlier that the set of zero points of $F$ has the same structure as the set of critical points of $\tilde{F}$. Therefore, the set of zero points of $F$ must also form an $n$-dimensional manifold. We will call this the **zero-manifold** of $F$. Most crucially, the zero-manifold of $F$ must have the same structure as the critical-point manifold of $\tilde{F}$.

Both the critical-point manifold and the zero-manifold are what we will call the **solution manifold** of their respective universal unfoldings.

In order to illustrate the relation between the two manifolds, let $\tilde{F}$ be the universal critical-point unfolding of the singularity $A_3$ of the cuspoid $t^4$; that is, the cusp catastrophe

$$\tilde{F}(t; b_1, b_2) = t^4 + b_1 t + b_2 t^2$$

We know that its critical-point manifold is the 2D surface shown on the left in Fig 9.2. Corresponding to $\tilde{F}$, the map $F$ would be the universal zero-unfolding of the singularity $A_2$ of the cuspoid $t^3$ (one step down). Notice that, in this case, the germ $t^3$ is that of the fold catastrophe. However, the unfolding is not the 1-dimensional critical-point unfolding known as the fold catastrophe, but the following 2-dimensional universal zero-unfolding:

$$F(t; a_0, a_1) = t^3 + a_0 + a_1 t$$

By the above discussion, the zero-manifold of $F$ would have the same structure as the critical-point manifold of $\tilde{F}$, and would therefore be as shown on the right in Fig 9.2.

The reader should carefully compare the terms in the above two equations. Furthermore, it is worth comparing the labels of the control-space axes in the two diagrams shown in Fig 9.2.

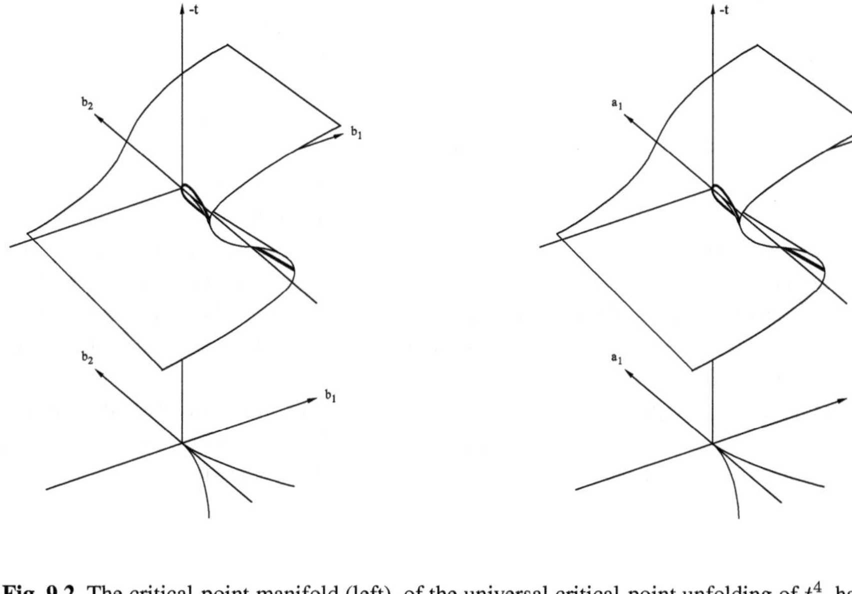


**Fig. 9.2.** The critical-point manifold (left), of the universal critical-point unfolding of $t^4$, has the same structure as the zero-manifold (right), of the universal zero-unfolding of $t^3$.

Now return to the general $n$-dimensional case of $\tilde{F}$ as given by expression (9.1) on p300. We know that any point on its critical-point manifold is a point $(t, b_1, \ldots b_n)$, in the total space $\mathbb{R} \times \mathbb{R}^n$, where $\partial\tilde{F}/\partial t = 0$. Let us now consider any point which is *degenerate* as a critical point. Such a point would be given by the condition that *both* its first derivative *and* its second derivative are zero; that is, $\partial\tilde{F}/\partial t = \partial^2\tilde{F}/\partial^2 t = 0$. In fact, this condition also means that the projection map, which sends the manifold onto the control space, would be singular at the point. We will call the set of degenerate critical points, the **degeneracy subset** of the critical-point manifold.

Correspondingly, return to the general $n$-dimensional case of $F$ as given by expression (9.2) on p300. We know that any point on the zero-manifold is a point $(t, a_0, \ldots a_{n-1})$, in the total space $\mathbb{R} \times \mathbb{R}^n$, where $F = 0$. Let us now consider any point which is *degenerate* as a zero. Such a point would be given by the condition that *both* $F$ *and* its first derivative are zero; that is, $F = \partial F/\partial t = 0$. In fact, this condition also means that the projection map, which sends the manifold onto the control space, would be singular at the point. We will call the set of zero-degenerate points, the **degeneracy subset** of the zero-manifold. Again, it would have the same structure as the degeneracy subset of the critical-point manifold of $\tilde{F}$.

To illustrate, return to Fig 9.2. The bold curve on the left manifold gives the degeneracy subset of the critical-point manifold of the universal critical-point unfolding of the singularity $A_3$ of the cuspoid $t^4$. The bold curve on the right manifold gives the degeneracy subset of the zero-manifold of the universal zero-unfolding of the singularity $A_2$ of the cuspoid $t^3$ (one step down). Something also to observe is that both the bold curve in the left manifold and the bold curve in the right manifold are where the tangent spaces to the manifold are vertical. Thus the projection of each manifold down onto its control space would be singular at each point on the bold curve.

Next, returning to the general $n$-dimensional case of $\tilde{F}$, we know that the critical-point separatrix of $\tilde{F}$ is the image of the projection of its degeneracy subset to the control space. Similarly, returning to the general $n$-dimensional case of $F$, we know that the zero-separatrix of $F$ is the image of the projection of its degeneracy subset to the control space. It is clear, from everything we have said, that the critical-point separatrix of $\tilde{F}$ is the same structure as the zero-separatrix of $F$.

Again, to illustrate: In Fig 9.2, the origin and the cusp-shaped curves in the control space of the left diagram give the critical-point separatrix of the universal critical-point unfolding of the singularity $A_3$ of the cuspoid $t^4$. Furthermore, the origin and the cusp-shaped curves in the control space of the right diagram give the zero-separatrix of the universal zero-unfolding of the singularity $A_2$ of the cuspoid $t^3$. We see illustrated the fact that the critical-point separatrix in the left diagram has the same structure as the zero-separatrix in the right diagram.

Now return to the general $n$-dimensional case; i.e., the universal critical-point unfolding of the singularity $A_{n+1}$ of the cuspoid $t^{n+2}$, and the universal zero-unfolding of the singularity $A_n$ of the cuspoid $t^{n+1}$. The discussion in this section has revealed that the following have the same structures in these two unfoldings:

(a) the dimension of the control space;
(b) the solution manifold;
(c) the degeneracy subset;
(d) the separatrix.

## 9.4 Comment on the Label $EA_n$

In the next section, we return to Interactive Singularity Theory. With respect to notation, the reader should note the following:

In Catastrophe Theory, the label $A_n$ is often used to denote both the singularity and the structure of its universal unfolding. That is, the literature often refers to these two entities, respectively, as the "singularity $A_n$" and the "catastrophe $A_n$". Similarly in Interactive Singularity Theory, we could use the label $EA_n$ to denote both the singularity, and the structure of its Interactive Unfolding. Therefore, these two entities could be referred to, respectively, as the "singularity $EA_n$" and the "Interactive Unfolding $EA_n$". However, to help the reader we will often state the latter as the "Interactive Unfolding *of* $EA_n$".

## 9.5 The Structures in Interactive Singularity Theory

The above discussion has given, in more detail, the structural similarities between the following two unfoldings:

(1) the *standard* universal critical-point unfolding of the singularity $A_{n+1}$ of the cuspoid $t^{n+2}$

(2) the universal zero-unfolding of the singularity $A_n$ of the cuspoid $t^{n+1}$

The purpose of the discussion has been to help the reader to better understand the organization of the zero-bifurcation that occurs in the Interactive Unfolding of $EA_n$. It should be noted that, in the unfolding (2), the index $n$ on the label $A_n$ is the same as the index $n$ on the label $EA_n$. However, the following should also be noted:

> **The conventional label $A_n$ is inadequate with respect to zero-unfolding, i.e., the contact relation to the domain axis $t$ of a function. The label $A_n$ means an $n$-fold degenerate critical point, but it does not tell us the contact relation of the critical point with respect to the $t$-axis. It is for this reason that, when one talks about a zero-unfolding of a singularity $A_n$, one has to give extra information concerning its initial contact status; e.g., in the unfolding (2) above, one has to specify that the singularity is in the cuspoid $t^{n+1}$, or at a contact-equivalent point in another function.**

**In contrast, the label $EA_n$ is adequate. It tells us that the singularity is an $n$-degenerate critical point and simultaneously has contact with the $t$-axis (in fact, contact of order $n+1$). Therefore, the label $EA_n$ does not require that we specify the type of function in which it occurs. The type of function is deducible from the label $EA_n$.**

**To emphasize: The label $A_n$ specifies its critical-point structure but not its contact structure with the domain axis. In contrast, the label $EA_n$ specifies *both* its critical-point structure and its contact structure with the domain axis.**

**Notice that this relates to our claim of the inadequacy of Catastrophe Theory with respect to describing spatial morphology. In Catastrophe Theory, the singularity $A_n$ is viewed with respect to its critical-point structure but not with respect to its range of alternative contact structures with the domain axis of a function.**

**It is the fact that $EA_n$ defines both its critical-point structure and its contact structure that enables it to model spatial morphology.**

Thus, according to our theory of spatial morphology, the following is a serious problem with the conventional labeling $A_n$ of a singularity: Consider again the universal zero-unfolding (9.2) page 300. The germ $t^{n+1}$ occurs when the control parameters are zero. In the conventional approach, the singularity in the germ $t^{n+1}$ is labeled $A_n$. However, now increase or decrease the control variable $a_0$ from zero, thus obtaining all the functions $t^{n+1} + a_0$ for non-zero $a_0$. The crucial fact is that, in the conventional labeling, the singularity does not change from $A_n$; that is, because the effect of the control variable $a_0$ is to raise and lower the graph of the function, the variable $a_0$ does not alter the type of critical point $A_n$ in the function.

It is a serious problem to give the singularity in a germ a label which does not disappear the moment one starts a universal unfolding of the germ. For example, this violates the view that the singularity in the germ should be regarded as unstable, and that the universal unfolding should have the minimal dimension needed to reveal the qualitative changes of the singularity induced by perturbations.

According to the Process Grammar, we need a system for identifying the *relationships* between the contact structure and the critical-point structure. As stated in section 9.1, our theory calls this the *Interactive Structure*; i.e., it defines the continuation operations and the way they organize the bifurcation operations. Interactive Singularity Theory provides the singularity structure needed to realize the Interactive Structure. Correspondingly, we need an expanded labeling of the singularities, as follows:

**NEW LABELS FOR SINGULARITIES**

**The singularity $A_n$, as used currently in singularity theory, will be called a** *conventional singularity.* **While it specifies the singularity to be an $n$-fold degenerate critical point, it does not specify the point's zero-structure, i.e., the contact relation between the graph of the function and the domain axis, at the point.**

**The singularity $EA_n$, of Interactive Singularity Theory, will be called a** *complete singularity.* **It is complete in the sense that it is an $n$-fold degenerate critical point, whose graph, at this point, is in $(n+1)$-point contact with the domain axis; the point being an $(n+1)$-fold degenerate zero.**

**The singularity $NA_n$, of Interactive Singularity Theory, will be called a** *non-zero singularity.* **It is an $n$-fold degenerate critical point, whose graph, at this point, has no contact with the domain axis.**

**Furthermore, the singularities $N^+A_n$ and $N^-A_n$, of Interactive Singularity Theory, distinguish between the positive and negative versions of $NA_n$.**

**Catastrophe Theory fails to model spatial morphology because it is concerned with only the conventional singularities $A_n$.**

**Interactive Singularity Theory gives the singularity basis of spatial morphology because it distinguishes between the complete singularities $EA_n$ and the non-zero singularities $N^-A_n$ and $N^+A_n$, and describes the** *Interactive Structure* **between them.**

Now, based on what we have said above, it is clear that we can *deduce* from the label $EA_n$ that it occurs at a point in a function that is both right-equivalent and contact-equivalent to the singularity at the origin of $t^{n+1}$; i.e., something that we cannot deduce from the conventional label $A_n$. Furthermore, based on our argument, the universal unfolding of $EA_n$ can be given as

$$F(t; a_0, \dots a_{n-1}) \quad = \quad t^{n+1} + a_0 + a_1 t + \dots + a_{n-1} t^{n-1} \tag{9.3}$$

That is, this unfolding not only gives the local critical-point structure (in local perturbations) of the germ $t^{n+1}$, but it raises and lowers this structure (via variable $a_0$), and therefore, unlike Catastrophe Theory, it provides the critical-point bifurcation operations in their fully-transversal form as given by the Process Grammar, and also provides the continuation operations of the Process Grammar that inter-relate those bifurcation operations.

Examining the equation in the unfolding (9.3) of $EA_n$, we see that it is exactly the same as the equation in the unfolding (9.2) page 300, which was specified as the universal zero-unfolding of the singularity $A_n$ in $t^{n+1}$. The crucial consequence is this:

**The** *space of functions* **defined by the canonical universal Interactive Unfolding of $EA_n$ is the same as the** *space of functions* **defined by the canonical universal zero-unfolding of the singularity $A_n$ of $t^{n+1}$.**

**However, while $EA_n$ uses the zero-structure given by the latter unfolding, it adds much more structure. This includes the solution manifold, degeneracy subset, and separatrix, of its critical-point structure, as well as the** *interaction* **of this structure with the solution manifold, degeneracy subset, and separatrix, of the zero-structure.**

**As we shall see, this** *Interactive Structure* **gives very powerful concepts.**

Thus, let us now begin to examine the Interactive Structure. In the following chapters, we will examine an essential part of this structure by examining the control space *volumes* that are bounded by *both* the critical-point separatrices and the zero-separatrices – i.e., the volumes defined by the *interaction*. These volumes are crucial to modeling the structure of the Process Grammar operations, because we will show that the operations go between those volumes.

In the remainder of the present chapter, we will examine the way these volumes are created.

Observe first that the critical-point manifold of the Interactive Unfolding of $EA_n$ must be the critical-point manifold of the catastrophe $A_n$ swept along the extra dimension $a_0$. This means that the critical-point degeneracy subset of the Interactive Unfolding of $EA_n$ must be the critical-point degeneracy subset of the catastrophe $A_n$ swept along $a_0$. This, in turn, means that the critical-point separatrix of the Interactive Unfolding of $EA_n$ must be the critical-point separatrix of the catastrophe $A_n$ swept along $a_0$.

We will now provide the following rules concerning the *intersections* between the critical-point structures and zero-structures of the Interactive Unfolding of $EA_n$. These intersections are crucial for understanding the *Interactive Structure*.

## POINTS OF INTERSECTION BETWEEN THE ZERO-MANIFOLD AND CRITICAL-POINT MANIFOLD OF THE INTERACTIVE UNFOLDING OF $EA_n$

**RULE 1:**
**2-fold zero-degenerate point:**

**This must be a point in the degeneracy subset of the zero-manifold where this manifold intersects the critical-point manifold, because a 2-fold degenerate zero is a critical point.**

**This is not a degenerate critical point, and therefore the point is not within the degeneracy subset of the critical-point manifold.**

**On one side of this intersection-point, the critical-point manifold must represent a positive minimum (or maximum), and on the other side, the critical-point manifold must represent a negative minimum (or maximum).**

**The projection of this intersection-point to the control space must be a point in the zero-separatrix but not the critical-point separatrix.**

**RULE 2:**
**$n$-fold zero-degenerate point for $n > 2$:**

**This must be a point where the degeneracy subset of the zero-manifold intersects the degeneracy subset of the critical-point manifold, because an n-fold degenerate zero must be an $n - 1$-fold degenerate critical-point.**

**The projection of this point to the control space must therefore be a point where the zero-separatrix intersects the critical-point separatrix.**

**CASE 3:** An additional type of intersection between the zero-separatrix and the critical-point separatrix can occur when there is a 2-fold degenerate zero at a point $t_1$ in the $t$-domain of a function and simultaneously a (non-zero) degenerate critical point that is located at another point $t_2$ in the $t$-domain of the same function. Note that this separatrix intersection will not be due to an intersection of the zero-manifold and critical-point manifold, but will be due to the fact that points $t_1$ and $t_2$ are on the $t$-domain of the same function, and are therefore projected onto the same point in the control space.

## 9.6 The Universal Structure of $Cm^+$

The most basic component of the Interactive Structure is the continuation operator $C$, because it relates the different sign versions of an extremum. In sections 8.5 – 8.7, its operation $Cm^+$ was modeled within the control space of the Interactive Unfolding of $EA_3$ by downward movement through the minima zero-separatrix. Its interactive role within this model is very important and will be investigated in detail in Chapter 10. However, in the present section, we will show that, independent of higher-order Interactive Unfoldings such as that of $EA_3$, the operation $Cm^+$ actually *contains* an Interactive Structure. That is, the *universal unfolding* defined by $Cm^+$, i.e., its *minimal* dimension unfolding, which is of dimension considerably smaller than that of $EA_3$, includes *within itself* an Interactive Structure. Furthermore, since that unfolding is universal, this will mean that its Interactive Structure will also be contained in the higher dimensional Interactive Unfoldings such as that of $EA_3$ and must therefore provide significant aspects of the Interactive Structure in those higher dimensional cases.

Thus, the purpose of the present section is to fully describe the Interactive Structure *within* the universal unfolding given by the Process Grammar operation

$$Cm^+ : m^+ \longrightarrow 0m^-0$$

We first note that we have already seen that the operation $Cm^+$ is given by a 2-fold zero-bifurcation of a positive minimum as the graph of the minimum is lowered through the $t$-domain axis.

However, a crucial fact, whose implications we must now deeply investigate, is that, *throughout the operation*, the minimum is non-degenerate as a critical point, and is therefore right-equivalent, *throughout the operation*, to the conventional singularity $A_1$ in $t^2$. Observe that the reason for this can be explained as follows: Recall from p237 that a non-degenerate critical point is right-equivalent to the critical point in $\pm t^2$. Recall also from p243 that the degree of degeneracy $i$ of a critical point corresponds to the index $i$ on the conventional singularity $A_i$ and is one number less then the power $i + 1$ of the canonical form $t^{i+1}$ at the critical point. Thus, in the operation $Cm^+$, the fact that the minimum is non-degenerate (1-fold degenerate) as a critical point, throughout the operation, means that, throughout the operation, it is right-equivalent to the singularity $A_1$ (i.e., the index 1 on the $A_1$ is the same as its degeneracy 1); and therefore, throughout the operation, this minimum is right-equivalent to the minimum in $t^2$ (the power 2 is one number more than the index 1 on $A_1$).

However, it is necessary now to use our crucial argument that right-equivalence, which is the qualitative equivalence used by Catastrophe Theory, is not enough. We must use contact-equivalence. By contact-equivalence, the state $m^+$ at the beginning of the operation $Cm^+$ is not qualitatively equivalent to the minimum in $t^2$, because the graph of $m^+$ is not in contact with the $t$-axis. The same argument applies also to the state $m^-$ at the end of the operation.

There is only one moment at which the minimum in $Cm^+$ is contact-equivalent to the minimum in $t^2$. That occurs in the transition-state. This is when the graph of the minimum, which is being lowered, touches the $t$-axis of the function.

At this stage, the minimum becomes the complete singularity $EA_1$. Note the subscript 1 on this singularity indicates that the singularity is non-degenerate as a critical point; but the label $E$ indicates that the graph of this critical point must be touching the $t$-axis. This singularity therefore gives the germ that is being unfolded by the operation $Cm^+$. Now, by what we have said in the previous sections, the universal zero-unfolding involved in the Interactive Unfolding of $EA_1$ is contact-equivalent to the universal zero-unfolding of the singularity $A_1$ of the germ $t^2$. That is, by requiring that the conventional singularity $A_1$ in the germ is not *any* right-equivalent example of $A_1$, but is the particular example of $A_1$ in $t^2$, we guarantee that the zero-unfolding is that required by $Cm^+$.

However, it is crucial to observe the following: The zero-unfolding of the singularity $EA_1$ does not require that we additionally specify what type of function $EA_1$ occurs in, because we can deduce the type of function from the singularity $EA_1$. That is, we can deduce that the singularity $EA_1$ must occur in a function that is locally contact-equivalent to $t^2$. In contrast, the zero-unfolding of the singularity $A_1$ does require that we additionally specify what type of function $A_1$ occurs in, because we cannot deduce the type of function from the singularity $A_1$. That is, in the $A_1$ case we are interested, we must *explicitly* specify that the function is contact-equivalent to $t^2$.

There is a further problem with the conventional label $A_1$. In the unfolding that moves the graph of the minimum away from contact with the $t$-axis, the label $A_1$ is preserved, because the minimum remains a non-degenerate critical point. Thus the label does not define what is happening to the singularity. In contrast, in Interactive Singularity Theory, what happens to the singularity, under the unfolding, is documented, because the singularity in the germ is defined as the complete singularity $EA_1$, and its unfolding changes it to a non-zero singularity $NA_1$.

**IMPORTANT ISSUE**

**Since the singularity in the germ of the Interactive Unfolding of $EA_1$ is non-degenerate as a critical point, it is not one of the singularities unfolded in Catastrophe Theory.**

**Yet, as shown in the Process Grammar, its unfolding is fundamental to spatial morphology.**

**In fact, since its unfolding provides the lowest operation on which much of the Interactive Structure of the Process Grammar is built, we will call the Interactive Unfolding of $EA_1$, the *Ground* Interactive Unfolding.**

Now let us turn to understanding the unfolding itself. The reader will recall that, generally, the dimension of the control space of the universal unfolding of $EA_n$ is the same as the index $n$ on the label $EA_n$. We can therefore deduce that the canonical universal unfolding of $EA_1$ is

$$F(t; a_0) \quad = \quad t^2 + a_0 \tag{9.4}$$

where the germ is $t^2$ and the control space is dimension 1, having only one control parameter $a_0$.

Now, by Method 2, the *zero*-structure given by this unfolding is the same as the *critical-point* structure given by the universal critical-point unfolding of the singularity $A_2$ (one step higher), which has the following canonical form:

$$\tilde{F}(t; b_1) = t^3 + b_1 t \tag{9.5}$$

where the germ is the cubic $t^3$ and the control dimension is also 1, but the control variable $b_1$ is the coefficient of $t$. Observe that this unfolding is the fold catastrophe.

The curve in the left diagram in Fig 9.3 shows the critical-point manifold of the fold catastrophe. Since this manifold is the collection of critical points of expression (9.5), it is the derivative $\partial\tilde{F}/\partial t$ of this expression, set to zero, which means that it must be a parabola.

Correspondingly, the curve in the right diagram in Fig 9.3 shows the zero-manifold of the canonical universal unfolding of $EA_1$. Since it is the collection of zero points of expression (9.4), it is given by setting this expression to zero, which means that the manifold is also a parabola. (Without loss of generality, we have done the obvious re-scaling of the $t$-axis to make the parabolas look the same size.)

Observe that both diagrams in Fig 9.3 show the axes of the total space of their respective unfoldings; i.e., in both cases, the total space is 2-dimensional, one axis being the $t$-axis and the other axis being the 1-dimensional control space. Furthermore, in both diagrams, the control space is shown as horizontal. For each point in the control space, there is a function whose domain axis (its $t$-axis) is the vertical line through that point in the control space.

It is important now to understand the following information that is given by the two manifolds in Fig 9.3, as follows:

Consider first the critical-point manifold shown in the left diagram in Fig 9.3. Fig 9.4 shows three representative functions from its control space $b_1$. The right function comes from a point in the positive $b_1$-axis; the middle function is the germ at the origin of the control space; and the left function comes from a point in the negative $b_1$-axis.

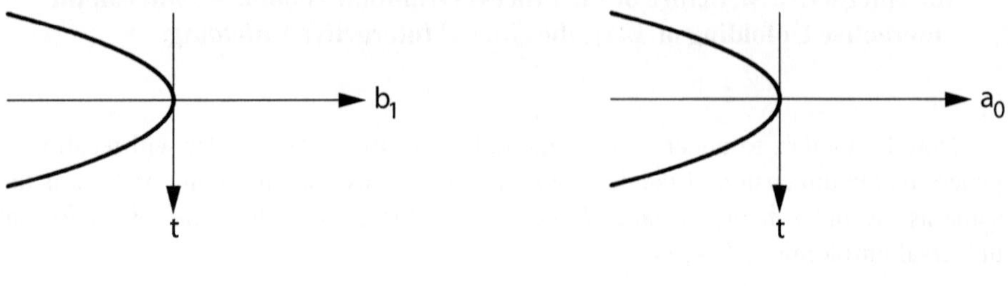


**Fig. 9.3.** Left: the critical-point manifold of the fold catastrophe $A_2$. Right: the zero-manifold of the Ground Interactive Unfolding $EA_1$.

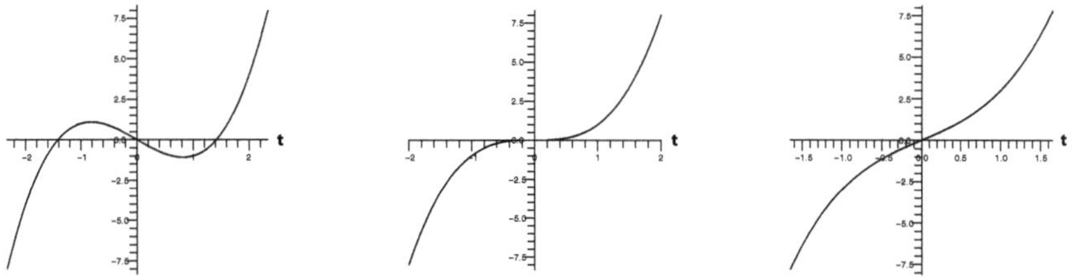


**Fig. 9.4.** The functions at points $b_1 = -2, 0, 2$ in the 1-dimensional control space $b_1$ of the fold catastrophe $A_2$.

Observe the following facts concerning the three functions in Fig 9.4: The right function has no critical point, i.e., no point of horizontal tangent; the middle function has one critical point, the point of horizontal tangency at the origin; and the left function has two critical points, a maximum and minimum. This progression is due to the "rotating-tangent effect", created by the linear term $b_1 t$ in the unfolding in expression (9.5). By clockwise rotating the tangent at the origin of the right function, i.e., by decreasing the value $b_1$ in the linear term $b_1 t$, a horizontal tangent appears at the origin of the middle function; and then, due to the further rotation of the tangent, a maximum and minimum bifurcate outwards from the origin, as illustrated in the left function.

This explains the structure of the critical-point manifold (parabola) shown in the left diagram in Fig 9.3. Because any function in the positive $b_1$-axis has no critical point, it contributes no point to the critical-point manifold, and therefore the critical-point manifold does not exist over the positive $b_1$-axis. Next, we have seen that the germ function at the origin of the control space has one critical point. Therefore the germ contributes a single point to the critical-point manifold. This point is the apex of the parabola which is the critical-point manifold. Then, as we move from the origin into the negative $b_1$-axis, we saw that the single critical point of the germ bifurcates into a maximum and a minimum. This bifurcation corresponds to the upper and lower branches, respectively, of the parabola. Thus, consider a vertical line slicing through a point on the negative $b_1$-axis. The line is the $t$-axis of the function located at this point on the $b_1$-axis. Observe that this line intersects the parabola at two points, one on the upper branch of the parabola, and the other on the lower branch. The upper point corresponds to the maximum in the function, and the lower point corresponds to the minimum.

Notice also the following: The *only* degenerate critical-point that occurs in the *entire* critical-point unfolding is the 2-fold degenerate critical point that occurs in the germ at the origin. Because it is degenerate, it satisfies the following two conditions: both the first derivative $\partial\tilde{F}/\partial t$ and the second derivative $\partial^2\tilde{F}/\partial^2 t$ are zero at this point. This means that the point must be on the critical-point manifold (parabola) where the tangent is *vertical*. Furthermore, it must be the only point of vertical tangency on the manifold. We clearly see this in the left diagram in Fig 9.3.

Now let us consider the zero-manifold shown in the right diagram in Fig 9.3. Fig 9.5 shows three representative functions from its control space $a_0$. The right function comes from a point in the positive $a_0$-axis; the middle function is the germ at the origin of the control space; and the left function comes from a point in the negative $a_0$-axis.

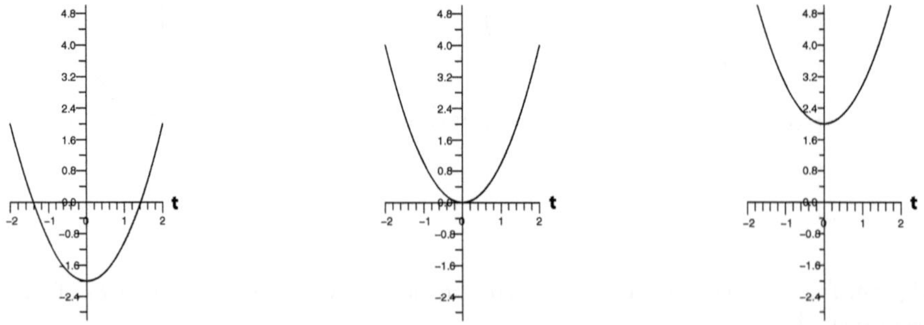

**Fig. 9.5.** The functions at points $a_0 = -2, 0, 2$ in the 1-dimensional control space $a_0$ of the Ground Interactive Unfolding $EA_1$.

Observe the following facts concerning the three functions in Fig 9.5: The right function has no zero; the middle function has one zero; and the left function has two zeros. The progression is due to the fact that the sequence of functions corresponds to the successive lowering of the corresponding graph.

This explains the structure of the zero-manifold (parabola) shown in the right diagram in Fig 9.3. Because any function in the positive $a_0$-axis has no zero, it contributes no point to the zero-manifold, and therefore the zero-manifold does not exist over the positive $a_0$-axis. Next, we have seen that the germ function at the origin of the control space has one zero. Therefore the germ contributes a single point to the zero-manifold. This point is the apex of the parabola which is the zero-manifold. Then, as we move from the origin into the negative $a_0$-axis, we saw that the single zero of the germ bifurcates into two zeros. This bifurcation corresponds to the upper and lower branches of the parabola. Thus, consider a vertical line slicing through a point in the negative $a_0$-axis. The line is the $t$-axis of the function located at this point on the $a_0$-axis. Observe that this line intersects the parabola at two points, one on the upper branch of the parabola, and the other on the lower branch. These two intersection-points are the two zeros of the function.

Notice also the following: The *only* degenerate zero that occurs in the *entire* zero-unfolding is the 2-fold degenerate zero that occurs in the germ at the origin. Because it is degenerate, it satisfies the following two conditions: both the function $F$ and its first derivative $\partial F/\partial t$ are zero at this point. This means that the point must be on the zero-manifold (parabola) where the tangent is *vertical*. Furthermore, it must be the only point of vertical tangency on the manifold. We clearly see this in the right diagram in Fig 9.3.

Now the reader will recall that, using Method 2, the zero-unfolding, given by the right diagram in Fig 9.3, can be regarded as the zero-unfolding of the singularity $A_1$ of the function $t^2$. The reason why our method considers this unfolding is that it is also the zero-unfolding in the Ground Interactive Unfolding $EA_1$. However, Interactive Singularity Theory considers not only the zero-structure of an unfolding, but also the critical-point structure of the same unfolding. Therefore, let us now consider the critical-point structure of this unfolding.

This can be seen clearly in the three functions in Fig 9.5. The entire critical-point structure is a single minimum that is preserved throughout the complete unfolding. Therefore, we can now give the critical-point manifold of this unfolding. It is the horizontal bold line in Fig 9.6. Thus, this figure shows *both* the zero-manifold (parabola) and the critical-point manifold (horizontal line) of the Interactive Unfolding of $EA_1$.

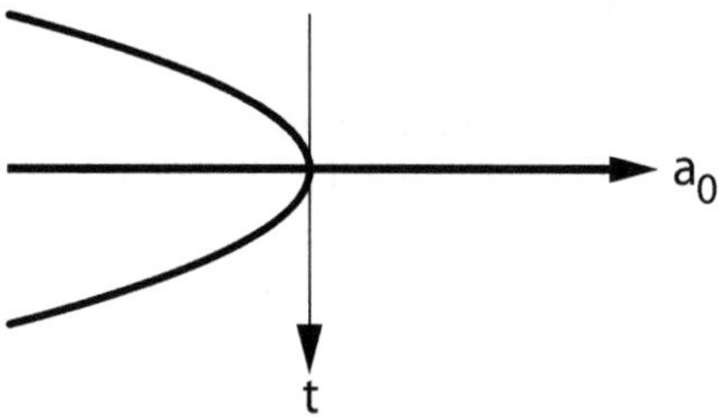


**Fig. 9.6.** The zero-manifold (parabola) and critical-point manifold (horizontal line) of the Ground Interactive Unfolding $EA_1$.

It is important to understand the *relationship* between the zero-manifold and critical-point manifold as follows: Consider again the three representative functions in Fig 9.5. We need to observe the relationship between the critical-point structure and zero-structure in the sequence of functions. The right function has one critical point, its minimum, and no zero. In the middle function (i.e., the germ), the single critical point (the minimum) has become simultaneously a zero, which is 2-fold degenerate. Finally, in the left function, the single critical point (the minimum) has remained in the center while the 2-fold degenerate zero has bifurcated into two zeros, one on each side of the critical point.

This explains the relationship between the critical-point manifold (horizontal line) and zero-manifold (parabola) in Fig 9.6. Consider a vertical line slicing through a point on the positive $a_0$-axis. The vertical line is the $t$-axis of the function at that point. We see that this line slices through the critical-point manifold (horizontal line) but not the zero-manifold (parabola). The reason is that any function in the positive $a_0$-axis has one minimum and no zeros. Now move this vertical line to the origin of the control space. At the origin, it slices through *both* the critical-point manifold (horizontal line) and the apex of the zero-manifold (parabola). Furthermore, the point through which it slices on the critical-point manifold is coincident with the point through which it slices on the zero-manifold. This is because the point is simultaneously the minimum and the zero in the germ function. Finally, move the vertical line into the negative $a_0$-axis. The line now slices through a single point on the critical-point manifold (horizontal line) and two separate points on the zero-manifold (parabola). This is because any function in the negative $a_0$-axis has a single minimum and two separate zeros, one on each side of the minimum.

We now therefore understand the relationship between the critical-point manifold and the zero-manifold of the Ground Interactive Unfolding $EA_1$.

Given this, let us now understand the *separatrix* structure given by these two manifolds. Recall that the separatrix given by a solution manifold is the projection of the manifold's degeneracy subset (subset of degenerate points) to the control space. Observe that the degeneracy subset of the zero-manifold (parabola) is only the apex of the parabola. This is the only point of a degenerate zero. The projection of this point to the control space is the origin of the control space. Thus, the zero-separatrix of the unfolding of $EA_1$ is only a single point, the origin of the control space. Now consider the degeneracy subset of the critical-point manifold (horizontal line). We know that there are no degenerate critical points in this unfolding, i.e., the minimum remains non-degenerate throughout the unfolding. Therefore there is no degeneracy subset of the critical-point manifold. Correspondingly, there is no critical-point separatrix in the control space.

Therefore, we see that the *full* separatrix of the Ground Interactive Unfolding $EA_1$ consists of only a zero-separatrix, and this consists of only a single point. In fact, we can also see this from the "vertical-tangent criterion". Recall that the degeneracy subset of a solution manifold is the subset of points where the tangent is vertical (when the $t$-axis is represented as vertical). We can see from Fig 9.6 that the zero-manifold has only one point where the tangent is vertical, and the critical-point manifold has no points where the tangent is vertical.

This situation therefore illustrates our Rule 1 (page 309) for the intersection between the zero-manifold and critical-point manifold of an Interactive Unfolding of $EA_n$. Rule 1 states that a 2-fold zero-degenerate point must be in the degeneracy subset of the zero-manifold where this manifold *intersects* the critical-point manifold. Furthermore, the rule says that the intersection-point is not within the degeneracy subset of the critical-point manifold. We can clearly visualize this in Fig 9.6 which gives the structure for $EA_1$. The intersection-point between the two manifolds is where the zero-manifold has a vertical tangent but where the critical-point manifold does not have a vertical tangent. Therefore the intersection-point is in the degeneracy subset of the zero-manifold but not in the degeneracy subset of the critical-point manifold, in accord with Rule 1.

Looking over the discussion of this section, and using Fig 9.6, we can see that the Ground Interactive Unfolding $EA_1$ gives a powerful model of the singularity structure of the Process Grammar operation

$$Cm^+ : m^+ \longrightarrow 0m^-0$$

The operation corresponds to moving within the control space $a_0$ for $EA_1$ from a point in the positive $a_0$-axis to a point in the negative $a_0$-axis. At the starting point, the vertical $t$-axis slices through only the critical-point manifold, and does so at a positive minimum $m^+$, which is the **domain of the operation $Cm^+$**. Moving the vertical $t$-axis along the control space $a_0$ towards the negative $a_0$-axis, we have seen that, at the origin, it slices through both the critical-point manifold and the apex of the zero-manifold. Therefore, at this position, the minimum, which remains non-degenerate as a critical point, has become a 2-fold degenerate zero, which therefore gives the **transition-state of the operation $Cm^+$**. Subsequently, on entering the negative $a_0$-axis, the vertical $t$-axis slices through both the critical-point manifold and the two branches of the zero-manifold, thus giving the singularity-configuration $0m^-0$ which is the **codomain of the operation $Cm^+$**.

The great advantage of modeling the singularity structure of $Cm^+$ by the Ground Interactive Unfolding $EA_1$ is that it gives the *universal* unfolding that corresponds to $Cm^+$. Recall that the term universal means that the control space has the minimal dimension needed to produce the singularity structure of an unfolding. This implies that $EA_1$ defines the singularity structure of $Cm^+$ even when it occurs in higher-order Interactive Unfoldings, i.e., for $EA_n$ with $n > 1$. Thus recall that, in section 8.7, we were studying the structure of the Interactive Unfolding of $EA_3$, and saw that the $Cm^+$ operation was given by a downward movement through any sheet of the minima zero-separatrix in the 3-dimensional control space. The value of the Ground Interactive Unfolding $EA_1$ is that it fully defines for us, in a minimal form, what is happening in this movement within the Interactive Unfolding of $EA_3$.

An important fact is that, despite the minimal nature of $EA_1$, it has an *Interactive* Structure *within* its own minimal total space. This Interactive Structure, as given by the form determined by Rule 1, is crucial for defining the powerful interactive role of $EA_1$ within a higher-order Interactive Unfolding such as that of $EA_3$. That is, the fact that Rule 1 gives the intersection between a zero-manifold and a critical-point manifold where the first manifold is 2-fold degenerate and the second is non-degenerate is what makes $EA_1$ give the relationship between the critical-point bifurcations in the Interactive Unfolding of $EA_3$, as we will see in Chapter 10. Thus the Interactive Structure *within* the Ground Interactive Unfolding $EA_1$ becomes the basis of significant Interactive Structure in higher-order Interactive Unfoldings such as that of $EA_3$.

As stated earlier, according to our theory, the latter Interactive Structure is fundamental to understanding spatial morphology. Furthermore, it is the failure of Catastrophe Theory to recognize this Interactive Structure that makes Catastrophe Theory unable to model spatial morphology.

Thus, we can see that Interactive Singularity Theory provides the singularity theory needed for spatial morphology. But what one also needs is the structure provided by my New Foundations to Geometry; i.e., the laws of memory storage, which lead to the Symmetry-Curvature Duality Theorem, the PISA definition of symmetry axis, and the causal compatibility of Interactive Singularity Theory with the causal structure provided by the PISA axes.

## 9.7 The Intersection Structure of the Interactive Unfolding of $EA_2$

In the last section, we created the Ground Interactive Unfolding $EA_1$. In the present section, we will create the next level Interactive Unfolding, which is the **Interactive Unfolding of $EA_2$**. This will become useful in Chapter 11, where we examine $EA_2$ in terms of the Interactive Structure of the pair-creation operators of the Process Grammar.

Before doing this, we will, in the present section, elaborate the *intersection structure* of the Interactive Unfolding of $EA_2$ by using the rules given on p309.

First observe this: In the fold catastrophe, the canonical germ is taken to be $t^3$, where it is unfolded with respect to only its critical-point structure. In the Interactive Unfolding of $EA_2$, we take the same canonical germ, but unfold it with respect to both its critical-point structure and its zero-structure. Therefore, Interactive Singularity Theory defines its canonical universal unfolding to be

$$F(t; a_0, a_1) = t^3 + a_0 + a_1 t \tag{9.6}$$

Using Method 2, the zero-manifold of this unfolding is the same structure as the critical-point manifold of the cusp catastrophe, and is shown as the left diagram in Fig 9.7.

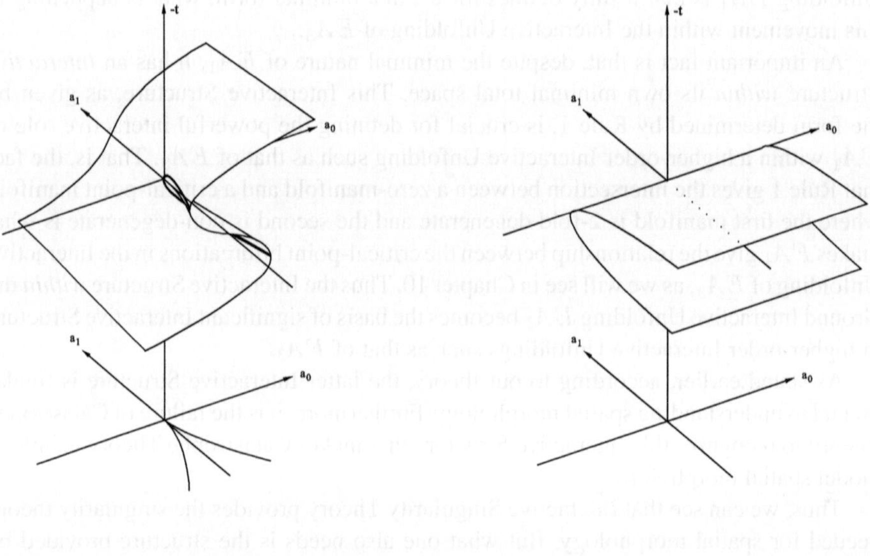


**Fig. 9.7.** The zero-manifold (left) and the critical-point manifold (right) of the Interactive Unfolding of $EA_2$.

Now let us turn to the critical-point manifold. As stated on p308, the critical-point manifold of the Interactive Unfolding of $EA_n$ must be the critical-point manifold of the catastrophe $A_n$ *swept along the extra dimension* $a_0$. As described in the previous section, the critical-point manifold of the fold catastrophe is a parabola. Therefore, to obtain the critical-point manifold of the Interactive Unfolding of $EA_2$, we sweep the parabola along the extra dimension $a_0$, thus obtaining the parabola surface shown as the right diagram in Fig 9.7.

Now let us use our intersection Rules 1 and 2, on p309, to understand how the zero-manifold and critical-point manifold of the Interactive Unfolding of $EA_2$ intersect; i.e., how the left and right diagrams in Fig 9.7 intersect. Our rules use the degeneracy subset of the zero-manifold to predict the types of intersection. Thus, let us first use Method 2 to find the degeneracy subset of the zero-manifold. In the discussion of the critical-point manifold of the cusp catastrophe, as illustrated in Fig 7.5 (page 247), we saw that the bold curve (degeneracy subset), labeled $\Delta_f$ in that figure, consists of a 3-fold degenerate critical point at the origin, and 2-fold degenerate critical points all along the rest of the curve. Therefore, the corresponding bold curve in the Interactive Unfolding of $EA_2$, as illustrated here in the left diagram of Fig 9.7, must consist of a 3-fold degenerate zero at the origin, and 2-fold degenerate zeros all along the rest of the curve. We will refer to the bold curve without the origin as the set of *non-origin points* on the bold curve.

Based on the above facts, let us now use our intersection rules (p309) to predict where the zero-manifold intersects the critical-point manifold.

First let us use Rule 1. This rule applies to any 2-fold degenerate zero. It says that such a point must be in the degeneracy subset of the zero-manifold where this manifold *intersects* the critical-point manifold; but the point is not within the degeneracy subset of the latter manifold.

Now we know that all the non-origin points on the bold curve in the zero-manifold (left diagram in Fig 9.7) represent 2-fold degenerate zeros. Therefore, applying Rule 1, we conclude that, at all these points, the zero-manifold must intersect the critical-point manifold (parabola surface in the right diagram).

Notice that all these non-origin points on the bold curve are points of vertical tangency on the zero-manifold (left diagram), but are points of non-vertical tangency on the critical-point manifold (right diagram). This means that they are all in the degeneracy subset of the zero-manifold but not in the degeneracy subset of the critical-point manifold. This is predicted by Rule 1.

We therefore see that, at any of these non-origin intersection-points, the critical-point manifold cuts through the zero-manifold transversally at the bold-curve in the zero-manifold. In fact, the upper sheet of the critical-point manifold (parabola surface) cuts the zero-manifold (left diagram) transversally along the left fold-curve; and the lower sheet of the critical-point manifold (parabola surface) cuts the zero-manifold (left diagram) transversally along the right fold-curve.

Now the crucial thing to observe is this: The intersection at any non-origin point on the bold curve is qualitatively exactly like that shown in Fig 9.6 (page 315), where the parabola in that figure corresponds to a fold in the zero-manifold in the left diagram of Fig 9.7; and the horizontal line in Fig 9.6 (page 315) corresponds to the upper or lower sheet of the critical-point manifold in the right diagram of Fig 9.7. That is, Fig 9.6 (page 315) illustrates the way the upper or lower sheet of the critical-point manifold in Fig 9.7 cuts through the left or right fold, respectively, in the zero-manifold in Fig 9.7.

The reason is this: Each non-origin point on the bold curve on the right fold of the zero-manifold of $EA_2$ (left diagram in Fig 9.7) is an example of the germ singularity that is unfolded in the Ground Interactive Unfolding $EA_1$, which we defined in the previous section. The Ground Interactive Unfolding $EA_1$ was illustrated in Fig 9.6 (page 315), and we have now seen that this unfolding occurs through each non-origin point on the right bold curve in the Interactive Unfolding of $EA_2$. Correspondingly, each non-origin point on the bold curve on the left fold corresponds to the dual of $EA_1$, which will be introduced later in the book, and has the same manifold structure as Fig 9.6 (page 315).

Having applied our intersection Rule 1, let us now use Rule 2. This rule concerns an $n$-fold zero-degenerate point where $n > 2$. The rule says that such a point must be where the degeneracy subset of the zero-manifold intersects the degeneracy subset of the critical-point manifold, because an n-fold degenerate zero must be an $n-1$-fold degenerate critical-point. Clearly, this rule applies to the origin of the zero-manifold in the left diagram of Fig 9.7. This point is a 3-fold degenerate zero. Consequently, it must be a 2-fold degenerate critical point, and this must therefore define the point at the origin of the critical-point manifold (the parabola surface in the right diagram). This is easily checked by the fact that the germ is located at the origin of the control space and is the function $t^3$, which, at the origin of its $t$-axis, is 3-fold degenerate as a zero and 2-fold degenerate as a critical point.

Note also the following: The $a_0$ control variable raises and lowers the graph of the function, and therefore, along the $a_0$-axis in the control space, the graph of the germ $t^3$ is raised and lowered, which means that the singularity at the origin of the $t$-axis stops being a zero but remains 2-fold degenerate as a critical point. This corresponds to the following facts: (1) the vertical edge of the parabola surface consists of the points where each of these raised and lowered versions of the germ has a 2-fold degenerate critical point; (2) the vertical edge of the parabola surface intersects the zero-manifold (left diagram) only at the origin, because this is the only position on the vertical edge where a 2-fold degenerate critical point has zero value.

The structure of the intersection at the origin is complex and subtle. In order to understand it, let us take the vertical planar slice along the $a_1$-axis through the total space; i.e., the slice defined by $a_0 = 0$. First, let us see what this slice shows about the zero-manifold (left diagram in Fig 9.7). By the fact that we are considering the zero-manifold, set $F$ to zero in the unfolding equation (9.6); and by the fact that we are only on this slice, also set $a_0$ to zero in this equation. Thus the structure of the zero-manifold on this slice is

$$0 = t^3 + a_1 t$$

which has two solutions: the line $t = 0$ and the inverted parabola $a_1 = -t^2$. These two solutions are illustrated by the bold straight line and parabola in the left diagram in Fig 9.8.

It is important to understand what this structure means in terms of the *full shape* of the zero-manifold (left diagram in Fig 9.7). First consider the straight line in this slice. This must correspond to the $a_1$-axis, which, as we can see in the left diagram in Fig 9.7, exists along the single sheet of the manifold behind the origin, then goes through the origin, and then goes along the *central sheet* of the manifold after the origin. Now consider the parabola. This is the vertical slice through the upper and lower sheet of the manifold together with the origin. The two branches of the parabola are the slice through the upper and lower sheet of the manifold, and the apex of the parabola is the origin. It is crucial to observe that the tangent to the apex of the parabola is *vertical*. This vertical tangency property, at the origin, can also be seen in the bold curve in the zero-manifold; i.e., while the projection of the bold curve down into the control space gives a curve with a cusp-point at the origin, the bold curve itself does not have a cusp-point at the origin, and is, in fact, smooth at the origin with a vertical tangent.

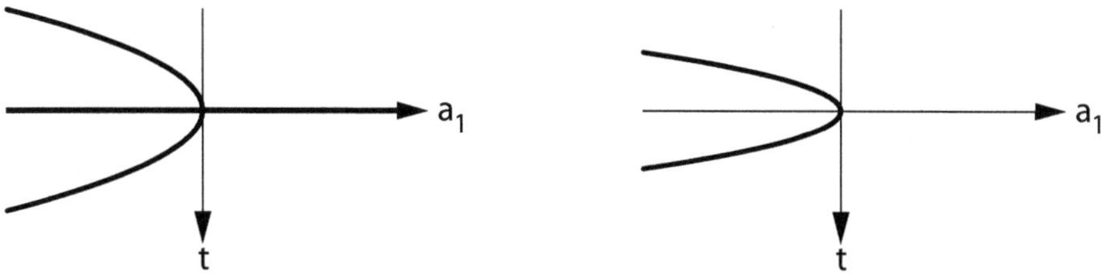


**Fig. 9.8.** For the Interactive Unfolding of $EA_2$: Left: the $a_0 = 0$ slice through its zero-manifold. Right: the $a_0 = 0$ slice through its critical-point manifold.

Now take the $a_0 = 0$ slice through the critical-point manifold (right diagram in Fig 9.7). This obviously also gives a parabola. However, it is a different scale. It will be useful to calculate the scale, as follows: Because we are now considering the critical-point manifold, the slice is given by setting the derivative of the unfolding equation (9.6) to zero, thus obtaining

$$0 = 3t^3 + a_1 t$$

which gives an inverted parabola $a_1 = -3t^2$ that is narrower than the one given by the zero-manifold. This narrower parabola is shown in the right diagram in Fig 9.8.

Thus, putting together the slice through the zero-manifold and the slice through the critical-point manifold, we obtain Fig 9.9. In this figure, the outer parabola and horizontal line come from the zero-manifold; and the inner parabola comes from the critical-point manifold. That is, the critical-point parabola is between the zero parabola and zero straight line.

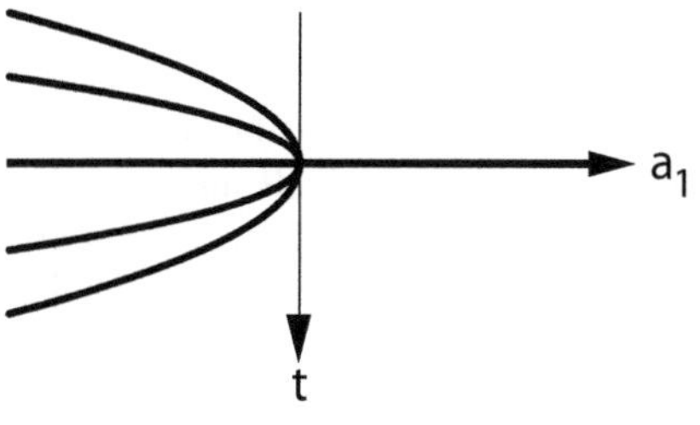


**Fig. 9.9.** The $a_0 = 0$ slice through *both* the zero-manifold and critical-point manifold of the Interactive Unfolding of $EA_2$.

We can see the logic of this structure by considering representative functions along the $a_1$-axis. These are shown in Fig 9.10. Note that, because the slice is along the $a_1$-axis, the unfolding, restricted to this slice, is the canonical unfolding of the fold catastrophe. However, Catastrophe Theory is not interested in what we are calling the *Interactive* Structure – the relation between the zeros and the critical-points. In fact, as we have pointed out, the $A_2$ singularity of the fold catastrophe does not specify any zero-structure, and thus, according to Catastrophe Theory, the set of functions shown in Fig 9.10 need not have the zero-structure shown there; i.e., the graphs could be higher or lower.

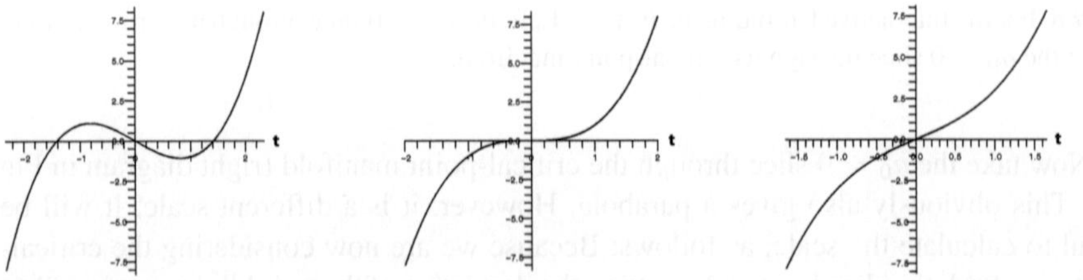

**Fig. 9.10.** The functions at points $(a_0, a_1) = (0, -2), (0, 0), (0, 2)$ in the 2D control space of the Interactive Unfolding of $EA_2$.

In contrast, because Interactive Singularity Theory specifies that the singularity in the germ is the complete singularity $EA_2$, rather than the conventional singularity $A_2$, we require the *Interactive Structure* that is seen in the functions in Fig 9.10.

Let us now examine that Interactive Structure. Observe the following facts concerning these three functions.The right function has one non-degenerate zero and no critical point; the middle function has one degenerate zero which is also a degenerate critical point; and the left function has three separate zeros that alternate with two separate critical points.

This explains the structure shown in Fig 9.9. Thus, consider a vertical line slicing through a point on the positive $a_1$-axis (in Fig 9.9). The vertical line is the $t$-axis of the function located at this point on the $a_1$-axis. Observe that this line intersects only the horizontal bold line, i.e., the zero-manifold. This corresponds to the single non-degenerate zero in the right function in Fig 9.10. Next consider the vertical $t$-axis slicing through the origin of the $a_1$-axis in Fig 9.9. It intersects the apex of large parabola and the horizontal line, which means that this intersection-point is a 3-fold degenerate zero; and it simultaneously intersects the apex of the smaller parabola which means that this intersection-point is also a 2-fold degenerate critical point. Next consider the vertical $t$-axis slicing through a point in the negative $a_1$-axis in Fig 9.9. It slices through the two branches of the larger parabola, as well as the horizontal line – thus giving 3 separate zeros – and it also slices through the two branches of the smaller parabola between these – thus giving 2 separate critical points between the 3 separate zeros. This 5-fold singularity-configuration, 3 zeros alternating with 2 critical points, is exactly the singularity-configuration that we see in the left-most function in Fig 9.10.

The wonderful thing about this example is that it illustrates an important feature of our intersection Rule 2, as follows:

**Because Rule 2 concerns an $n$-fold zero-degenerate point where $n > 2$, it predicts that this point must be where the degeneracy subset of the zero-manifold intersects the degeneracy subset of the critical-point manifold.**

**From this prediction, we deduce that the tangent to *both* manifolds, at this intersection-point, must be "vertical" in the sense that the projection map from both manifolds to the control space must be singular at this intersection-point.**

**This is exactly what we see in the two-parabola example of $EA_2$ shown in Fig 9.9. At their intersection-point, the two parabolas are tangential to each other, and share a vertical tangent at that point, thus illustrating the above fundamental fact about Rule 2.**

## 9.8 The Intersection Structure of the Interactive Unfolding of $EA_3$

The total space of the Interactive Unfolding of $EA_3$ is 4-dimensional, and therefore it will not be possible to draw its critical-point manifold and zero-manifold. Nevertheless, as we have seen in Chapter 8, we are able to draw its critical-point separatrix and zero-separatrix, because these are 2-dimensional surfaces in the 3-dimensional control space. Furthermore, we will be able to use the discoveries in the previous two sections, concerning the critical-point manifolds and zero-manifolds of the lower-order Interactive Unfoldings $EA_1$ and $EA_2$, to predict important aspects of the structure of the critical-point manifold and zero-manifold for $EA_3$, because the latter contains structures from $EA_1$ and $EA_2$.

To understand this, let us begin by considering the zero-separatrix for $EA_3$, shown again here as Fig 9.11. In section 8.7, we used Method 2 to create a correspondence between the zero-separatrix for $EA_3$ and the critical-point separatrix of the swallowtail catastrophe $A_4$. In the critical-point separatrix for $A_4$, the origin obviously corresponds to the germ for $A_4$. Furthermore, we saw the following concerning that critical-point separatrix: Each point on the back and diagonal sheets is locally right-equivalent to the germ of the fold catastrophe $A_2$; each point on the upward bottom arch sheet is locally right-equivalent to the germ of the dual-fold catastrophe $A_{-2}$; each point on the right cuspidal edge is locally right-equivalent to the germ of the cusp catastrophe $A_3$; and each point on the left cuspidal edge is locally right-equivalent to the germ of the dual-cusp catastrophe $A_{-3}$. In particular, Fig 8.19 (page 289), which showed a vertical planar slice through the front part of the critical-point separatrix for $A_4$, labeled the separatrix components by the germs $A_{\pm k}$ to which the points on the components are locally right-equivalent.

Using Method 2, we can now understand the germs to which the points on the zero-separatrix for $EA_3$ correspond. We know already that the origin corresponds to the

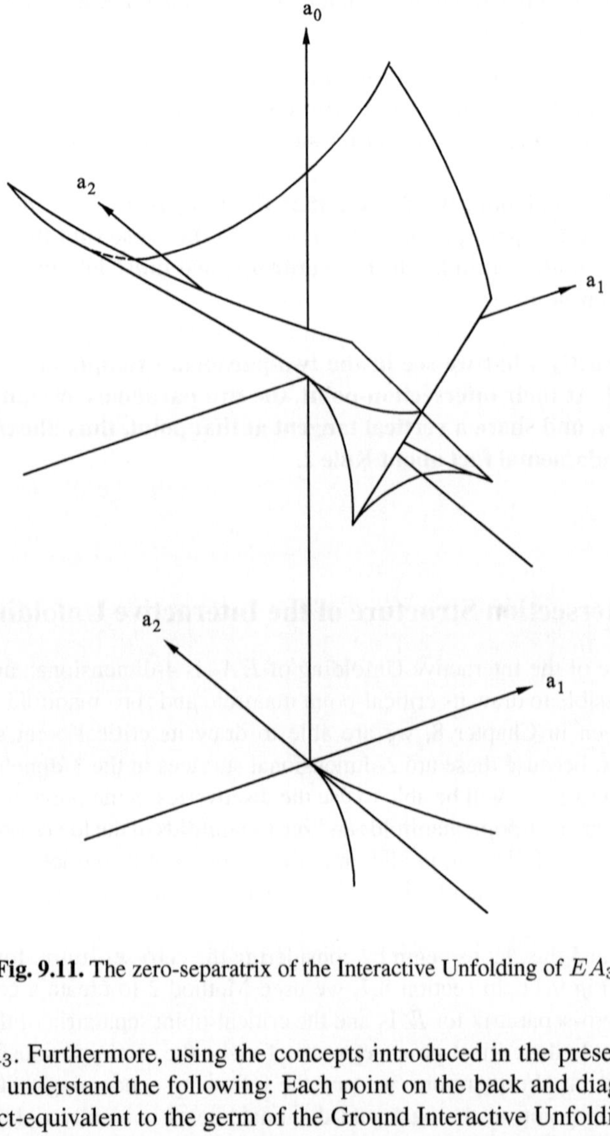


**Fig. 9.11.** The zero-separatrix of the Interactive Unfolding of $EA_3$.

germ for $EA_3$. Furthermore, using the concepts introduced in the present chapter, we now can also understand the following: Each point on the back and diagonal sheets is locally contact-equivalent to the germ of the Ground Interactive Unfolding $EA_1$; each point on the upward bottom arch is locally contact-equivalent to the germ of the *Dual* Ground Interactive Unfolding $EA_{-1}$ (this dual will be defined later in the book); each point on the right cusped edge is locally contact-equivalent to the germ of the Interactive Unfolding of $EA_2$; and each point on the left cusped edge is locally contact-equivalent to the germ of the dual, the Interactive Unfolding of $EA_{-2}$. In particular, Fig 9.12 shows a vertical planar slice through the front part of the zero-separatrix for $EA_3$, and labels the separatrix components by the germs $EA_{\pm k}$ to which the points on the components are locally contact-equivalent.

Fig 9.13 shows representative functions in relation to this vertical slice. This diagram was shown earlier on the bottom half of p290; but here we discuss it in relation to the

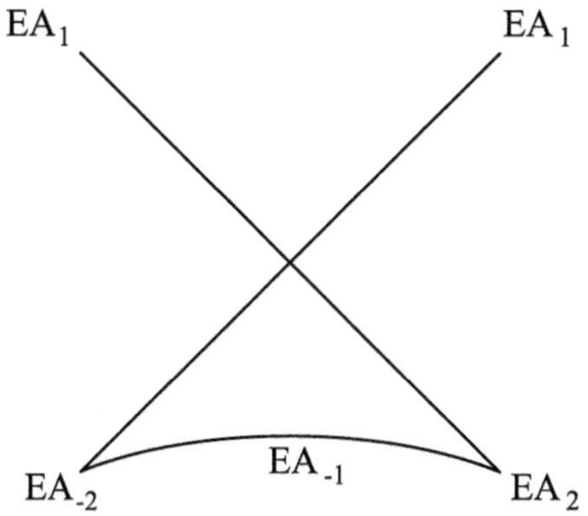


**Fig. 9.12.** The local contact-equivalence of points in the zero-point separatrix for $EA_3$ with germs in lower ordered Interactive Unfoldings.

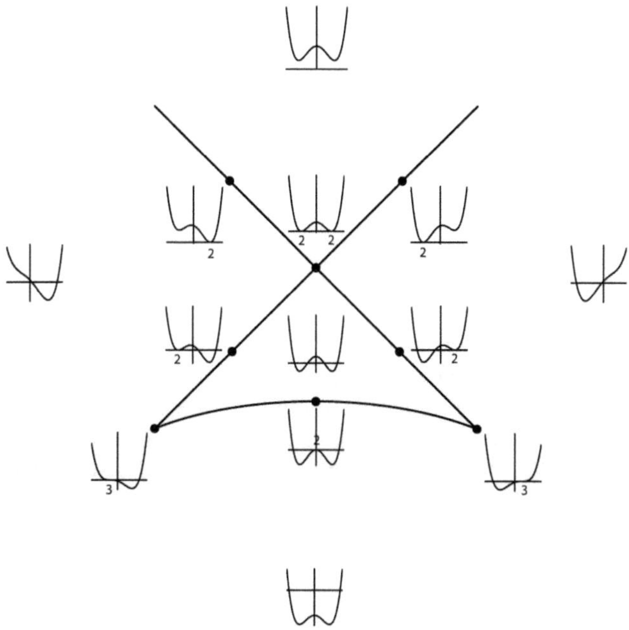

**Fig. 9.13.** Representative functions in relation to a vertical slice through the zero-separatrix of the Interactive Unfolding of $EA_3$.

new diagram above it, Fig 9.12. Note that any degenerate zero in any of the functions in Fig 9.13 is labeled with a number indicating the degree of its degeneracy as a zero. Observe that this number $n$ is one number more than the number $n-1$ of the subscript on the corresponding germ $EA_{n-1}$ in Fig 9.12. The reason is that the singularity in the corresponding germ $EA_{n-1}$ is an $n-1$-fold degenerate critical point but simultaneously an $n$-fold degenerate zero. We observe that each singularity $EA_{n-1}$ is therefore a complete singularity, in contrast to the corresponding singularity $A_{n-1}$ in Catastrophe Theory.

In relation to our discussion in the previous two sections, observe this: Consider the two diagonal sheets in Fig 9.13. As shown in this diagram, any function on these sheets contains a zero that is labeled with the number 2; that is, the zero is 2-fold degenerate as a zero. This means that this zero is contact-equivalent to the zero in the germ of the Ground Interactive Unfolding $EA_1$, as shown by the labels on the two diagonal sheets in Fig 9.12. The crucial consequence is that a path which downward crosses either of these two diagonal sheets, or the back sheet in Fig 9.11, must be equivalent to the 1-dimensional control space, the $a_0$-axis, of the Ground Interactive Unfolding $EA_1$ as illustrated in Fig 9.6 (page 315). That is, as illustrated by the *functions* in Fig 9.13, such a path has the effect of lowering a non-degenerate minimum through the $t$-axis of the function. Thus, the zero-manifold and critical-point manifold of this lowering effect has the structure given by the zero-manifold and critical-point manifold shown in Fig 9.6 (page 315). In particular, at any point on the diagonal sheets or the back sheet of the zero-separatrix for $EA_3$ in Fig 9.11, the zero-manifold and critical-point manifold must intersect as shown in Fig 9.6 (page 315), in accord with our intersection Rule 1. The reader will recall that this intersection-point corresponds to the single-point separatrix for $EA_1$. We can therefore state the following concerning the continuation operation $Cm^+$:

**CONTINUATION OPERATION $Cm^+$**
**IN THE**
**INTERACTIVE UNFOLDING OF $EA_3$**

**The Process Grammar operation $Cm^+$, which describes the continuation of a squashing force or process till it indents, can be modeled by a downward movement through a point in the minima zero-separatrix of the Interactive Unfolding of $EA_3$, and corresponds to movement along the 1-dimensional control space of the Ground Interactive Unfolding $EA_1$, through the single-point separatrix for $EA_1$.**

In relation to our discussion in the previous section, observe this: Consider the right cuspidal edge in Fig 9.11. As illustrated in Fig 9.13, a function at any point on this edge contains a zero that is labeled with the number 3; that is, the zero is 3-fold degenerate as a zero. This means that this zero is contact-equivalent to the *complete singularity* in the germ of the Interactive Unfolding of $EA_2$, as shown by the label at the right cusp-point in Fig 9.12. The crucial consequence is that a path that maximally crosses the cusp-point (i.e., from the right into the central volume) must be equivalent to the $a_1$-axis of the Interactive Unfolding of $EA_2$ as illustrated in Fig 9.9 (page 321). Thus, the structure of the zero-manifold and critical-point manifold at the cusp-point is as we discussed in the previous section and is given by our intersection Rule 2.

In addition to this 1-dimensional path, the reader will recall that the control space for $EA_2$ is 2-dimensional, and has the zero-manifold *surface* and critical-point manifold *surface* shown in Fig 9.7 (page 318). In Chapter 11 we will see how these surfaces relate to $EA_3$ and contribute to its Interactive Structure.

The full Interactive Structure of the zero-separatrix and critical-point separatrix for $EA_3$ is enormously powerful, and in the next three chapters we will be investigating this Interactive Structure and seeing how it organizes the interactions of the Process Grammar operations.

# 10

# Operators $B$ and $C$ in the Interactive Unfoldings of $EA_3$ and $EA_{-3}$

## 10.1 Introduction

To realize the set of operations of the Level 3 Process Grammar (page 72), we need not only the Interactive Unfolding of $EA_3$ but also its dual, the Interactive Unfolding of $EA_{-3}$, which is obtained by inverting the functions. The Level 3 Process Grammar operations are not only distributed across these two Interactive Unfoldings, but *inter-relate* these two Interactive Unfoldings in interesting ways, as we shall see. To help guide the reader through the following sections, we first note this:

**The two operations $Bm^+$ and $Bm^-$ occur in the Interactive Unfolding of $EA_3$, because they are the bifurcation of a *minimum*.**

**The two operations $BM^+$ and $BM^-$ occur in the Interactive Unfolding of $EA_{-3}$, because they are the bifurcation of a *maximum*.**

**The two continuation operations $Cm^+$ and $CM^-$ occur in *both*:
the Interactive Unfolding of $EA_3$ and the Interactive Unfolding of $EA_{-3}$.**

We begin by looking at the operations within the Interactive Unfolding of $EA_3$, and then turn to the Interactive Unfolding of $EA_{-3}$.

## 10.2 Operations $Bm^+$ and $Bm^-$ in the Interactive Unfolding of $EA_3$

First, we need to emphasize the following important facts, concerning the *inferred processes* of the *shapes*, as given by our theory, and as illustrated in Figs 10.1 and 10.2, and not causally understood in conventional singularity theories of shape:

**Both the bifurcation operations $Bm^+$ and $Bm^-$ are applied to a single** *minimum.*

**In $Bm^+$, the inferred force** *causing* **the initial minimum is** *compressive.*

**In $Bm^-$, the inferred force** *causing* **the initial minimum is** *penetrative.*

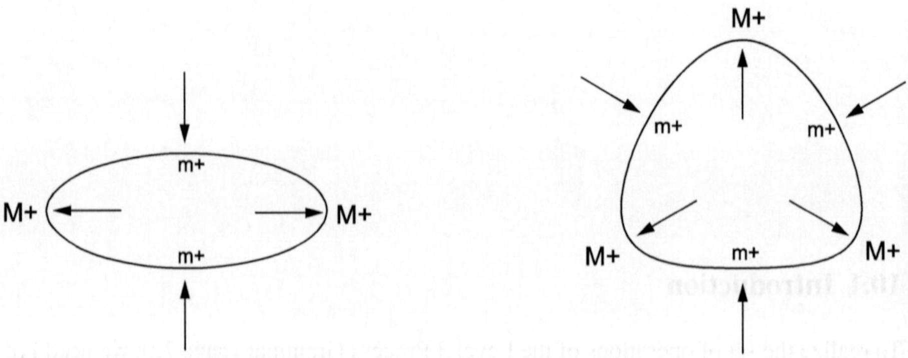


**Fig. 10.1.** Process-bifurcation B at $m^+$.

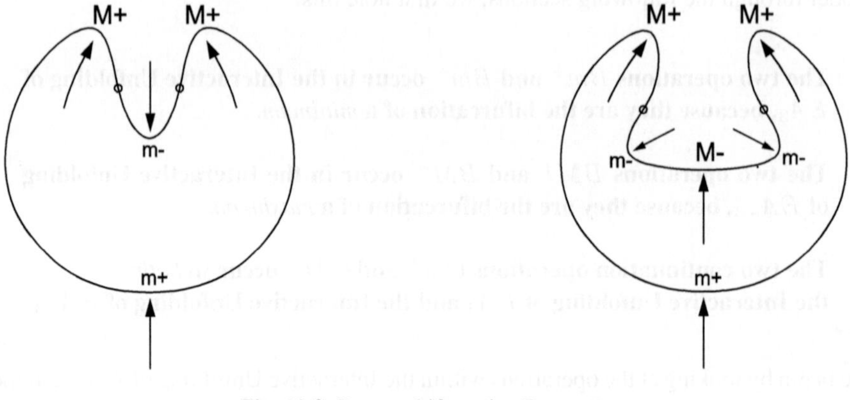


**Fig. 10.2.** Process-bifurcation B at $m^-$.

Next, we need to emphasize the following important *distinction*, concerning the corresponding sets of curvature *functions*, as illustrated in the distinction between Fig 10.3 and Fig 10.4, and not distinguished in Catastrophe Theory:

$Bm^+$ **is applied to a** *positive* **minimum, resulting in a** *positive* **minimum, maximum, minimum.**

$Bm^-$ **is applied to a** *negative* **minimum, resulting in a** *negative* **minimum, maximum, minimum.**

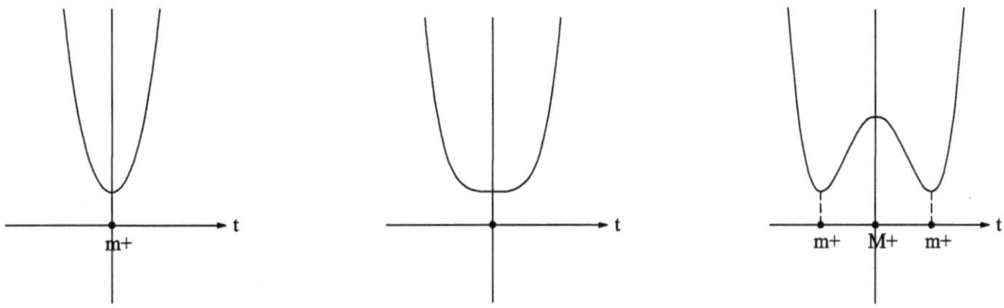


**Fig. 10.3.** Curvature functions given by the bifurcation operation $Bm^+ : m^+ \longrightarrow m^+M^+m^+$.

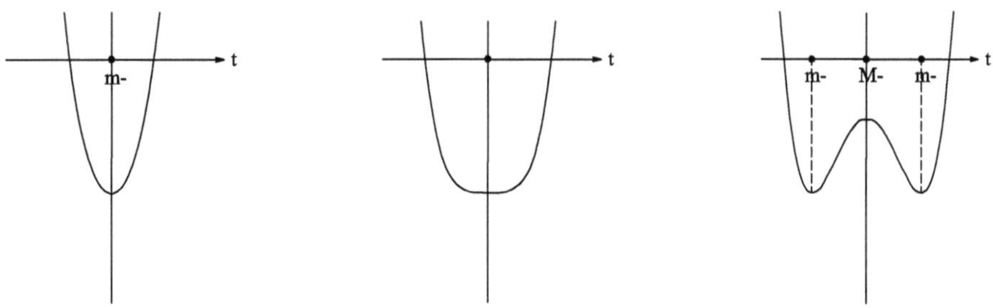


**Fig. 10.4.** Curvature functions given by the bifurcation operation $Bm^- : m^- \longrightarrow m^-M^-m^-$.

### 10.2.1 Relation of $Bm^+$ and $Bm^-$ to Critical-Point Manifolds

The operations $Bm^+$ and $Bm^-$ each occur within critical-point manifolds, which are based on distinctions that Catastrophe fails to give, and which are parts of what our theory defines as an Interactive Folding. The occurrence of operations $Bm^+$ and $Bm^-$ within such manifolds is illustrated in the figures on the next two pages, Fig 10.5 and Fig 10.6. In the case of $Bm^+$, the manifold consists entirely of positive critical points; and in the case of $Bm^-$, the manifold consists entirely of negative critical points – again, a distinction which Catastrophe Theory fails to give.

With respect to the 2D *subspace* $(a_1, a_2)$ of the 3D control space of $EA_3$, both operations go from a point in the region surrounding the critical-point separatrix of the subspace (e.g., point 1), through the origin of the subspace (point 2), to a point in the region between the separatrix curves (e.g., point 3). In the two figures, this is illustrated, without loss of generality, with points along the $a_2$-axis.

In the figures, the bifurcating effect is shown in a corresponding-sign critical-point manifold illustrated above the 2D control *subspace*. The non-degenerate extremum, located on the manifold above point 1, moves along the manifold to the origin, where it becomes degenerate as a critical point, and then separates into three extrema, on the three sheets of the manifold, above point 3. A crucial fact is this:

> **The operations $Bm^+$ and $Bm^-$ have the same structural relationship to their respective critical-point manifolds, but their manifolds are of opposite signs from each other, and these signs are crucial to their *causal* explanations, and the manifolds are related by the Interactive Structure.**

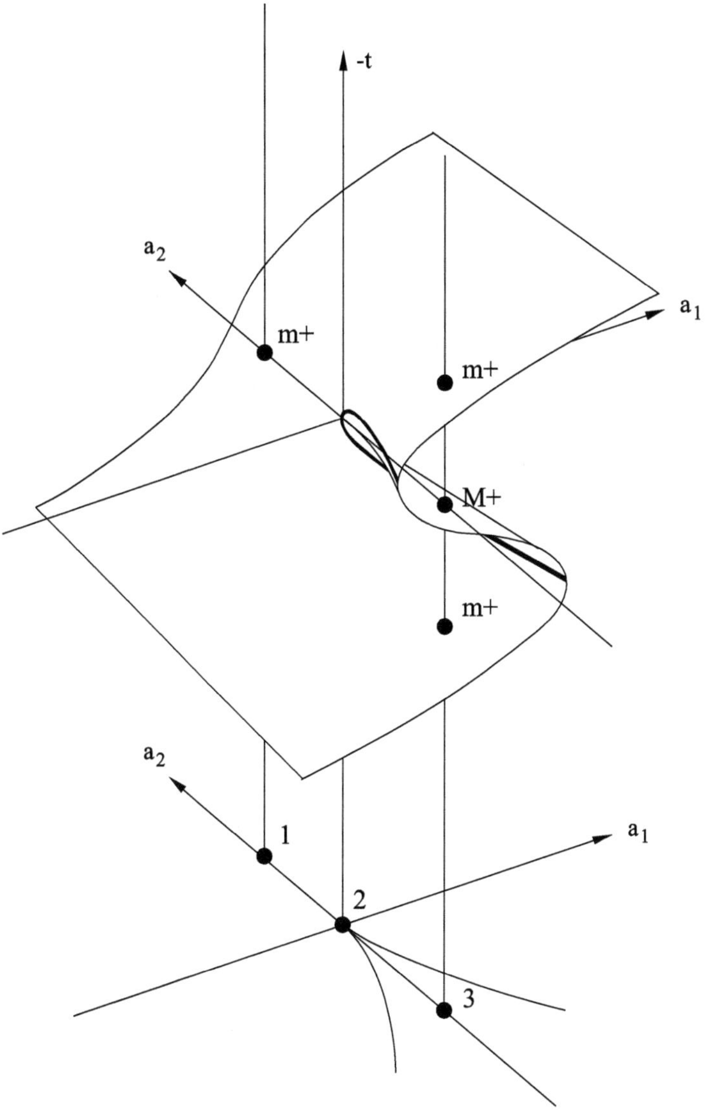


**Fig. 10.5.** The effect of $Bm^+$ within its corresponding critical-point manifold.

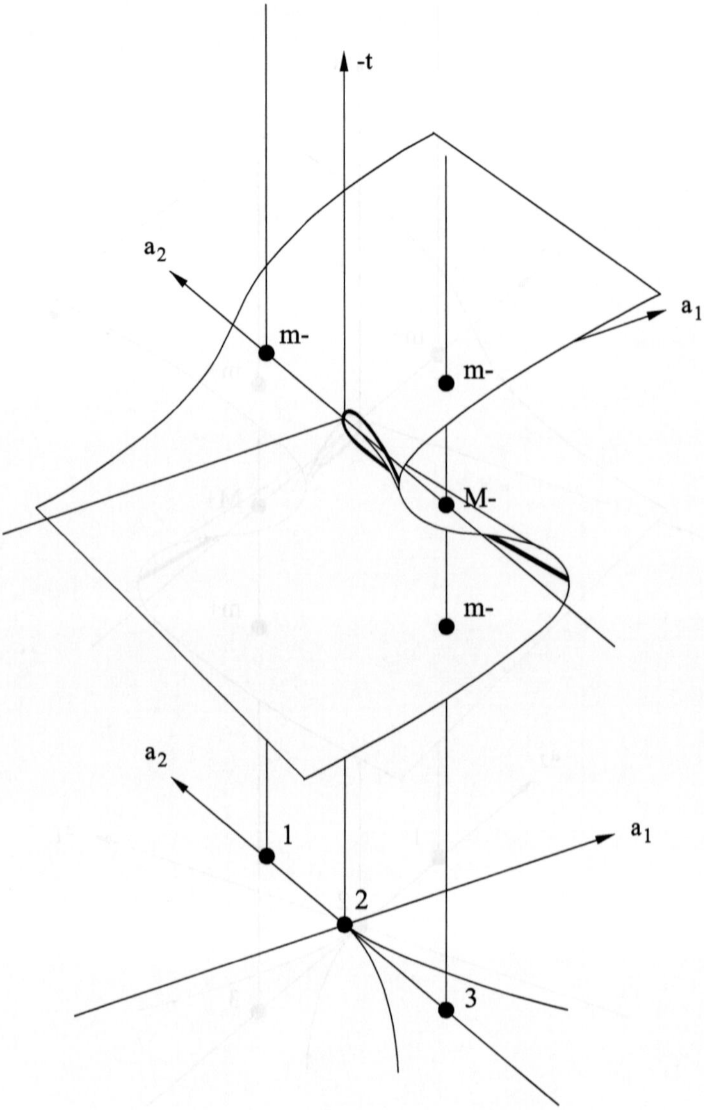


**Fig. 10.6.** The effect of $Bm^-$ within its corresponding critical-point manifold.

### 10.2.2 Relation of $Bm^+$ and $Bm^-$ to the Full Separatrix for $EA_3$

While the operations $Bm^+$ and $Bm^-$ act within their respective critical-point manifolds, they map the singularity-configuration given by the functions in one part of the 3D control space for $EA_3$ to the singularity-configuration given by the functions in another part.

To analyze this, the trajectories given in the previous two diagrams for $Bm^+$ and $Bm^-$ are illustrated respectively by the upper and lower trajectories marked $1 \to 2 \to 3$ in the 3D control space for $EA_3$, illustrated in Fig 10.7.

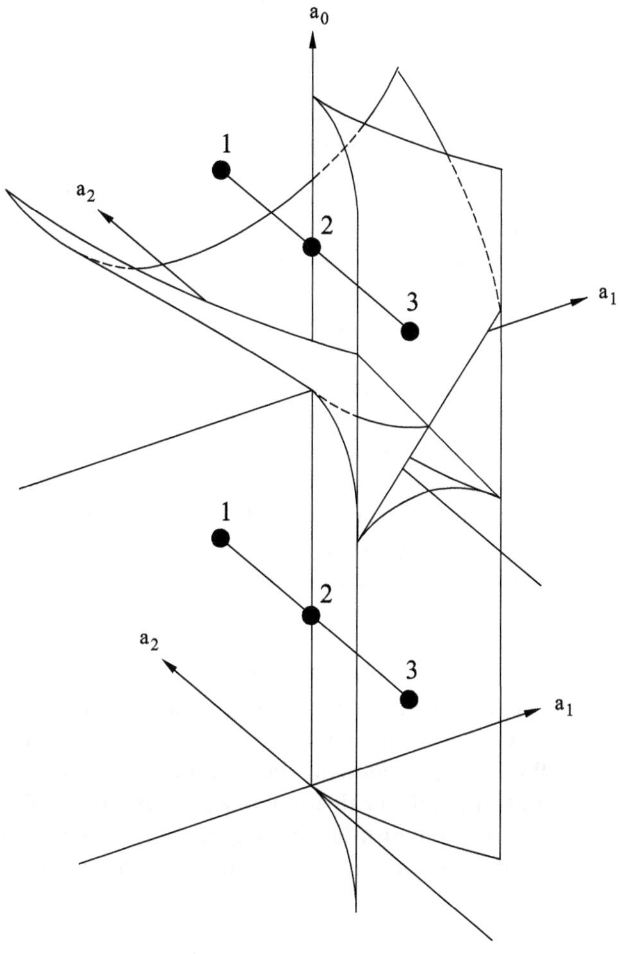


**Fig. 10.7.** Operations $Bm^+$ and $Bm^-$, as shown in the previous two figures, are shown here, respectively, as the upper and lower sequence $1 \to 2 \to 3$, in the 3D control space for $EA_3$.

Important concepts are as follows: In the control space of an Interactive Unfolding, the full separatrix is defined as the set of functions that have degenerate critical points with respect to either or both of our degeneracy criteria. The full separatrix divides the control space into a set of multi-dimensional volumes containing non-degenerate functions. (In the case of $EA_3$, these volumes are 3-dimensional.) Such a volume will be called a **volume of non-degeneracy** or simply a **volume**. The functions in a volume all have the same singularity-configuration, which will be called the **singularity-configuration** *belonging to* **the volume**. The volume is connected. Also, the full separatrix itself is partitioned into connected components, where all the functions in a component have the same singularity-configuration. Such a connected component will be called a **component** of the full separatrix. The singularity-configuration of the functions in a component will be called the **singularity-configuration** *belonging to* **the component**.

## VOLUMES AND THE INTERACTIVE STRUCTURE IN $Bm^+$ AND $Bm^-$

**Fig 10.7 shows how the operations $Bm^+$ and $Bm^-$ map between volumes of the 3D control space for $EA_3$. We see that these volumes represent a crucial aspect of what our theory calls the *Interactive Structure*; i.e., these volumes are crucially bounded by *both* the critical-point separatrix and zero-separatrix, as follows:**

> **$Bm^+$ must start at a point in the volume that is above the zero-separatrix and behind the critical-point separatrix (e.g., point 1); then go through a cusp-point in the critical-point separatrix above the zero-separatrix (e.g., point 2); and then go to a point in the volume that is above the zero-separatrix and in front of the critical-point separatrix (e.g., point 3). $Bm^+$ sends the singularity-configuration belonging to the volume containing point 1 to the singularity-configuration belonging to the volume containing point 3.**
>
> **$Bm^-$ must start at a point in the volume that is below the zero-separatrix and behind the critical-point separatrix (e.g., point 1); then go through a cusp-point in the critical-point separatrix below the zero-separatrix (e.g., point 2); and then go to a point in the volume that is below the zero-separatrix and in front of the critical-point separatrix (e.g., point 3). $Bm^-$ sends the singularity-configuration belonging to the volume containing point 1 to the singularity-configuration belonging to the volume containing point 3.**

Given what we have just said about the volumes corresponding to these operations, we can see that these volumes also determine important aspects of the critical-point manifolds corresponding to the operations, i.e., the manifolds illustrated in Figs 10.5 and 10.6. These figures showed the critical-point manifolds over 2D slices of the 3D control space. However, we now understand that these slices must occur through the corresponding volumes, as follows: Define a $(a_1, a_2)$**-slice** of the 3D control space to be a 2-manifold that is either entirely above or entirely below the zero-separatrix, and is such that the canonical projection of the slice onto the $(a_1, a_2)$-plane is non-singular, thus giving the slice, natural $(a_1, a_2)$-coordinates. It is clear that in Figs 10.5 and 10.6 the $(a_1, a_2)$-planes shown in the figures were slices in this sense, rather than actual planes (i.e., of $a_0$ constant value). For example, if the $(a_1, a_2)$ slice shown in Fig 10.5 had actually been a plane, then the zero-separatrix (which bends upward) would have cut through this slice, and the critical-point manifold would have changed from positive extrema to negative extrema where this intersection occurred. Thus, what is depicted in this figure, as the $(a_1, a_2)$ control space, must actually be a slice that bends upward in the 3D control space in Fig 10.7, to avoid intersecting with the zero-separatrix. A corresponding requirement must also be true for the $(a_1, a_2)$ control space shown Fig 10.6, so that it avoids the lowest arch of the zero-separatrix shown in Fig 10.7.

Finally, it is important to note that point 1, above the zero-separatrix in Fig 10.7, is a *compressive* extremum. In contrast, point 1, below the zero-separatrix, is a *penetrative* extremum. Thus the diagram shows how the bifurcation operator $B$ relates to the full separatrix in the application of $B$ to compressive vs. penetrative extrema.

## 10.3 Operations $Cm^+$ and $CM^-$ in the Interactive Unfolding of $EA_3$

While the minimum-based bifurcation operations $Bm^+$ and $Bm^-$ both exist in the Interactive Unfolding of $EA_3$, the maximum-based bifurcation operations $BM^+$ and $BM^-$ do not. The reason is that there is no example of 3-fold bifurcation of a maximum in the Interactive Unfolding of $EA_3$. For this, we will need the Interactive Unfolding of the dual $EA_{-3}$, to be discussed in section 10.5.

However, remarkably, although the maximum-based bifurcation operations $BM^+$ and $BM^-$ do not exist in the Interactive Unfolding of $EA_3$, the maximum-based continuation operation $CM^-$ does. We have already seen, in Chapters 8 and 9 how the minimum-based continuation operation $Cm^+$ exists in the Interactive Unfolding of $EA_3$, and therefore the conclusion will be that *both* continuation operations, $Cm^+$ and $CM^-$, exist in the Interactive Unfolding of $EA_3$.

In this paragraph, let us briefly recall the structure of $Cm^+$ in the control space for $EA_3$. As we said, it is given by any downward path through the minima zero-separatrix (i.e., the back sheet or two diagonal sheets of the zero-separatrix). Recall also from section 9.6 that the *universal unfolding* corresponding to $Cm^+$ is the Ground Interactive Unfolding $EA_1$. Therefore, as shown in section 9.8, any downward path through the minima-zero separatrix for $EA_3$ corresponds to the 1-dimensional control space, the $a_0$-axis, for $EA_1$. In particular, any point on the minima zero-separatrix for $EA_3$ corresponds to the single-point full separatrix for $EA_1$.

We now turn to the structure of the *dual* operation $CM^-$ in the Interactive Unfolding of $EA_3$. Consider the *lower* point marked 3 in Fig 10.7 (p335). It corresponds to the right-hand graph in Fig 10.8. Suppose we raise the right-hand graph upwards, relative to its $F$-axis, till the *maximum* has crossed the $t$-axis, but before the minima have crossed the $t$-axis. Then the transition is clearly described by the operation $CM^- : M^- \longrightarrow 0M^+0$. That is, the maximum, which is initially negative, becomes zero and then positive, while that zero undergoes a 2-fold bifurcation. Also, note that the maximum is non-degenerate throughout the raising, because the function is within Region 3 of the control subspace $(a_1, a_2)$.

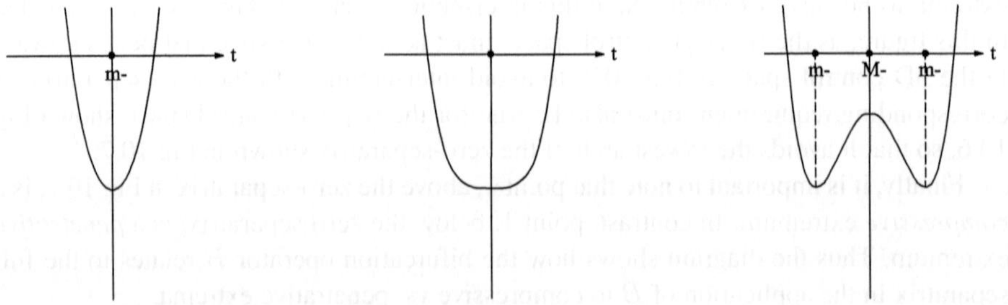


**Fig. 10.8.** Curvature functions given by the bifurcation operation $Bm^- : m^- \longrightarrow m^-M^-m^-$.

Observe now that raising the function in this way is given by moving the function from the lower point 3 in Fig 10.7, *upwards*, till it crosses the lower (arched) sheet of the zero-separatrix. In our Interactive Singularity Theory, this sheet is called the **maxima zero-separatrix**, because it is the set of functions that have a maximum of value zero.

Now, since the operation $CM^-$ is the dual of $Cm^+$, and the universal unfolding corresponding to $Cm^+$ is given by the Ground Interactive Unfolding $EA_1$, we conclude that the universal unfolding corresponding to $CM^-$ must be the inversion of the functions in the Ground Interactive Unfolding $EA_1$ relative to their codomain axes, i.e., the inversion of the functions illustrated in Fig 9.5 (page 314). The universal unfolding for $CM^-$ will be called the **Dual Ground Interactive Unfolding $EA_{-1}$**. Notice that, just as the control space for $EA_1$ is 1-dimensional, i.e., the $a_0$-axis illustrated in Fig 9.6 (page 315), the control space for $EA_{-1}$ must also be 1-dimensional, which we will also label as the $a_0$-axis. Furthermore, since the functions in the control space for $EA_{-1}$ are exact inversions of the functions in the control space for $EA_1$, the singularities (zeros and critical-points) occur in exactly the same positions on the $t$-axis (even though the minimum is changed to a maximum). This means that the structure of the critical-point manifold and zero-manifold, shown in Fig 9.6 (page 315) for $EA_1$, is exactly the same as the structure of the critical-point manifold and zero-manifold for $EA_{-1}$. Furthermore, just as the full separatrix for $EA_1$ is the single point at the origin of its control space, the full separatrix for $EA_{-1}$ is the single point at the origin of its control space.

With this in mind, return to the Interactive Unfolding of $EA_3$. Consider any *upward path* that crosses the lower arched sheet in the zero-separatrix for $EA_3$, for example, starting from point 3 in Fig 10.7 (page 335). Based on the above discussion, we conclude that any such upward path must, locally to the maximum in the changing function, be equivalent to the 1-dimensional control space $a_0$ of $EA_{-1}$. Furthermore, the function at any point on the lower arch must be locally equivalent (at its maximum) to the germ of $EA_{-1}$. It is for this reason that the label $EA_{-1}$ was assigned to the points on the lower arch in Fig 9.12 (page 325).

Thus, summarizing what we said in section 9.8 and the present section:

**OPERATIONS $Cm^+$ AND $CM^-$**
**IN THE**
**INTERACTIVE UNFOLDING OF $EA_3$**

**The Process Grammar operation $Cm^+$ can be realized by a downward movement through a point in the minima zero-separatrix of the Interactive Unfolding of $EA_3$, and corresponds to movement along the 1-dimensional control space of the Ground Interactive Unfolding $EA_1$, through the single-point separatrix for $EA_1$.**

**The Process Grammar operation $CM^-$ can be realized by an upward movement through a point in the maxima zero-separatrix of the Interactive Unfolding of $EA_3$, and corresponds to movement along the 1-dimensional control space of the Dual Ground Interactive Unfolding $EA_{-1}$, through the single-point separatrix for $EA_{-1}$.**

We will see later that these two operations are realized also in the Interactive Unfolding of the *dual* $EA_{-3}$, where they "switch places". The difference between the occurrence of these operations in the control space for $EA_3$ and the control space for $EA_{-3}$ can be seen as follows: In the control space for $EA_3$, the operation $CM^-$ is realized by raising a function of the form shown on the right in Fig 10.8; that is, a function that not only has a maximum but two minima. In the control space for $EA_{-3}$, the same operation $CM^-$ can be realized by raising a function with only a maximum. The reverse relationship exists for the operation $Cm^+$ in the two respective spaces.

## Causal Compatibility

With respect to the operations $Cm^+$ and $CM^-$, we now demonstrate the crucial fact that Interactive Singularity Theory realizes fundamentally important causal aspects of morphology:

**CAUSAL COMPATIBILITY OF OPERATIONS $Cm^+$ AND $CM^-$ IN THE INTERACTIVE UNFOLDING OF $EA_3$**

**It is a fundamental fact that Interactive Singularity Theory is a singularity theory of shape that is causally compatible with morphology, whereas conventional singularity theories of shape are not.**

**In the control space for $EA_3$, the operations $Cm^+$ and $CM^-$ are compatible with plausible forces and processes, as follows:**

**Consider the control space for $EA_3$, shown in Fig 10.7 (p335). The transition from the upper point 1 to the lower point 1 is given by the operation $Cm^+$. The upper point is a $m^+$ extremum, which, by our process-inference rules, represents a *squashing* process. The lower point is a $m^-$ extremum, which, by our process-inference rules, represents an *indenting* process. Thus the movement down from the upper point to the lower point represents the following process-structure: *squashing continues till it indents.***

**Consider the transition from the lower point 3 upward across the maxima zero-separatrix. This is given by the operation $CM^-$. The lower point gives a $M^-$ extremum, which, by our process-inference rules, represents a *resistance* process. The upper point gives a $M^+$ extremum, which, by our process-inference rules, represents a *protruding* process. Thus the movement from the lower point, upward across the maxima zero-separatrix, represents the following process-structure: *resistance continues till it protrudes.***

## 10.4 Comment on $EA_1$ and $EA_{-1}$

While the canonical form of the interactive singularity $EA_1$ is $t^2$, and the canonical form of the interactive singularity $EA_{-1}$ is $-t^2$, it does not matter whether the single control variable $a_0$, in their universal unfoldings, has a positive or negative sign in front of it, since changing the sign simply means reversing the direction of the $a_0$-axis in the unfolding. This fact becomes relevant when we look at $EA_1$ and $EA_{-1}$ within the Interactive Unfolding of $EA_3$, and shows that it does not reduce the powerful causal compatibility of the Interactive Unfolding of $EA_3$. Thus, recall from section 9.6 that we gave the canonical universal unfolding of $EA_1$ as

$$F(t; a_0) = t^2 + a_0 \tag{10.1}$$

and that this gave the direction of the $a_0$-axis to be that shown in Fig 9.6 (page 315). Notice also that if we choose the canonical universal unfolding of $EA_{-1}$ to be

$$F(t; a_0) = -t^2 - a_0 \tag{10.2}$$

then this will give exactly the same diagram with the $a_0$-axis going in the same direction.

Now, if we switch the sign of the control variable $a_0$, in each of these two unfoldings, i.e., if we choose the canonical universal unfolding of $EA_1$ to be

$$F(t; a_0) = t^2 - a_0 \tag{10.3}$$

and if we choose the canonical universal unfolding of $EA_{-1}$ to be

$$F(t; a_0) = -t^2 + a_0 \tag{10.4}$$

then the diagram would be exactly the same, except that, in both cases, the direction of the $a_0$-axis would be reversed.

A crucial fact is that changing the sign of the control variable does not alter the property that the equation is a universal unfolding of its germ. What cannot, of course, be altered is the sign of the germ; i.e., the germ with its sign describes whether we are modeling a non-degenerate minimum or a non-degenerate maximum.

We will call versions (10.1) and (10.2), the *first* canonical universal unfoldings of $EA_1$ and $EA_{-1}$, respectively. And we will call versions (10.3) and (10.4), the *second* canonical universal unfoldings of $EA_1$ and $EA_{-1}$, respectively.[1] Notice that, in the first version for both $EA_1$ and $EA_{-1}$, the 2-fold zero-bifurcation occurs by moving along the $a_0$-axis in the direction of *decreasing* $a_0$. In contrast, in the second version for both $EA_1$ and $EA_{-1}$, the 2-fold zero-bifurcation occurs by moving along the $a_0$-axis in the direction of *increasing* $a_0$.

Now we have seen that the change in sign of the control variable does not alter the manifold structure; i.e., the manifold structure remains as shown in Fig 9.6 (page 315), that is, a parabola zero-separatrix intersected at its apex by a straight-line critical-point separatrix. This is crucial because it means that both versions for $EA_1$ model the singularity structure of the Process Grammar operation $Cm^+$, and both versions for $EA_{-1}$ model the singularity structure of the Process Grammar operation $CM^-$.

The final thing to observe is this: Within $EA_3$, the unfolding that is used for $EA_1$ is the first version, and the unfolding that is used for $EA_{-1}$ is the second version. We can see this by looking at Fig 10.7 (page 335). That is, because a 2-fold zero-bifurcation of a minimum is given by a downward movement through the minima zero-separatrix, it moves along the $a_0$-axis in the direction of *decreasing* $a_0$ (i.e., the first version for $EA_1$). In contrast, because a 2-fold zero-bifurcation of a maximum is given by an upward movement through the maxima zero-separatrix, it moves along the $a_0$-axis in the direction of *increasing* $a_0$ (i.e., the second version for $EA_{-1}$). The corresponding switch will occur within $EA_{-3}$.

**Most crucially, in the different versions, the *causal* structure is preserved.**

[1] Generally, we can define the first version of a canonical universal unfolding of $EA_n$ and $EA_{-n}$ to be the case where, in the equation of the unfolding, all the control variables are given the same sign as the germ; and we can define the second version to be the case where all the control variables are given the opposite sign of the germ. In most of the cases in the book, it will not be necessary to explicitly distinguish between the first and second versions, because they will be obvious. However, we will distinguish here between the first and second versions for $EA_1$ and $EA_{-1}$ in order to initially introduce the reader to this concept.

## 10.5 Operations $BM^-$ and $BM^+$ in the Interactive Unfolding of $EA_{-3}$

First, we need to emphasize the following important facts, concerning the *inferred processes* of the *shapes*, as given by our theory, and as illustrated in Figs 10.9 and 10.10, and not causally understood in conventional singularity theories of shape:

**Both the bifurcation operations $BM^-$ and $BM^+$ are applied to a single *maximum*.**

**In $BM^-$, the inferred force *causing* the initial maximum is *compressive*.**

**In $BM^+$, the inferred force *causing* the initial maximum is *penetrative*.**

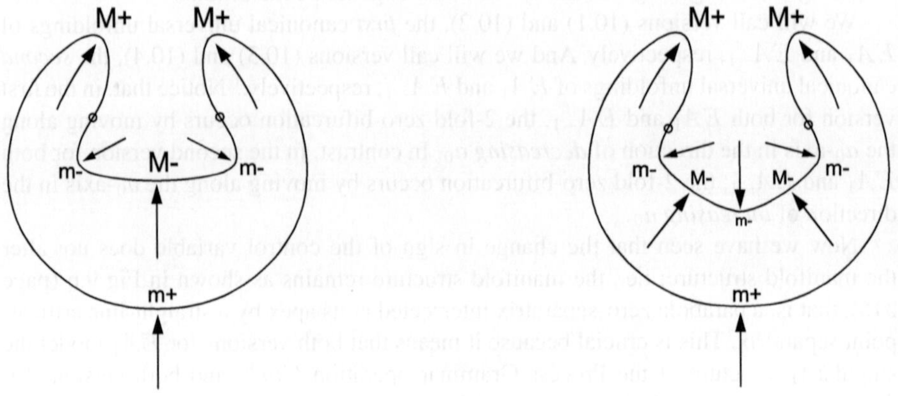


**Fig. 10.9.** Process-bifurcation B at $M^-$.

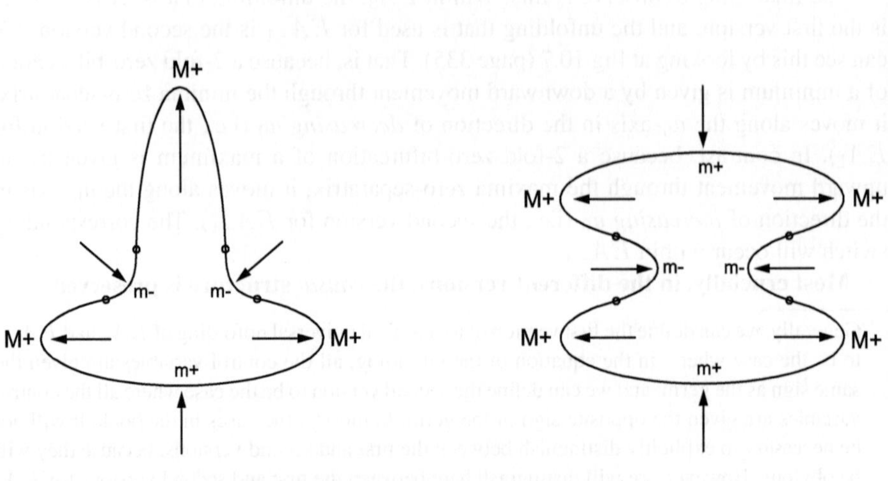


**Fig. 10.10.** Process-bifurcation B at $M^+$.

Next, we need to emphasize the following important *distinction*, concerning the corresponding sets of curvature *functions*, as illustrated in the distinction between Fig 10.11 and Fig 10.12, and not distinguished in Catastrophe Theory:

**$BM^-$ is applied to a *negative* maximum, resulting in a *negative* maximum, minimum, maximum.**

**$BM^+$ is applied to a *positive* maximum, resulting in a *positive* maximum, minimum, maximum.**

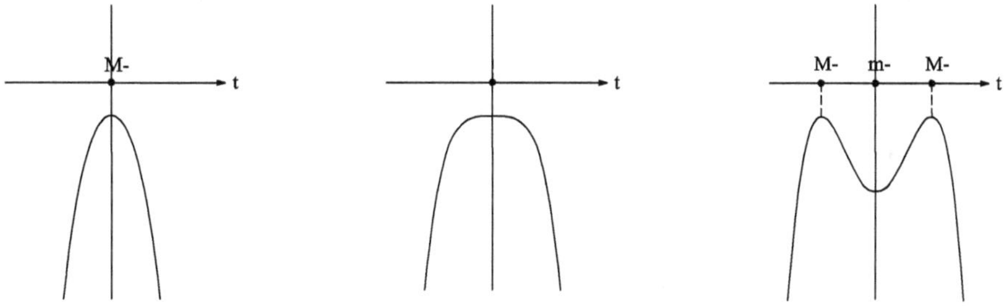


**Fig. 10.11.** Curvature functions given by the bifurcation operation $BM^- : M^- \longrightarrow M^-m^-M^-$.

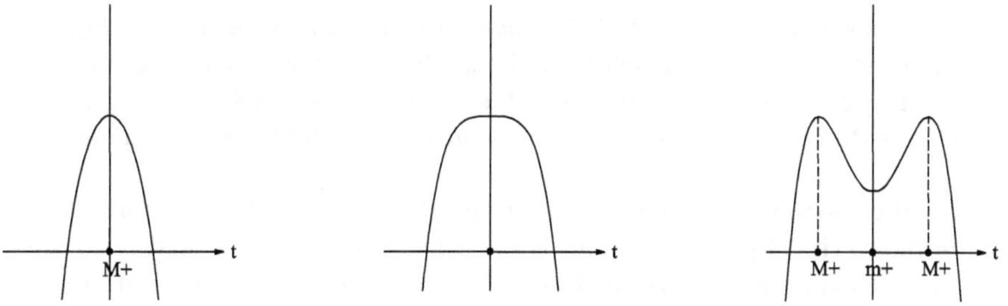


**Fig. 10.12.** Curvature functions given by the bifurcation operation $BM^+ : M^+ \longrightarrow M^+m^+M^+$.

### 10.5.1 Relation of $BM^-$ and $BM^+$ to the Critical-Point Manifolds

To model the operations $BM^+$ and $BM^-$ within their critical-point manifolds we need the Interactive Unfolding of the *dual* $EA_{-3}$. To define this, it is worth looking first at the dual cusp catastrophe $A_{-3}$, from Catastrophe Theory. This is obtained by simply inverting the unfolding function of the cusp catastrophe, thus:

$$F(t; a_1, a_2) \quad = \quad -t^4 \; - \; a_1 t \; - \; a_2 t^2 \tag{10.5}$$

Therefore, given any point $(a_1, a_2)$ in the control space, the function located at that point is simply the inverted version of the function located at the same point $(a_1, a_2)$ in the control space of the cusp catastrophe; i.e., minima become maxima, etc. Notice that, despite this change of minima to maxima, etc., the *positions* of the extrema, on the $t$-axis of the function, are preserved; and thus the critical-point manifold is exactly congruent with, and in the same position as, the critical-point manifold of the cusp catastrophe, in the space of variables $(a_1, a_2, t)$. The only change is that the middle sheet of the manifold is now a sheet of minima, and the remaining sheet is now a sheet of maxima. Also, the separatrix, in the control space $(a_1, a_2)$, is exactly congruent with, and in the same position as, the separatrix of the cusp catastrophe.

Now, as with the cusp catastrophe, the standard unfolding function (10.5) of the dual cusp catastrophe is not compatible with plausible causal explanations given by our system of process-inference rules. The reason is the same as before: the failure of Catastrophe Theory to consider both transversality conditions, and thus contact-equivalence in addition to right-equivalence. Therefore, we need to add, to the unfolding in expression (10.5), a term $-a_0$ that unfolds the germ also with respect to our second transversality condition – which separates extrema $M^-$ from $M^+$. Thus we obtain the unfolding function:

$$F(t; a_0, a_1, a_2) \quad = \quad -t^4 \; - \; a_0 \; - \; a_1 t \; - \; a_2 t^2 \tag{10.6}$$

which we will call the **Interactive Unfolding of the Dual $EA_{-3}$**.

This allows the operations $BM^-$ and $BM^+$ to each occur within their own critical-point manifolds, as illustrated in the figures on the next two pages. In the case of $BM^-$, the manifold consists entirely of negative critical points; and in the case of $BM^+$, the manifold consists entirely of positive critical points – again, a distinction which Catastrophe Theory fails to give. We note the following:

> **The operations $BM^-$ and $BM^+$ have the same structural relationship to their respective critical-point manifolds, but, their manifolds are of opposite signs from each other, and these signs are crucial to their *causal* explanations, and the manifolds are related by the Interactive Structure.**

> **Furthermore, the structural relationship that $BM^-$ and $BM^+$ both have to their critical-point manifolds is the same as that of $Bm^+$ and $Bm^-$ to their critical-point manifolds. But each of these four operations correspond to a different *causal* structure, and this is why they are distinguished and separated by our Interactive Singularity Theory.**

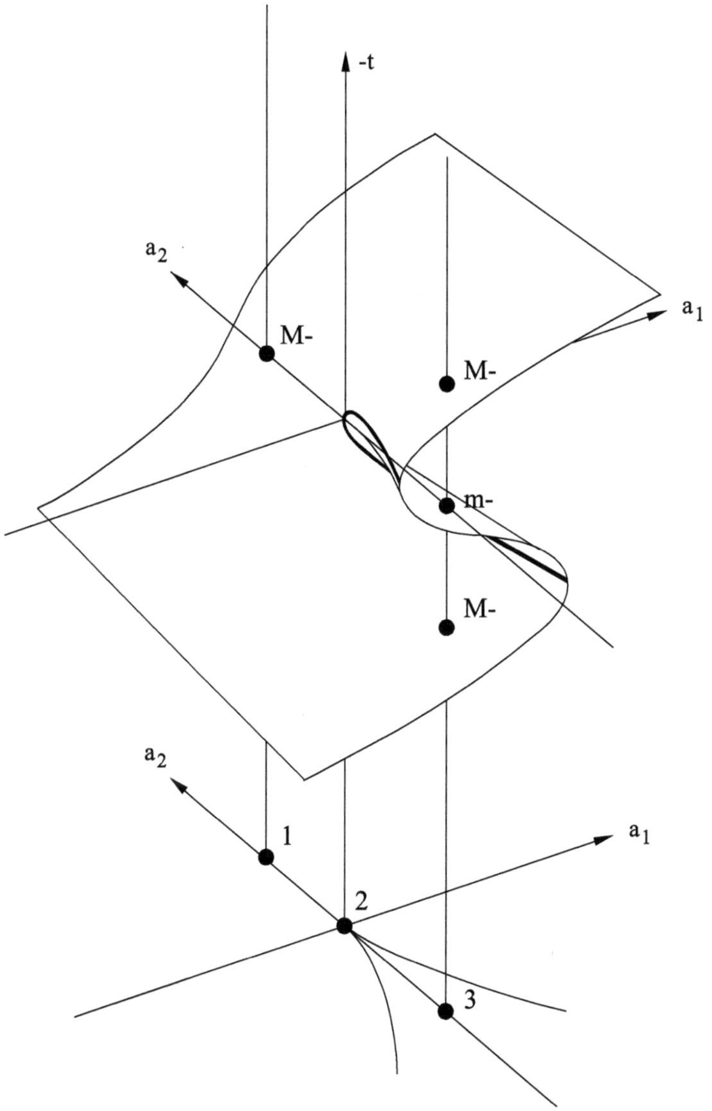


**Fig. 10.13.** The effect of $BM^-$ within its corresponding critical-point manifold.

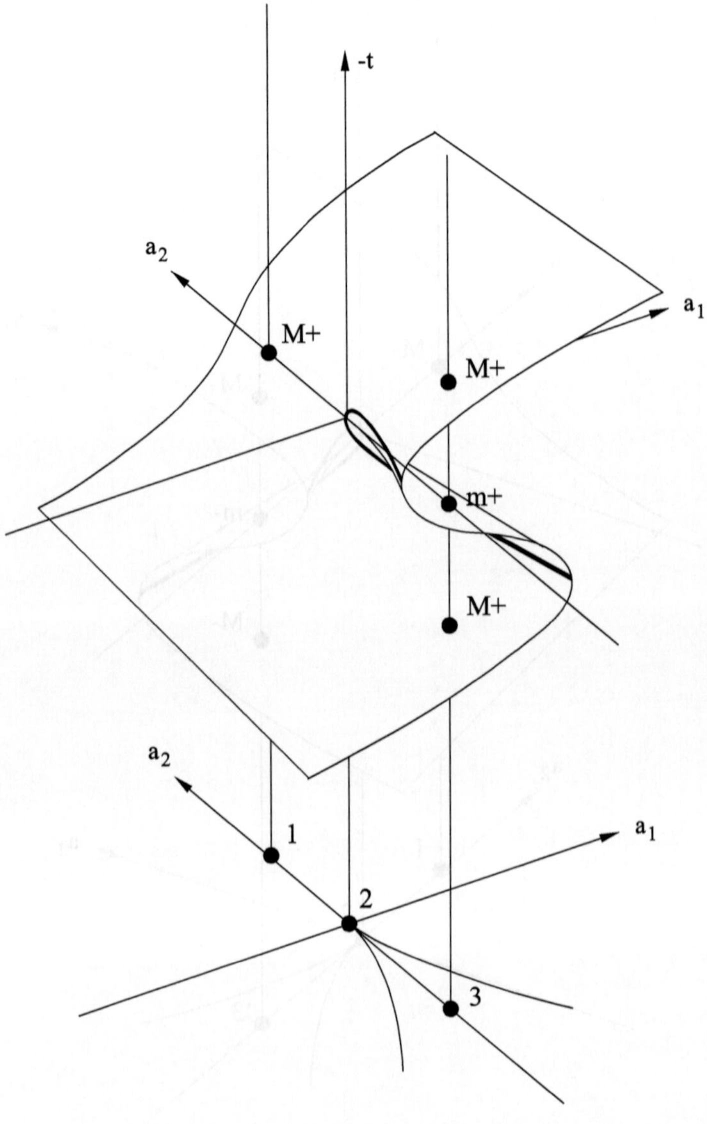


**Fig. 10.14.** The effect of $BM^+$ within its corresponding critical-point manifold.

### 10.5.2 Relation of $BM^-$ and $BM^+$ to the Full Separatrix for $EA_{-3}$

For reasons to be given soon, it will be convenient to draw the full separatrix for $EA_{-3}$ as an inversion of the full separatrix for $EA_3$. Thus, the full separatrix for $EA_{-3}$ will be drawn as shown in Fig 10.15. Notice that the vertical axis is now $-a_0$ rather than $a_0$. The figure also shows the location of the trajectories given in the previous two diagrams, for $BM^-$ and $BM^+$, as, respectively, the *lower* and *upper* sequences marked $1 \rightarrow 2 \rightarrow 3$ in the control space. Clearly, the operations map between volumes in the Interactive Structure (i.e., bounded by *both* the critical-point separatrix and zero-separatrix) in a way that corresponds to that described on p336 for $EA_3$.

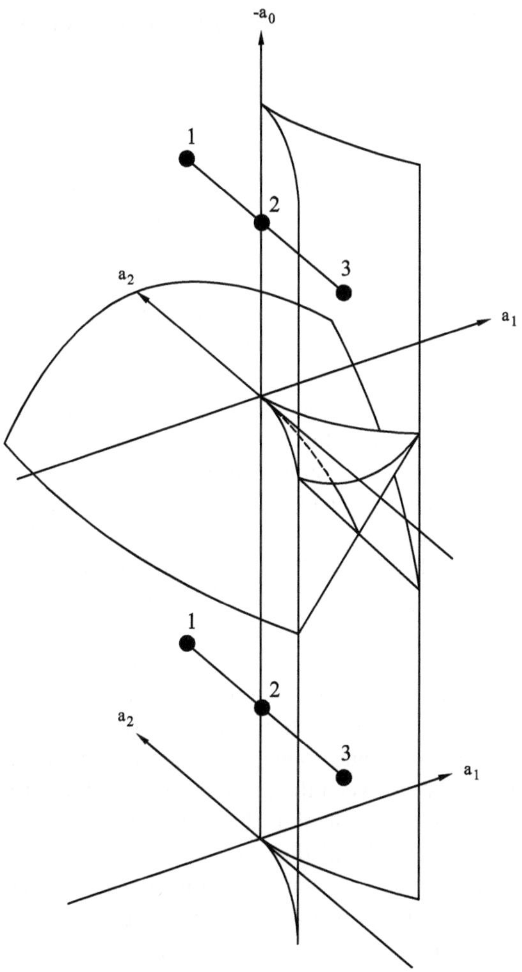


**Fig. 10.15.** Operations $BM^-$ and $BM^+$, as shown in the previous two figures, are shown here, respectively, as the lower and upper sequence $1 \rightarrow 2 \rightarrow 3$, in the 3D control space for $EA_{-3}$.

It is worth comparing the non-dual case Fig 10.7 (p335) with the present dual case Fig 10.15. In the former case, the bifurcation operation $Bm^+$, which acts on a *compressive* extremum, was given by a trajectory *above* the zero-separatrix; and the bifurcation operation $Bm^-$, which acts on a *penetrative* extremum, was given by a trajectory *below* the zero-separatrix. In the present case, the situation is inverted: The bifurcation operation $BM^-$, which acts on a *compressive* extremum, is given by a trajectory *below* the zero-separatrix; and the bifurcation operation $BM^+$, which acts on a *penetrative* extremum, is given by a trajectory *above* the zero-separatrix.

## 10.6 Operations $CM^-$ and $Cm^+$ in the Interactive Unfolding of $EA_{-3}$

The minimum-based bifurcation operations, $Bm^+$ and $Bm^-$, exist in the Interactive Unfolding of $EA_3$, but not in the Interactive Unfolding of $EA_{-3}$; and the maximum-based bifurcation operations, $BM^+$ and $BM^-$, exist in the Interactive Unfolding of $EA_{-3}$, but not in the Interactive Unfolding of $EA_3$. However, the minimum-based continuation operation, $Cm^+$, and the maximum-based continuation operation, $CM^-$, exist both in the Interactive Unfolding of $EA_3$ and in the Interactive Unfolding of $EA_{-3}$. We have seen how the two continuation operations move in the Interactive Unfolding of $EA_3$. However, the way they move in the Interactive Unfolding of $EA_{-3}$ raises issues that are more complex. Therefore we need to study this in greater detail.

Before doing so, it is first necessary to compare this Interactive Unfolding with unfolding in Catastrophe Theory, to further show the inadequacy of Catastrophe Theory for morphology. As an important example, it is necessary to understand some issues concerning the two points labeled 1 in Fig 10.15. Consider the vertical line going between them. Obviously, it intersects the zero-separatrix at a point. In the particular way that the points 1 were drawn in Fig 10.15, the vertical line between them intersects the zero-separatrix at a point on the positive $a_2$-axis of the *2D control space* of the dual cusp catastrophe. Since the function at this control point in the *non-dual* cusp catastrophe has a non-degenerate minimum which touches its $t$-axis, the function at this control point in the *dual* cusp catastrophe must have a non-degenerate maximum which touches its $t$-axis, as illustrated by the middle function in Fig 10.16.

Again, a crucial failure of Catastrophe Theory, in morphology, is that Catastrophe Theory does not distinguish between the positive, zero, and negative, values of a maximum; i.e., does not make the distinction shown in Fig 10.16. In contrast, Interactive Singularity Theory does make this distinction.

Thus, moving from the point we have just considered in the zero-separatrix of the Interactive Unfolding of $EA_{-3}$, downwards in the 3D control space, e.g., to the lower point labeled 1 in Fig 10.15, means that we have increased the value of $a_0$ (see the vertical axis). By the unfolding function for $EA_{-3}$

$$F(t; a_1, a_2) \quad = \quad -t^4 \; - \; a_0 \; - \; a_1 t \; - \; a_2 t^2 \qquad (10.7)$$

increasing $a_0$ means that we have lowered the function relative to its own $F$-axis.

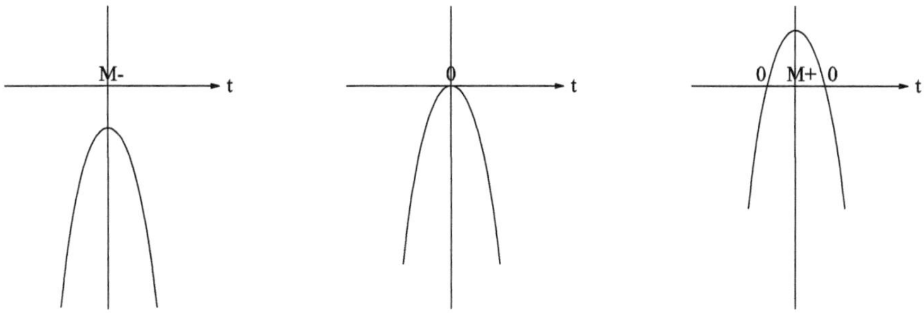


**Fig. 10.16.** Movement in the *decreasing* $a_0$ direction in the control space for $EA_{-3}$ *raises* the graph of the function.

Therefore, the function at the lower point 1 in Fig 10.15 must be below its own $t$-axis, as illustrated by the left-most function in Fig 10.16, and is crucially distinguished in Interactive Singularity Theory, from the other two functions shown in Fig 10.16.

Furthermore, moving from the point we considered in the zero-separatrix, upwards in the 3D control space, e.g., to the upper point labeled 1 in Fig 10.15, must *raise* the function relative to its $F$-axis. Thus the function at this higher point 1 is as illustrated by the right-most function in Fig 10.16, and is crucially distinguished, in Interactive Singularity Theory, from the other two functions shown in Fig 10.16.

This distinction, which Catastrophe Theory fails to make, gives one of the fundamentally important morphological transitions defined by the Process Grammar, the operation $CM^-$. Thus, we have seen that, in the control space for $EA_{-3}$, moving vertically in Fig 10.15, from the lower point 1, through the zero-separatrix, to the higher point 1, is given by the sequence of three *distinguished* functions in Fig 10.16, and this sequence shows that the movement corresponds to the Process Grammar operation $CM^- : M^- \longrightarrow 0M^+0$. That is, a negative non-degenerate maximum has been raised to become zero and then positive, the maximum remaining non-degenerate throughout, while the zero underwent a 2-fold bifurcation. This corresponds to a crucial morphology defined by our PISA axis: a *process-continuation*, which all other morphological theories, e.g., those based on the Medial axis, completely fail to model.

In the remainder of this section, we will discuss the zero-separatrix for $EA_{-3}$ in relation to Method 1 and Method 2 for producing it, because the two methods emphasize different information.

We shall soon see that the zero-separatrix, as drawn in Fig 10.15, was produced by using Method 1. However, it is now worthwhile to get the particular information that comes more easily by using Method 2. Recall that Method 2 obtains the zero-separatrix of an Interactive Unfolding by understanding that separatrix as having the same structure as the critical-point separatrix of the next higher catastrophe. In this way, the zero-separatrix for $EA_{-3}$ is understood as having the same structure as the critical-point separatrix of the dual swallowtail catastrophe $A_{-4}$. However, Catastrophe Theory does not recognize the existence of $A_{-4}$, because it is right-equivalent to $A_4$, due to the even-order degeneracy,

4, of its critical point. Therefore, Catastrophe Theory does not recognize the existence of the dual swallowtail catastrophe $A_{-4}$. In contrast, $A_{-3}$ is not right-equivalent to $A_3$, due to the odd-order degeneracy, 3, of its critical point. Therefore, Catastrophe Theory does recognize the existence of the dual cusp catastrophe $A_{-3}$. However, in Interactive Singularity Theory, we need to recognize $A_{-4}$, and its unfolding, as different from $A_4$, and its unfolding, for a number of reasons. One of these reasons becomes evident very soon.

Now, we saw earlier that the critical-point separatrix for $A_3$, in its 2D control space $(a_1, a_2)$, is exactly congruent with, and in the same position as, the critical-point separatrix for $A_{-3}$ in its 2D control space $(a_1, a_2)$. This is because the functions in the respective spaces are inversions of each other, which means that, despite the change of minima to maxima, etc., the *positions* of the critical points, on the $t$-axis, are preserved. The same argument shows that the critical-point separatrix for $A_4$ is exactly congruent with, and in the same position as, the critical-point separatrix for $A_{-4}$, in the 3D space of control variables.

From the arguments in the previous two paragraphs, we conclude that the full separatrix for the Interactive Unfolding of the Dual $EA_{-3}$, in its 3D control space $(a_0, a_1, a_2)$, must be exactly congruent with, and in the same position as, the full separatrix for the Interactive Unfolding of $EA_3$, in its 3D control space $(a_0, a_1, a_2)$.

Therefore, exactly the same drawing used for the full separatrix for $EA_3$, in Fig 8.22 (p296), can now be used for the full separatrix for $EA_{-3}$. Thus Fig 10.17 now uses this drawing for the full separatrix for $EA_{-3}$.

Also notice that Fig 10.15, which was the first diagram we gave for $EA_{-3}$, has exactly the same structure as the new diagram, Fig 10.17, for $EA_{-3}$. The two figures have simply been drawn up-side-down relative to each other. In fact, the former comes from Method 1, and the latter comes from Method 2. This latter version is useful for making a *direct* comparison between the control space for $EA_3$ and the control space for $EA_{-3}$.

Thus let us directly compare Fig 8.22 (p296) of the control space for $EA_3$ with Fig 10.17 of the control space for $EA_{-3}$. The figures are absolutely identical. However, the crucial fact is that the *functions* in the two control spaces are different.

For example, consider a point anywhere in the volume above the zero-separatrix and behind the critical-point separatrix, e.g., the point shown as 1 in Fig 10.17. In the case of $EA_3$, a function in this volume would have a single critical point that is a positive minimum $m^+$. However, at the same control point, in the case of $EA_{-3}$, the function is inverted, and therefore has a single critical point that is a negative maximum $M^-$, as illustrated by the left function in Fig 10.16. In fact, the three successive points $1 \rightarrow 2 \rightarrow 3$, shown in Fig 10.17, correspond to the three successive functions from left to right in Fig 10.16. Therefore, this succession corresponds to the use of the Process Grammar operation $CM^- : M^- \longrightarrow 0M^+0$.

Notice that this uses the fact that point 2, in Fig 10.17, is in a component of the zero-separatrix in which the functions have a 2-fold zero-degenerate point. The advantage of using Method 2 for obtaining the zero-separatrix for $EA_{-3}$ is that it is easy to quickly visualize the fact that all components of this zero-separatrix have the same degeneracies as the corresponding components of the zero-separatrix for $EA_3$ as discussed in section

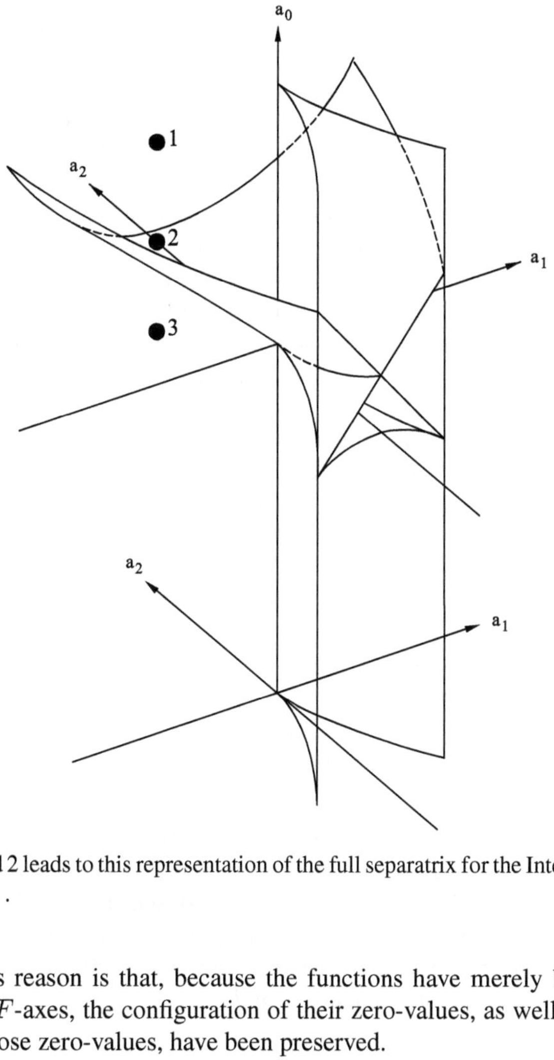


**Fig. 10.17.** Method 2 leads to this representation of the full separatrix for the Interactive Unfolding of the Dual $EA_{-3}$ .

8.7. The obvious reason is that, because the functions have merely been inverted in relation to their $F$-axes, the configuration of their zero-values, as well as the extent of degeneracy of those zero-values, have been preserved.

To investigate this further, recall that Fig 8.20 (page 290) shows representative functions in a vertical slice through the front part of the critical-point separatrix of the *non-dual* swallowtail catastrophe $A_4$. In particular, recall the following fact concerning that figure: Using the fact that the functions in the separatrix are the only functions in that slice that have a degenerate critical point, the degeneracy of that critical point has been indicated in the figure by the number we have put at that point within the function. Then, using Method 2, Fig 8.21 (page 290) showed the corresponding functions, i.e., the derivatives, in the same slice through the zero-separatrix in the Interactive Unfolding of $EA_3$. Therefore, in this figure, the functions in the separatrix exhibited the same numbers at the same positions within the functions, but this time the positions were zeros, instead of critical points, and the numbers indicated the degeneracy of those zeros.

Based on this, recall that Fig 9.12 (page 325) showed the type of *complete* singularity $EA_{\pm k}$that occurs within each function in the zero-separatrix (within the slice).

Now, in the *dual* swallowtail catastrophe $A_{-4}$, the same slice will have the inversion of the representative functions shown in Fig 8.20 (page 290). This of course will mean that the numbers, indicating the critical-point degeneracy of points within the functions in the critical-point separatrix, will be the same and occur at the same positions within those functions, even though the functions are inverted. Correspondingly, in the Interactive Unfolding of $EA_{-3}$, the same slice will have the inversion of the functions shown in Fig 8.21 (page 290). This will mean that the numbers, indicating the zero-degeneracy of points within the functions in the zero-separatrix, will be the same and occur at the same positions within those functions, even though the functions are inverted.

As a result of this, the labels shown in Fig 9.12 (page 325), for the *complete* singularities $EA_{\pm k}$ that occur within the functions in the zero-separatrix (within the slice) for $EA_3$, will simply have, for $EA_{-3}$, their indices $\pm k$ switch sign from positive to negative and vice versa.

In particular, observe that the singularity that occurs within any function in the two diagonal sheets, and also within any function in the back sheet (in Fig 10.17), will therefore be the *complete* singularity $EA_{-1}$. That is, the singularity is a zero that is a non-degenerate maximum. Thus, the two diagonal sheets and the back sheet will be called the **maxima zero-separatrix** for $EA_{-3}$. We can therefore state the following concerning the continuation operation $CM^-$ in the Interactive Unfolding of $EA_{-3}$:

### CONTINUATION OPERATION $CM^-$ IN THE INTERACTIVE UNFOLDING OF THE DUAL $EA_{-3}$

**The Process Grammar operation $CM^-$, which describes the continuation of a resistance force or process till it protrudes, can be modeled by a movement through a point in the maxima zero-separatrix of the Interactive Unfolding of the Dual $EA_{-3}$, and corresponds to movement along the 1-dimensional control space of the Dual Ground Interactive Unfolding $EA_{-1}$, through the single-point separatrix for $EA_{-1}$.**

It is worthwhile now considering a relationship that we have not so far examined between $EA_3$ and $EA_{-3}$. For this we turn from Method 2 to Method 1 for constructing the zero-separatrix of the Interactive Unfolding. Whereas Method 2 is based on using the *critical-point* separatrix of the *next higher* catastrophe, Method 1 is based on using the *critical-value* surface of the *same level* catastrophe.

Recall that the critical-value surface of the cusp catastrophe $A_3$ has the structure we now show again in Fig 10.18. This is interpreted as follows: Given a position $(a_1, a_2)$ in the control plane, the vertical line through that position is the $F$-axis of the function located at that position; and the points where that vertical line intersects the surface are the points where that function has critical *values*. For example, at the bottom of Fig 10.18, we have drawn the graph of a function that is located at a position on the negative $a_2$-axis. This function has two negative minima of the same value as each other, and a maximum of zero value. Therefore the vertical line, at this position of the control space, cuts through the surface above where two minima sheets of the surface intersect (i.e., representing the two minima of the same value), and also cuts through the maxima sheet (the upper arched sheet) where that sheet has $F$ value zero; i.e., representing the function's maximum of zero value.

Now consider Fig 10.19. This shows the critical-value surface of the *dual* cusp catastrophe $A_{-3}$. The axes $a_1$, $a_2$, and $F$, are all oriented in the same direction as in the previous figure. However, the critical-value surface is now inverted. The obvious reason is that the functions, at all the points in the control space $(a_1, a_2)$, have been inverted in relation to their $F$-axes. For example, at the bottom of the figure, we have drawn the graph of the function located at the same position as that considered at the bottom of the previous figure. It is therefore an inverted version of the previous function. Thus, it has two positive maxima of the same value as each other, and a minimum of zero value. Therefore the vertical line, at this position of the control space, cuts through the surface where two maxima sheets of the surface intersect (i.e., representing the two maxima of the same value), and also cuts through the minima sheet (the lower arched sheet) where that sheet has $F$ value zero; i.e., representing the function's minimum of zero value.

Now Method 1 uses the following procedure for obtaining the zero-separatrix of the Interactive Unfolding of $EA_3$ and the Interactive Unfolding of the *Dual* $EA_{-3}$. Consider any function $F(t; a_1, a_2)$, from the corresponding catastrophe control space $(a_1, a_2)$. Suppose that a critical point of that function has level $c$ on its $F$-axis. Then, in order to raise (or lower) the function so that this critical point has level zero on the $F$-axis, it is necessary merely to add $-c$ to the function.

Thus, to obtain the zero-separatrix of the corresponding Interactive Unfolding, one has to invert the critical-value surface, and replace the inverted $F$-axis by the term which, in the unfolding function for the Interactive Unfolding, raises (or lowers) the function $F(t; a_1, a_2)$. In the case of $EA_3$, this term is $a_0$. In the case of $EA_{-3}$, this term is $-a_0$, as can be seen from the unfolding function in expression (10.7) page 348.

Therefore, Fig 10.20 and Fig 10.21 show the way Method 1 represents the zero-separatrices for $EA_3$ and $EA_{-3}$ respectively. Notice that the two surfaces are inversions of each other, because the critical-value surfaces, from which they were derived, were inversions of each other. Also observe that, in Fig 10.20, the $a_0$ axis is vertically upward; whereas, in Fig 10.21, the $-a_0$ axis is vertically upward.

This reinforces the point made earlier, that the zero-separatrices for $EA_3$ and $EA_{-3}$ are congruent to each other, and are in exactly the same position with respect to the three control variables $(a_0, a_1, a_2)$. That is, they both depict the same information as the zero-separatrix in Fig 10.17 (p351), which, as we said, is a single diagram that works for both $EA_3$ and $EA_{-3}$.

However, there is an advantage of seeing the two separatrices drawn in the inverted relation to each other shown in Figs 10.20 and 10.21, as obtained from Method 1. The advantage is this:

Recall that the control space for $EA_3$ actually contains the operation $CM^- : M^- \longrightarrow 0M^+0$ as the movement upward through the lowest surface, the upward arched surface at the front of Fig 10.20. A negative maximum $M^-$ below this arch is raised to become a positive maximum $M^+$ above the arch.

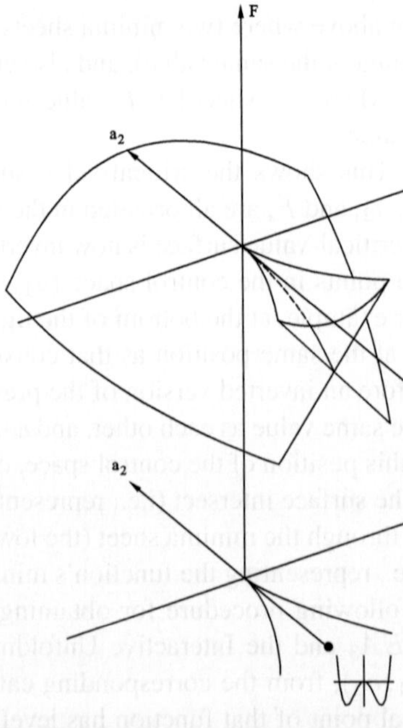


**Fig. 10.18.** The critical-value surface of the cusp catastrophe, showing a function on the negative $a_2$-axis.

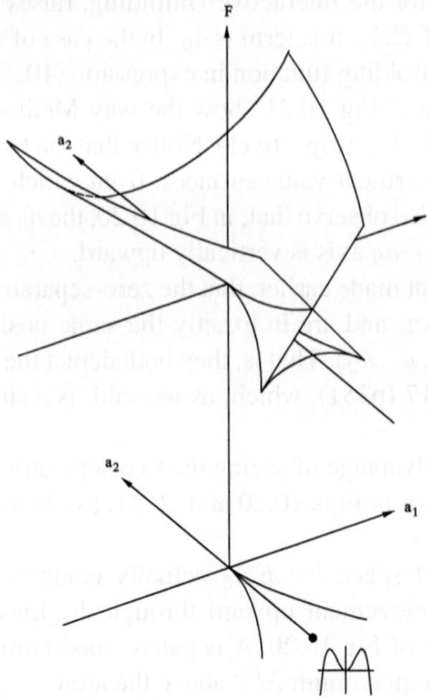


**Fig. 10.19.** The critical-value surface of the *dual* cusp catastrophe, showing a function on the negative $a_2$-axis.

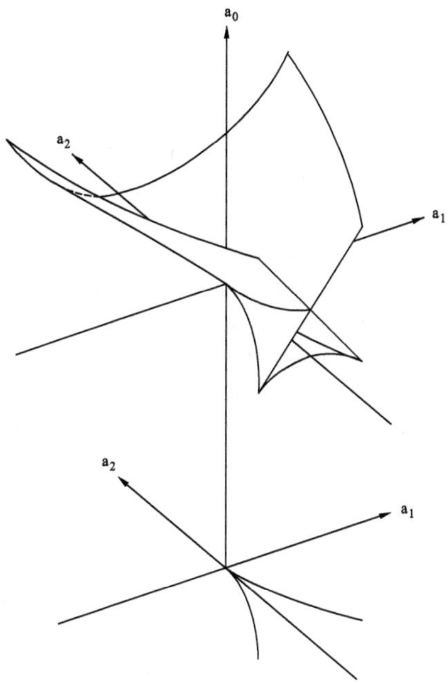


**Fig. 10.20.** Method 1 leads to this representation of the zero-separatrix for $EA_3$.

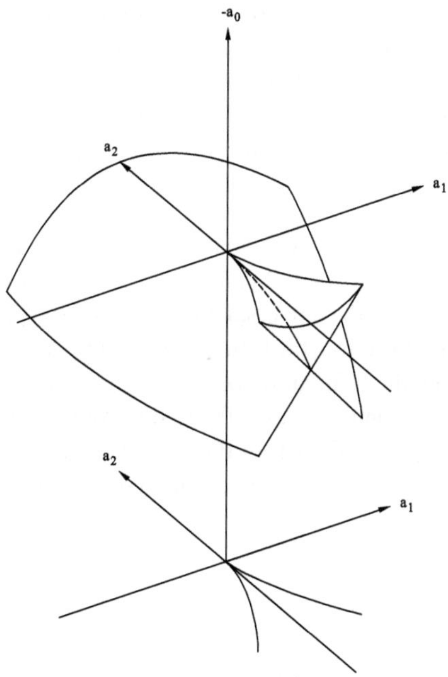


**Fig. 10.21.** Method 1 leads to this representation of the zero-separatrix for $EA_{-3}$.

Remarkably, this upward arched surface for $EA_3$, at the *front* of Fig 10.20, corresponds to the upward arched surface for $EA_{-3}$, at the *back* of Fig 10.21. Furthermore, in relation to the latter arch, the operation $CM^-$ is realized in exactly the same way: A negative maximum $M^-$ below the arch is raised to become a positive maximum $M^+$ above the arch. Notice that the upward arched surface in Fig 10.20 is the *maxima zero-separatrix* for $EA_3$, and the upward arched surface in Fig 10.21 is part of the *maxima zero-separatrix* for $EA_{-3}$.

In exactly the same way, the downward arched surface at the top-front of Fig 10.21 corresponds to the downward arched surface at the back of Fig 10.20. In both cases, the operation $Cm^+$ crosses from above the arch to below it. That is, a positive minimum $m^+$ above the arch is lowered through the arch to become a negative minimum $m^-$ below the arch. Notice that the downward arched surface in Fig 10.21 is the *minima zero-separatrix* for $EA_{-3}$, and the downward arched surface in Fig 10.20 is part of the *minima zero-separatrix* for $EA_3$.

Thus, representing the zero-separatrices in the inverted relation to each other, as in Fig 10.20 and Fig 10.21, allows us to match the arched surfaces between the separatrices, and also allows us to match the directions of the continuation operations, $Cm^+$ and $CM^-$, as follows:

**OPERATIONS $Cm^+$ AND $CM^-$ IN**
**<u>BOTH</u>**
**THE INTERACTIVE UNFOLDING OF $EA_3$ AND**
**THE INTERACTIVE UNFOLDING OF THE DUAL $EA_{-3}$**

**The 3D control space for $EA_3$, and the 3D control space for $EA_{-3}$ (using the Method 1 representation), both model the continuation operation $Cm^+ : m^+ \longrightarrow 0m^-0$ by movement downward across the minima zero-separatrix; and model the continuation operation $CM^- : M^- \longrightarrow 0M^+0$ by movement upward across the maxima zero-separatrix.**

However, observe that, despite this matching, the operations are in the opposite directions relative to the $a_0$-axes for $EA_3$ and $EA_{-3}$, and therefore they switch between the first and second versions of the universal unfoldings of $EA_1$ and $EA_{-1}$, as defined on p341. Recall that in the first version, the bifurcation occurs in the *decreasing* $a_0$ direction; and in the second version, the bifurcation occurs in the *increasing* $a_0$ direction. Thus, in the case of $EA_3$, the operation $Cm^+$ uses the first version for $EA_1$, and $CM^-$ uses the second version for $EA_{-1}$. In contrast, in the case of $EA_{-3}$, the operation $Cm^+$ uses the second version for $EA_1$, and $CM^-$ uses the first version for $EA_{-1}$.

## 10.7 The Preservation Scenario

This section will describe a scenario, for the bifurcation operation $B$, that demonstrates that Interactive Singularity Theory, with the use of our process-inference rules, has the power to give causal explanations that have valuable, highly-plausible, details.

It is necessary, at this point, to consider not only the qualitative aspects of Interactive Singularity Theory, but also the quantitative aspects.

We begin by considering three successive positions *within* the $a_2$-axis of the 3D control space for $EA_3$. They will be (1) a position in the positive $a_2$-axis, (2) the origin, and (3) a position in the negative $a_2$-axis. The functions at these three successive positions are illustrated from left to right in Fig 10.22. The following qualitative property should be observed about these three functions: In the transition from the first to the second function, the value of the minimum is *preserved.* However, the moment the bifurcation happens, the value of the minimum is no longer preserved. In fact, as illustrated by the third function, the two minima, emerging from the bifurcation, move downwards.

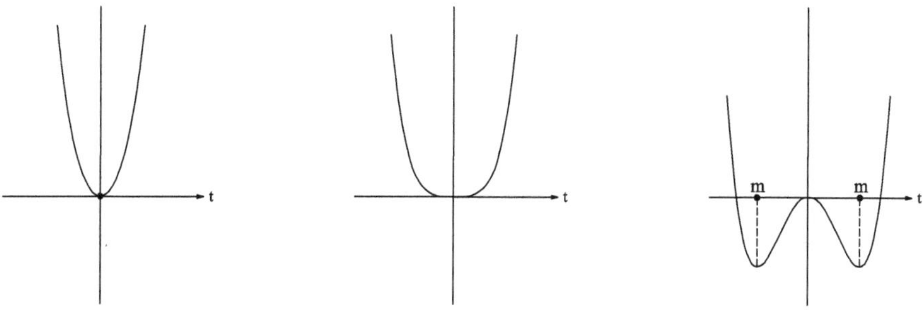


**Fig. 10.22.** Functions *within* the $a_2$-axis. From left to right: (1) at a position in the positive $a_2$-axis, (2) at the origin, and (3) at a position in the negative $a_2$-axis.

Now raise all three functions by a single constant $c$, so that the functions are all above the $t$-axis, as shown in Fig 10.23. Obviously, the particular qualitative property described in the previous paragraph still holds. In the present case, it is expressed in this way: The value $c$ of the minimum in the first function is preserved in going to the second function, but decreases after that.

It is important to see that, because all three functions have been raised by the same constant, they are positioned in the 3D control space of the Interactive Unfolding of $EA_3$ in the way that is illustrated by the three points 1, 2, 3, in Fig 10.24. That is, they are within a *horizontal straight line* directly above the $a_2$-axis. Furthermore, the height of this line on the $a_0$-axis is $c$, the value by which the functions were raised with respect to their $F$-axes.

Now return to the qualitative fact that was pointed out about the three functions in Fig 10.23: the value of the minimum is preserved in going from the first to the second function, but is lowered in the third. Based on this, we can now make the following statement concerning the 3D control space in Fig 10.24: In a horizontal straight line, directly above the $a_2$-axis, movement along the positive $a_2$-axis preserves the value

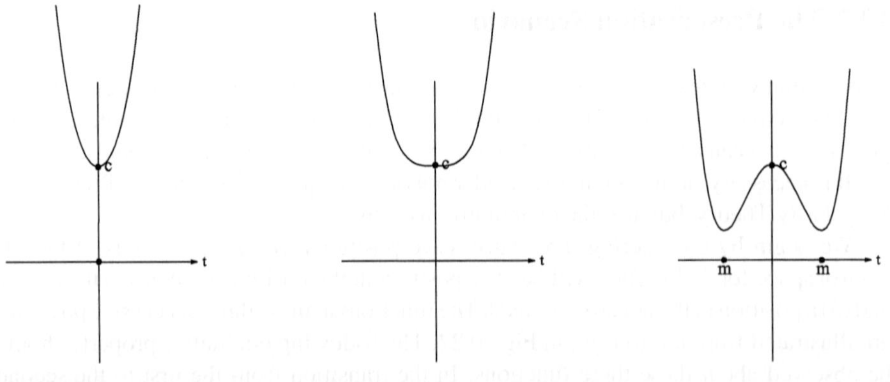


**Fig. 10.23.** The functions in the previous figure have all been raised by a single constant c, to place them above the $t$-axis.

of the minimum, but movement past the critical-point separatrix cusp-point (point 2) decreases the value of the minimum.

This fact will be important to know, because the trajectory $1 \to 2 \to 3$ shown in Fig 10.24 is an example of the bifurcation operation $Bm^+ : m^+ \longrightarrow m^+M^+m^+$.

The next crucial concepts are as follows: As we let point 3 move further along the same straight line in the control space, then the minima in the third function of Fig 10.23 will descend further. Therefore, it is clear from Fig 10.23 that, eventually, the minima will move down across the $t$-axis and become negative. Thus the function at this stage would no longer correspond to the co-domain $m^+M^+m^+$ of the operation $Bm^+$. In fact, what has happened is this: The movement of the two minima, down across the $t$-axis, corresponds to the horizontal line $1 \to 2 \to 3$, in the control space, crossing the upward curve, marked $d$ in Fig 10.24, where the two diagonal minima zero-separatrix sheets intersect. This of course violates our *separatrix* condition on $Bm^+$: That is, $Bm^+$ must occur in a trajectory above the zero-separatrix.

It is clear therefore that, if we allow the point 3 to be sufficiently far in the negative $a_2$ direction, then, to maintain the operation $Bm^+$, the trajectory $1 \to 2 \to 3$ must "curl upwards" in the 3D control space.

Now, this is *one* of the reasons why the trajectory must curl upwards. However, our theory also claims that some circumstances require an even stronger constraint, that automatically satisfies the requirement just given, but has an extra condition that is useful to impose because it captures important aspects of the causal explanation involved: The stronger constraint is this:

**PRESERVATION SCENARIO**

> **In this scenario, the curvature of an extremum is preserved throughout the bifurcation of that extremum, i.e., at its copies emerging from the bifurcation.**

This scenario does not always occur, but it does occur with sufficient frequency and causal plausibility, that it is worth defining. Also, it can be the basis of a useful operator in Computer-Aided Design (CAD).

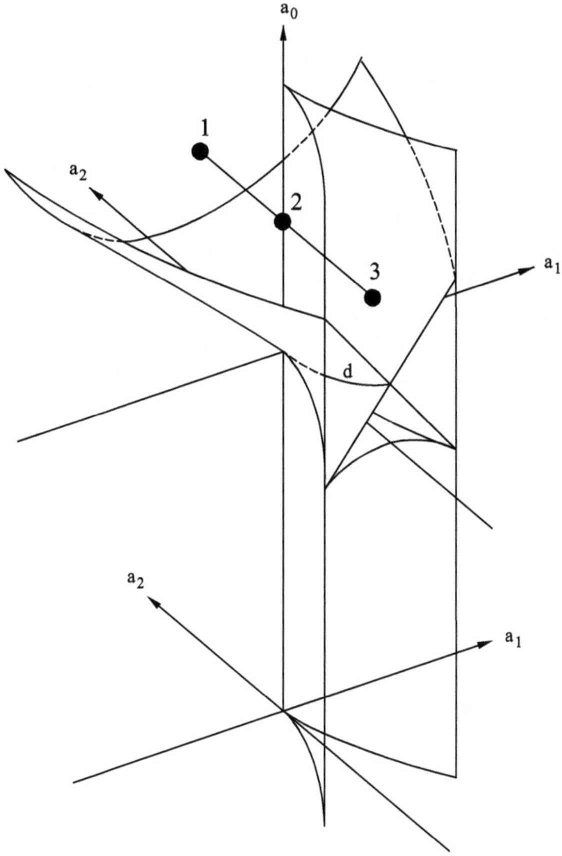


**Fig. 10.24.** Illustration of the operation $Bm^+$ along a *horizontal straight line* directly above the $a_2$-axis, in the control space for $EA_3$.

We have seen that preservation of the minimum curvature value occurs along the horizontal straight line in Fig 10.24 from points 1 to 2, but not past 2. It is clear that, to ensure the preservation scenario, we have to use the straight line as far as point 2, but then we have to continue on a particular upward curve, to be defined as follows: First consider the continuous line consisting of the positive $a_2$-axis, the origin, and the intersection curve $d$. Along that line, the initial minimum has value zero, which means zero curvature, and this zero value for curvature is preserved through the bifurcation; i.e., in the two emerging minima. The reason is that the positive $a_2$-axis is on the minima zero-separatrix; and the component $d$, where the minima zero-separatrix crosses itself, must be the set of functions in which both minima emerging from the bifurcation are zero. Therefore we will call this entire line, the **zero-preservation line**. Now if we simply raise this entire line by the amount $c$, then the initial minimum must be of value $c$, and must bifurcate, after the critical-point separatrix cusp-point, into two minima that are of value $c$. Therefore, we will call this line the $c$**-preservation line**

A useful way to understand this is to consider a vertical line upward through a point $a_2$ on the negative $a_2$-axis. The function located at this point is as illustrated by the third graph in Fig 10.22; i.e., the minima have a negative value, which we will notate using:

$$F_{m,a_2} \;=\; F\text{-value of the minimum } m \text{ in the function located at the point } \textit{within} \text{ the } a_2\text{-axis of the control space, at position } a_2$$

Then traveling upward on the vertical line in the control space, the point at which that line intersects the curve $d$ is the function that results from raising that previous graph so that the minima are now of value zero. This means that we have taken the previous function and raised it by the positive amount $-F_{m,a_2}$. Finally, traveling further upward on the vertical line, by the amount $c$, moves the minima of the graph up to value $c$.

Let us now calculate the unfolding function that defines the functions along the $c$-preservation line.

First note that, given the following argument, it is easy to *build* the unfolding function that produces the functions along the $c$-preservation line. To do so, first consider the unfolding function that produces the functions *within* the $a_2$-axis:

**$a_2$-Axis Unfolding:**

$$F(t; 0, 0, a_2) \;=\; t^4 \;+\; a_2 t^2 \tag{10.8}$$

Now, to convert this into the 0-preservation line, we raise each point $a_2$ in the $a_2$-axis by the amount $-F_{m,a_2}$, thus:

**Zero-Preservation Unfolding:**

$$F_{m=0}(t; a_2) \;=\; t^4 \;+\; a_2 t^2 \;-\; F_{m,a_2} \tag{10.9}$$

And, finally, to convert this into the $c$-preservation line, we raise the entire zero-preservation line by the constant $c$, thus:

**$c$-Preservation Unfolding:**

$$F_{m=c}(t; a_2) \;=\; t^4 \;+\; a_2 t^2 \;-\; F_{m,a_2} \;+\; c \tag{10.10}$$

Notice that, for any function *within* the *positive* $a_2$-axis, and for the germ, the value $F_{m,a_2}$ of the minimum is zero. It is only for functions within the *negative* $a_2$-axis that $F_{m,a_2}$ becomes non-zero. This is coded in the fact that the line $d$ curls upward away from the negative $a_2$-axis.

Expression (10.10) for the $c$-preservation line is useful because it exhibits the three-stage logic by which the line is derived. It order to see further aspects of our Interactive Singularity Theory, it is valuable to understand the relation of this logic to Method 1, as follows.

First let us calculate the actual value of the term $F_{m,a_2}$. Expression (10.8) gives the form of any function *within* the $a_2$-axis, and therefore the first derivative of any such function is

$$dF/dt = 2t(2t^2 + a_2)$$

which, when equated to 0 gives the minima as occurring at positions $t = \pm\sqrt{-a_2/2}$ for $a_2 \leq 0$. Substituting these positions into $F(t; 0, 0, a_2)$, in expression (10.8), gives the value of the minima to be $-a_2^2/4$. Therefore the intersection curve in the *critical-value surface* in Fig 10.18 (p354) must be a *downward semi-parabola*. Using Method 1, we invert this to obtain the intersection curve $d$ in Fig 10.24 (p359) as an *upward semi-parabola*, given by the equation $a_0 = a_2^2/4$. Therefore, Method 1 gives the following expression for (10.10):

**$c$-Preservation Unfolding:**

$$F_{m=c}(t; a_2) = t^4 + \begin{cases} a_2 t^2 + c & a_2 > 0 \\ a_2 t^2 + a_2^2/4 + c & a_2 \leq 0 \end{cases} \tag{10.11}$$

As the result of applying this raising process to the three functions in Fig 10.22, one obtains the three functions in Fig 10.25, where we see that the minima all now have value $c$. Notice that the amount by which the first two functions were raised was the same, that is $c$; but the amount by which the third function was raised must have been more, that is $a_2^2/4 + c$.

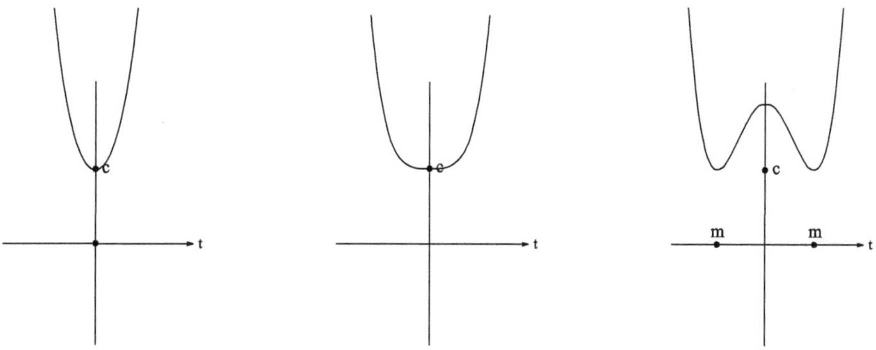


**Fig. 10.25.** The sequence of three functions, in Fig 10.22, raised in such a way that the chosen value $c$ of the minimum is preserved along the sequence.

Now, the illustration in Fig 10.25 is for the operation $Bm^+$. And the same unfolding function works for $Bm^-$, where one chooses $c$ to have a negative value. In the 3D control space, this means that the positive $a_2$-axis and the $d$-curve are simply translated downward by $c$. However, immediately, this shows us that something extra happens for a preserved negative $c$-value. Although, the lowered $d$-curve is still an upward semi-parabola, it will now cut upwards through the maxima zero-separatrix sheet in Fig 10.24 (p359), which is the lowest arched sheet at the front of the figure. However, the upward parabola will *never* reach the $d$-curve itself because it is parallel to the latter. Thus the two minima in the function can never cross their $t$-axis, because this can happen only when the function crosses the $d$-curve. Note that the $d$-curve is on the *minima* zero-separatrix.

In summary: Although the semi-parabola crosses the *maxima* zero-separatrix, it does not cross the *minima* zero-separatrix.

It is worth comparing what we have described in the control space, with an illustration of the functions themselves. Fig 10.26 illustrates the operation $Bm^-$ along the $c$-preservation line we have just discussed, where the preserved value $c$ of the minimum is negative; i.e., note that the minima in all the functions have the same negative value $c$.

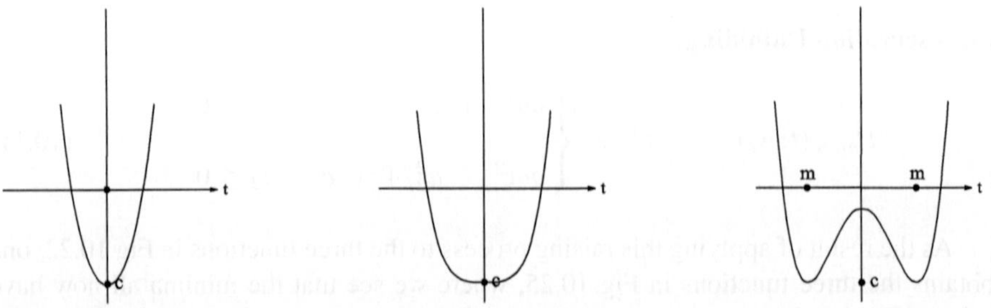


**Fig. 10.26.** The sequence of three functions, in Fig 10.22, lowered in such a way that the chosen value $c$ of the minimum is preserved along the sequence.

But observe that the maximum in the third function in Fig 10.26 is also negative. This corresponds to the fact that the third function occurs at a point on the upward semi-parabola, *before* the semi-parabola has reached the maxima zero-separatrix in the 3D control space of Fig 10.24.

We know that what has happened is this: The maximum started at value $c$ in the second function in Fig 10.26 where it was coincident with the minimum. It then moved upward relative to the minimum, reaching its value shown in the third function. Generally, we know that, as one goes in the negative $a_2$-direction in the control space, the value-difference between the minima and the maximum, in a function, increases.

The crucial consequence is this: In continuing along the semi-parabola, the minima in the third function in Fig 10.26 will remain at the negative value $c$, but the maximum, in moving upwards, will eventually cross the $t$-axis. This corresponds to the semi-parabola crossing the maxima zero-separatrix in the control space in Fig 10.24. The crossing is captured by the Process Grammar operation $CM^- : M^- \longrightarrow 0M^+0$, applied to the maximum emerging from the preceding operation $Bm^- : m^-M^-m^-$.

This gives a fundamental difference between $c$-preservation for a positive minimum and $c$-preservation for a negative minimum. However, this difference is not a disadvantage. In fact, it captures the *causal* structure of the respective situations, as will now be shown.

## 10.8 Causal Structure

We shall now see that the relation between the $c$-preservation line and the separatrices corresponds to the *causal* explanation of the shape. Recall from section 3.24 that one of the crucial properties of our process-inference rules is that the rules have the capacity to causally explain the bifurcations, in the following way:

> **Each of the $B$ bifurcations is *causally explained* by the process inferred from the central extremum that appears between the two copies of the bifurcating extremum. We therefore call this central process, the *bifurcation-causing process*.**
>
> **This bifurcation-causing process causes the bifurcation by *opposing* the pre-existing process that splits in the bifurcation; i.e., its opposition to that pre-existing process causes that pre-existing process to split.**

In this section, we are going to show that the preservation scenario has a profound consequence on the bifurcation-causing process. We will first show this in the case of a preserved positive minimum, and then we will show this in the case of a preserved negative minimum.

When we discussed the preservation scenario for a positive minimum, we saw that the $c$-preservation line, in the control space, does not cross another separatrix sheet after crossing the critical-point separatrix (at a cusp-point). We will now show that this corresponds to our theory's plausible *causal explanation* of the shape.

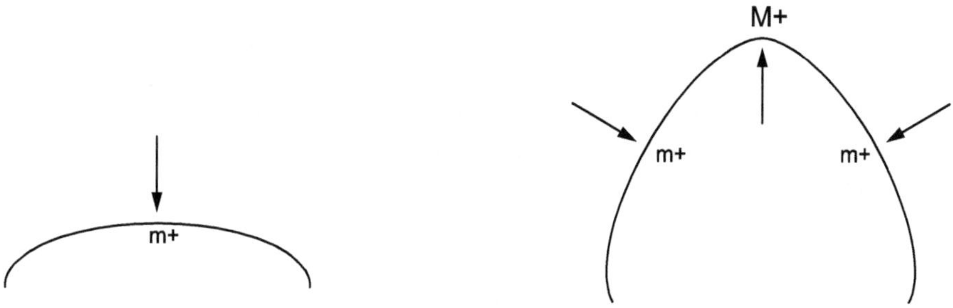


**Fig. 10.27.** $Bm^+$ with preservation of the curvature of $m^+$.

First consider the causal explanation corresponding to crossing the critical-point separatrix at a cusp-point. For a positive minimum, crossing this separatrix is the use of the $Bm^+$ operation. Fig 10.27 illustrates the $Bm^+$ operation in the case where the curvature of the positive minimum is preserved in the bifurcation. That is, the central $m^+$ extremum, at the top of the first shape, bifurcates into the two side $m^+$ extrema, in the second shape, and these two side $m^+$ extrema have the same curvature as the $m^+$ extremum in the first shape; i.e., the curvature of the initial $m^+$ has been *preserved* in its two bifurcated copies.

Note that, in Fig 10.27, as always occurs in the $Bm^+$ operation, a new $M^+$ appears between the two copies of the bifurcated $m^+$. In section 3.21, we gave our theory's causal explanation of the operation $Bm^+$. Let us first recall that causal explanation, in order to demonstrate the causal compatibility of Interactive Singularity Theory, and also in order to then show how the preservation scenario effects a property of the causal explanation. Thus, first recall from section 3.21 that our causal explanation for $Bm^+$ is this: Our inference rules say that the inferred process, at the created $M^+$, *caused* the bifurcation; i.e., in Fig 10.27, the upward protruding process, shown at the $M^+$ in the second shape, was the process that opposed the downward squashing process in the first shape, and *broke-through* that squashing process, causing the squashing process to split to the two sides of the protrusion created in the second shape by the upward opposing process.

Now, since the $c$-preservation line crosses the critical-point separatrix at a cusp-point, what we have just stated is the *causal explanation* that our theory gives for that separatrix-crossing which occurs on the $c$-preservation line.

A further crucial fact is that, being on the $c$-preservation line, the value of the minimum is preserved, and this has the following additional consequence: Since, after crossing the critical-point separatrix, the $c$-preservation line is directly above the negative $a_2$-axis, the value-difference between the minima and the maximum is increasing. Therefore, since the minima are being preserved, the maximum is increasing. Thus, the shape curvature given by the maximum is increasing. The consequence is that the extremum corresponding to the maximum is *sharpening*. That is, we have the following resulting conclusion:

**In the case of the Process Grammar operation $Bm^+$ that accords with the preservation scenario for its $m^+$, i.e., that occurs along the $c$-preservation line for its $m^+$ in the Interactive Unfolding of $EA_3$, there is the following important causal consequence: The process that causes the bifurcation also causes a continuous sharpening of its maximum after the bifurcation.**

This corresponds, in the right shape in Fig 10.27, to the central $M^+$ extremum moving upward with increasing curvature, i.e., sharpening.

Another crucial fact is that the $c$-preservation line for a positive minimum $m^+$ does not cross any of the zero-separatrix sheets.

Now let us turn to the preservation scenario for a negative minimum. In discussing this case previously, we saw that the $c$-preservation line successively crosses *two* separatrices: first the critical-point separatrix at a cusp-point, and then the maxima zero-separatrix. We will now show that this corresponds to the plausible *causal explanation* of the shape.

First consider the causal explanation corresponding to crossing the critical-point separatrix at a cusp-point. For a negative minimum, crossing this separatrix is the use of the $Bm^-$ operation. Fig 10.28 illustrates the $Bm^-$ operation in the case where the curvature of the negative minimum is preserved in the bifurcation. That is, the central

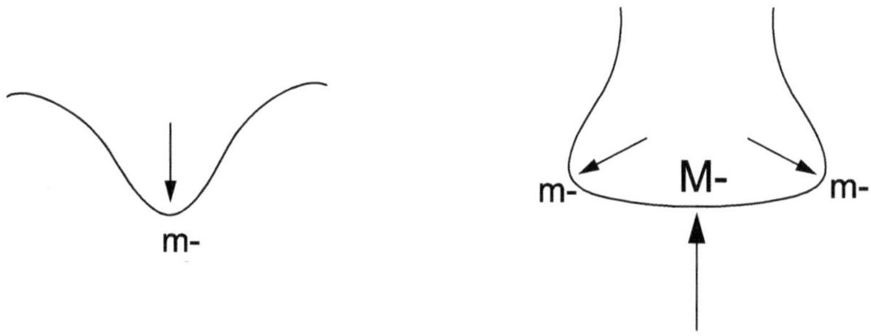


**Fig. 10.28.** $Bm^-$ with preservation of the curvature of $m^-$.

$m^-$ extremum, in the first shape, bifurcates into the two side $m^-$ extrema, in the second shape, and these two side $m^-$ extrema have the same curvature as the $m^-$ extremum in the first shape; i.e., the curvature of the initial $m^-$ has been *preserved* in its two bifurcated copies.

Note that, in Fig 10.28, as always occurs in the $Bm^-$ operation, a new $M^-$ appears between the two copies of the bifurcated $m^-$. In section 3.19, we gave our theory's causal explanation of the operation $Bm^-$. Let us first recall that causal explanation, in order to demonstrate the causal compatibility of Interactive Singularity Theory, and also in order to then show how the preservation scenario effects a property of the causal explanation. Thus, first recall from section 3.19 that our causal explanation for $Bm^-$ is this: Our inference rules say that the inferred process, at the created $M^-$, *caused* the bifurcation; i.e., in Fig 10.28, the upward resisting process, shown at the $M^-$ in the second shape, was the process that opposed the downward indenting process in the first shape, causing the indenting process to split to the two ends of the bay created in the second shape by the upward opposing process.

Now, since the $c$-preservation line crosses the critical-point separatrix at a cusp-point, what we have just stated is the *causal explanation* that our theory gives for that separatrix-crossing which occurs on the $c$-preservation line.

A further crucial fact is that, being on the $c$-preservation line, the value of the minimum is preserved, and this has the following additional consequence: Since, after crossing the critical-point separatrix, the $c$-preservation line goes along the vertical half-plane whose middle line is the negative $a_2$-axis, the value-difference between the minima and the maximum is increasing. Therefore, since the minima are being preserved on the $c$-preservation line, the value of the maximum is increasing. Thus, the shape curvature given by the maximum is increasing in the *positive curvature direction* on the shape.

In fact, as we said previously, the $c$-preservation line for a negative minimum will eventually cross the maxima zero-separatrix, and this crossing corresponds to the application of Process-Grammar continuation operation $CM^-$. That is, the central resistance process will continue till protrudes, as illustrated in the transition from the left to right

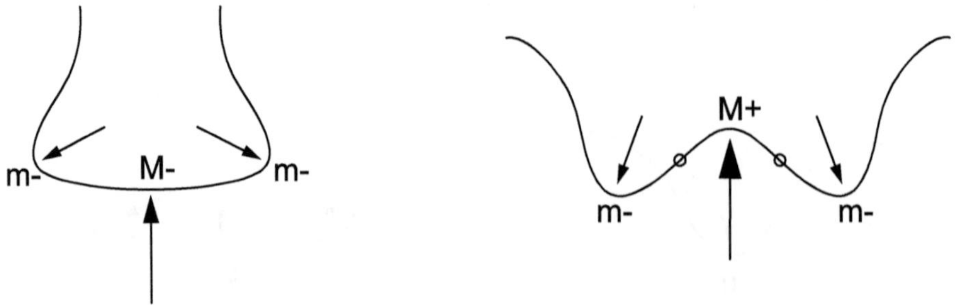


**Fig. 10.29.** $CM^-$ with preservation of the curvature of $m^-$.

shape in Fig 10.29. This is a consequence of the fact that the shape curvature given by the maximum increases in the *positive curvature direction* on the shape.

As a result of this, we have seen that, in *both* the positive $c$ and negative $c$ cases, the *bifurcation-causing process* increases the curvature of the central maximum in the positive curvature direction. It is important to understand the full reason we have demonstrated for this, as follows:

**In the Interactive Unfolding of $EA_3$:**
**The 3-fold bifurcation-causing process in a minimum-preservation scenario (in both the positive and negative case) necessarily causes the curvature of the central created maximum to increase, in the positive curvature direction, due to a combination of two reasons:**

**(1) Since, in the control space, we are moving within the vertical half-plane whose middle line is the negative $a_2$-axis, and we are moving in the direction of increasing negativity of $a_2$, the value-difference between the minima and the maximum increases.**

**(2) Since we are moving along the $c$-preservation line, the value of the minima is preserved, and therefore the value of the maximum is forced to increase.**

In other words, reason (1) *alone* is not sufficient for the central maximum to increase. In fact, movement *within* the negative $a_2$-axis actually preserves the maximum.

As a result of the above discussion, we therefore see again that our Interactive Singularity Theory is compatible with plausible causal explanations offered by our process-inference rules. To summarize the causal structure described in the present section, we have the following:

**CAUSAL STRUCTURE**
**ALONG A**
**$c$-PRESERVATION LINE**
**IN THE**
**INTERACTIVE UNFOLDING OF $EA_3$**

**For a positive minimum $m^+$, movement along its $c$-preservation line corresponds to the continuous application of a single force: First through the critical-point separatrix, the force creates a protrusion which splits a squashing process on the shape, and then along the upward semi-parabola of the $c$-preservation line, the force persists, causing the protrusion to extend and sharpen.**

**For a negative minimum $m^-$, movement along its $c$-preservation line corresponds to the continuous application of a single force: First through the critical-point separatrix, the force resists an indenting process, causing it to split, and then along the semi-parabola of the $c$-preservation line, and through the maxima zero-separatrix, the resistance force continues till it protrudes.**

Observe that this important causal situation is not represented in Catastrophe Theory because the dimension of the control space for the same germ $t^4$, in Catastrophe Theory, is not large enough to contain the $c$-preservation line. Furthermore the Catastrophe Theory control space does not contain what our theory calls the *full separatrix*, and therefore cannot model the *interactive causal relation* between what our theory defines as the process-bifurcation operator $B$ and the process-continuation operator $C$. Also, this important causal structure is not given by conventional symmetry analyses such as the Medial axis because, using the Medial axis, the causal force would, at the zero-separatrix, switch sides on the shape, which would be causally meaningless, as demonstrated in section 3.25.

Finally, we note that the above discussion demonstrated the causal compatibility of our $EA_3$ model with respect to the Process Grammar operations $Bm^+$ and $Bm^-$. Exactly the dual statements show the causal compatibility of our $EA_{-3}$ model with respect to the Process Grammar operations $BM^-$ and $BM^+$.

## 10.9 Quantitative Example of $Bm^+$

This section gives a *quantitative* example of $Bm^+$ with the preservation scenario. Throughout the bifurcation, the minima will have the following value:

$$\text{preserved curvature} \quad c = 0.1$$

The three figures, on the next three pages, i.e., Figs 10.30 – 10.32, show the functions at three successive points along the $c$-preservation line. Each function, as a curvature function, is shown together with its resulting unit-speed parameterized curve, which is determined by our usual condition that the curve-point with parameter value $t = 0$ is placed at the origin of the $(x, y)$-plane and has velocity vector corresponding to the point $(1, 0)$ in that plane. Since the curvature function accords with the $c$-preservation unfolding equation

$$F_{m=c}(t; a_2) \;=\; t^4 + a_2 t^2 + c - F_{m,a_2} \tag{10.12}$$

the parametric equations of the corresponding curve are:

$$\begin{aligned} x(t) &= \int_0^t \cos(\frac{1}{5}s^5 + \frac{1}{3}a_2 s^3 + [c - F_{m,a_2}]s)\,ds \\ y(t) &= \int_0^t \sin(\frac{1}{5}s^5 + \frac{1}{3}a_2 s^3 + [c - F_{m,a_2}]s)\,ds \end{aligned} \tag{10.13}$$

Now, recall that $F_{m,a_2}$, which is the value of the minimum of the function at point $a_2$ within the $a_2$-axis, has value 0 for $a_2 \geq 0$, and value $-a_2^2/4$ for $a_2 < 0$. Therefore it is a trivial matter to calculate the functions in each of the three following figures. The three successive functions are on the $c$-preservation line at positions that are, respectively:

(1) before the cusp-point,
(2) at the cusp-point,
(3) after the cusp-point; i.e., on the semi-parabola.

The specific values chosen for these positions are:

$a_2 = 2$
$a_2 = 0$
$a_2 = -2.3$

Using the $c$-preservation unfolding equation (10.12), the three respective functions are:

$$\begin{aligned} F_{m=c}(t;\ 2) &= t^4 + 2t^2 + 0.1 \\ F_{m=c}(t;\ 0) &= t^4 + 0.1 \\ F_{m=c}(t; -2.3) &= t^4 - 2.3t^2 + 0.1 + 1.3 \end{aligned}$$

The text now continues in the captions of the three figures, which the reader is recommended to read.

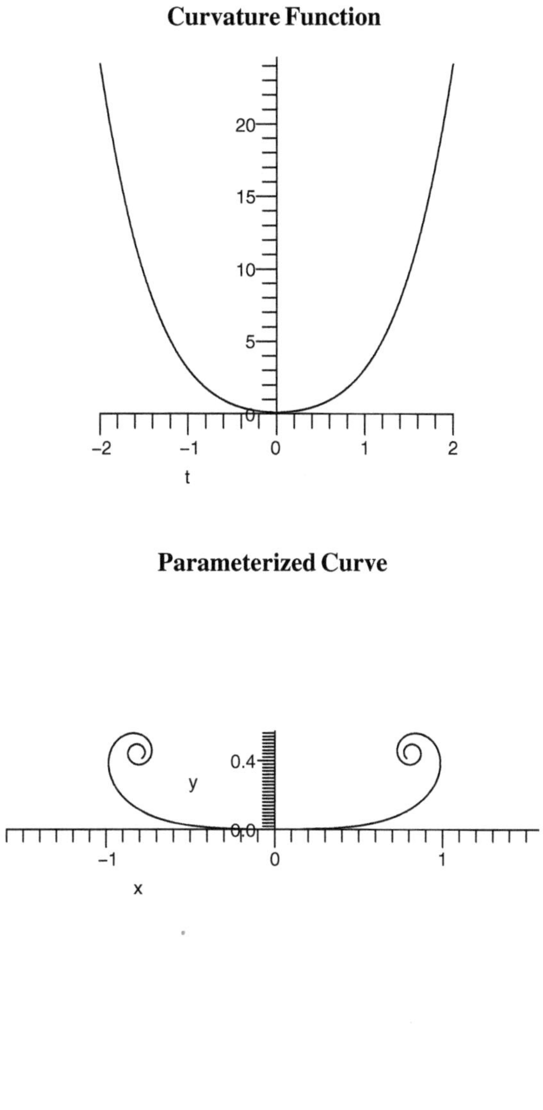


**Fig. 10.30.** Pre-Bifurcation: The upper figure shows the curvature function $t^4 + 2t^2 + 0.1$ which is at the position $(a_0, a_1, a_2) = (0.1,\ 0,\ 2)$ in the control space of the Interactive Unfolding of $EA_3$. It is therefore above the zero-separatrix in the control space, but before the cusp-point. In fact, it is on the straight-line component of the $c$-preservation line. Thus, the height $a_0$ of the function in the 3D control space is the same as the preserved curvature $0.1$. The function has a single non-degenerate minimum whose value is that preserved curvature (although this value $0.1$ is a little difficult to see on the $F$-axis, due to the scaling of the axis). The lower figure shows the resulting unit-speed parameterized curve. The solid is above the curve and the empty space below the curve, due to the way the curve is parameterized in correspondence to the function. The curvature extremum $m^+$ is at the center of the curve, directly below the minimum in the function.

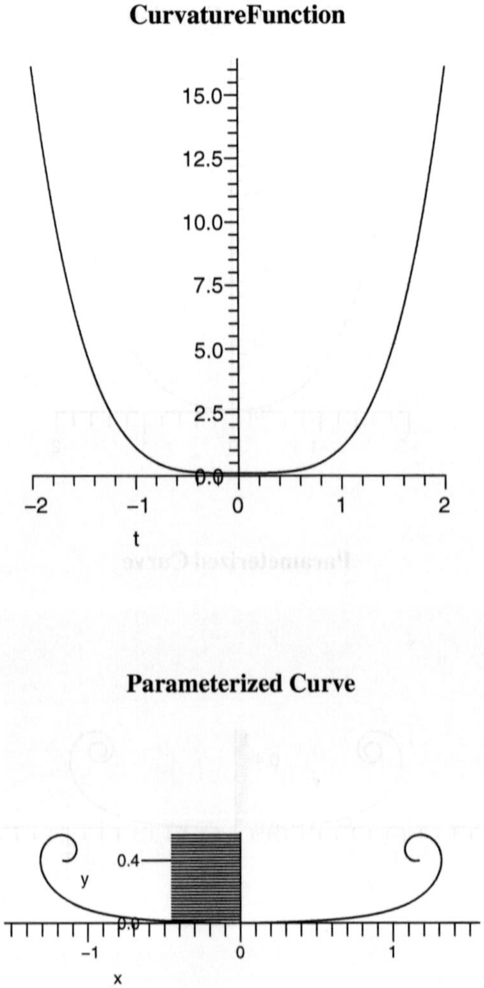


**Fig. 10.31.** Cusp-Point: The upper figure shows the curvature function $t^4 + 0.1$ which is at the position $(a_0, a_1, a_2) = (0.1,\ 0,\ 0)$ in the control space of the Interactive Unfolding of $EA_3$. This means that the function is located at a cusp-point of the critical-point separatrix, but above the zero-separatrix. In fact, the height $a_0$ of the function in the 3D control space is the same as the preserved curvature 0.1. The function has a single degenerate minimum whose value is that preserved curvature. The lower figure shows the resulting unit-speed parameterized curve. Again, the solid is above the curve and the empty space below the curve, due to the way the curve is parameterized in correspondence to the function. The curvature extremum is at the center of the curve, directly below the minimum in the function.

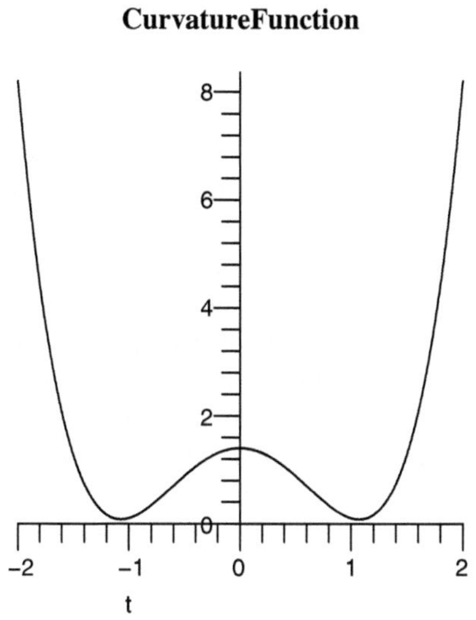


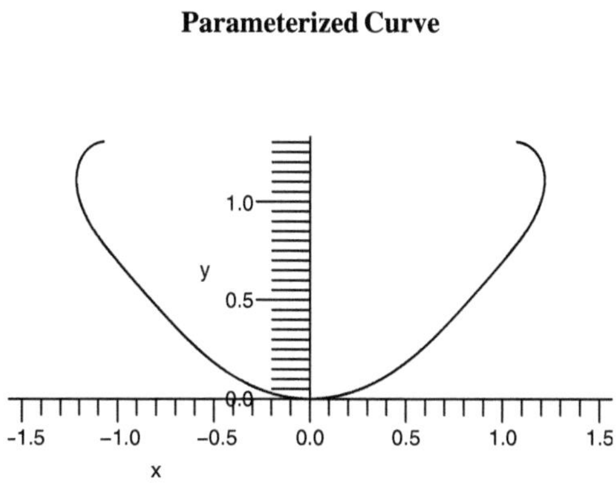


**Fig. 10.32.** Post-Bifurcation: The upper figure shows the curvature function $t^4 - 2.3t^2 + 1.4$ which is at the position $(a_0, a_1, a_2) = (1.4,\ 0,\ -2.3)$ in the control space of the Interactive Unfolding of $EA_3$. It is therefore after the cusp-point of the critical-point separatrix, but above the zero-separatrix; in fact, on the upward semi-parabola of the $c$-preservation line. Thus the height $a_0$ of the function in the 3D control space has now moved considerably above the preserved curvature value $0.1$. Nevertheless, that upward movement in the 3D control space has preserved that curvature value $0.1$ in the function. The function now has three critical points: two minima and one maximum – where the minima have the preserved curvature value. The lower figure shows the resulting unit-speed parameterized curve. Again, the solid is above the curve and the empty space below the curve, due to the way the curve is parameterized in correspondence to the function. The maximum $M^+$ is at the center of the curve, and the two minima $m^+$ are on the sides of the curve.

The sequence of figures on the previous three pages clearly realizes the bifurcation operation

$$Bm^+ : m^+ \longrightarrow m^+M^+m^+$$

from the Process Grammar. That is, Fig 10.30 exhibits the single starting $m^+$, then Fig 10.31 shows the degenerate transition function, and Fig 10.32 exhibits the post-bifurcation triple $m^+M^+m^+$.

Now let us look quantitatively at the causal structure defined in section 10.8. The four successive curves in Fig 10.33 show the effect of moving along the semi-parabola of the $c$-preservation line, starting at the cusp-point. The four curves are at the four successive points $a_2 = 0, \ -1.1, \ -2, \ -2.3$. Observe that the curvature of the central minimum, in the first curve, is preserved in the two side minima that occur in each of the following three curves. The central maximum, which appears in the second curve, increases in curvature in the following curves. From this maximum, our rules infer the force which caused both, the bifurcation, and the subsequent increase in curvature of the maximum.

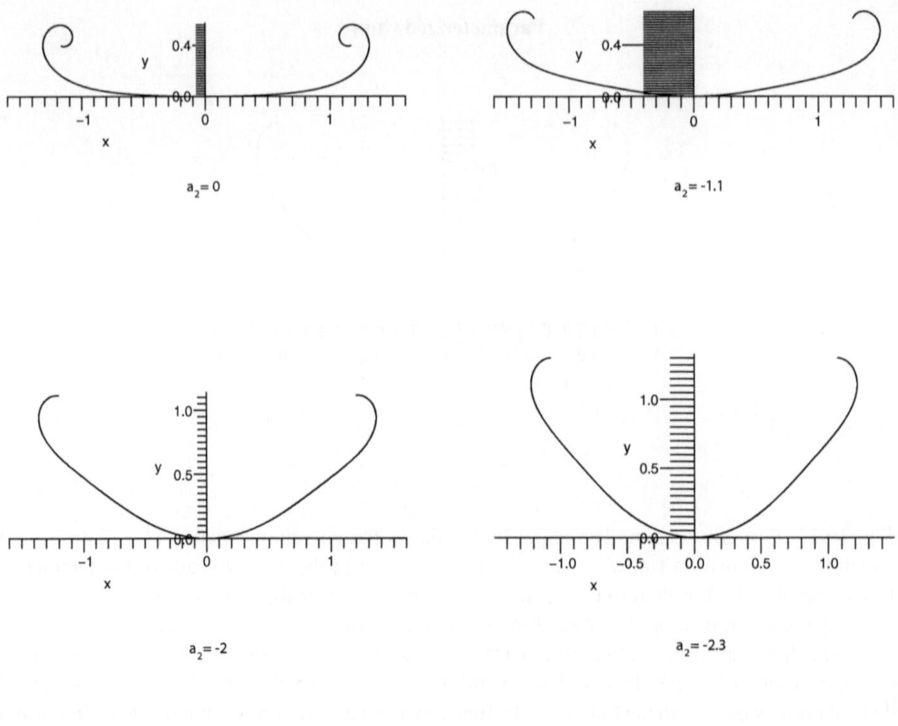


**Fig. 10.33.** Four successive curves along the $c$-preservation line, starting at the cusp-point.

## 10.10 Quantitative Example of $Bm^-$ and Subsequent $CM^-$

Recall the following fact about the Interactive Unfolding of $EA_3$:

Whereas a $c$-preservation line for positive $c$ crosses only the critical-point separatrix, i.e., corresponding to only the bifurcation operation $Bm^+$, a $c$-preservation line for negative $c$ crosses the critical-point separatrix and subsequently the zero-separatrix, i.e., corresponding to the bifurcation operation $Bm^-$ followed by the continuation operation $CM^-$.

This section will give a quantitative example of this sequence of two operations along a negative $c$-preservation line. Throughout the successive operations, the minima will have the following value:

$$\text{preserved curvature} \qquad c = -2.5$$

The three figures, on the next three pages, i.e., Figs 10.34 – 10.36, show the functions at three successive points along the $c$-preservation line:

(1) before the critical-point separatrix,
(2) after the critical-point separatrix but before the zero-separatrix,
(3) after the zero-separatrix.

Their positions on the $c$-preservation line are given respectively by the values:

$$a_2 = 2.3$$
$$a_2 = -3.1$$
$$a_2 = -3.5$$

Using the preservation unfolding equation

$$F_{m=c}(t;\ a_2) \quad = \quad t^4\ +\ a_2t^2\ +\ c\ -\ F_{m,a_2}$$

and the fact that $F_{m,a_2} = 0$ for positive $a_2$-value, and $F_{m,a_2} = -a_2^2/4$ for negative $a_2$-value, the functions at the three respective points are:

$$\begin{aligned} F_{m=c}(t;\ 2.3) &= t^4 + 2.3t^2 - 2.5 \\ F_{m=c}(t;-3.1) &= t^4 - 3.1t^2 - 2.5 + 2.4 \\ F_{m=c}(t;-3.5) &= t^4 - 3.5t^2 - 2.5 + 3 \end{aligned}$$

Each function, as a curvature function, is shown, in the three figures, together with its resulting unit-speed parameterized curve, computed by our usual method. Notice that, in each case, the solid is above the curve, and the empty space below the curve, due to the way the curve is parameterized in correspondence to the function.

The text now continues in the captions of the three figures, which the reader is recommended to read.

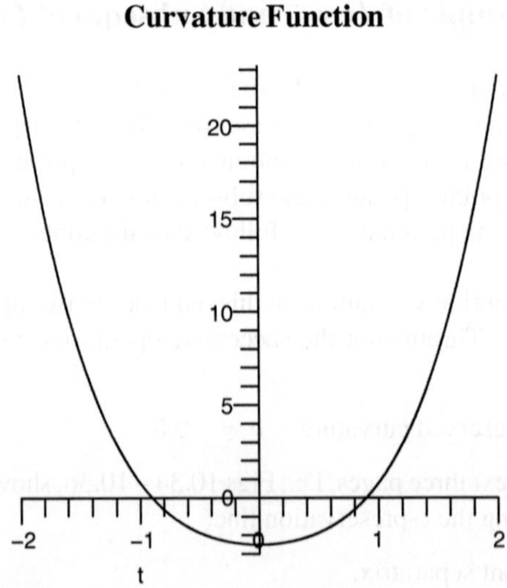


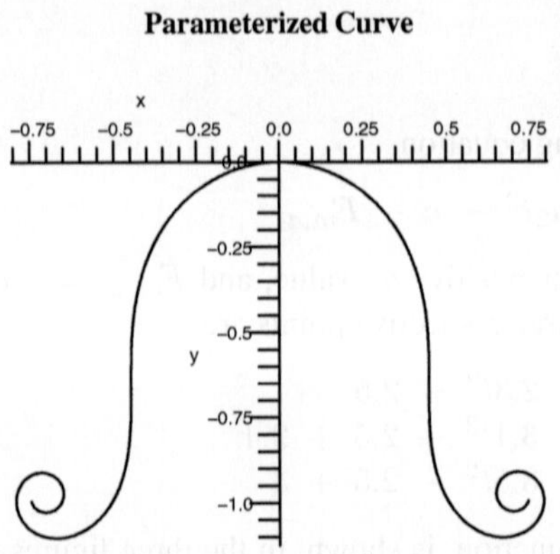


**Fig. 10.34.** Before the critical-point separatrix: The upper figure shows the curvature function $t^4 + 2.3t^2 - 2.5$ which is at the position $(a_0, a_1, a_2) = (-2.5,\ 0,\ 2.3)$ in the control space of the Interactive Unfolding of $EA_3$. It is therefore below the zero-separatrix in the control space, but before the cusp-point. In fact, it is on the straight-line component of the $c$-preservation line. Thus, the height $a_0$ of the function in the 3D control space is the same as the preserved curvature $-2.5$. The function, as seen above, has a single non-degenerate minimum whose value is that preserved curvature. The lower figure shows the resulting unit-speed parameterized curve. The curvature extremum $m^-$ is at the center of the curve, directly below the minimum in the function. Note that the curvature function crosses its $t$-axis at two zeros. Consequently, we see two flat points directly below, on the sides of the parameterized curve.

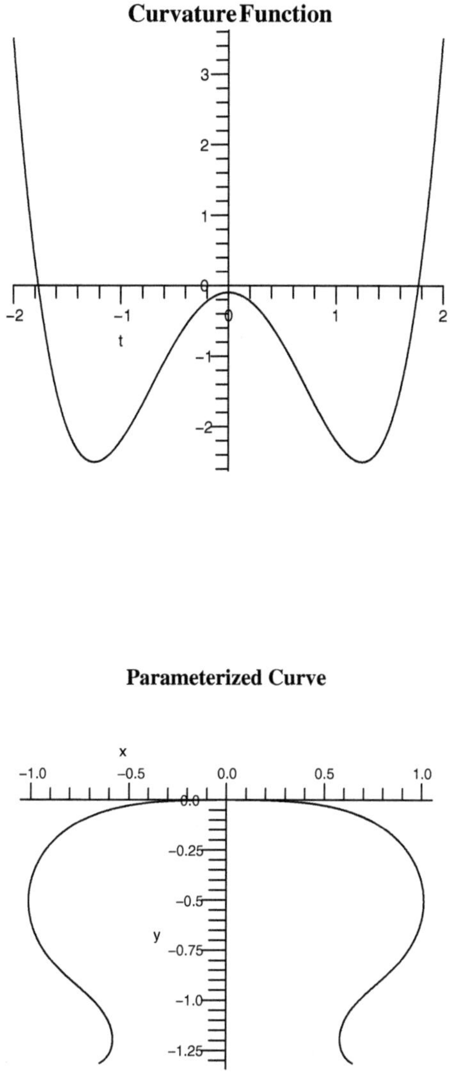


**Fig. 10.35.** After the critical-point separatrix but before the zero-separatrix: The upper figure shows the curvature function $t^4 - 3.1t^2 - 0.1$ which is at the position $(a_0, a_1, a_2) = (-0.1,\ 0,\ -3.1)$ in the control space of the Interactive Unfolding of $EA_3$. The minimum $m^-$ of the previous function has bifurcated into two minima and a maximum; i.e., the sequence $m^-M^-m^-$. A further crucial fact is that the value $-2.5$ of the previous minimum has been preserved in the present minima. The maximum, which at the moment of bifurcation-transition must have been the same value as the minima, has risen, but not yet crossed the $t$-axis, which means that the present function is located before the maxima zero-separatrix in the control space. The lower figure shows the unit-speed parameterized curve determined by the curvature function. We can see the maximum in the center and the two minima at the sides. Observe that the curve therefore has the shape of a *bay*; i.e., it is given by the sequence $m^-M^-m^-$.

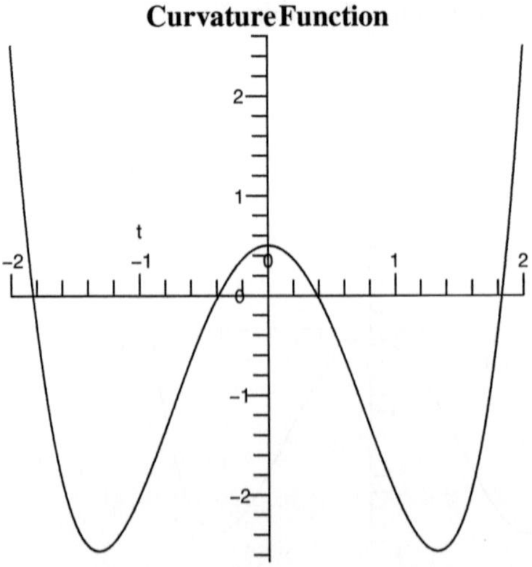


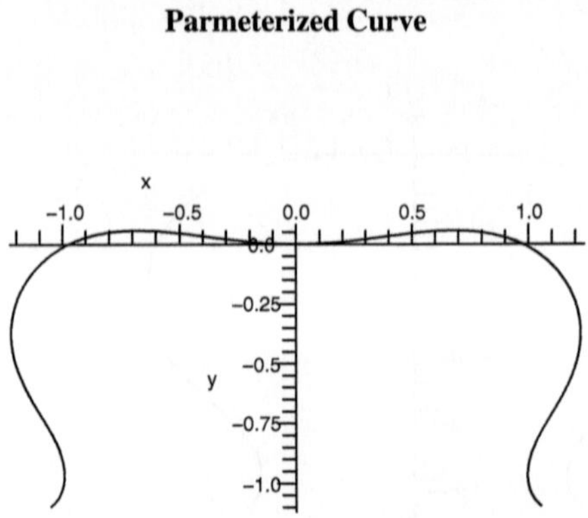


**Fig. 10.36.** After the zero-separatrix: The upper figure shows the curvature function $t^4 - 3.5t^2 + 0.5$, which is at the position $(a_0, a_1, a_2) = (0.5, 0, -3.5)$ in the control space of the Interactive Unfolding of $EA_3$. The negative maximum $M^-$ of the previous function moved upwards, in the function, across the $t$-axis to the positive value 0.5, with two zeros bifurcating to the sides of the maximum; thus producing the sequence $0M^+0$. A further crucial fact is that the minima are still the same value $-2.5$ as previously. Thus, in the control space, the function has moved along the $c$-preservation line across the maxima zero-separatrix. The lower figure shows the unit-speed parameterized curve determined by the curvature function. We can see the positive maximum at the center, with the two new zeros on the sides of the maximum.

The sequence of figures on the previous three pages clearly illustrates the fact that movement along the $c$-preservation line, for negative $c$, realizes the succession of two operations:

$$Bm^- : m^- \longrightarrow m^-M^-m^-$$

followed by

$$CM^- : M^- \longrightarrow 0M^+0$$

from the Process Grammar. That is, it is an example of what our singularity theory calls the *Interactive Structure*.

To show the shape development involved, Fig 10.37 puts together the three successive parameterized curves from the previous three figures. The curvature extrema are labeled here on the shapes, and the arrows are shown, inferred from the extrema by our process-inference rules. The sequence of shapes show the remarkable power of these rules to causally explain the shape evolution, as follows: From the negative minimum $m^-$ in the first shape, our rules infer an upward arrow creating that minimum, as shown. The arrow is therefore an *indenting* process – the solid being above the curve and the empty space below it, with the causal force coming upward from the empty side, pushing the curve inwards into the solid region. In the second shape, the downward resistance force, inferred by our rules from the central $M^-$, has opposed the indenting process and forced it to bifurcate to the two sides, thus creating what our theory calls a *complete bay*, that is, $m^-M^-m^-$. Finally, in the third shape, the downward resistance force has continued till it protruded into the empty space below the bay.

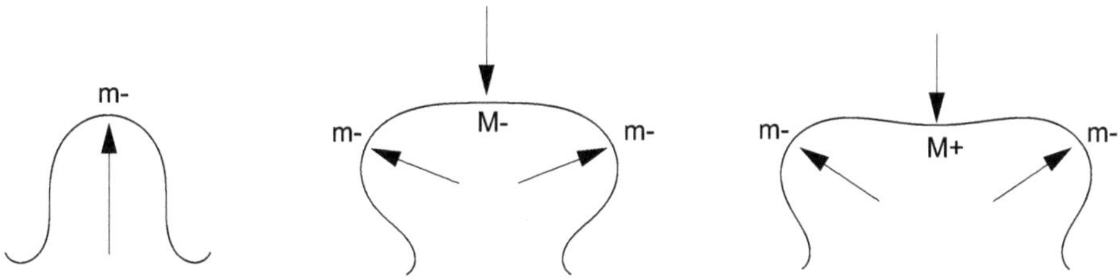


**Fig. 10.37.** The three successive curves from the previous three figures, showing the processes inferred by our rules.

# 11

# Operator $\pi$ in the Interactive Unfoldings of $EA_3$ and $EA_{-3}$

## 11.1 Introduction

In the previous chapter, we saw that the Process Grammar operators $B$ and $C$, which cannot be modeled in the cusp catastrophes $A_3$ and $A_{-3}$, can be modeled in the Interactive Unfoldings of $EA_3$ and $EA_{-3}$. We shall now see, similarly, that the Process Grammar operator $\pi$ fails to be modeled in the cusp catastrophes $A_3$ and $A_{-3}$, but can be modeled in the Interactive Unfoldings of $EA_3$ and $EA_{-3}$. This will lead to the discovery of additional interesting structure in Interactive Singularity Theory.

If the reader needs a quick review of the pair-creation operator $\pi$ and its operations, then this can be obtained simply by reading the statements in bold print in Chapter 4.

## 11.2 Pair-Creation and the Cusp Catastrophes $A_3$ and $A_{-3}$

Consider again Fig 7.4 (page 245), which shows representative functions in the 2D control space $(a_1, a_2)$ of the cusp catastrophe $A_3$. The reader will recall that, in Region 1 (the region surrounding the separatrix), all the functions have only one critical point: a non-degenerate minimum. In contrast, in Region 3 (between the separatrix curves), all the functions have three critical points: two non-degenerate minima and a non-degenerate maximum. Crossing from a function in Region 1 to a function in Region 3, via the germ at the origin, corresponds to a *3-fold* bifurcation of the initial minimum; i.e., when reaching the germ, the initial minimum becomes 3-fold degenerate, and then after this, the minimum bifurcates simultaneously into the non-degenerate minimum, maximum, minimum. In contrast, crossing from the same function in Region 1 to the same function in Region 3, via a point in one of the separatrix curves, corresponds to a *2-fold* bifurcation; i.e., the initial minimum remains non-degenerate throughout, and a non-critical point in

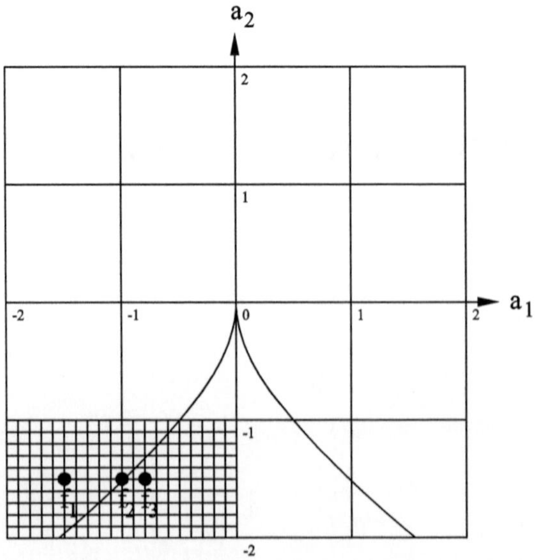


**Fig. 11.1.** Crossing the left separatrix curve in the control space of the cusp catastrophe.

the initial function becomes a critical inflection-point (horizontal inflection) within the graph of the function at the crossing point in the separatrix curve, and this inflection then bifurcates into a newly created maximum and minimum, after the separatrix.

We need to look at this 2-fold bifurcation in more detail. Consider Fig 11.1, which shows three successive points $f_1$, $f_2$, $f_3$, across the left separatrix curve (point $f_2$ being on the curve itself). As the reader can see from the figure, the specific positions $(a_1, a_2)$, chosen for these three points, are:

$$\begin{aligned} f_1 &= (-1.5, \; -1.5) \\ f_2 &= (-1, \quad -1.5) \\ f_3 &= (-0.8, \; -1.5) \end{aligned}$$

Substituting these coordinates into the general form $F(t; a_1, a_2) = t^4 + a_1 t + a_2 t^2$ of any function in the control space of the cusp catastrophe, we see that the functions located at these three points are, respectively:

$$\begin{aligned} f_1(t) &= t^4 - 1.5t - 1.5t^2 \\ f_1(t) &= t^4 - \quad t - 1.5t^2 \\ f_1(t) &= t^4 - 0.8t - 1.5t^2 \end{aligned}$$

The graphs of these three functions are shown in Fig 11.2. Observe that, in the positive $t$-axis, of each of these functions, there is a large *negative* minimum. This minimum is non-degenerate throughout. In contrast, consider what happens on the negative $t$-axis throughout. In the first function, $f_1$, there is a point in the negative $t$-axis, where the

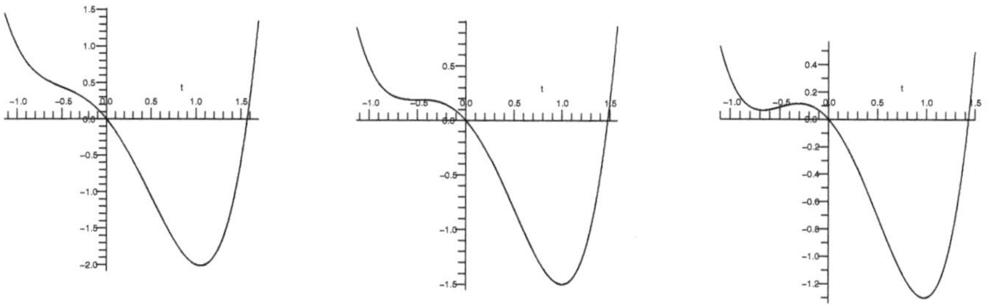

**Fig. 11.2.** The functions at the positions $f_1$, $f_2$, $f_3$, in Fig 11.1.

gradient of the graph has rotated towards the horizontal but has not reached it. Next, in $f_2$, the gradient at that point has actually reached the horizontal; in fact, that point is now a critical inflection-point. Finally, in $f_3$, this inflection-point has bifurcated into a positive minimum and positive maximum. Both of these are small, compared with the large minimum on the positive $t$-axis. The reason is that $f_3$ has been chosen to be near to the separatrix curve in the control space, as shown in Fig 11.1. Nevertheless, $f_3$ is in Region 3, and therefore, as shown in the third graph of Fig 11.2, it has the critical-point structure of any function in Region 3: a non-degenerate minimum, maximum, minimum.

Now we know, from the discussion of Fig 8.5 (p267), that if we continue to move $f_3$ further along the same horizontal line in the control space, towards the center of Region 3, the left minimum in the function will become larger and the right minimum will become smaller. The two minima will reach the same size when the function reaches the center of Region 3. At this point, the maximum will be of value 0, and will be located at the origin of the function's $t$-axis.

It is now worth noting this: Crossing the left separatrix curve, from Region 1 to 3, but remaining sufficiently close to that curve, realizes a *particular application* of the Process Grammar operation,

$$\pi\emptyset^{\downarrow+} \; : \; \emptyset^{\downarrow+} \quad \longrightarrow \quad m^+M^+$$

Recall, from section 4.2, that the empty-set symbol $\emptyset$ indicates that the point of application of the operation is a non-critical point in the first function in Fig 11.2. The downward arrow on the symbol $\emptyset^{\downarrow+}$ indicates that the gradient of this function at the application-point is downward; and the $+$ sign on $\emptyset^{\downarrow+}$ indicates that the application-point has positive value in the function. Therefore, since, in crossing the separatrix curve, the anchor-extremum is a negative minimum, the application of the operation is to only one of its application-types: a point in an opposite-sign tail of the anchor-extremum.

Now let us take the reflectionally opposite route in the control space (the reflection being about the $a_2$-axis). That is, let us cross the *right* separatrix curve from Region 1 to 3, using the reflectionally opposite positions to $f_1$, $f_2$, $f_3$ in Fig 11.1. The functions at these three positions are shown in Fig 11.3. We can see that, over the positive $t$-axis of these functions, they realize a *particular application* of the following Process Grammar operation:

$$\pi\emptyset^{\uparrow+} \; : \; \emptyset^{\uparrow+} \quad \longrightarrow \quad M^+m^+$$

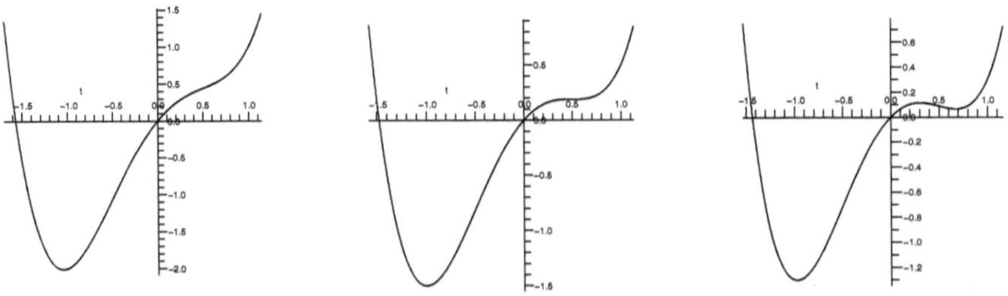

**Fig. 11.3.** The functions at the reflectionally opposite positions to $f_1$, $f_2$, $f_3$, in Fig 11.1.

where the symbol $\emptyset^{\uparrow +}$ means that, at the application-point $\emptyset$, the first function has upward gradient and +ve curvature. Therefore, since, in crossing the separatrix curve, the anchor-extremum is a negative minimum, the application of the operation is to only one of its application-types: a point in an opposite-sign tail of the anchor-extremum.

Now consider the dual cusp catastrophe $A_{-3}$, obtained by simply inverting the unfolding function of the cusp catastrophe $A_3$ with respect to its $F$-axis. Recall that, since the individual functions in the 2D control space are inverted, the extrema must be in the same positions on their $t$-axes, with maxima converted to minima and vice versa. Therefore, the separatrix in the control space $(a_1, a_2)$ of $A_{-3}$ must be congruent with, and in the same position as, the separatrix in the control space $(a_1, a_2)$ of $A_3$. Thus, by making exactly the same movements across the separatrix of $A_{-3}$, as we did across the separatrix of $A_3$, we obtain the duals of the above operations. This means that, taking the same three points, $f_1$, $f_2$, $f_3$, as before, gives us the functions shown in Fig 11.4; and this realizes a *particular application* of the Process Grammar operation:

$$\pi\emptyset^{\uparrow -} : \emptyset^{\uparrow -} \quad \longrightarrow \quad M^- m^-$$

where the symbol $\emptyset^{\uparrow -}$ means that, at the application-point $\emptyset$, the first function has upward gradient and -ve curvature. Therefore, since, in crossing the separatrix curve, the anchor-extremum is a positive maximum, the application of the operation is to only one of its application-types: a point in an opposite-sign tail of the anchor-extremum.

Correspondingly, crossing the right separatrix curve, at the reflectionally opposite points to $f_1, f_2, f_3$, gives us the functions shown in Fig 11.5; and this realizes a *particular application* of the Process Grammar operation:

$$\pi\emptyset^{\downarrow -} : \emptyset^{\downarrow -} \quad \longrightarrow \quad m^- M^-$$

where the symbol $\emptyset^{\downarrow -}$ means that, at the application-point $\emptyset$, the first function has downward gradient and -ve curvature. Therefore, since, in crossing the separatrix curve, the anchor-extremum is a positive maximum, the application of the operation is to only one of its application-types: a point in an opposite-sign tail of the anchor-extremum.

From the above discussion, we conclude that the cusp catastrophe $A_3$ and the dual cusp catastrophe $A_{-3}$ contain only one particular application of each of the non-zero pair-creation operations. Therefore, again Catastrophe Theory is not sufficient to model morphology. In fact, with respect to pair-creation, Catastrophe Theory has *two* problems that will force us to use the *Interactive Unfoldings* of $EA_3$ and $EA_{-3}$.

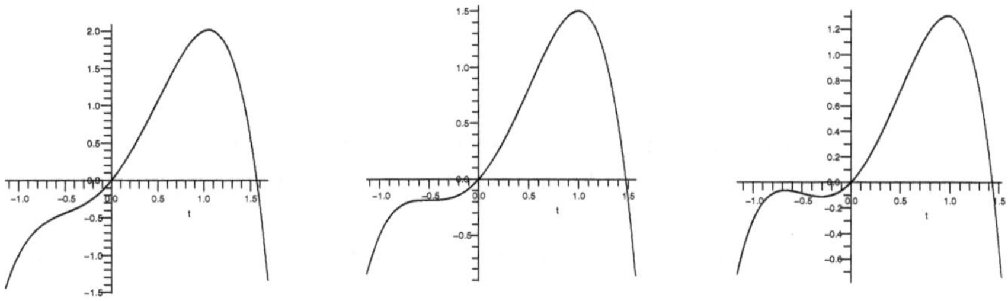

**Fig. 11.4.** In the *dual* cusp catastrophe, the functions at the positions $f_1$, $f_2$, $f_3$.

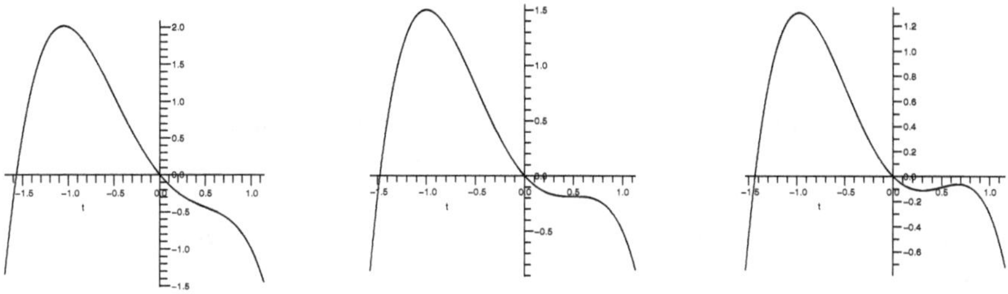

**Fig. 11.5.** In the *dual* cusp catastrophe, the functions at the reflectionally opposite positions to $f_1$, $f_2$, $f_3$

The first problem is this: Using the graphs in Fig 11.2 and Fig 11.3, we saw that, in the cusp catastrophe $A_3$, the anchor-extremum of the application of the positive $\pi$ operations, $\pi\emptyset^{\downarrow +}$ and $\pi\emptyset^{\uparrow +}$, was a negative minimum $m^-$. Correspondingly, using the graphs in Fig 11.4 and Fig 11.5, we saw that, in the *dual* cusp catastrophe $A_{-3}$, the anchor-extremum of the application of the negative $\pi$ operations, $\pi\emptyset^{\uparrow -}$ and $\pi\emptyset^{\downarrow -}$, was a positive maximum $M^+$.

However, in Table 4.1 (page 105), we saw that, in the Process Grammar, each of the non-zero pair-creation operations has *three* possible anchor-extrema. This means that the catastrophes $A_3$ and $A_{-3}$ are highly restrictive with respect to the use of the pair-creation operations: They allow *only one* of the three anchor-extrema for each operation.

This problem is listed in the following table, where we also define a second problem with modeling the $\pi$ operations in the catastrophes $A_3$ and $A_{-3}$.

**OPERATOR $\pi$**
**AND THE CUSP CATASTROPHES $A_3$ AND $A_{-3}$**

---

**Problem 1: While the positive and negative $\pi$ operations occur in the catastrophes $A_3$ and $A_{-3}$, their applications in those catastrophes is very limited:**

**(1) The positive $\pi$ operations are restricted to only a $m^-$ anchor-extremum.**

**(2) The negative $\pi$ operations are restricted to only a $M^+$ anchor-extremum.**

---

**Problem 2: In moving the codomain function of a positive $\pi$ operation, from its position in Region 3 near a separatrix curve, towards the center of Region 3, the following two things happen in the function:**

**(1) The small minimum increases as a minimum and moves downward across the $t$-axis of the function.**

**(2) The maximum moves downward and becomes tangent to the $t$-axis of the function, when the function reaches the negative $a_2$-axis.**

**In our theory, the above two statements imply, respectively, the following:**

**(1) There must be a** *minima zero-separatrix* **in the control space, between the critical-point separatrix curve and the negative $a_2$-axis.**

**(2) The negative $a_2$-axis must be in a** *maxima zero-separatrix.*

**The dual statements hold for the negative $\pi$ operations.**

---

The additional separatrices, mentioned in Problem 2 above, do not exist in catastrophe theory, but are needed by our requirement of compatibility with causal explanation.

The two problems, described above, are solved by going to Interactive Singularity Theory, as we do in the next section.

In particular, Interactive Singularity Theory supplies what we call the *Interactive Structure*, i.e., the *interaction* between the zero-structure and critical-point structure. Furthermore, Interactive Singularity Theory defines the crucial structure of the volumes that are bounded by the two separatrices; i.e., the volume structure that our theory claims is essential for defining spatial *morphology*.

## 11.3 Pair-Creation in the Interactive Unfoldings of $EA_3$ and $EA_{-3}$

This section shows that Interactive Singularity Theory realizes, in a powerful way, the pair-creation operations of the Process Grammar. This will lead to the discovery of additional interesting structure in the control space and separatrices for $EA_3$ and $EA_{-3}$.

We begin with further examination of the 3D control space for $EA_3$ using Fig 11.6. Consider the two points $p$ and $q$, in the $(a_1, a_2)$ plane at the bottom of the figure. While, points $p$ and $q$, in Fig 11.6, are in Region 3, to the left and right, respectively, of the negative $a_2$-axis, a crucial fact is this: They no longer have to be near the left and right separatrix curves, as was required by the point $f_3$ and its reflection, in the last section. The importance of this fact will be seen as we examine the functions associated with these two points.

Now, the vertical line through point $p$ will be called the $p$-line, and the vertical line through point $q$ will be called the $q$-line. The diagram shows representative functions on these two lines. All the functions have the same *critical-point* structure – that belonging to Region 3 – i.e., two minima and a maximum. For the functions on the $p$-line, the left minimum is smaller than the right minimum; and for functions on the $q$-line, the right minimum is smaller than the left minimum. (The reader will recall that the functions on a vertical line are simply raised and lowered versions of each other.)

Let us first consider the $p$-line: Going down this line, the qualitative structure of the functions in the successive *volumes* are illustrated, as follows: The top function shown is above all the zero-separatrix sheets, and therefore its maximum and both its minima are above its $t$-axis. The middle function shown on the $p$-line is below only the upper minima zero-separatrix sheet, and therefore the larger minimum in the function is now negative, but the smaller minimum and maximum are still positive. The lowest function shown on the $p$-line is below the entire zero-separatrix, and therefore both its minima are now negative, and also its maximum is now negative.

The functions down the $q$-line are reflected versions of the functions down the $p$-line.

Notice that, in both lines, we have not illustrated a function in the intermediate volume: below both minima zero-separatrices and above the maxima zero-separatrix. This is the triangular volume in the center of the figure, and we will call it the **enclosed volume**. A function in this volume is not the co-domain of a positive or negative pair-creation operation, whereas all the functions illustrated are. We will, for several sections, be concentrating on the positive and negative pair-creation operations.

Thus, let us study how their pair-creating effect is realized in this 3D control space.

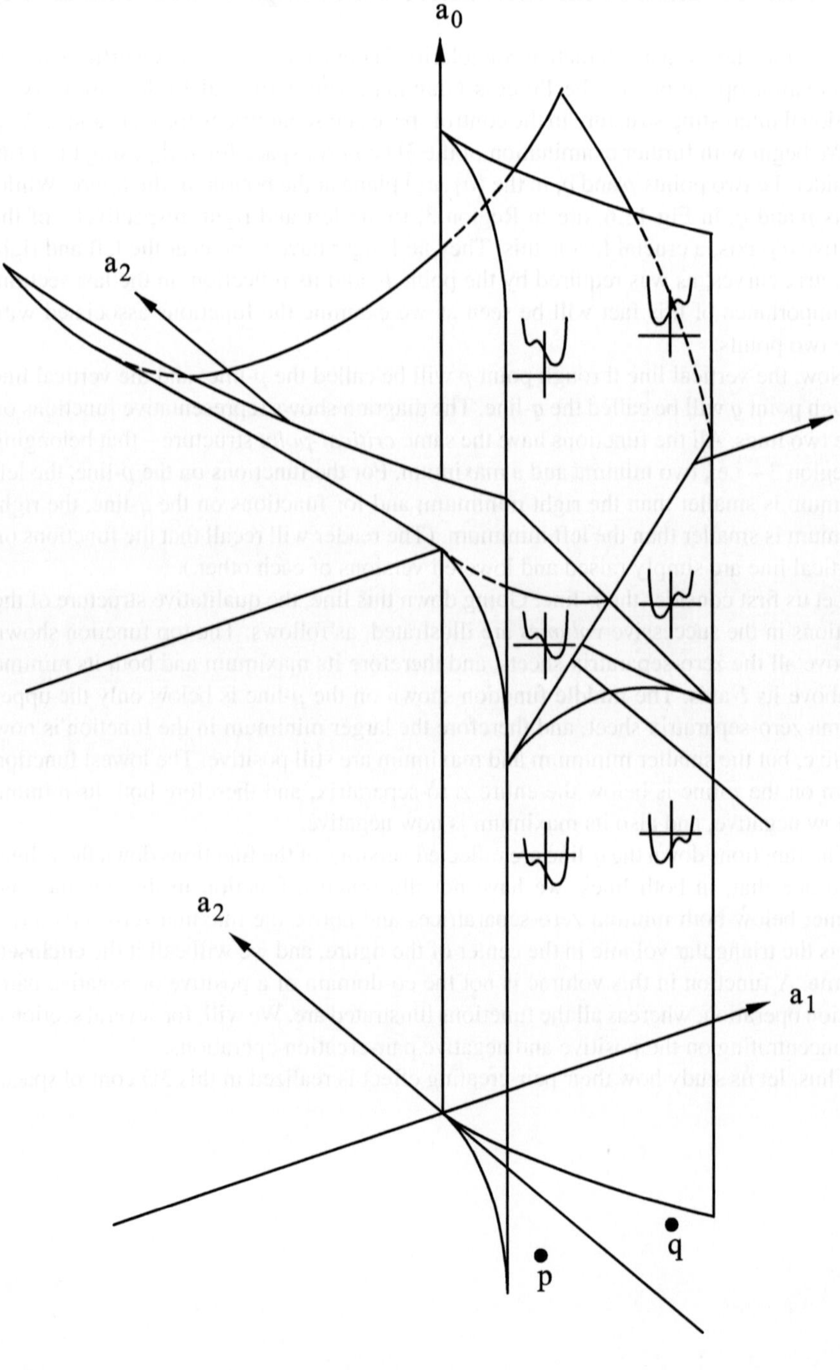


**Fig. 11.6.** The 3D control space for $EA_3$: Representative functions in vertical lines through points $p$ and $q$.

The next figure, Fig 11.7, shows the same three functions on the $p$-line. In addition, it shows three functions that are on the same level as those on the $p$-line, but are on the vertical line through a point $r$, which is to the left of the left-separatrix curve in the $(a_1, a_2)$ plane at the bottom of the figure.

Movement from a function on the $r$-line, to its corresponding function on the $p$-line, illustrates the use of the pair-creation operator. It is worth studying the logic of this in detail because it reveals new information about the *Interactive* Structure in the control space. To do so, it is necessary first to set up some new definitions:

**Definition 11.1.** *In the 3D control space for $EA_3$ or $EA_{-3}$, the separatrix sheet corresponding to one of the two side-curves of the separatrix of $A_3$ or $A_{-3}$ (i.e., not including the cusp point), will be called a* **side-sheet of the critical-point separatrix**.

**Definition 11.2.** *A line of intersection between a side-sheet from the critical-point separatrix and a sheet from the zero-separatrix will be called a* **bi-valent line**.

To emphasize: the term *bi-valent* refers to the fact that sheets of *two* types cross at the line: a sheet from the *critical-point* separatrix, and a sheet from the *zero* separatrix. The concept of a bi-valent line becomes crucial in understanding how the pair-creation operations are organized in the control space of the Interactive Unfolding of $EA_3$ and the control space of the Interactive Unfolding of $EA_{-3}$. To see this, observe first that each side-sheet intersects the zero-separatrix at two bi-valent lines, which we will define as follows:

**Definition 11.3.** *The* **bi-valent line, e,** *is where a side-sheet of the critical-point separatrix intersects a single sheet of the zero-separatrix.*

**Definition 11.4.** *The* **bi-valent line, g,** *is where a side-sheet of the critical-point separatrix is coincident with the curve where the minima zero-separatrix meets the maxima zero-separatrix.*

Fig 11.7 illustrates the bi-valent lines $e$ and $g$ on the left side-sheet of the control space of the Interactive Unfolding of $EA_3$. We can see, from this illustration, that the bi-valent lines decompose the side-sheet into three regions. Thus, we have:

**Definition 11.5.** *In the Interactive Unfoldings of $EA_3$ and $EA_{-3}$, the bi-valent lines divide each side-sheet of the critical-point separatrix into three regions, which will be called the* **bi-valent bounded regions** *of the side-sheet. They are defined as follows:*

*(1) The* **e-bounded region**: *this is bounded by the single bi-valent line $e$ and unbounded in the positive $a_0$-direction from that line.*
*(2) The* **e&g-bounded region**: *this is bounded by the bi-valent lines $e$ and $g$.*
*(3) The* **g-bounded region**: *this is bounded by the single bi-valent line $g$ and unbounded in the negative $a_0$-direction from that line.*

We will now study the logical role of the bi-valent lines, and the bi-valent bounded regions, with respect to the pair-creation operator. To illustrate, we will begin by looking at the effect of the operator in crossing each of the three successive bi-valent bounded regions down the left side-sheet of the control space of the Interactive Unfolding of $EA_3$.

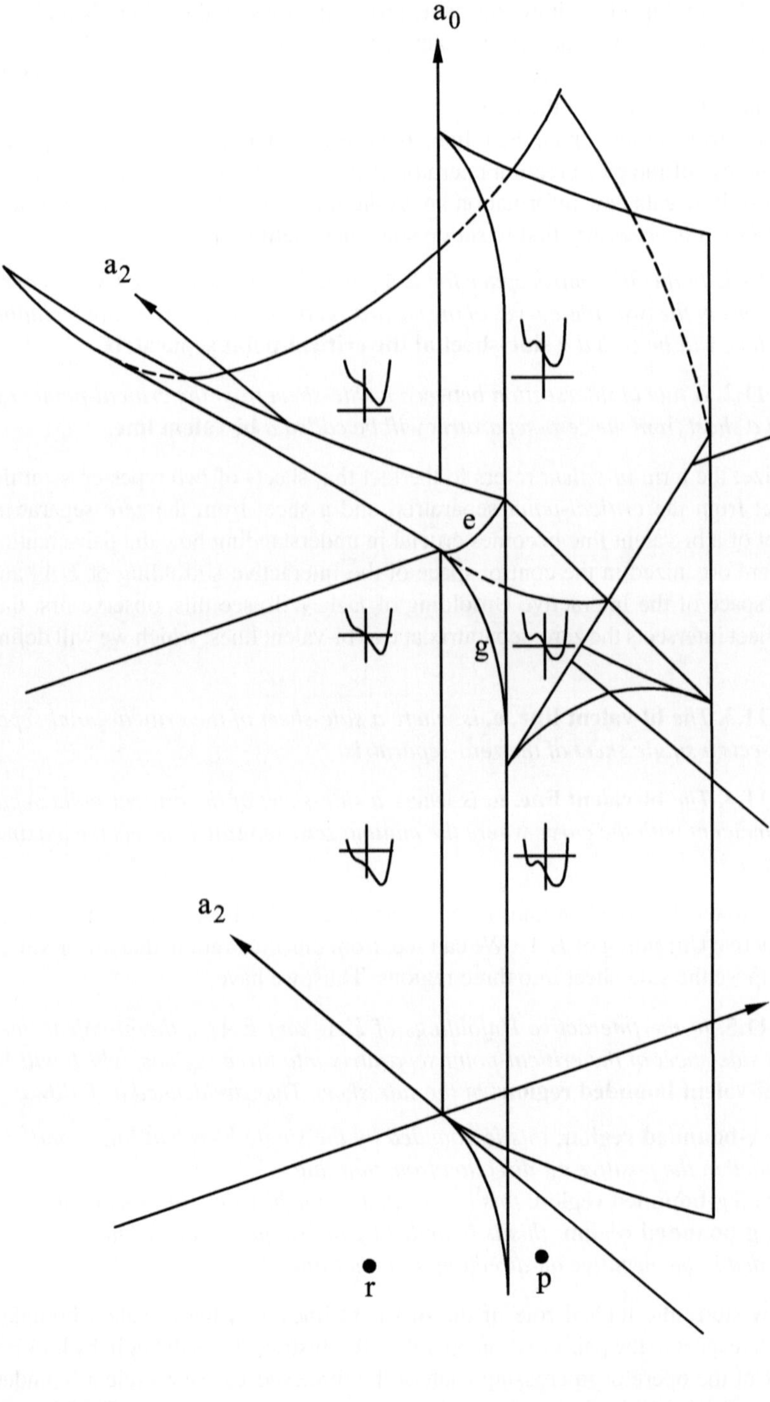


**Fig. 11.7.** The 3D control space for $EA_3$: Representative functions in vertical lines through points $r$ and $p$.

**(1) Crossing the e-bounded region.**
Consider the top function on the $r$-line, in Fig 11.7. It is in the volume that is around the critical-point separatrix and *above* the minima zero-separatrix. Therefore, the graph of this function is entirely above its own $t$-axis, and has only one critical point, a positive minimum. Moving the function horizontally, through the left side-sheet of the critical-point separatrix, to its corresponding function on the $p$-line, has the effect of creating the pair, $m^+M^+$, in the left side of the original minimum in the function. Therefore, this movement corresponds to the use of the pair-creation operation

$$\pi\emptyset^{\downarrow+} : \emptyset^{\downarrow+} \quad \longrightarrow \quad m^+M^+$$

on the left side of an anchor-extremum $m^+$. Furthermore, since $m^+$ is a compressive extremum, this crossing corresponds to application-type 1 in Table 4.2 (page 106).

**(2) Crossing the e&g-bounded region.**
Consider the middle function on the $r$-line, in Fig 11.7. It is in the volume that is around the critical-point separatrix and *below* the minima zero-separatrix. Therefore, the graph of this function is partially below its own $t$-axis, and has only one critical point, a negative minimum. Now, moving this function horizontally, through the left side-sheet of the critical-point separatrix, to its corresponding function on the $p$-line, also has the effect of creating the pair, $m^+M^+$, in the positive region of the left side of the original minimum in the function. Therefore, this movement corresponds to using the same pair-creation operation as the previous case:

$$\pi\emptyset^{\downarrow+} : \emptyset^{\downarrow+} \quad \longrightarrow \quad m^+M^+$$

However, in the present case, the anchor-extremum is $m^-$, not $m^+$, as previously. Furthermore, since $m^-$ is a penetrative extremum, and the pair-creation occurs in an opposite-sign tail of the extremum, this crossing must correspond to application-type 2 in Table 4.2 (page 106).

**(3) Crossing the g-bounded region.**
Consider the bottom function on the $r$-line, in Fig 11.7. Like the previous function, it is in the volume that is around the critical-point separatrix and *below* the minima zero-separatrix. Therefore, the graph of this function is again partially below its own $t$-axis, and has only one critical point, a negative minimum. However, moving this function horizontally, through the left side-sheet of the critical-point separatrix, to its corresponding function on the $p$-line, has the effect of creating the pair, $m^-M^-$, in the negative region of the left side of the original minimum in the function. Therefore, this movement corresponds to the use of a different pair-creation operation:

$$\pi\emptyset^{\downarrow-} : \emptyset^{\downarrow-} \quad \longrightarrow \quad m^-M^-$$

In this case, the anchor-extremum is $m^-$, as in the previous case, but the application-point has changed from an opposite-sign tail to a same-sign component of the anchor-extremum. We therefore conclude that crossing the left side-sheet, through the $g$-bounded region, corresponds to application-type 3 in Table 4.2 (page 106).

The above three cases have therefore demonstrated the following correspondence, which can easily be shown to be true for all the side-sheets of the Interactive Unfoldings of $EA_3$ and $EA_{-3}$:

**CORRESPONDENCE BETWEEN BI-VALENT BOUNDED REGIONS AND APPLICATION-TYPES**

| | | |
|---|---|---|
| **e-bounded region** | $\longleftrightarrow$ | **application-type 1** |
| **e&g-bounded region** | $\longleftrightarrow$ | **application-type 2** |
| **g-bounded region** | $\longleftrightarrow$ | **application-type 3** |

## 11.4 Logical Relation between Bi-Valent Lines and Pair-Creation Operations

It is important to show the logical role of the bi-valent lines in distinguishing between applications of the pair-creation operations. This logic will be illustrated by considering again the left side-sheet of the Interactive Unfolding of $EA_3$.

**The logic of crossing the bi-valent line $e$ from the $e$-bounded region to the $e$&$g$-bounded region**: We saw that crossing the line $e$ downward in the left side-sheet of Fig 11.7 preserves the pair-creation operation $\pi\emptyset^{\downarrow+}$, but changes its anchor-extremum from $m^+$ to $m^-$. This is explained by the fact that the bi-valent line $e$ is where a *single* sheet of the zero-separatrix intersects the side-sheet. Therefore crossing the line $e$ downwards must affect only one extremum in the function. However, there is only one extremum in the function on the side-sheet, and therefore this must be the anchor-extremum of the pair-creation operator. Thus crossing the bi-valent line $e$ downward, which means crossing the *zero-separatrix* on the side-sheet, must move the anchor-extremum in the function from above to below the $t$-axis of the function. Now, above the line $e$, the anchor-extremum, being $m^+$, is *compressive*. This is a general situation in the Interactive Unfoldings of $EA_3$ and $EA_{-3}$: In the $e$-bounded region, the anchor-extremum is compressive. Furthermore, crossing from the $e$-bounded region to the $e$&$g$-bounded region changes the compressive extremum to a penetrative extremum in exactly the way prescribed by the continuation operator $C$ of the Process Grammar, i.e., by continuing a compressive process till it penetrates. Notice also that, even though the sign of the anchor-extremum changes, in this way, one nevertheless preserves the sign of the application-point. This is achieved by ensuring that the application-point, which had the same sign as the compressive extremum prior to continuation, is on the *opposite-sign tail* of the penetrative extremum after continuation.

**The logic of crossing the bi-valent line $g$ from the $e$&$g$-bounded region to the $g$-bounded region**: We saw that crossing the line $g$ downward in the left side-sheet of Fig 11.7 changes the pair-creation operation from $\pi\emptyset^{\downarrow+}$ to $\pi\emptyset^{\downarrow-}$. This is explained by the following argument: The bi-valent line $g$ is where the minimum *and* maximum zero-separatrix sheets meet each other, and simultaneously touch the side-sheet. Note that, in a function, a point where a minimum and maximum meet, i.e., coalesce, is a critical inflection-point. Now, *every* function on the side-sheet has a critical inflection-point. However, the positions on the side-sheet where the minima zero-separatrix and maxima zero-separatrix meet each other must be where the critical inflection-point in a function has value *zero*. Thus, any function on the bi-valent line $g$ has an critical inflection-point of value zero. Furthermore, given the role of the $a_0$-axis in the control space, the critical inflection-point must become positive for functions on the side-sheet above the line $g$, and negative for functions on the side-sheet below the line $g$. Thus crossing the bi-valent line $g$ downward on the side-sheet must move the critical inflection-point in the function from above to below the $t$-axis of the function. However, the critical inflection-point in a function on the side-sheet is the point from which the maximum and minimum *bifurcate*, in crossing the side-sheet; i.e., undergo pair-creation. It is from this that we see why crossing the bi-valent line $g$, downward on the side-sheet, must correspond to *changing the operation* from a positive pair-creation operation $\pi\emptyset^{\downarrow +}$ to a negative pair-creation operation $\pi\emptyset^{\downarrow -}$.

## 11.5 The Configuration of Positive and Negative Pair-Creation Operations in the Interactive Unfoldings of $EA_3$ and $EA_{-3}$

All the positive and negative pair-creation operations share the same three application-types. These were listed in Table 4.2 (page 106), and were illustrated in Fig 4.8 (page 107). Now each of the positive and negative pair-creation operations realizes those three *application-types* in three specific *applications* which are different from the three applications of each of the other operations. The three applications for each operation were listed in Table 4.1 (page 105). We now need to see how the specific applications are distributed throughout the control spaces for $EA_3$ and $EA_{-3}$.

Recall that the three applications of an operation correspond to the three anchor-extrema allowable by that operation. However, an individual side-sheet permits only two alternative anchor-extrema. Therefore, the three application-types on a side-sheet cannot correspond to the three applications of a *single* operation. In fact, two of the regions on a side-sheet, the $e$-bounded region and the $e\&g$-bounded region, correspond to two of the applications of one operation $\pi_1$, and the remaining region, the $g$-bounded region, corresponds to an application of another operation $\pi_2$. As we shall see, the third application of $\pi_1$ corresponds to the $g$-bounded region on the opposite side-sheet of the dual of this Interactive Unfolding; and the other two applications of $\pi_2$ correspond to the $e$-bounded region and the $e\&g$-bounded region, also on the opposite side-sheet of the dual.

Section 11.4 developed the logic of how the bi-valent lines relate to the positive and negative pair-creation operations. Using this logic, we find that the configuration of the applications of these operations in the control spaces for $EA_3$ and $EA_{-3}$ are as shown in Table 11.1.

| $\pi$ | Application to Function | Region of Side-Sheet Crossed in Control Space |
|---|---|---|
| $\pi\emptyset^{\downarrow+}$ | | |
| | left side of $m^+$ | $e$-bounded region on left side-side sheet of $EA_3$. |
| | $+$ve left side of $m^-$ | $e\&g$-bounded region on left side-side sheet of $EA_3$. |
| | $+$ve right side of $M^+$ | $g$-bounded region on right side-side sheet of $EA_{-3}$. |
| $\pi\emptyset^{\uparrow+}$ | | |
| | right side of $m^+$ | $e$-bounded region on right side-side sheet of $EA_3$. |
| | $+$ve right side of $m^-$ | $e\&g$-bounded region on right side-side sheet of $EA_3$. |
| | $+$ve left side of $M^+$ | $g$-bounded region on left side-side sheet of $EA_{-3}$. |
| $\pi\emptyset^{\uparrow-}$ | | |
| | left side of $M^-$ | $e$-bounded region on left side-side sheet of $EA_{-3}$. |
| | $-$ve left side of $M^+$ | $e\&g$-bounded region on left side-side sheet of $EA_{-3}$. |
| | $-$ve right side of $m^-$ | $g$-bounded region on right side-side sheet of $EA_3$. |
| $\pi\emptyset^{\downarrow-}$ | | |
| | right side of $M^-$ | $e$-bounded region on right side-side sheet of $EA_{-3}$. |
| | $-$ve right side of $M^+$ | $e\&g$-bounded region on right side-side sheet of $EA_{-3}$. |
| | $-$ve left side of $m^-$ | $g$-bounded region on left side-side sheet of $EA_3$. |

**Table 11.1.** Correspondence between the different applications of a non-zero pair-creation operation and the regions crossed on the side-sheets.

To illustrate each of the entries in the right-hand column of this table, one can use Fig 11.6 for $EA_3$ and Fig 11.8 for the dual $EA_{-3}$. The following comment should be made about diagrammatic issues: In these two figures, as well as most figures in the book, the $e$-lines will be omitted because they are additional cusp-shaped lines that can visually confuse the reader, because several cusp-shaped lines emerge from the origin. However, their presence can be judged in the following way: Observe that, in both Fig 11.6 and Fig 11.8, there are two vertical straight lines at the front of the figure, which represent the cross-sections of the side-sheets. Each of these lines is divided into three intervals, which are obviously the cross-sections of the three bi-valent bounded regions of the side-sheet.

Now, in both figures, three functions are shown next to these three intervals. They represent the co-domains of the different applications of the pair-creation operations. Notice that the middle function, in each case, is in a triangle within the front planar cross-section of the figure. The triangle represents the *volume* of functions which have the same singularity-configuration as that of the function shown in the triangle. The vertical side-edge of the triangle is the cross-section of the $e\&g$ bounded region on the

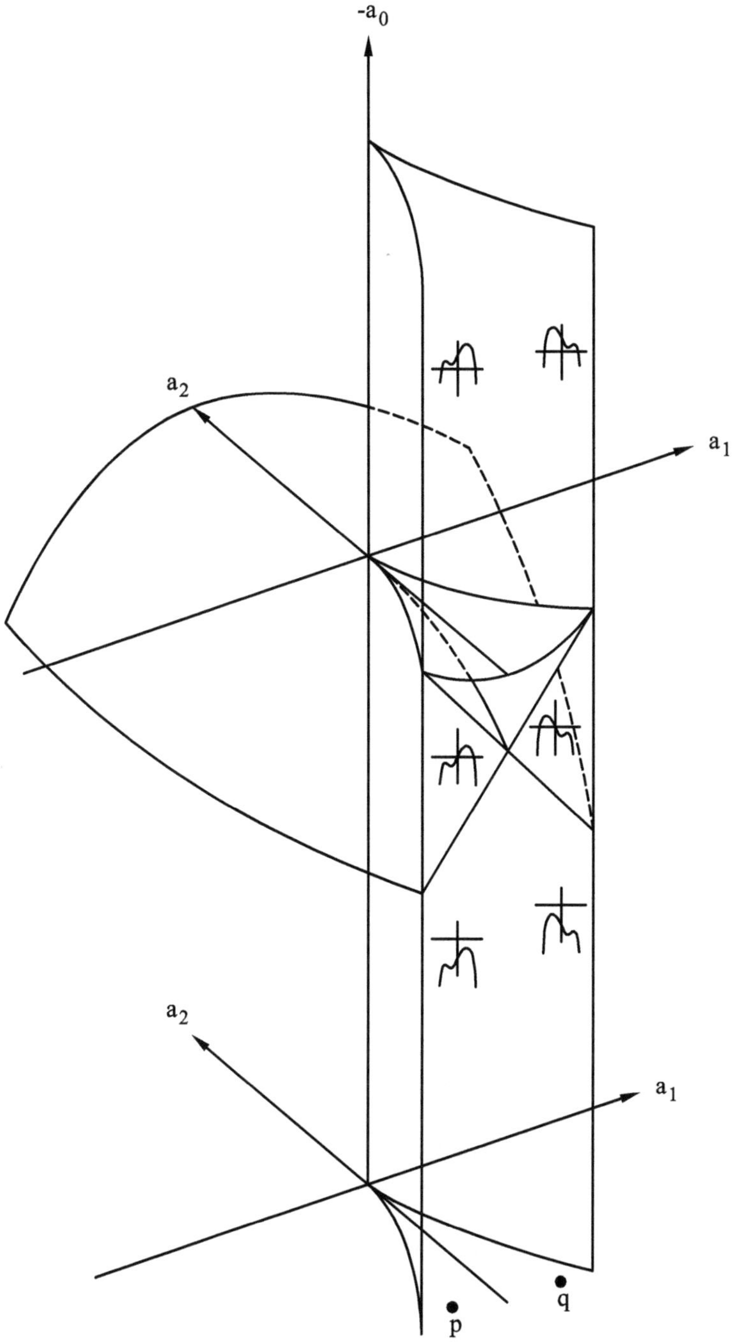


**Fig. 11.8.** The 3D control space for $EA_{-3}$: Representative functions in vertical lines through points *p* and *q*.

side-sheet. Finally, notice that the $e$-bounded region is *above* the $e\&g$ bounded region in $EA_3$ (Fig 11.6), but *below* the $e\&g$ bounded region in $EA_{-3}$ (Fig 11.8), because, in accord with Definition 11.5 (page 387), the $e$-bounded region corresponds to an interval on the *positive* $a_0$-axis. Conversely, the $g$-bounded region is *below* the $e\&g$ bounded region in $EA_3$, but *above* the $e\&g$ bounded region in $EA_{-3}$, because, in accord with Definition 11.5, the $g$-bounded region corresponds to an interval on the *negative* $a_0$-axis.

With these comments, the reader can now see how the right-hand entries in Table 11.1 are realized in Fig 11.6 for $EA_3$ and Fig 11.8 for $EA_{-3}$.

The following additional facts are worth observing with respect to these figures. First recall, from page 384, what we called Problem 2 in using the catastrophes $A_3$ and $A_{-3}$ to model the *pair-creation operations*; i.e., moving from the codomain of the operations to the $a_2$-axis. This problem lead to two conclusions, which were stated for $A_3$ as follows:

> Unlike Catastrophe Theory, in the 2D control space of the cusp catastrophe $A_3$:
>
> (1) There should be a *minima zero-separatrix,* between the critical-point separatrix curve and the negative $a_2$-axis.
>
> (2) The negative $a_2$-axis should be understood as a *maxima zero-separatrix.*

We can now see how Fig 11.6, for the Interactive Unfolding of $EA_3$, correctly possesses these valuable morphological properties: The 2D control space of the cusp catastrophe $A_3$ is the central horizontal plane in that figure. Now observe that this central plane cuts each of the two *minima zero-separatrix* sheets at a curve between the critical point separatrix and the negative $a_2$-axis. This corresponds to statement (1) above. Second, observe that the central plane contains the negative $a_2$-axis, which, as can be seen in this figure, is part of the *maxima zero-separatrix*. This corresponds to statement (2) above.

The dual statements define the same issue with the dual catastrophe $A_{-3}$, and show the correctness of the Interactive Unfolding of the Dual $EA_{-3}$ as seen in Fig 11.8.

## 11.6 The $c$-Preservation Surface

Section 10.7 defined the concept of $c$-preservation, i.e., the preservation of the curvature $c$ of an extremum through a bifurcation. In that section, the curvature $c$ being preserved was that of an extremum that was undergoing 3-fold bifurcation. In the present section and the following sections, we are going to be interested in preserving the curvature $c$ of the *anchor-extremum*, while 2-fold bifurcation is occurring elsewhere in the curvature function. Nevertheless, several aspects of the logic are the same as for the previous case.

However, in section 10.7, to help the reader, we illustrated the logic by considering a movement purely along the $a_2$-axis. In that case, we showed that $c$-preservation occurs along a line consisting of the positive $a_2$-axis, the origin, and the intersection line $d$, all raised by $c$ in the direction of the $a_0$-axis in the 3D control space. A more general version, of 3-fold bifurcation, would have meant that the starting point could have been anywhere over Region 1 in the 2D control subspace $(a_1, a_2)$, and the finishing point could have been anywhere over Region 3, provided that the extremum preserved the value $c$ through the bifurcation. This means that, instead of having a $c$-preservation *line*, we would have a $c$-preservation *surface*:

$c$-**PRESERVATION SURFACE**

**The $c$-preservation surface is the zero-separatrix in the 3D control space, raised by the amount $c$ in the direction of the $a_0$-axis.**

**A function at a point on this surface has an extremum of value $c$. This is because, in the function at the corresponding point on the zero-separatrix, the same extremum had value zero.**

Exactly the same surface is used for $c$-preservation of the anchor-extremum in pair-creation. The next three sections will illustrate this by giving quantitative examples of the pair-creation operator going, respectively, through the three successive bi-valent bounded regions down the left side-sheet of the critical-point separatrix for $EA_3$. That is, the three successive quantitative examples, in the next three sections, will illustrate the three application-types with the condition of $c$-preservation.

## 11.7 Quantitative Example of $\pi\emptyset^{\downarrow +}$ with a $c$-Preserved Anchor-Extremum $m^+$

This section will give a quantitative illustration of the operation

$$\pi\emptyset^{\downarrow +} \; : \; \emptyset^{\downarrow +} \quad \longrightarrow \quad m^+M^+$$

with anchor-extremum $m^+$. Therefore, the operation will be modeled within the Interactive Unfolding of $EA_3$ as going through the left side-sheet of the critical-point separatrix, above the bi-valent line $e$. It will thus be a quantitative illustration of application-type 1 of a non-zero pair-creation operation. Furthermore, we will require that the initial curvature $c$ of the anchor-extremum is preserved, which means that the function will move on the $c$-preservation surface.

Throughout the bifurcation, the anchor-extremum will have the following value:

$$\text{preserved curvature} \qquad c = 0.1$$

The quantitative illustration of the bifurcation will choose three successive points, whose positions with respect to the $(a_1, a_2)$-plane are shown as $p_1, p_2, p_3$, in Fig 11.9. The $(a_1, a_2)$ coordinates of these points are:

$$\begin{aligned} p_1 &= (-1.3, -0.7) \\ p_2 &= (-0.7607, \; -1.25) \quad \text{separatrix point} \\ p_3 &= (-0.4, \; -1.6) \end{aligned}$$

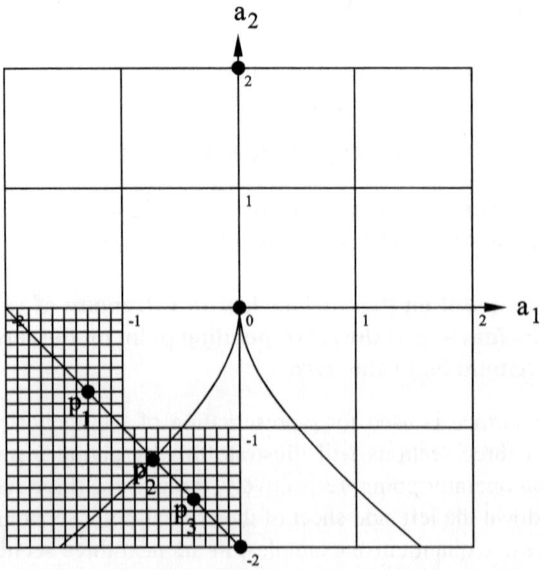


**Fig. 11.9.** The sequence of points $p_1, p_2, p_3$ are the $(a_1, a_2)$-positions of the three functions that we use to illustrate the pair-creation operation $\pi\emptyset^{\downarrow+}$ with anchor-extremum $m^+$.

Substituting these coordinates into the general form $F(t; 0, a_1, a_2) = t^4 + a_1 t + a_2 t^2$ of any function in the $(0, a_1, a_2)$-plane, we see that these three points in that plane are, respectively, the three functions:

$$p_1(t) = t^4 - 1.3t - 0.7t^2$$
$$p_2(t) = t^4 - 0.7607t - 1.25t^2$$
$$p_3(t) = t^4 - 0.4t - 1.6t^2$$

Furthermore, to fully illustrate the effect of the applied force on the shape, the sequence will be preceded by the most symmetric shape for the initial state of the anchor-extremum: that given by a point over the positive $a_2$-axis. For illustration, the $(a_1, a_2)$ coordinates of this point will be chosen to be $s_1 = (0,\ 2)$, shown as the central top point of Fig 11.9. Furthermore, the sequence will be followed by the most symmetric post-bifurcation state: that given by a point over the negative $a_2$-axis. For illustration, the $(a_1, a_2)$ coordinates of this point will be chosen to be $s_2 = (0,\ -2)$, shown as the central bottom point of Fig 11.9. Therefore, the entire evolution will be given by a sequence of five points, whose $(a_1, a_2)$-positions are $s_1, p_1, p_2, p_3, s_2$.

Given the preserved value $c$ for the anchor-extremum, these five $(a_1, a_2)$-positions determine five unique positions on the $c$-preservation surface. These positions will be labeled $\tilde{s}_1, \tilde{p}_1, \tilde{p}_2, \tilde{p}_3, \tilde{s}_2$.

Now, in the 3D control space for $EA_3$, let us take a $(a_1, a_2)$-*slice* (defined in the last chapter) containing these five points. Its corresponding critical-point manifold is shown

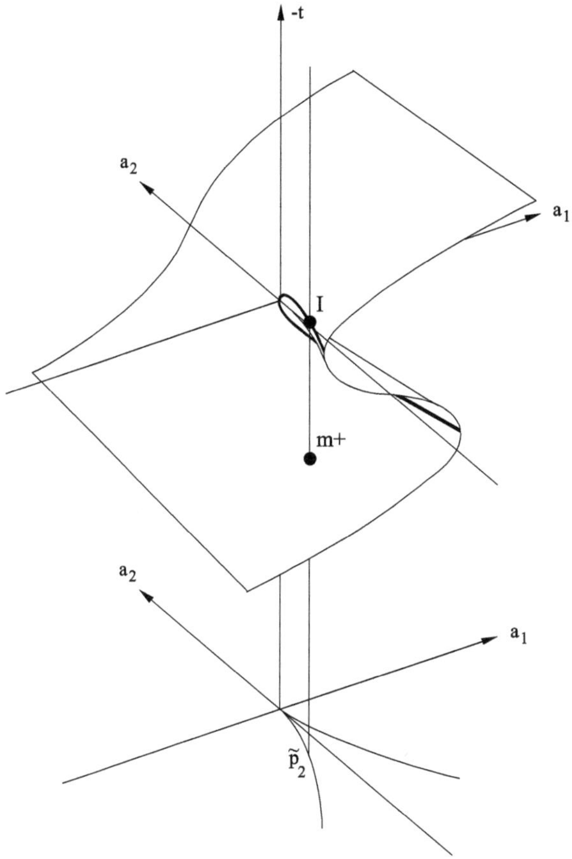


**Fig. 11.10.** The two positions $m^+$ and $I$, where the $t$-axis of the function at $\tilde{p}_2$ intersects the critical-point manifold.

in Fig 11.10. In this figure, the position $\tilde{p}_2$ is shown on the $(a_1, a_2)$-slice at the bottom of the figure. It is on the left separatrix-curve.

The vertical line through this point represents the $t$-axis of the function $\tilde{p}_2$, and the diagram shows that this axis intersects the critical-point manifold at two points. The lower point is the extremum $m^+$ in the function $\tilde{p}_2$, and the upper point $I$ is the critical inflection-point in that function. To understand this, the qualitative structure of the function is illustrated in Fig 11.11, and the critical points of the function are marked $m^+$ and $I$ on the $t$-axis. Taking this $t$-axis, together with those marked points, $m^+$ and $I$, and placing that $t$-axis vertically provides, in Fig 11.10, the vertical line shown with the marked points $m^+$ and $I$.

Now consider, in the control space, the two positions $\tilde{p}_1$ and $\tilde{p}_3$, which correspond to the two positions $p_1$ and $p_3$ in Fig 11.9. Qualitatively, their functions correspond to the two functions at the top of Fig 11.7 (page 388). As can be seen, the left top function in that figure has a single critical point, a positive minimum $m^+$; and the right top function

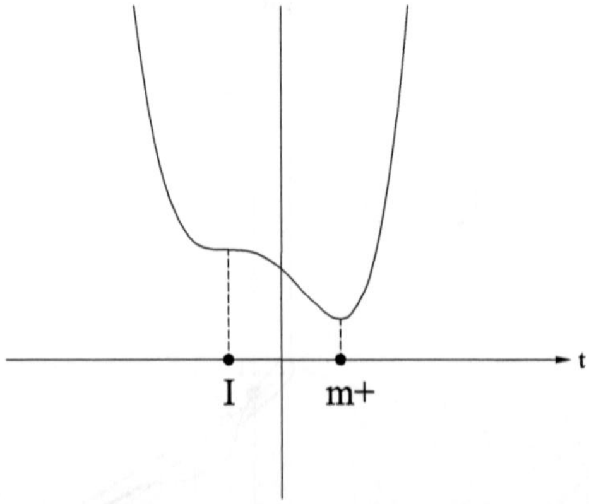


**Fig. 11.11.** Illustrating the qualitative structure of the function $\tilde{p}_2$.

has two additional critical points, a positive maximum $M^+$ and a positive minimum $m^+$.

Now the first position $\tilde{p}_1$ is a point to the left of the left separatrix-curve in Fig 11.10, and its vertical $t$-axis, would intersect the critical manifold at only one point, which would be on the lower surface. This intersection point is the extremum $m^+$ in the function, and is the anchor-extremum. On reaching position $\tilde{p}_2$ in Fig 11.10, the critical inflection-point $I$ appears in the function, because the $t$-axis hits the side of the upper fold. The point $I$ is a 2-fold degenerate critical point. Then, as we move past the left separatrix-curve, to position $\tilde{p}_3$ in Region 3, the point $I$ bifurcates into a positive minimum $m^+$ on the top surface, and a positive maximum $M^+$ on the middle surface. This therefore models the pair-creation operation $\pi\emptyset^{\downarrow +}$. Notice that the anchor-extremum persists throughout: It is the point $m^+$ on the lower surface where the $t$-axis of each of the three successive points, $\tilde{p}_1, \tilde{p}_2, \tilde{p}_3$, intersects the critical-point manifold.

The five figures, on the next five pages, i.e., Figs 11.12 – 11.16, show the functions at the five successive points, $\tilde{s}_1, \tilde{p}_1, \tilde{p}_2, \tilde{p}_3, \tilde{s}_2$, on the $c$-preservation surface. The text now continues in the captions of the five figures, which the reader is recommended to read.

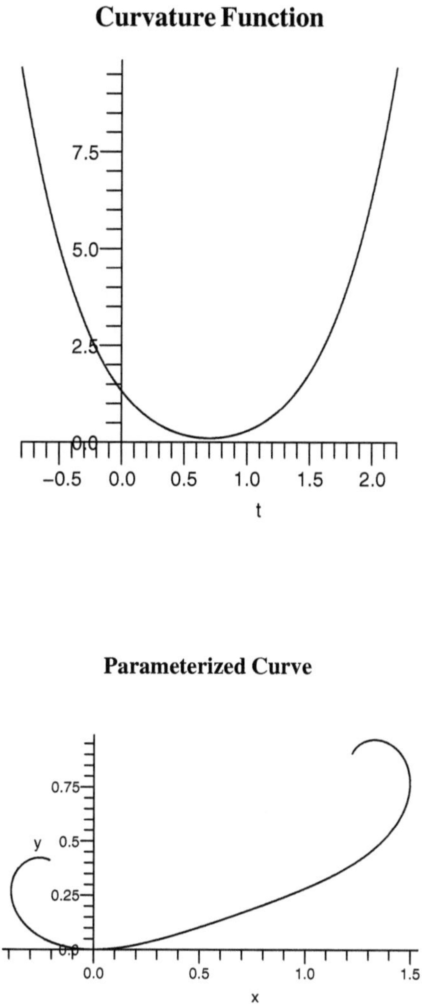


**Fig. 11.12.** First Pre-Bifurcation Stage: The curvature function at the first point $(a_0, a_1, a_2) = (0.1,\ 0,\ 2)$, in the control space of the Interactive Unfolding of $EA_3$, is the function $\tilde{s}_1(t) = t^4 + 2t^2 + 0.1$. However, we will translate the graph of this function along its $t$-axis, by the amount 0.7 in order to best align its single minimum $m^+$ with its position in the subsequent functions. Thus we obtain the function $\tilde{s}_1'(t) = (t - 0.7)^4 + 2(t - 0.7)^2 + 0.1$, which is shown as the curvature function above. The lower figure shows the resulting unit-speed parameterized curve, which has undergone rotation, due to the translation of the curvature function. The minimum $m^+$ is at the center of this curve, the flattest point, directly below its central position in the function above. This minimum has the preserved curvature value 0.1, and will be the anchor-extremum throughout the subsequent bifurcation.

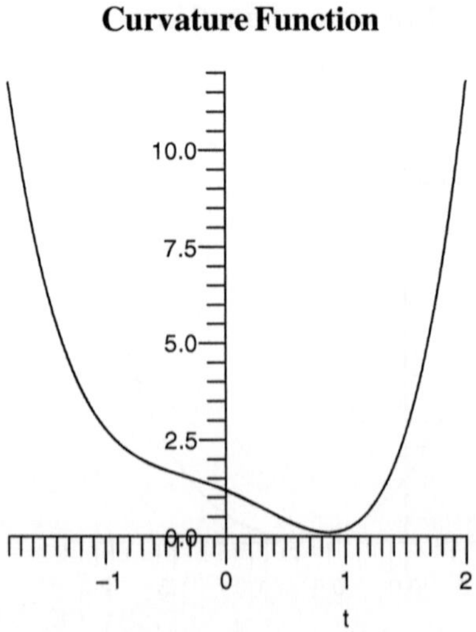


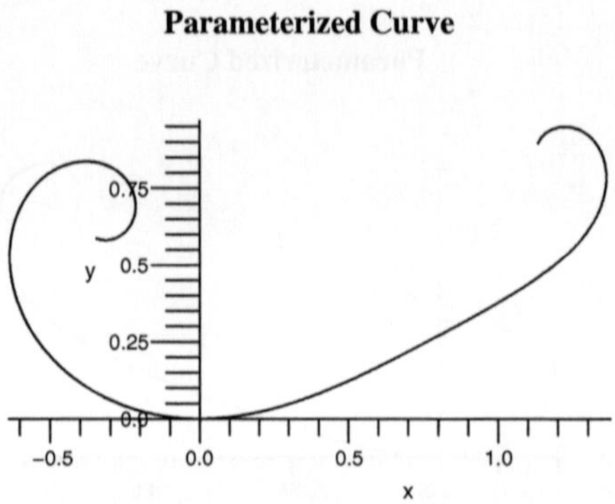


**Fig. 11.13.** Second Pre-Bifurcation Stage: The upper figure shows the curvature function $\tilde{p_1}(t) = t^4 + 1.18 - 1.3t - 0.7t^2$ at the position $(a_0, a_1, a_2) = (1.18, \ -1.3, \ -0.7)$ in the control space of the Interactive Unfolding of $EA_3$. It is on the $c$-preservation surface, to the left of the left side-sheet of the critical-point separatrix. Thus the height $a_0 = 1.18$ of the function in the 3D control space has now moved considerably above the preserved curvature value $0.1$. This is because the $c$-preservation sheet bends upward as one moves away from the $a_2$-axis in the space. Nevertheless, the curvature of the minimum is still $0.1$. The lower figure shows the resulting unit-speed parameterized curve. The flattening, which is beginning to occur in the *function* to the left of the minimum, corresponds to the broadening on the left side of the parameterized *curve*. The curve still has only one extremum: the minimum $m^+$, the flattest point on the right, corresponding to the minimum in the function above. This minimum is the anchor-extremum of the subsequent bifurcation, and exhibits the preserved curvature $0.1$.

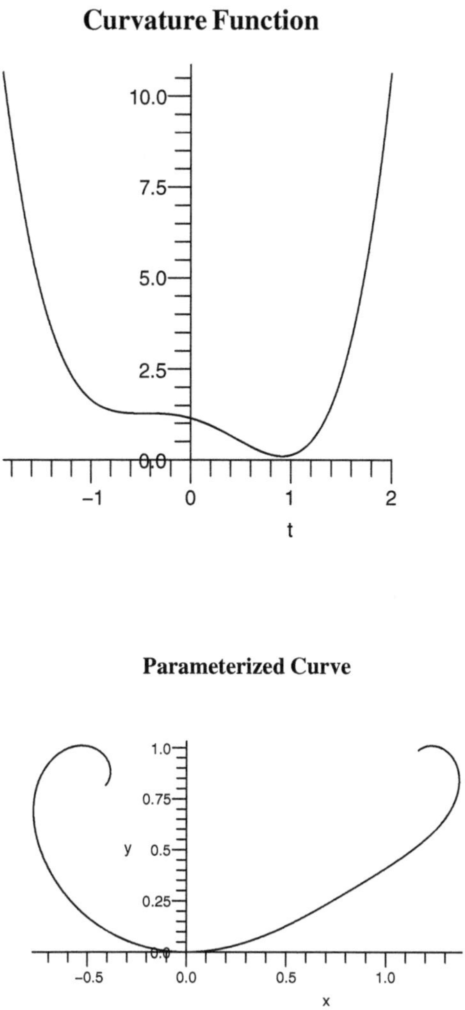


**Fig. 11.14.** Transition-Point: The upper figure shows the curvature function $\tilde{p_2}(t) = t^4 + 1.14 - 0.7607t - 1.25t^2$ at the position $(a_0, a_1, a_2) = (1.14, \ -0.7607, \ -1.25)$ in the control space of the Interactive Unfolding of $EA_3$. It is on the intersection between the $c$-preservation surface and the left side-sheet of the critical-point separatrix. Furthermore, this intersection is within the $e$-bounded region of the left side-sheet; thus making this crossing an example of application-type 1 of a non-zero pair-creation operation. Because we are now on the left side-sheet, a critical inflection-point has appeared in the left side of the curvature function shown above. The lower figure shows the resulting unit-speed parameterized curve. To the left of the origin on the curve, there is an approximately circular region, because the inflection-point, to which it corresponds in the function, creates an approximately constant region, i.e., thus giving a region of approximately constant curvature on the curve. The curve still has only one extremum: the minimum $m^+$, the flattest point on the right, corresponding to the minimum in the function above. This minimum is the anchor-extremum of the subsequent bifurcation, and exhibits the preserved curvature 0.1.

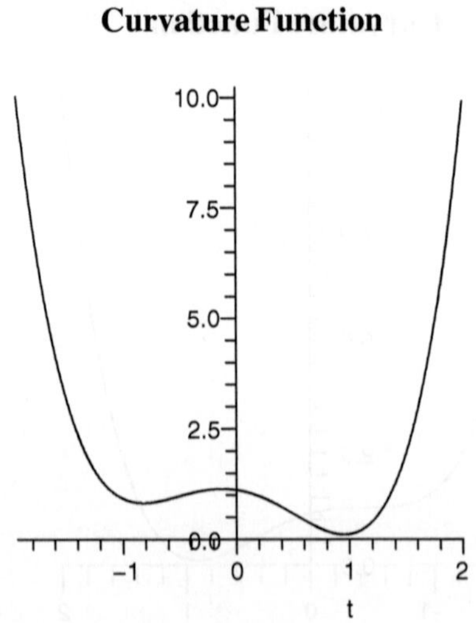


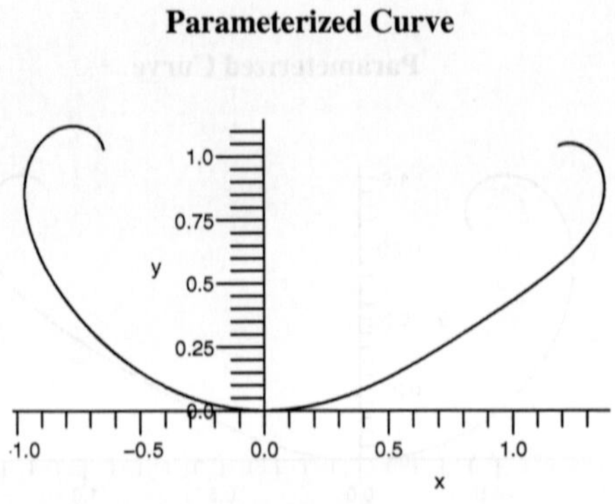


**Fig. 11.15.** First Post-Bifurcation Stage: The upper figure shows the curvature function $\tilde{p_3}(t) = t^4 + 1.1095 - 0.4t - 1.6t^2$ at the position $(a_0, a_1, a_2) = (1.1095, \ -0.4, \ -1.6)$ in the control space of the Interactive Unfolding of $EA_3$. It is on the $c$-preservation surface, and to the right of the left side-sheet of the critical-point separatrix. Therefore, having crossed the $e$-bounded region of the left side-sheet, the inflection-point in the previous function has bifurcated into a positive minimum $m^+$ and positive maximum $M^+$. Thus the function has the critical-point configuration $m^+M^+m^+$ of functions that are between the two side-sheets but above the zero-separatrix. The lower figure shows the resulting unit-speed parameterized curve. The new $m^+$ is the flattest point on the left side of the curve, and the new $M^+$ is the sharpest point at the bottom of the curve. As previously, the right side of the curve has a $m^+$ extremum, the flattest point, which is the anchor-extremum of the bifurcation, and which exhibits the preserved curvature 0.1 of the anchor.

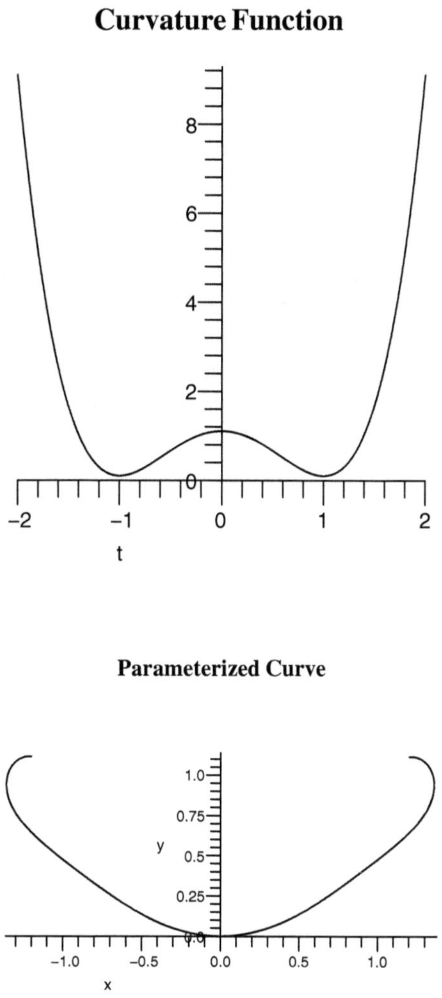


**Fig. 11.16.** Second Post-Bifurcation Stage: The upper figure shows the curvature function $\tilde{s_2}(t) = t^4 + 1.1 - 2t^2$ at the position $(a_0, a_1, a_2) = (1.1,\ 0,\ -2)$ in the control space of the Interactive Unfolding of $EA_3$. It is on the $c$-preservation surface, and over the negative $a_2$-axis. Therefore, it has the same critical-point configuration $m^+M^+m^+$ as the previous function, but is now symmetric. In the parameterized curve, shown below the function, the left minimum (flattest point on the left) now has the same curvature as the right minimum (flattest point on the right), which is the anchor-extremum, i.e., they both have curvature 0.1. The extremum at the bottom of the curve is $M^+$. The three extrema, $m^+$, $M^+$, $m^+$, on the curve, are positioned directly below the three corresponding critical points on the function. The vertical axis through the bottom extremum $M^+$, on the curve, is now the symmetry axis of the entire curve.

The five figures just given illustrate the shape evolution involved in the pair-creation operation $\pi\emptyset^{\downarrow+}$ with a $c$-preserved anchor-extremum $m^+$. The five parameterized curves, in those figures, are now shown together in Fig 11.17. Going through these curves successively, we see the following effects: In the transition from the first to the second curve, the left side of the curve is broadened. In both these curves, the left side remains a *spiral* (purely monotonically decreasing curvature). Then, in the transition from the second to the third curve, which is on the critical-point separatrix, part of that spiral becomes almost circular, due to the appearance of the 2-fold critical inflection-point; i.e., the beginning of pair-creation. Then, in the transition from the third to the fourth curve, the 2-fold critical-point bifurcation causes a flattening extremum $m^+$ to appear for the first time on the left, and a sharpening extremum $M^+$ to its right. The curve now has the extremum-structure $m^+M^+m^+$. Finally, in the transition from the fourth to the fifth curve, we obtain the fully symmetric double-shield structure. Notice that, because the pair $m^+M^+$ was created on a spiral side of a *compressive* anchor-extremum, this has been an example of application-type 1 of a non-zero pair-creation operation.

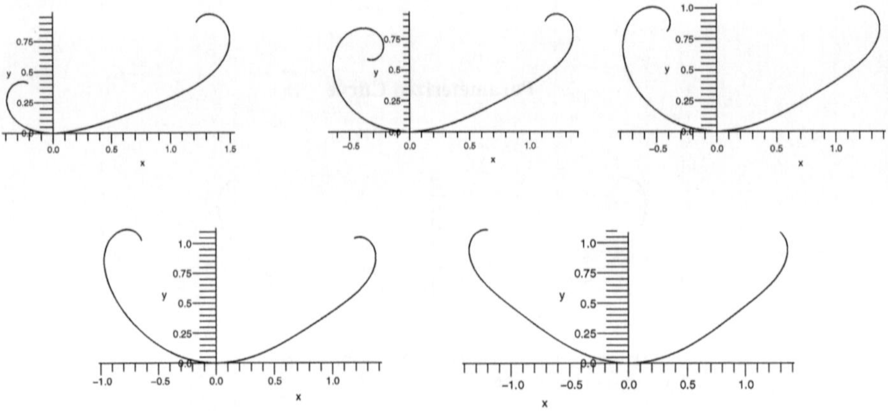


**Fig. 11.17.** The five successive curves used to illustrate the shape evolution involved in pair-creation $\pi\emptyset^{\downarrow+}$ with $c$-preserved anchor-extremum $m^+$.

It is worth reminding the reader of an important aspect of this bifurcation operation. The transition

$$\pi\emptyset^{\downarrow +} \ : \ m^+ \ \longrightarrow \ m^+M^+m^+$$

unambiguously means that the extremum $m^+$, in the domain, is *added to* by the pair $m^+M^+$, that appears in the co-domain. The operation is illustrated in Fig 11.18, where the $m^+$ extremum in the center of the first shape is *added to* by the two extrema to its left in the second shape.

Therefore, the **double-shield** structure $m^+M^+m^+$, of this result, is produced differently from the same double-shield structure $m^+M^+m^+$, that results from the operation $Bm^+$. The latter operation is illustrated in Fig 11.19. Here the double-shield is not created by *adding to* the initial $m^+$. The double-shield comes *entirely out of* the initial $m^+$. That is, the initial $m^+$ is pushed to both the left and the right, becoming the two sides of the resulting double-shield, and the $M^+$ emerges between them.

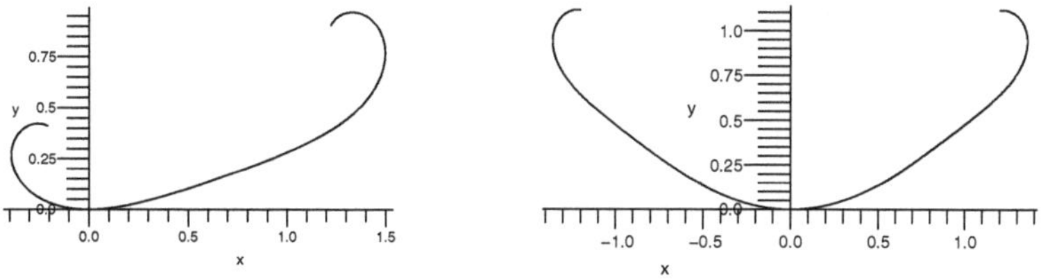


**Fig. 11.18.** Double-shield created by pair-creation $\pi\emptyset^{\downarrow +}$ applied to the left side of the initial $m^+$.

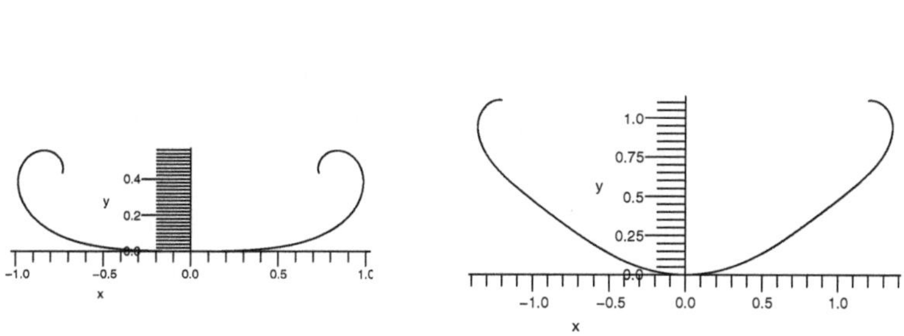


**Fig. 11.19.** Double-shield created by 3-fold bifurcation $Bm^+$ applied to the initial $m^+$.

## 11.8 Quantitative Example of $\pi\emptyset^{\downarrow+}$ with a $c$-Preserved Anchor-Extremum $m^-$

The previous section gave a quantitative example of the pair-creation operator $\pi$ applied *above* the bi-valent line $e$ on the left side-sheet for $EA_3$; that is, across the $e$-bounded region of that sheet. For the present section, we move down the same sheet, and what we do is apply the operator *below* the bi-valent line $e$ but *above* the bi-valent line $g$; that is, across the $e\&g$-bounded region of that sheet. Thus, the applied operation

$$\pi\emptyset^{\downarrow+} \;:\; \emptyset^{\downarrow+} \quad\longrightarrow\quad m^+M^+$$

is the same as in the previous section, but the *anchor-extremum* has been *changed* from $m^+$ to $m^-$. Therefore, this is an example of application-type 2 of a non-zero pair-creation operation.

We will require again that the initial curvature $c$ of the anchor-extremum is preserved. The chosen value will be:

$$\text{preserved curvature} \qquad c = -4.4$$

and thus the movement will take place on the corresponding $c$-preservation surface.

The quantitative illustration of the bifurcation will select three successive points, whose positions with respect to the $(a_1, a_2)$-plane are shown as $q_1$, $q_2$, $q_3$, in Fig 11.20. The $(a_1, a_2)$ coordinates of these points are:

$$\begin{aligned} q_1 &= (-3.4, \;-2.7) \\ q_2 &= (-2.9279972, \;-3.07) \quad \text{separatrix point} \\ q_3 &= (-2, \;-3.8) \end{aligned}$$

Substituting these coordinates into the general form $F(t; 0, a_1, a_2) = t^4 + a_1 t + a_2 t^2$ of any function in the $(0, a_1, a_2)$-plane, we see that these three points in that plane are, respectively, the three functions:

$$\begin{aligned} q_1(t) &= t^4 - 3.4t - 2.7t^2 \\ q_2(t) &= t^4 - 2.9279972t - 3.07t^2 \\ q_3(t) &= t^4 - 2t - 3.8t^2 \end{aligned}$$

Furthermore, to fully illustrate the effect of the applied force on the shape, the sequence will be preceded by the most symmetric shape for the initial state of the anchor-extremum: that given by a point below the positive $a_2$-axis. For illustration, the $(a_1, a_2)$ coordinates of this point will be chosen to be $v_1 = (0, \; 8)$.

However, unlike the previous section, the sequence cannot be followed by a symmetric version of the post-bifurcation state. To see this, first consider Fig 11.21, which shows the zero-separatrix. There are two surfaces that we need to consider: that marked 1, which is the upper minima sheet on the left, and that marked 2, which is the lower minima sheet on the left. Now the $c$-preservation surface will be a lowered version of the entire zero-separatrix surface. Let us denote the lowered version of sheet 1 by the symbol L1, and the lowered version of sheet 2 by the symbol L2. It is important for the reader to keep this notation in mind as we now return to Fig 11.7 (page 388), which shows both the zero-separatrix and the critical-point separatrix. The pair-creation operation we are

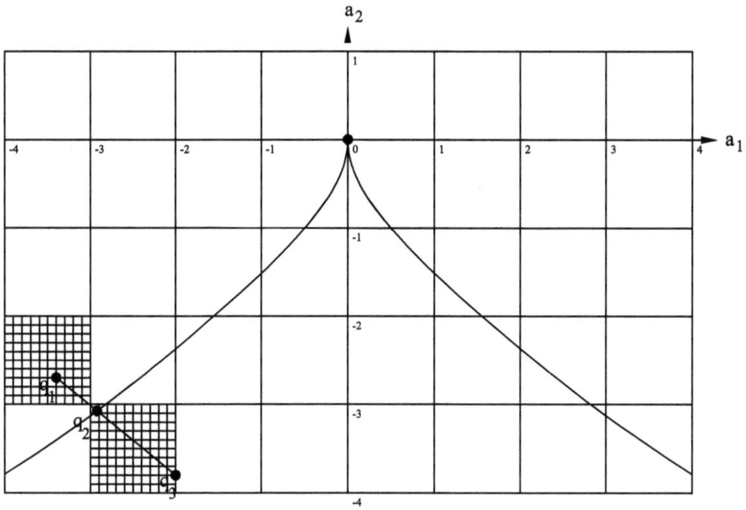


**Fig. 11.20.** The sequence of points $q_1, q_2, q_3$ are the $(a_1, a_2)$-positions of the three functions that we use to illustrate the pair-creation operator $\pi\emptyset^{\downarrow +}$ with anchor-extremum $m^-$.

applying corresponds to the transition, qualitatively, between the left and right graphs on the *middle* level of the figure. This creates, in the right graph, the left minimum and maximum (both positive). Again, we will be moving within the $c$-preservation surface, which is a lowered version of the zero-separatrix surface. Observe that L1, the lowered version of sheet 1, must necessarily cross the $e\&g$-bounded region of the left side-sheet, because the pair-creation operation, we are applying, crosses that region. Now, we can see from the structure of the diagram that a consequence, of the fact that L1 crosses the $e\&g$-bounded region, is that L1 must also cross sheet 2 in the zero-separatrix. But crossing sheet 2 would mean that the left minimum in the function (see the middle right graph in Fig 11.7) would descend through the $t$-axis of the function, i.e., one would be applying the operation $Cm^+$ to the left minimum in the function. This means that, after applying the pair-creation operation, i.e., crossing the left side-sheet, if one tried to go along the $c$-preservation sheet all the way to the center of the space, i.e., to make the function symmetric, we would need to apply the extra operation $Cm^+$. That is, there is no way of producing a symmetric function by using *only* the pair-creation operation through the $e\&g$-bounded region and maintaining $c$-preservation. This contrasts with the last section, where we were able to use *only* the pair-creation operation to produce a symmetric function; i.e., due to the fact that the $c$-preservation surface, in the previous case, was above the zero-separatrix.

Thus, in the present section, while we can begin the trajectory with a symmetric function, we cannot end the trajectory with a symmetric function, by applying only the pair-creation operator. As a consequence, in the present section, we have a sequence of

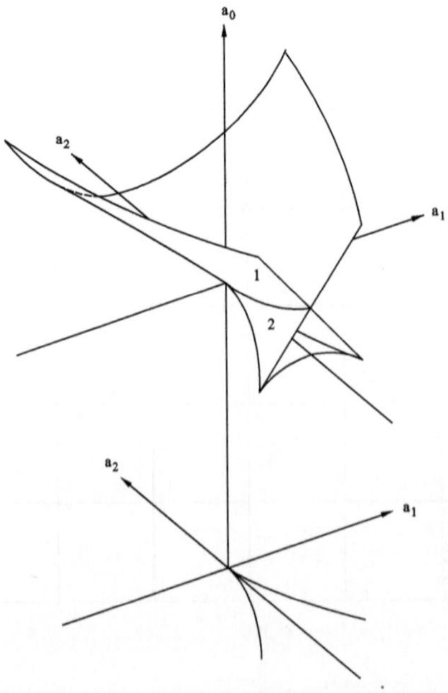


**Fig. 11.21.** Sheets 1 and 2 in the zero-separatrix.

*four* points, as opposed to *five* points, the total used in the previous section; i.e., the final symmetric function is missing.

The $(a_1, a_2)$-positions for the four points are $v_1$, $q_1$, $q_2$, $q_3$, defined above. Given the preserved value $c$ for the anchor-extremum, these four $(a_1, a_2)$-positions determine four unique positions on the corresponding $c$-preservation surface. The positions will be labeled $\tilde{v}_1$, $\tilde{q}_1$, $\tilde{q}_2$, $\tilde{q}_3$.

The four figures, on the next four pages, i.e., Figs 11.22 – 11.25, show the functions at the four successive points $\tilde{v}_1$, $\tilde{q}_1$, $\tilde{q}_2$, $\tilde{q}_3$, on the $c$-preservation surface. The text now continues in the captions of the four figures, which the reader is recommended to read.

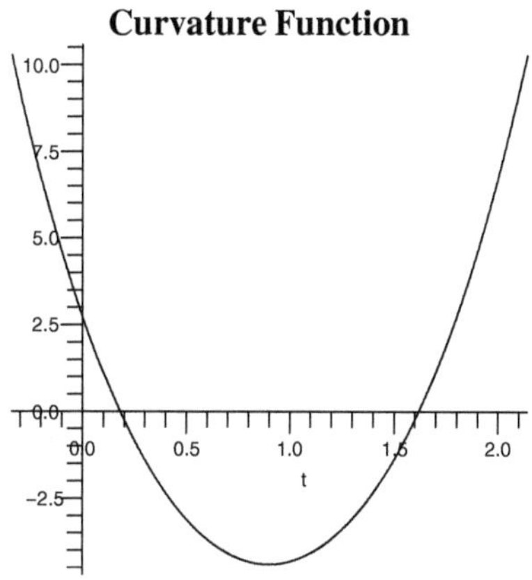


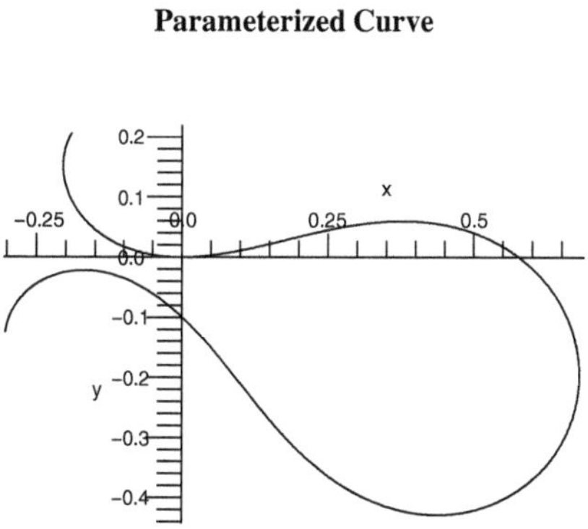


**Fig. 11.22.** First Pre-Bifurcation Stage: The curvature function at the first point $(a_0, a_1, a_2) = (-4.4,\ 0,\ 8)$, in the control space of the Interactive Unfolding of $EA_3$, is the function $\tilde{v}_1(t) = t^4 + 8t^2 - 4.4$. However, we will translate the graph of this function along its $t$-axis, by the amount 0.9, in order to best align its single minimum $m^-$ with its position in the subsequent functions. Thus we obtain the function $\tilde{v}_1'(t) = (t - 0.9)^4 + 8(t - 0.9)^2 - 4.4$, which is shown as the curvature function above. The lower figure shows the resulting unit-speed parameterized curve, which has undergone rotation, due to the translation of the curvature function. The left zero-value of the *function* corresponds to the point of zero-curvature directly below it in the upper branch of the *curve*. The pair-creation operator will be applied at a point to the left of this; i.e., in the *opposite-sign tail* of the anchor-extremum. The anchor-extremum, which is the minimum $m^-$, is at the center of the curve, the sharpest point on the right, and has the preserved curvature value $-4.4$.

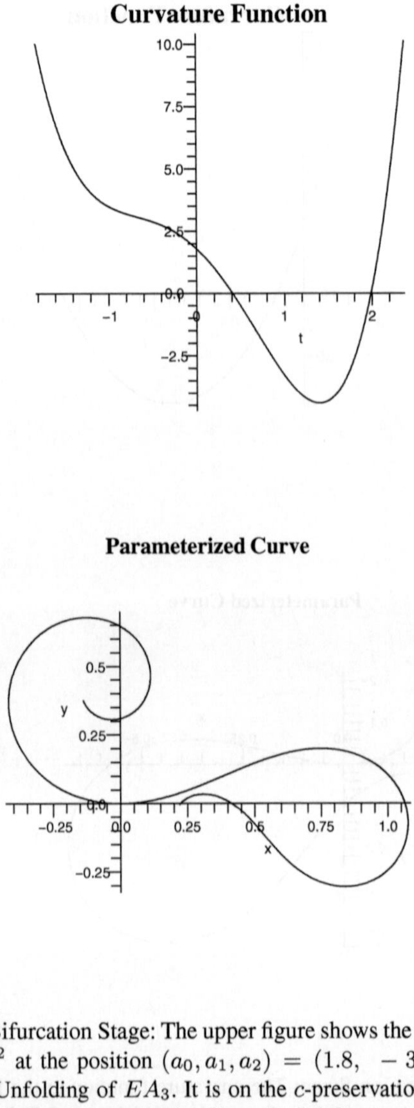


**Fig. 11.23.** Second Pre-Bifurcation Stage: The upper figure shows the curvature function $\tilde{q_1}(t) = t^4 + 1.8 - 3.4t - 2.7t^2$ at the position $(a_0, a_1, a_2) = (1.8, \ -3.4, \ -2.7)$ in the control space of the Interactive Unfolding of $EA_3$. It is on the $c$-preservation surface, to the left of the left side-sheet of the critical-point separatrix. Thus the height $a_0 = 1.8$ of the function in the 3D control space has now moved considerably above the preserved curvature value $-4.4$. This is because the $c$-preservation sheet bends upward as one moves away from the $a_2$-axis in the space. Nevertheless, the curvature of the minimum is still $-4.4$. The lower figure shows the resulting unit-speed parameterized curve. The left zero-value in the *function* is located directly above the point of zero-curvature in the upper branch of the *curve* below. The flattening, which is beginning to occur to the left of that zero-value in the *function*, corresponds to the broadening to the left of that point of zero-curvature in the parameterized *curve*; i.e., in the *opposite-sign tail* of the anchor-extremum. The anchor-extremum is still the only extremum in the curve. It is the $m^-$, the sharpest point on the right; and exhibits the preserved curvature $-4.4$.

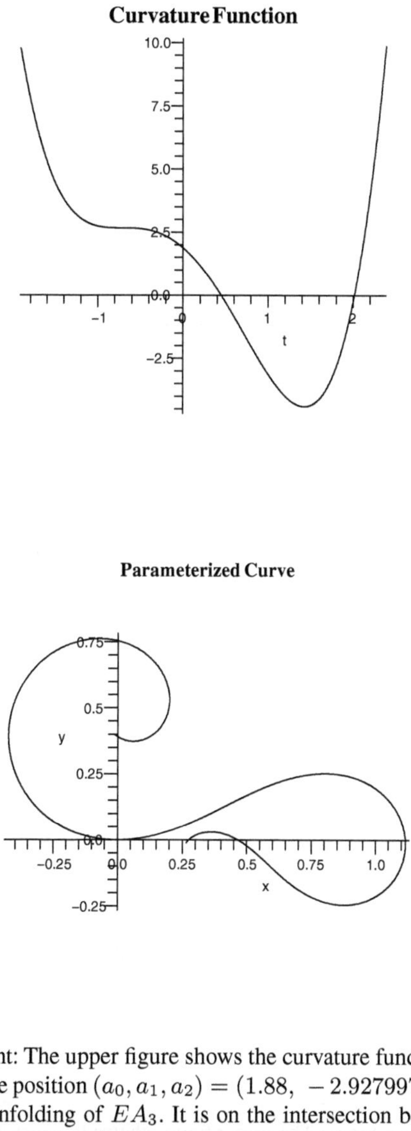


**Fig. 11.24.** Transition-Point: The upper figure shows the curvature function $\tilde{q}_2(t) = t^4 + 1.88 - 2.9279972t - 3.07t^2$ at the position $(a_0, a_1, a_2) = (1.88, \; -2.9279972, \; -3.07)$ in the control space of the Interactive Unfolding of $EA_3$. It is on the intersection between the $c$-preservation surface and the left side-sheet of the critical-point separatrix. Furthermore, it is in the $e\&g$ bounded region within the left side-sheet. Therefore, a critical inflection-point has appeared in the function to the left of its left zero-value; i.e., above the $t$-axis, whereas the anchor-extremum, the minimum, is below the $t$-axis. The lower figure shows the resulting unit-speed parameterized curve. The left zero-value in the *function* is located directly above the point of zero-curvature in the upper branch of the *curve*. The critical inflection-point, in the *function*, corresponds to the approximately circular region to the left of that point of zero-curvature in the parameterized *curve*; i.e., in the *opposite-sign tail* of the anchor-extremum. The anchor-extremum is still the only extremum in the curve. It is the $m^-$, the sharpest point on the right; and exhibits the preserved curvature $-4.4$.

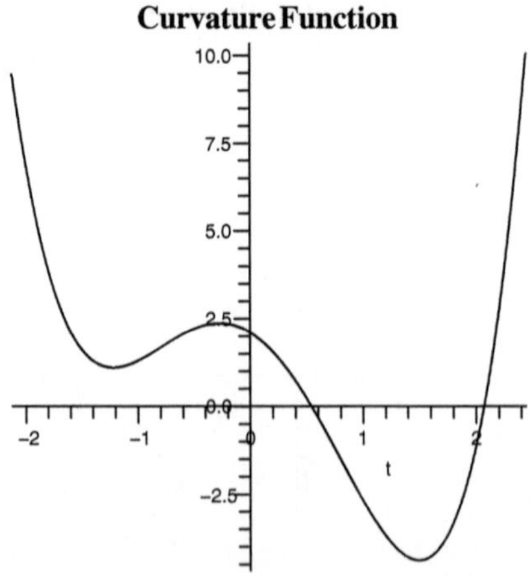


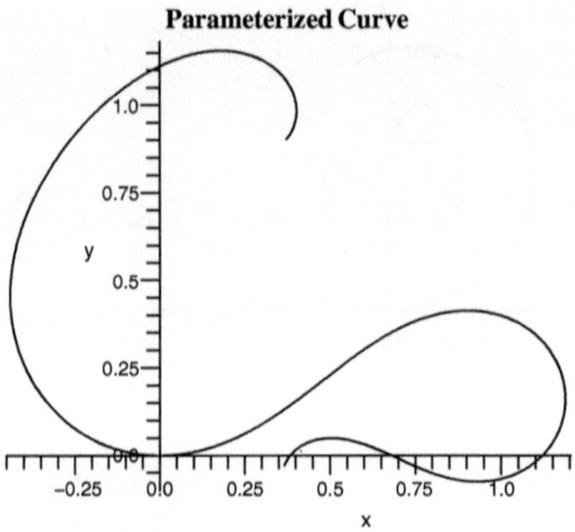


**Fig. 11.25.** Post-Bifurcation Stage: The upper figure shows the curvature function $\tilde{q}_3(t) = t^4 + 2.1 - 2t - 3.8t^2$ at the position $(a_0, a_1, a_2) = (2.1, \ -2, \ -3.8)$ in the control space of the Interactive Unfolding of $EA_3$. It is on the $c$-preservation surface, and to the right of the left side-sheet of the critical-point separatrix. Therefore, having crossed the $e\&g$-bounded region of the left side-sheet, the inflection-point in the previous function has bifurcated into a positive minimum $m^+$ and a positive maximum $M^+$; that is, both are above the $t$-axis in the function, whereas the anchor-extremum is below the $t$-axis. The lower figure shows the resulting unit-speed parameterized curve. The new $m^+$ is the flattest point on the left side of the curve; and the new $M^+$ is the sharpest point at the bottom-left of the curve; i.e., the downward protrusion. Both new extrema were created in the *opposite-sign tail* of the anchor-extremum. The anchor-extremum is still the $m^-$, the sharpest point on the right; and exhibits the preserved curvature $-4.4$.

The four figures, just given, illustrate the shape evolution involved when the *positive* pair-creation operation $\pi\emptyset^{\downarrow+}$ is applied in an opposite-sign tail of the *negative* extremum $m^-$, which is a penetrative extremum. Fig 11.26 shows the first and last curves in the sequence. The reader can see that the pair-creation operation has created the broad *shield* within the upper tail of the shape. This is an example of application-type 2 of a non-zero pair-creation operation. All labels have been omitted from the curves so that the reader can fully appreciate the powerful beauty of this transition.

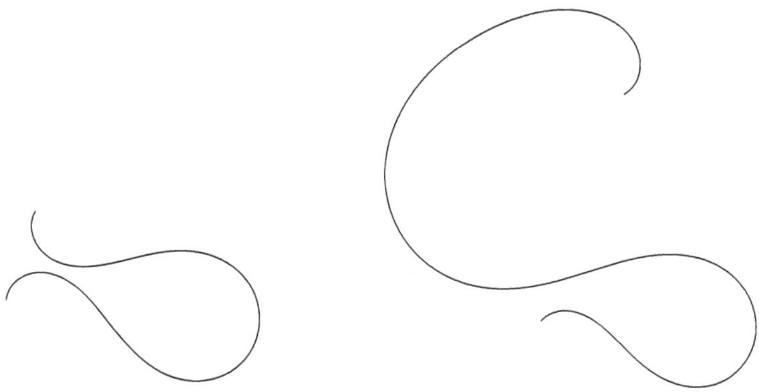

**Fig. 11.26.** The transition between the first and last curves in the previous sequence of figures.

## 11.9 Quantitative Example of $\pi\emptyset^{\downarrow-}$ with a $c$-Preserved Anchor-Extremum $m^-$

The previous section gave a quantitative example of the pair-creation operator $\pi$ applied *between* the bi-valent lines $e$ and $g$ on the left side-sheet for $EA_3$; that is, across the $e\&g$-bounded region of that sheet. For the present section, we move down the same sheet, and what we do is apply the operator *below* the bi-valent line $g$; that is, across the $g$-bounded region of that sheet. Thus, the applied operation is

$$\pi\emptyset^{\downarrow-} \; : \; \emptyset^{\downarrow-} \quad \longrightarrow \quad m^- M^-$$

Since the anchor-extremum $m^-$ has remained the same as the previous section, the application-point is no longer in the opposite-sign tail of that extremum but is now in the same-sign component of the extremum. Therefore, this is an example of application-type 3 of a non-zero pair-creation operation.

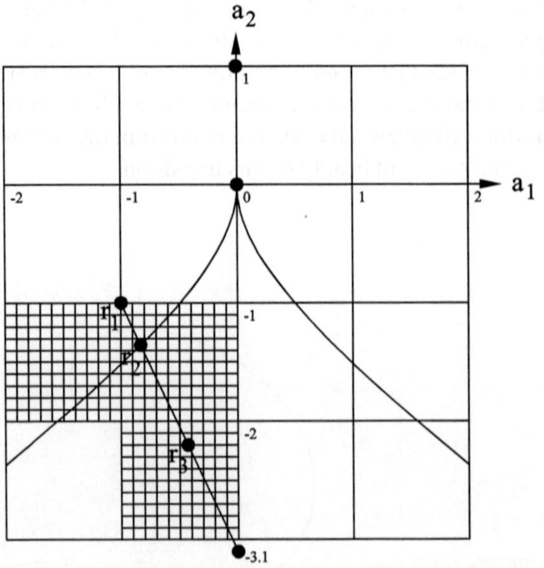


**Fig. 11.27.** The sequence of points $r_1, r_2, r_3$ are the $(a_1, a_2)$-positions of the three functions that we use to illustrate the pair-creation operator $\pi\emptyset^{\downarrow-}$ with anchor-extremum $m^-$.

We will require again that the initial curvature $c$ of the anchor-extremum is preserved. The chosen value will be:

$$\text{preserved curvature} \quad c = -2.5$$

and thus the movement will take place on the corresponding $c$-preservation surface.

The quantitative example of the bifurcation will select three successive points, whose positions with respect to the $(a_1, a_2)$-plane are shown as $r_1$, $r_2$, $r_3$, in Fig 11.27. The $(a_1, a_2)$ coordinates of these points are:

$$\begin{aligned} r_1 &= (-1, -1) \\ r_2 &= (-0.8538149, \ -1.35) \quad \text{separatrix point} \\ r_3 &= (-0.43, \ -2.2) \end{aligned}$$

Substituting these coordinates into the general form $F(t; 0, a_1, a_2) = t^4 + a_1 t + a_2 t^2$ of any function in the $(0, a_1, a_2)$-plane, we see that these three points in that plane are, respectively, the three functions:

$$\begin{aligned} r_1(t) &= t^4 - t - t^2 \\ r_2(t) &= t^4 - 0.8538149t - 1.35t^2 \\ r_3(t) &= t^4 - 0.43t - 2.2t^2 \end{aligned}$$

Furthermore, to fully illustrate the effect of the applied force on the shape, the sequence will be preceded by the most symmetric shape for the initial state of the anchor-extremum: that given by a point below the positive $a_2$-axis. For illustration, the $(a_1, a_2)$-coordinates of this point will be chosen to be $w_1 = (0,\ 1)$, shown as the central top point of Fig 11.27. Furthermore, unlike the previous section, the sequence can be followed by a fully symmetric post-bifurcation state, because the $c$-preservation surface, being below the bi-valent line $g$, can now reach the negative $a_2$-axis without crossing any sheet of the zero-separatrix. For illustration, the $(a_1, a_2)$-coordinates of the symmetric final function will be chosen to be $w_2 = (0,\ -3.1)$, shown as the central bottom point of Fig 11.27. Therefore, the entire evolution will be given by a sequence of five points, whose $(a_1, a_2)$-positions are $w_1, r_1, r_2, r_3, w_2$.

Given the preserved value $c$ for the anchor-extremum, these five $(a_1, a_2)$-positions determine five unique positions on the corresponding $c$-preservation surface. These positions will be labeled $\tilde{w}_1, \tilde{r}_1, \tilde{r}_2, \tilde{r}_3, \tilde{w}_2$.

The five figures, on the next five pages, i.e., Figs 11.28 – 11.32, show the functions at the five successive points, $\tilde{w}_1, \tilde{r}_1, \tilde{r}_2, \tilde{r}_3, \tilde{w}_2$, on the $c$-preservation surface. The text now continues in the captions of the five figures, which the reader is recommended to read.

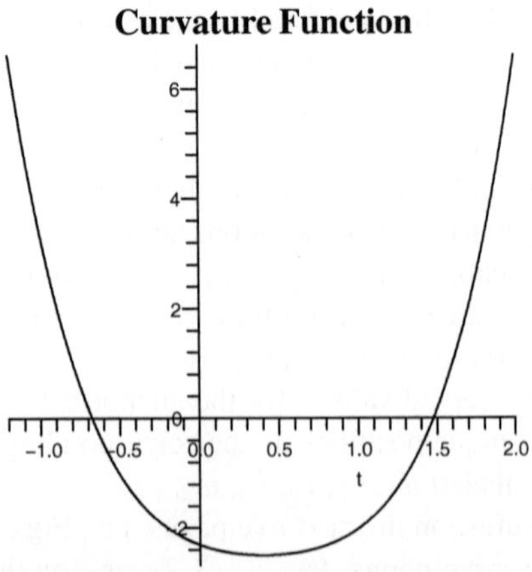


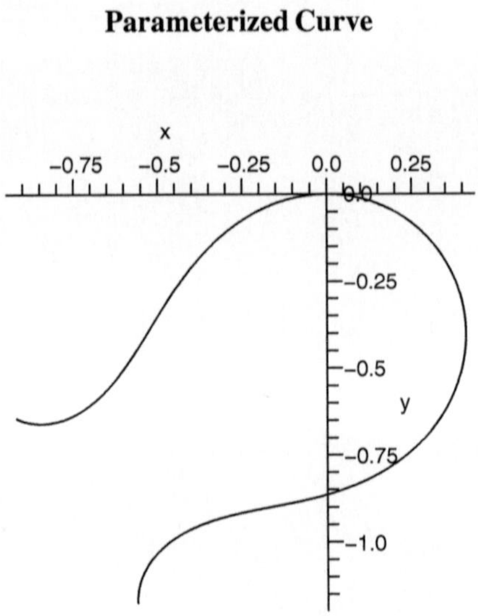


**Fig. 11.28.** First Pre-Bifurcation Stage: The curvature function at the first point $(a_0, a_1, a_2) = (-2.5,\ 0,\ 1)$, in the control space of the Interactive Unfolding of $EA_3$, is the function $\tilde{w}_1(t) = t^4 + t^2 - 2.5$. However, we will translate the graph of this function along its $t$-axis, by the amount $0.4$, in order to best align its single minimum $m^-$ with its position in the subsequent functions. Thus we obtain the function $\tilde{w}_1'(t) = (t - 0.4)^4 + (t - 0.4)^2 - 2.5$, which is shown as the curvature function above. The lower figure shows the resulting unit-speed parameterized curve, which has undergone rotation, due to the translation of the curvature function. The minimum $m^-$ is at the center of this curve, the sharpest point. It has the preserved curvature value $-2.5$, and will be the anchor-extremum throughout the subsequent bifurcation.

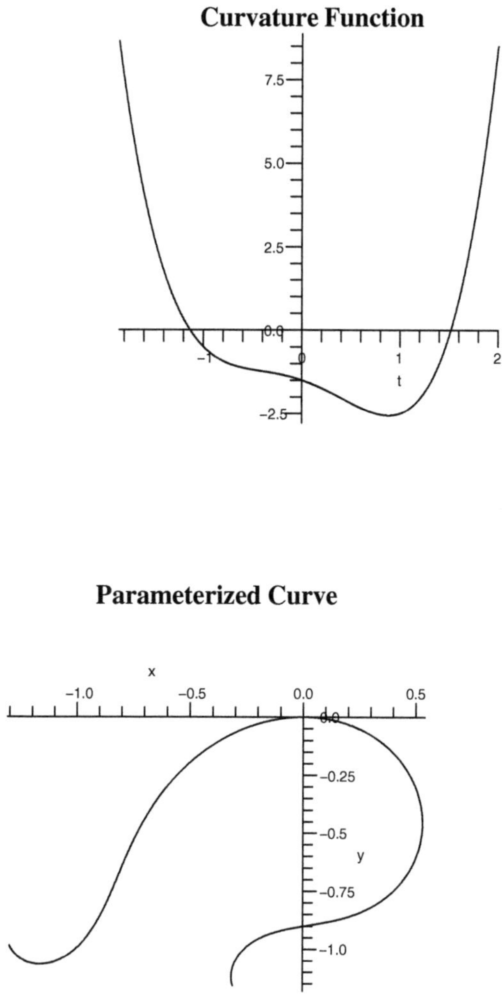


**Fig. 11.29.** Second Pre-Bifurcation Stage: The upper figure shows the curvature function $\tilde{r_1}(t) = t^4 - 1.5 - t - t^2$ at the position $(a_0, a_1, a_2) = (-1.5, \ -1, \ -1)$ in the control space of the Interactive Unfolding of $EA_3$. It is on the $c$-preservation surface, to the left of the left side-sheet of the critical-point separatrix. Thus the height $a_0 = -1.5$ of the function in the 3D control space has now moved considerably above the preserved curvature value $-2.5$. This is because the $c$-preservation sheet bends upward as one moves away from the $a_2$-axis in the space. Nevertheless, the curvature of the minimum is still $-2.5$. The lower figure shows the resulting unit-speed parameterized curve. The flattening, which is beginning to occur in the *function* between the left zero-value and the minimum, corresponds to the broadening in the upper branch of the parameterized *curve*, between the left point of zero curvature and the anchor-extremum; i.e., in the *same-sign component* of the anchor-extremum. The anchor-extremum is still the only extremum in the curve. It is the $m^-$, the sharpest point on the right; and exhibits the preserved curvature $-2.5$.

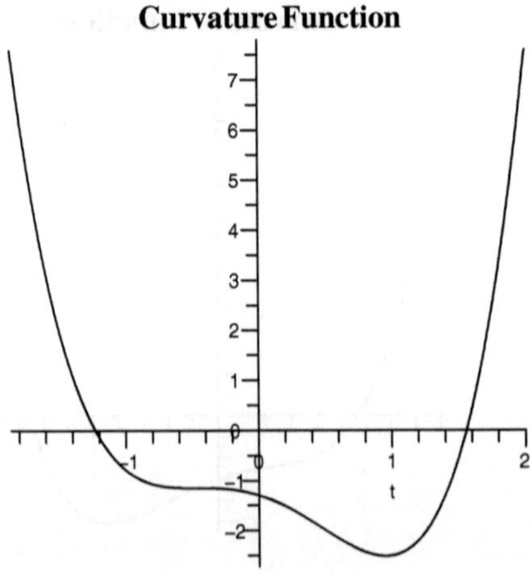


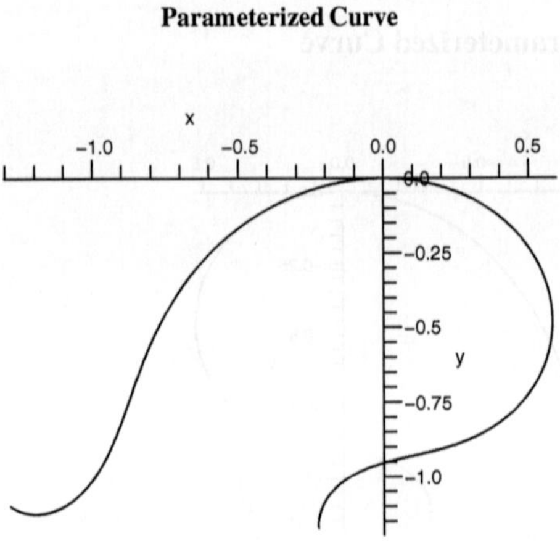


**Fig. 11.30.** Transition-Point: The upper figure shows the curvature function $\tilde{r_2}(t) = t^4 - 1.3 - 0.8538149t - 1.35t^2$ at the position $(a_0, a_1, a_2) = (-1.3, \ -0.8538149, \ -1.35)$ in the control space of the Interactive Unfolding of $EA_3$. It is on the intersection between the $c$-preservation surface and the left side-sheet of the critical-point separatrix. Furthermore, this part of the intersection is within the $g$-bounded region of the left side-sheet; thus making this crossing an example of application-type 3 of a non-zero pair-creation operation. Because we are now on the left side-sheet, a critical inflection-point has appeared in the left side of the curvature function shown above. The lower figure shows the resulting unit-speed parameterized curve. Since the critical inflection-point occurs in the function *between* the left zero-value and the minimum, it corresponds to the approximately circular region in the upper branch of the parameterized curve *between* the left point of zero curvature and the anchor-extremum; i.e., in the *same-sign component* of the anchor-extremum. The anchor-extremum is still the only extremum in the curve. It is the $m^-$, the sharpest point on the right; and exhibits the preserved curvature $-2.5$.

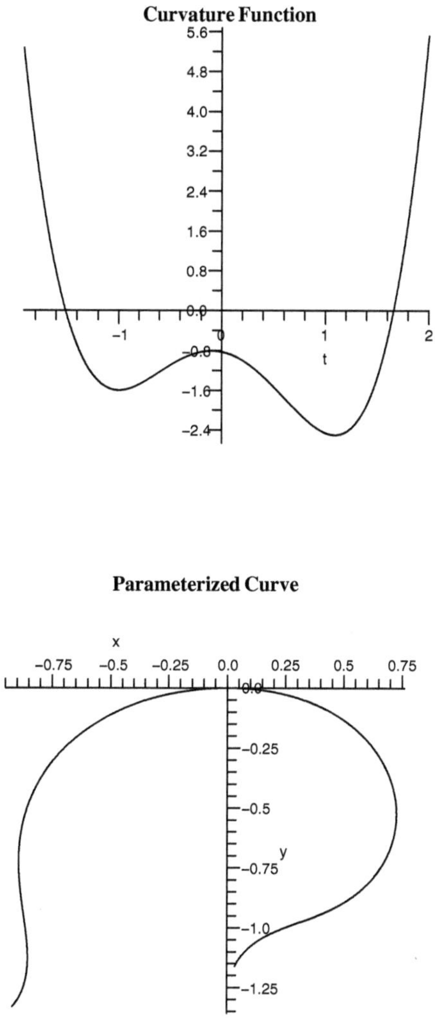


**Fig. 11.31.** First Post-Bifurcation Stage: The upper figure shows the curvature function $\tilde{r_3}(t) = t^4 - 0.83 - 0.43t - 2.2t^2$ at the position $(a_0, a_1, a_2) = (-0.83, -0.43, -2.2)$ in the control space of the Interactive Unfolding of $EA_3$. It is on the $c$-preservation surface, and to the right of the left side-sheet of the critical-point separatrix. Therefore, having crossed the $g$-bounded region of the left side-sheet, the critical inflection-point in the previous function has bifurcated into a negative minimum $m^-$ and negative maximum $M^-$. Thus the function has the critical-point configuration $m^-M^-m^-$ possessed by functions that are between the two side-sheets but below the zero-separatrix. The lower figure shows the resulting unit-speed parameterized curve. The new $m^-$ is the sharpest point on the left side of the curve, and the new $M^-$ is the flattest point at the top of the curve. Both new extrema were created in the *same-sign component* of the anchor-extremum. The anchor-extremum is still the $m^-$, the sharpest point on the right; and exhibits the preserved curvature $-2.5$.

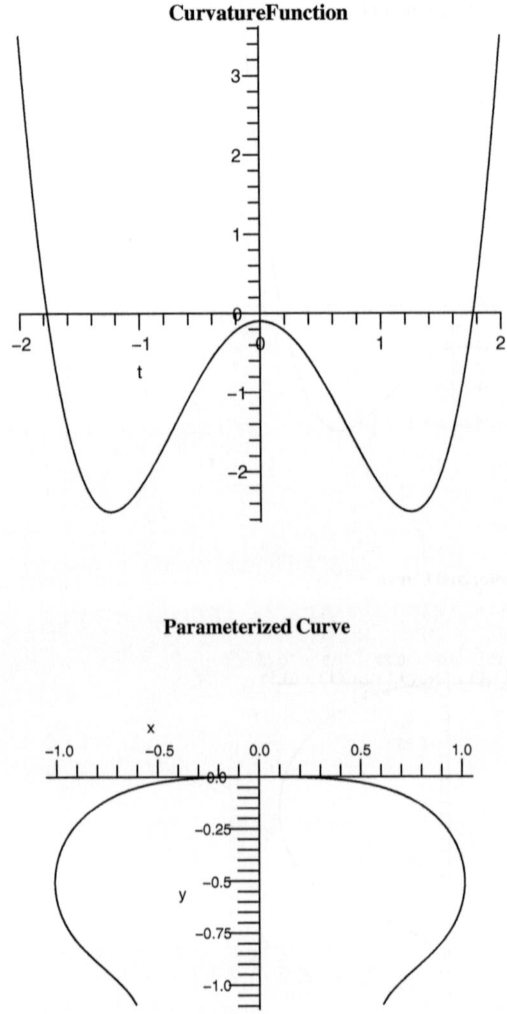


**Fig. 11.32.** Second Post-Bifurcation Stage: The upper figure shows the curvature function $\tilde{\omega_2}(t) = t^4 - 0.1 - 3.1t^2$ at the position $(a_0, a_1, a_2) = (-0.1,\ 0,\ -3.1)$ in the control space of the Interactive Unfolding of $EA_3$. It is on the c-preservation surface, and below the negative $a_2$-axis. Therefore, it has the same critical-point configuration $m^- M^- m^-$ as the previous function, but is now symmetric. In the parameterized curve, shown below the function, the left minimum (sharpest point on the left) now has the same curvature as the right minimum (sharpest point on the right), which is the anchor-extremum, i.e., they both have curvature $-2.5$. The extremum at the top of the curve is $M^-$. The curve is now a fully symmetric complete bay.

The five figures just given illustrate the shape evolution involved in the pair-creation operation $\pi\emptyset^{\downarrow-}$ with a $c$-preserved anchor-extremum $m^-$. The five parameterized curves, in those figures, are now shown together in Fig 11.33. Going through these curves successively, we see the following effects: In the transition from the first to the second curve, the upper side of the curve is broadened. In both these curves, the left side remains a spiral (purely monotonically increasing curvature). Then, in the transition from the second to the third curve, which is on the critical-point separatrix, part of that spiral becomes almost circular, due to the appearance of the critical inflection-point; i.e., the beginning of pair-creation. Then, in the transition from the third to the fourth curve, the 2-fold critical-point bifurcation causes a sharpening extremum $m^-$ to appear for the first time on the left of the shape, and a flattening extremum $M^-$ towards the top. The curve now has the extremum structure $m^-M^-m^-$. Finally, in the transition from the fourth to the fifth curve, we obtain a fully symmetric complete bay. Notice that, because the created pair $m^-M^-$ are the same sign as the penetrative anchor-extremum $m^-$, this has been an example of application-type 3 of a non-zero pair-creation operation.

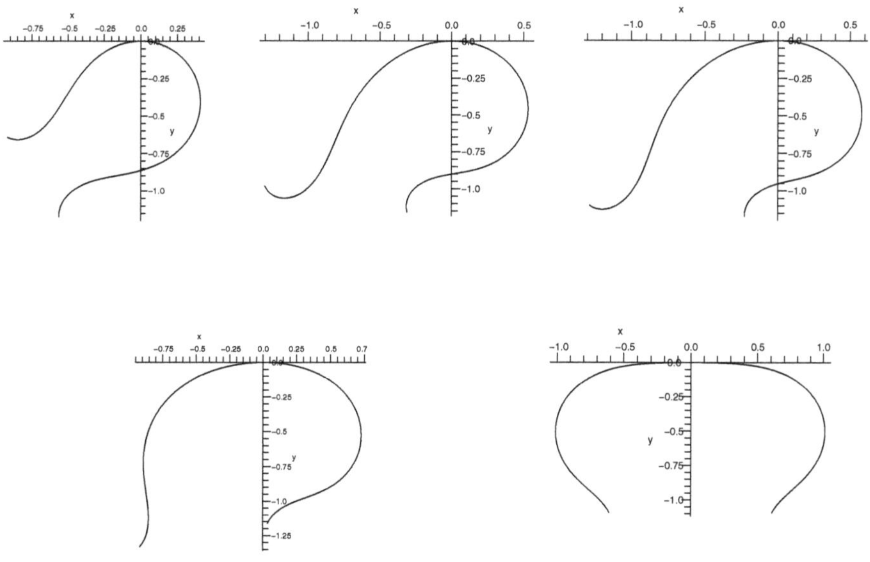


**Fig. 11.33.** The five successive curves used to illustrate the shape evolution involved in pair-creation $\pi\emptyset^{\downarrow-}$ with $c$-preserved anchor-extremum $m^-$

It is worth reminding the reader of an important aspect of this bifurcation operation. The transition

$$\pi\emptyset^{\downarrow -} \; : \; m^- \; \longrightarrow \; m^- M^- m^-$$

unambiguously means that the extremum $m^-$, in the domain, is *added to* by the pair $m^- M^-$ that appears in the co-domain. The operation is illustrated in Fig 11.34, where the $m^-$ extremum in the center of the first shape is *added to* by the two extrema to its left in the second shape.

Therefore, the bay structure $m^- M^- m^-$, of the result, is produced differently from the same bay structure $m^- M^- m^-$, that results from the operation $Bm^-$. The latter operation is shown in Fig 11.35. Here the bay structure is not created by *adding to* the initial $m^-$. The bay comes *entirely out of* the initial $m^-$. That is, the initial $m^-$ is pushed to both the left and the right, becoming the two sides of the final bay, with the $M^-$ emerging between them.

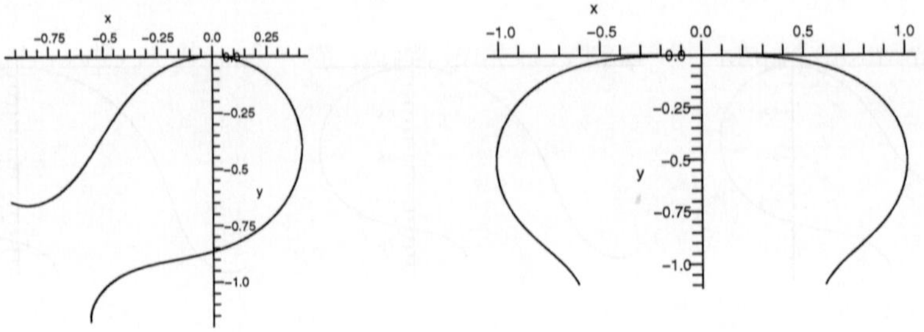


**Fig. 11.34.** Bay created by pair-creation $\pi\emptyset^{\downarrow -}$ applied to the left side of the initial $m^-$.

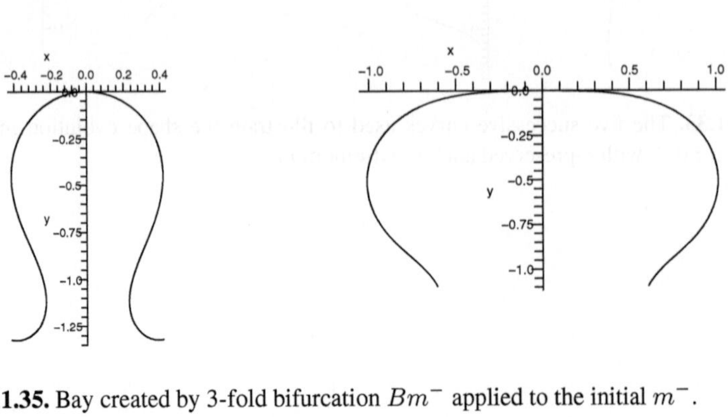


**Fig. 11.35.** Bay created by 3-fold bifurcation $Bm^-$ applied to the initial $m^-$.

## 11.10 Zero Pair-Creation in the Interactive Unfoldings of $EA_3$ and $EA_{-3}$

As the reader has seen, the purpose of the present chapter is to show that the pair-creation operations can be modeled in the Interactive Unfoldings of $EA_3$ and $EA_{-3}$. So far, this chapter has dealt with the *positive* and *negative* pair-creation operations; i.e., the *non-zero* pair-creation operations. Now we deal with the remaining case: the *zero* pair-creation operations.

To do so, and to understand their *interactive* role in the Interactive Unfoldings of $EA_3$ and $EA_{-3}$, we will need the Interactive Unfoldings of $EA_2$ and $EA_{-2}$, which were invented in section 9.7, and we will need to work out how the Interactive Unfoldings of $EA_2$ and $EA_{-2}$ are embedded in the Interactive Unfoldings of $EA_3$ and $EA_{-3}$. These *Interactive Structures* will also give us additional understanding of how the *non-zero* pair-creation operators are realized in the Interactive Unfoldings of $EA_3$ and $EA_{-3}$.

To model the zero pair-creation operations, let us begin by examining the fold and dual-fold catastrophes $A_2$ and $A_{-2}$ with respect to these operations. Even though we have previously described $A_2$ and $A_{-2}$, we did not consider them with respect to these Process Grammar operations. Thus it is worth briefly reviewing the structure of $A_2$ and $A_{-2}$ in order to see what they specifically imply with respect to these operations.

Thus, in this paragraph, let us briefly review the fold catastrophe $A_2$. Recall that its canonical universal unfolding is

$$F(t; a_1) \quad = \quad t^3 \quad + \quad a_1 t \tag{11.1}$$

where $t^3$ is the germ, and $a_1$ gives the 1-dimensional control space. This control space is shown as the horizontal dotted line in Fig 11.36. It has two Morse regions, labeled Region 0 and Region 2 in the figure. The labeling is due to the fact that, in Region 0, the functions have no *degenerate* critical points (e.g., the function illustrated at the right of the figure); and in Region 2, the functions have two critical points (e.g., the function illustrated at the left of the figure). As noted previously, since the two Morse regions are respectively the positive and negative $a_1$-axes, the critical-point separatrix consists of only a single point: the origin, which is shown as the dot at the center of the figure.

With this brief review of the fold catastrophe $A_2$, let us now examine it with respect to the corresponding Process Grammar operation. Examining the graphs in Fig 11.36,

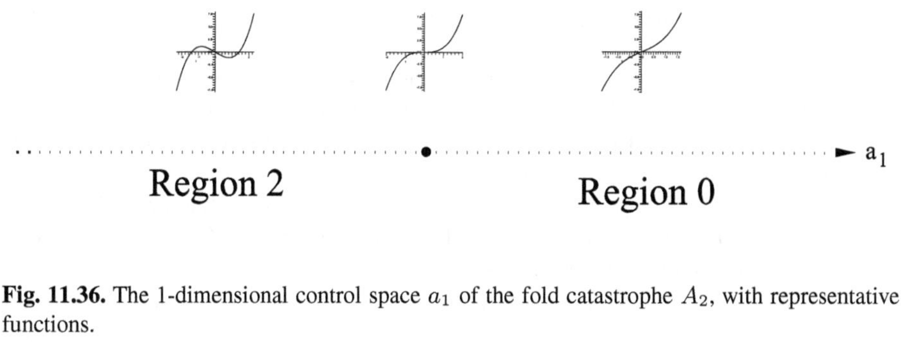


**Fig. 11.36.** The 1-dimensional control space $a_1$ of the fold catastrophe $A_2$, with representative functions.

we see that the singularity configuration of a function in Region 0 of the *canonical* unfolding is a single non-degenerate zero, and the singularity configuration of a function in Region 2 is the 5-fold singularity configuration $0M^+0m^-0$. Therefore, in the canonical unfolding, the transition from the singularity-configuration of a function in Region 0, to the singularity-configuration of a function in Region 2, is given by the following *zero* pair-creation operation from the Process Grammar, which was introduced in section 4.5:

$$[C(3)\pi]0^{\uparrow} \;:\; 0^{\uparrow} \quad\longrightarrow\quad 0M^+0m^-0$$

However, recall that there is a fundamental problem with the Catastrophe Theory view: Catastrophe Theory defines the singularity in the germ to be $A_2$, which means that the singularity can be right-equivalent to any 2-fold degenerate critical point, including such a point that is not of zero value and thus will not undergo the above 5-fold bifurcation defined by the Process Grammar operation $[C(3)\pi]0$.

It is for this reason that Interactive Singularity Theory demands that the singularity in the germ is the *complete* singularity $EA_3$. Furthermore, the singularity must undergo a universal unfolding not only with respect to its critical-point structure, as in expression (11.1), but with respect to its zero-structure as follows:

$$F(t; a_0, a_1) \;=\; t^3 \;+\; a_0 \;+\; a_1 t \tag{11.2}$$

Clearly, the corresponding argument applies to the dual fold catastrophe $A_{-2}$. It has exactly the same control space $a_1$, but the functions in the space are simply inverted. Therefore, in Catastrophe Theory's *canonical* unfolding, crossing the single-point separatrix (the origin) from Region 0 to Region 2 is given, in the Process Grammar, by the other *zero* pair-creation operation, which was also introduced in section 4.5:

$$[C(3)\pi]0^{\downarrow} \;:\; 0^{\downarrow} \quad\longrightarrow\quad 0m^-0M^+0$$

However, again Catastrophe Theory would not require the singularity in the germ to have zero value, and thus the singularity would not have to undergo the above 5-fold bifurcation.

It is for this reason that Interactive Singularity Theory demands that the singularity in the germ is the *complete* singularity $EA_{-3}$. Furthermore, the singularity must undergo a universal unfolding not only with respect to its critical-point structure, but with respect to its zero-structure, and thus be of the form:

$$F(t; a_0, a_1) \;=\; -\,t^3 \;-\; a_0 \;-\; a_1 t \tag{11.3}$$

An additional problem with Catastrophe Theory is that, because it allows the singularities $A_{\pm 2}$ to have any value, it would not recognize the distinction between the zero, positive, and negative pair-creation operations. In contrast, this distinction does exist in the Interactive Unfoldings of $EA_{\pm 2}$.

Now, in section 9.7, we have already studied the structure of the Interactive Unfolding of $EA_2$. However, we have not studied it with respect to the Process Grammar operations. This is what we will do in the present section.

The Process Grammar operations correspond to crossing the separatrices in the control space of an Interactive Unfolding, between regions or volumes defined by the

Interactive Structure of the separatrices. Therefore we must understand the separatrices for $EA_2$.

Recall from section 9.7 that the zero-manifold and critical-point manifold for $EA_2$ were shown, respectively, as the left and right diagrams in Fig 9.7 (page 318). Consider the zero-manifold, i.e., the left diagram. The *zero-separatrix* for $EA_2$ must be the projection of the *vertical edge* of this manifold (i.e., the bold curve) onto the 2-dimensional control space $(a_0, a_1)$; i.e., the zero-separatrix must be the side-curves of the cusp and the cusp-point shown in the bottom plane in the left diagram. In contrast, consider the critical-point manifold, i.e., the right diagram. The *critical-point separatrix* for $EA_2$ must be the projection of the *vertical edge* of this manifold onto the 2-dimensional control space $(a_0, a_1)$; i.e., the critical-point separatrix must be the $a_0$-axis in the control space.

Therefore, putting the zero-separatrix and critical-point separatrix together, we obtain the full separatrix for $EA_2$, which is shown here in Fig 11.37. That is, this figure exhibits the 2-dimensional control space $(a_0, a_1)$ for $EA_2$. The *zero-separatrix* consists of the cusp side-curves and the origin. The *critical-point separatrix* is the $a_0$-axis. Thus the set of *solid lines* in Fig 11.37 illustrate the *full separatrix* for $EA_2$. The reader should try to imagine the zero-manifold and critical-point manifold hanging over this space, with their respective vertical edges directly over the solid lines in this figure.

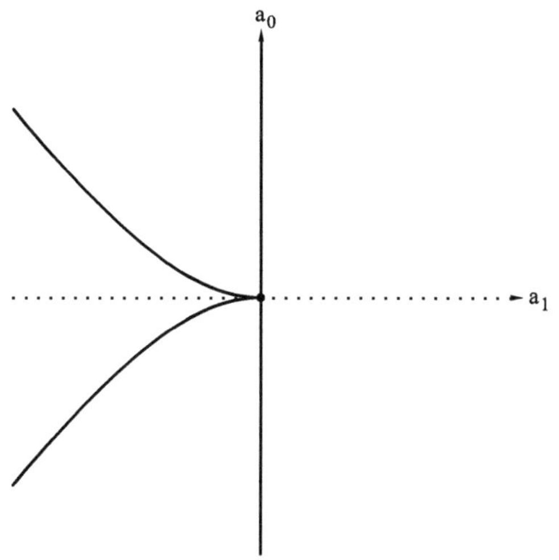


**Fig. 11.37.** The full separatrix of the Interactive Unfolding of $EA_2$.

Before deducing the Process Grammar operations, let us first notice the following: Fig 11.36 (page 423) illustrates functions along the $a_1$-axis of the full 2-dimensional control space shown in Fig 11.37. The remainder of the functions in the full control space are obtained simply by raising and lowering the graphs of those functions with respect to their own $F$-axes; i.e., this is the role of the $a_0$-axis in the control space. That is, movement up the $a_0$-axis in Fig 11.37 will raise the graphs of the functions with respect to their own $F$-axes; and movement down the $a_0$-axis will lower the graphs of the functions with respect to their own $F$-axes.

Now we turn to the Process Grammar operations. The operations realized in the Interactive Unfolding of $EA_2$ will cross the separatrices between the regions defined by the *Interactive Structure*. However, to fully understand these Process Grammar operations, it is useful to consider how they arise from the Interactive Structure between the two solution manifolds, as follows:

First observe that the *lower sheet* of the critical-point manifold (the right diagram in Fig 9.7 page 318) consists of non-degenerate *minima* because the lower sheet is in the positive $t$-axis and we see from Fig 11.36 (page 423) that the only critical points that occur in the positive $t$-axis are non-degenerate minima; i.e., as illustrated by the left representative function.

Now, as stated in section 9.7, the lower sheet of the critical-point manifold (the right diagram in Fig 9.7 page 318) intersects the *lower fold* in the zero-manifold (the left diagram) through the *bold curve* on this fold. Notice that the lower fold is in the positive region of the $a_0$-axis of the control space. Thus, in the control space, consider the separatrix curve corresponding to the lower fold. Also within the control space, consider a path that crosses this separatrix curve in the decreasing $a_0$ direction. The path will define a trajectory on the lower sheet of the critical-point manifold, and this trajectory will therefore slice through the lower fold of the zero-manifold through a non-origin point on the bold curve. Notice that this trajectory on the lower sheet of the critical-point manifold will be a non-degenerate minimum that is positive until it reaches the bold curve, where it will become a 2-fold degenerate zero, and after this, as it continues on the lower sheet of the critical-point manifold, it will become a negative non-degenerate minimum, while, simultaneously, the 2-fold degenerate zero bifurcates into two non-degenerate zeros, one on the upper side and the other on the lower side of this fold of the zero-manifold. Therefore, the crucial thing to observe is that the path, which we are considering in the control space, defines a slice through the lower sheet of the critical-point manifold and the lower fold of the zero-manifold, that together on this slice has the same organization as the manifold structure shown in Fig 9.6 (page 315) for the Ground Interactive Unfolding $EA_1$. Furthermore, recall that the Ground Interactive Unfolding $EA_1$ is the universal unfolding that corresponds to the Process Grammar operation $Cm^+ : m^+ \longrightarrow 0m^-0$. That is, moving from right to left in Fig 9.6 (page 315), the horizontal bold line is a non-degenerate minimum that goes from positive to negative, and the parabola is a two-fold degenerate zero at its apex that bifurcates into two non-degenerate zeros along its two branches.

Now observe this: Since the path we have taken in the control space for $EA_2$ is in the positive region of the $a_0$-axis, through the separatrix curve, in the decreasing $a_0$ direction, this path corresponds, in Fig 11.37, to crossing the upper separatrix curve downward. We therefore conclude that the downward crossing of the upper separatrix curve, in Fig 11.37, corresponds to the Process Grammar operation $Cm^+$. We also conclude that each point on the upper separatrix curve corresponds to the singularity $EA_1$.

Also observe that, because, in this space, the application of $Cm^+$ starts with a function above the upper separatrix curve, such a function has a *positive* maximum and a *positive* minimum and *one* zero; i.e., is a raised version of the left-most representative function in Fig 11.36 (page 423). The application of $Cm^+$ results in a function that has a *positive* maximum and a *negative* minimum and *three* zeros; i.e., a function that is qualitatively equivalent to the left-most representative function in Fig 11.36. Such a function is a *maximally bifurcated* function in the control space; i.e., the maximally bifurcated functions are between the two zero-separatrix curves.

Let us now consider the *upper sheet* of the critical-point manifold (the right diagram in Fig 9.7 page 318). Observe that this consists of non-degenerate *maxima* because the upper sheet is in the negative $t$-axis and we see from Fig 11.36 (page 423) that the only critical points that occur in the negative $t$-axis are non-degenerate maxima; i.e., as illustrated by the left representative function.

Recall, as stated in section 9.7, the upper sheet of the critical-point manifold (the right diagram in Fig 9.7 page 318) intersects the *upper fold* in the zero-manifold (the left diagram) through the *bold curve* on this fold. Notice that the upper fold is in the negative region of the $a_0$-axis of the control space. Thus, in the control space, consider the separatrix curve corresponding to the upper fold. Also within the control space, consider a path that crosses this separatrix curve in the increasing $a_0$ direction. The path will define a trajectory on the upper sheet of the critical-point manifold, and this trajectory will therefore slice through the upper fold of the zero-manifold through a non-origin point on the bold curve. Notice that this trajectory on the upper sheet of the critical-point manifold will be a non-degenerate maximum that is negative until it reaches the bold curve, where it will become a 2-fold degenerate zero, and after this, as it continues on the upper sheet of the critical-point manifold, it will become a positive non-degenerate maximum, while, simultaneously, the 2-fold degenerate zero bifurcates into two non-degenerate zeros, one on the upper side and the other on the lower side of this fold of the zero-manifold. Therefore, the crucial thing to observe is that the path, which we are considering in the control space, defines a slice through the upper sheet of the critical-point manifold and the upper fold of the zero-manifold, that together on this slice has the same organization as the manifold structure shown in Fig 9.6 (page 315) for the Ground Interactive Unfolding $EA_1$, except that, in the present case, this structure is created by the *Dual* Ground Interactive Unfolding $EA_{-1}$. Also, the present case uses what we defined on p341 as the second version of the universal unfolding of $EA_{-1}$, that is, where the $a_0$-axis is in the reverse direction from that shown in Fig 9.6 (page 315). Most crucially, recall that the Dual Ground Interactive Unfolding $EA_{-1}$ is the universal unfolding that corresponds to the Process Grammar operation $CM^- : M^- \longrightarrow 0M^+0$, which we are seeing here.

Now observe this: Since the path we have taken in the control space for $EA_2$ is in the negative region of the $a_0$-axis, through the separatrix curve, in the increasing $a_0$ direction, this path corresponds, in Fig 11.37, to crossing the lower separatrix curve upward. We therefore conclude that the upward crossing of the lower separatrix curve, in this diagram, corresponds to the Process Grammar operation $CM^-$. We also conclude that each point on the lower separatrix curve corresponds to the singularity $EA_{-1}$.

Also observe that, because, in this space, the application of $CM^-$ starts with a function below the lower separatrix curve, such a function has a *negative* minimum and a *negative* maximum and *one* zero; i.e., is a lowered version of the left-most representative function in Fig 11.36 (page 423). The application of $CM^-$ results in a function that has a *negative* minimum and a *positive* maximum and *three* zeros; i.e., a function that is qualitatively equivalent to the left-most representative function in Fig 11.36. Again, such a function is a *maximally bifurcated* function in the control space, as are all the functions between the two zero-separatrix curves.

At this stage, let us give some additional logic concerning the zero-separatrix in the control space; in particular, the cusp-curves in Fig 11.37. The purpose is to show how Method 1 relates to the argument we have just given concerning the intersection of the critical-point manifold and zero-manifold. The first stage of Method 1 is to construct the critical-*value* manifold of the fold catastrophe $A_2$. Inspection of Fig 11.36 (page 423) shows us what that manifold must look like: We see that, from the bifurcation, the *value* of the maximum moves upward in the graph as one moves from the origin of the $a_1$ control space in the negative $a_1$-direction. Furthermore, we see that, from the bifurcation, the *value* of the minimum moves simultaneously downward in the graph. Therefore, the maximum and minimum branch away from each other as one moves away from the origin of the $a_1$ control space in the negative $a_1$-direction. This gives, to the critical-*value* manifold, the branching cusp-curve structure we see in the zero-separatrix.

The second stage in Method 1 is to vertically invert the critical-*value* manifold. Thus the maxima component of the manifold will branch downwards, and the minima component of the manifold will branch upwards. In conclusion from this, the lower branch, in Fig 11.37, must be the *maxima zero-separatrix*, and the upper branch must be the *minima zero-separatrix*.

Notice the following important logic in this situation: The minima zero-separatrix is the set of functions that have a minimum at level 0 on their $F$-axis. When we return to Fig 11.36, and examine the left function, we see that this function would have to be *raised* by a *positive* amount $a_0$, to make the minimum have value 0. This is why the minima zero-separatrix bends *upwards* in Fig 11.37. That is, the positive $a_0$-axis in this figure points upwards and gives the amount that a function in the $a_1$-axis has been raised to create the function at a point above it in the space.

Correspondingly, the maxima zero-separatrix is the set of functions that have a maximum at level 0 on their $F$-axis. When we return to Fig 11.36, and examine the left function, we see that this function would have to be *lowered* to make its maximum have value 0. This is why the maxima zero-separatrix bends *downwards* in Fig 11.37.

Now observe the following: The logic we have just given is reinforced by the argument given earlier that, in the Interactive Unfolding of $EA_2$, the surface of minima in the critical-*point* manifold intersects the zero-manifold at the fold edge that projects onto the upper separatrix curve; and the surface of maxima of the critical-*point* manifold intersects the zero-manifold at the fold edge that projects onto the lower separatrix curve.

So far, we have established the Process Grammar operations that cross the two curves of the zero-separatrix for $EA_2$. Let us now establish the Process Grammar operations that cross the critical-point separatrix for $EA_2$. To do so, we begin by considering the critical-point manifold, i.e., the right diagram in Fig 9.7 page 318. Recall that each point in the upper sheet is a non-degenerate maximum, and each point in the lower sheet is a non-degenerate minimum. Therefore, any point on the vertical edge of the manifold consists of the *coalescence* of a maximum and a minimum, i.e., a 2-fold degenerate critical inflection-point. Notice that, since the critical-point manifold does not enter the positive $a_1$ half-plane of the control space, any point in that half-plane is a function that does not have a critical point.

Thus, consider the following type of path in the control space: The path starts in the positive $a_1$ half-plane, then crosses the $a_0$-axis, and then enters the negative $a_1$ half-plane. The $t$-axis, that moves along the path, will initially not intersect the critical-point manifold, and therefore the function will initially not contain a critical point; then, the $t$-axis will touch the vertical-edge of the critical-point manifold, and therefore the function will contain a 2-fold degenerate critical inflection-point; and finally, after crossing the $a_0$-axis, the $t$-axis will simultaneously slice through the top and bottom sheet of the critical-point manifold, and the function will therefore contain a non-degenerate maximum and non-degenerate minimum. Thus, the path corresponds to a pair-creation operation of the Process Grammar.

Observe that, if the $a_0$ value along the path is positive, then the crossing of the $a_0$-axis will correspond to the upward positive pair-creation operation:

$$\pi\emptyset^{\uparrow+} \;:\; \emptyset^{\uparrow+} \quad\longrightarrow\quad M^{+}m^{+}$$

Correspondingly, if the $a_0$ value along the path is negative, then the crossing of the $a_0$-axis will correspond to the upward negative pair-creation operation:

$$\pi\emptyset^{\uparrow-} \;:\; \emptyset^{\uparrow-} \quad\longrightarrow\quad M^{-}m^{-}$$

Furthermore, if the $a_0$ value along the path is zero, then the crossing of the $a_0$-axis will correspond to the upward zero pair-creation operation:

$$[C(3)\pi]0^{\uparrow} \;:\; 0^{\uparrow} \quad\longrightarrow\quad 0M^{+}0m^{-}0$$

It is important to observe that, in this third case, when the $t$-axis crosses the vertical edge of the critical-point manifold, it simultaneously crosses the vertical edge of the zero-manifold, in the way described in section 9.7 and illustrated in Fig 9.9 (page 321). Recall that this figure is the $a_0 = 0$ slice through the zero-manifold and critical-point manifold; the horizontal line and larger parabola being the slice through the zero-manifold, and the smaller parabola being the slice through the critical-point manifold. Thus, moving the $t$-axis along the $a_0 = 0$ slice, from the positive to negative $a_1$-axis, implies that the $t$-axis will, on the negative $a_1$-axis, slice only the zero-manifold at a single point; then, at the origin, the $t$-axis will slice the apex of the critical-point manifold where it is coincident with the apex of the zero-manifold as well as a point on the horizontal line of the zero-manifold; i.e., this intersection-point is simultaneously a 2-fold degenerate critical point and a 3-fold degenerate zero, i.e., the *complete* singularity $EA_2$; and finally, on entering the negative $a_1$-axis, the $t$-axis will simultaneously slice the two branches of the critical-point manifold and the two branches and horizontal line of the zero-manifold, i.e., the complete singularity $EA_2$ will have undergone the 5-fold bifurcation into two critical points that alternate with three zeros. The total effect will therefore correspond to the zero pair-creation operation given above.

We have now elaborated the complete set of Process Grammar operations in the control space of the Interactive Unfolding of $EA_2$. With this complete set, it is useful to take our diagram of the control space with its *full* separatrix, as shown in Fig 11.37, and map out the entire action of the Process Grammar operations in this space. This is shown in Fig 11.38. First, the diagram exhibits the fact that the zero pair-creation operation $[C(3)\pi]0^{\uparrow}$ crosses not only the cusp-point of the zero-separatrix, thus giving a 3-fold *zero*-bifurcation, but simultaneously crosses the critical-point separatrix (the central vertical line), thus giving a 2-fold *critical-point* bifurcation. Also the diagram shows that crossing the central vertical line, above and below the origin, corresponds, respectively, to the positive and negative pair-creation operations, $\pi\emptyset^{\uparrow+}$ and $\pi\emptyset^{\uparrow-}$. And finally, the diagram shows that crossing the upper and lower zero-separatrix curves corresponds, respectively, to the two continuation operations $Cm^{+}$ and $CM^{-}$.

**Fig 11.38 will be called the**

**Process Grammar operation-diagram for the Interactive Unfolding of $EA_2$.**

The reader will note that the three pair-creation operations across the $a_0$-axis are all *upward* pair-creation operations, i.e., as indicated by the upward arrow on each. The reason is this: As shown in Fig 11.36 (page 423), the representative functions, from right to left, undergo the bifurcation at a point which initially has an upward gradient.

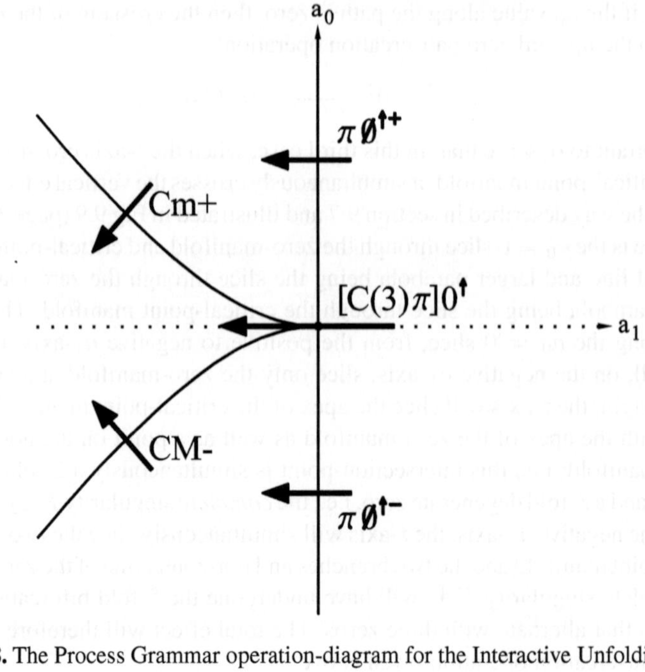


**Fig. 11.38.** The Process Grammar operation-diagram for the Interactive Unfolding of $EA_2$.

Furthermore, since the other functions in the full 2D control space are only raised or lowered versions of the functions in the $a_1$-axis, the gradient direction does not change.

Now recall from the bold-printed statements on page 119 and the figure on page 120, that there are three upward pair-creation operations and three downward pair-creation operations. We have therefore seen that *all* the upward operations are realized in the Interactive Unfolding of $EA_2$, as shown in Fig 11.38. By the corresponding argument, we also conclude that all the three *downward* operations act in the corresponding positions in the Interactive Unfolding of the Dual $EA_{-2}$ because the functions are inverted in that space. Thus we can state:

**The three *upward* pair-creation operations $\pi\emptyset^{\uparrow+}$, $[C(3)\pi]0^{\uparrow}$, $\pi\emptyset^{\uparrow-}$ cross the critical-point separatrix of the Interactive Unfolding of $EA_2$.**

**The three *downward* pair-creation operations $\pi\emptyset^{\downarrow+}$, $[C(3)\pi]0^{\downarrow}$, $\pi\emptyset^{\downarrow-}$ cross the critical-point separatrix of the Interactive Unfolding of the Dual $EA_{-2}$.**

Now, the middle operation, in each of these two sets, is an example of the *zero* pair-creation operator, $[C(3)\pi]$, which was defined in the bold print on page 112. When we originally explained the notation $[C(3)\pi]$ for this operator in section 4.5, we said that it was due to the operator being the simultaneous application of two operators: the 3-fold zero-bifurcation operator $C(3)$ and the 2-fold extremum-bifurcation operator $\pi$. Now, using Fig 11.38 as illustration, we can see that the notation $[C(3)\pi]$ can be justified also from the point of view of Interactive Singularity Theory, as follows:

---

**The *zero* pair-creation operator, $[C(3)\pi]$, is labeled by *two operators*:**

**$C(3)$ and $\pi$**

**because it simultaneously crosses *two separatrices*:**

**the zero-separatrix and the critical-point separatrix.**

---

Notice, in Fig 11.38, that the zero pair-creation operation is the *only* operation that simultaneously crosses two separatrices. This is because, the zero pair-creation operation is the only operation that crosses a point of vertical tangency on the zero-manifold that is simultaneously a point of vertical tangency on the critical-point manifold – i.e., in accord with our intersection Rule 2. All other operations in Fig 11.38 cross only one separatrix. That is, the continuation operations, $Cm^+$ and $CM^-$, cross only the zero-separatrix;

and the positive and negative pair-creation operations, $\pi\emptyset^{\uparrow+}$ and $\pi\emptyset^{\uparrow-}$, cross only the critical-point separatrix. This is because the continuation operations, $Cm^+$ and $CM^-$, cross a point of vertical tangency of only the zero-manifold, while nevertheless being on the critical-point manifold, not at a point of vertical tangency, in accord with our intersection Rule 1; and the positive and negative pair-creation operations, $\pi\emptyset^{\uparrow+}$ and $\pi\emptyset^{\uparrow-}$, cross a point of vertical tangency of only the critical-point manifold, while this point is not an intersection point at all.

Having worked out the structures of the Interactive Unfoldings of $EA_2$ and $EA_{-2}$, we shall now see that they are embedded in the Interactive Unfoldings of $EA_3$ and $EA_{-3}$. To illustrate, consider Fig 11.39 which shows the 3D control space for $EA_3$. In this figure, an arbitrary point $g_i$ has been chosen in the bi-valent line $g$ on the right side-sheet. Also a rectangle has been drawn around this point to represent a "vertical planar neighborhood" of $g_i$ (the quotation marks indicate that one can choose a neighborhood that is qualitatively the same). Notice that this neighborhood slices the full separatrix (zero-separatrix and critical-point separatrix) for $EA_3$ in a set of lines. One can see that these lines have the same qualitative structure as the full separatrix for $EA_2$ as illustrated in Fig 11.37.

It is necessary to carefully understand this by comparing the individual separatrix components in Fig 11.37 for $EA_2$ with the individual separatrix components within the "vertical planar neighborhood" of $g_i$ in Fig 11.39 for $EA_3$, as follows: The cusp-point in the separatrix in Fig 11.37 for $EA_2$ corresponds to the cusp-point $g_i$ in Fig 11.39 for $EA_3$. The upper separatrix curve in Fig 11.37 is the minima zero-separatrix for $EA_2$, and corresponds to the curve which is the slice through a diagonal sheet of the minima zero-separatrix in Fig 11.39 for $EA_3$. The lower separatrix curve in Fig 11.37 is the maxima zero-separatrix for $EA_2$, and corresponds to the curve that is the slice through the maxima zero-separatrix sheet in Fig 11.39 for $EA_3$. Finally, the vertical line which is the critical-point separatrix in Fig 11.37 for $EA_2$ corresponds to the vertical line which is the slice through the right sheet of the critical-point separatrix in Fig 11.39 for $EA_3$.

Now recall that Fig 11.38 exhibited the Process Grammar operation-diagram for the Interactive Unfolding of $EA_2$. The important thing to observe is that, corresponding to what we said in the previous paragraph, the Process Grammar operation-diagram for the Interactive Unfolding of $EA_2$ is the same as the Process Grammar operation-diagram, in the Interactive Unfolding of $EA_3$, for a "vertical planar neighborhood" of any point $g_i$ in the bi-valent line $g$ on the right side-sheet. Thus, as we have seen previously in the Interactive Unfolding of $EA_3$, the positive pair-creation operation, $\pi\emptyset^{\uparrow+}$, crosses through the right side-sheet *above* the bi-valent line $g$; and the negative pair-creation operation, $\pi\emptyset^{\uparrow-}$, crosses through the right side-sheet *below* the bi-valent line $g$. Again, as we have seen previously in the Interactive Unfolding of $EA_3$, the continuation operation $Cm^+$ crosses a minima zero-separatrix sheet downwards; and the continuation operation $CM^-$ crosses the maxima zero-separatrix sheet upwards.

There is, in the Interactive Unfolding of $EA_2$, only one remaining operation which we have not previously assigned in the Interactive Unfolding of $EA_3$: the zero pair-creation operation $[C(3)\pi]0^{\uparrow}$. In the Interactive Unfolding of $EA_2$, this operation maximally crosses the cusp-point of the zero-separatrix as well as the critical-point separatrix

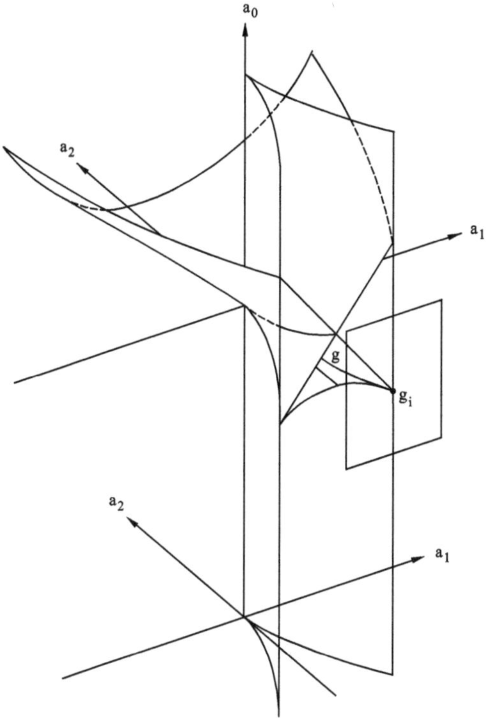


**Fig. 11.39.** A "vertical planar neighborhood" of an arbitrary point $g_i$ in the bi-valent line $g$ on the right side-sheet, in the Interactive Unfolding of $EA_3$.

simultaneously. Correspondingly, in the Interactive Unfolding of $EA_3$, this operation maximally crosses the right cuspidal edge of the zero-separatrix as well as the right side-sheet of the critical-point separatrix simultaneously. That is, the operation crosses the *right bi-valent line* $g$, into the central triangular volume which we have called the *enclosed volume*. This is worthwhile understanding in detail, as follows:

Observe first that each of the operations in the Process Grammar operation-diagram for the Interactive Unfolding of $EA_2$ are applied to functions that do not have an anchor-extremum, but are applied in the Interactive Unfolding of $EA_3$ to functions that have an anchor-extremum, a minimum; and they are applied to the right side of that extremum. Thus, whereas the functions in Fig 11.36 (page 423) illustrated the zero pair-creation operation $[C(3)\pi]0^{\uparrow}$ in the Interactive Unfolding of $EA_2$, the corresponding functions in the Interactive Unfolding of $EA_3$, for the same operation, necessarily have a minimum, as an anchor-extremum to the left of the application-point of the operation, as illustrated by the functions in Fig 11.40. This figure shows a cross-section of the full separatrix for $EA_3$, where the cross-section is given by the plane $a_2 = c$ for a negative constant $c$.

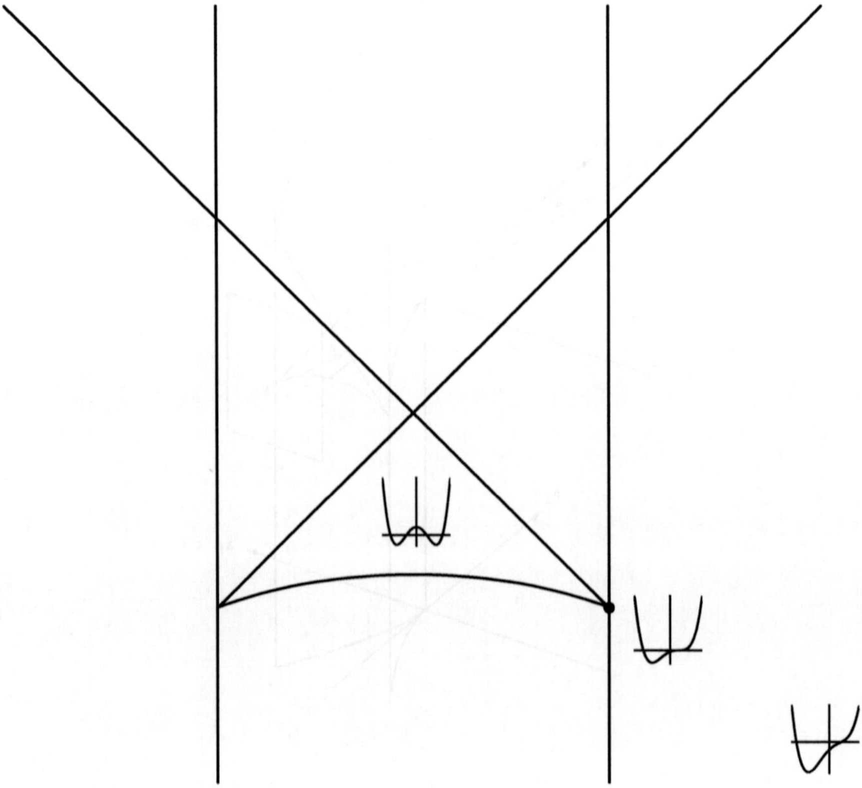

**Fig. 11.40.** The upward zero pair-creation operation $[C(3)\pi]0^{\uparrow}$ in the Interactive Unfolding of $EA_3$.

The figure is drawn large so that the reader can clearly see the singularity-configuration of each of the three functions.

It is worth carefully understanding Fig 11.40 in the following way: The domain function for $[C(3)\pi]0^{\uparrow}$ is illustrated as the right-most function in this figure. It is located to the right of the right side-sheet, and below the single zero-separatrix sheet that occurs to the right of the right side-sheet. Observe that this function has only one critical point (horizontal tangent), its negative minimum $m^{-}$, which is the anchor-extremum. The zero which will eventually undergo bifurcation is the right zero in this function. Currently, it is non-degenerate. Next, the transition function of the operation is illustrated by the middle function in Fig 11.40. This function is located on the bi-valent line $g$; that is, it is not only on the right side-sheet of the critical-point separatrix, but is also on a right cusp-point of the zero-separatrix. Because it is on the side-sheet, it has a 2-fold degenerate critical point, a horizontal inflection; and because it is on the cusp-point, this inflection is in tangential contact with its own $t$-axis and is a 3-fold degenerate zero.

Finally, the codomain of the operation is illustrated by the left-most function in Fig 11.40. This function is in the enclosed volume of the zero-separatrix. We observe that, in this function, the 2-fold critical inflection that occurred in the transition function has now bifurcated into a positive maximum and negative minimum. Thus the critical-point structure of the function is negative minimum, positive maximum, negative minimum. All functions in the enclosed volume have this critical-point structure; i.e., none have a positive minimum or a negative maximum. The other thing to observe is that, in this function, the 3-fold degenerate zero that occurred in the transition function has now bifurcated into 3 zeros. Thus, there are a total of 4 zeros in the function. All functions in the enclosed volume have 4 zeros. Thus the singularity-configuration of any function in the enclosed volume is $0m^-0M^+0m^-0$.

Therefore, the transition between the singularity-configuration of the first function and the singularity-configuration of the third function is given by the following application of the operation $[C(3)\pi]0^\uparrow$

$$[C(3)\pi]0^\uparrow \ : \ 0m^-0 \quad \longrightarrow \quad 0m^-0M^+0m^-0 \tag{11.4}$$

Note that this expression is non-ambiguous, because the upward arrow on the operation $[C(3)\pi]0^\uparrow$ means that it must be applied to the right 0 in the domain-triple $0m^-0$, and it must bifurcate, from this 0, the sequence $0M^+0m^-0$ in the codomain, in accord with the definition of $[C(3)\pi]0^\uparrow$ in expression (4.5) on page 111.

It is also worthwhile seeing how our terminology fits with explaining expression (11.4). The component $C(3)$ in the operator $[C(3)\pi]$ is responsible for the 3-fold zero-bifurcation that occurs in the transition; and the component $\pi$ in the operator $[C(3)\pi]$ is responsible for the pair-creation that simultaneously occurs in the transition. We have seen that the simultaneous occurrence of the two operators $C(3)$ and $\pi$ corresponds to the fact that they are simultaneously crossing the zero-separatrix and the critical-point separatrix. Now recall that the term *bi-valent* line was defined as a line where the two separatrices meet. In particular, the *bi-valence* of the line $g$ therefore corresponds to the simultaneous occurrence of the two operators $C(3)$ and $\pi$.

Expression (11.4) shows the form of application of the *upward* zero pair-creation operation $[C(3)\pi]0^\uparrow$ in the Interactive Unfolding of $EA_3$. The *downward* zero pair-creation operation $[C(3)\pi]0^\downarrow$ occurs in the Interactive Unfolding of $EA_3$ by maximally crossing the bi-valent line $g$ on the *left* side-sheet into the same enclosed volume. This is illustrated by the functions shown in Fig 11.41. These illustrating functions are simply the reflectional opposites of the illustrating functions shown in Fig 11.40 for the upward zero pair-creation operation $[C(3)\pi]0^\uparrow$.

Notice that the singularity-configuration of the *codomain* function is therefore the same as before, because the codomain is in the same volume of the control space as before. Furthermore, the singularity-configuration of the *domain* function is also the same as before, because the domain is in the same volume as it was before. To understand this, observe that, in Fig 11.40, the volume in which the domain function was drawn, i.e., to the right of the right side-sheet and below the single zero-separatrix sheet, *surrounds*

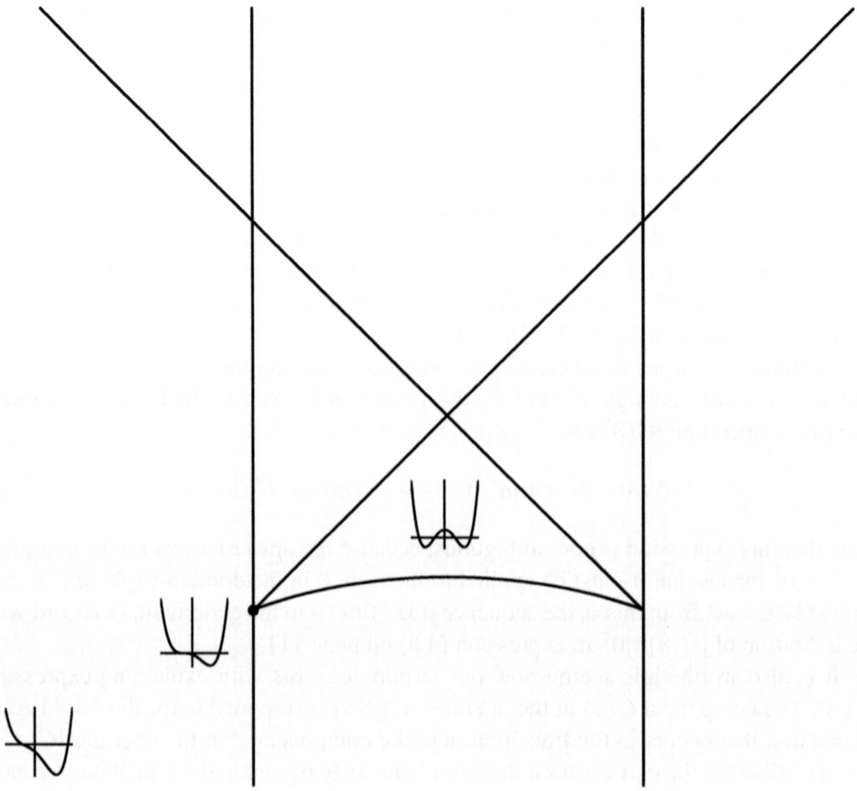

**Fig. 11.41.** The downward zero pair-creation operation $[C(3)\pi]0^{\downarrow}$ in the Interactive Unfolding of $EA_3$.

the critical-point separatrix below the single zero-separatrix sheet (as illustrated in Fig 8.22 page 296), and is therefore also on the left of the left side-sheet, below the single zero-separatrix sheet. That is, the domain function for both operations $[C(3)\pi]0^{\uparrow}$ and $[C(3)\pi]0^{\downarrow}$ in the Interactive Unfolding of $EA_3$ can be chosen from anywhere in this entire volume. The singularity-configuration of any function in this volume is $0m^-0$.

The following expression therefore gives the transition between the singularity-configuration of the domain and the singularity-configuration of the codomain of the downward zero pair-creation operation $[C(3)\pi]0^{\downarrow}$ in the Interactive Unfolding of $EA_3$:

$$[C(3)\pi]0^{\downarrow} \ : \ 0m^-0 \quad \longrightarrow \quad 0m^-0M^+0m^-0 \tag{11.5}$$

Notice that the domain and codomain are the same as in expression (11.4) for the upward zero pair-creation operation. However, in the present case, the downward arrow on the operation $[C(3)\pi]0^{\downarrow}$ means that it must be applied to the left 0 in the domain-triple $0m^-0$, and it must bifurcate from this 0, the sequence $0m^-0M^+0$ in the codomain – and this accords with the definition of $[C(3)\pi]0^{\downarrow}$ in (4.6) on page 111.

It is clear that this operation comes from the Interactive Unfolding of the Dual $EA_{-2}$. Thus, to summarize what we have said so far in this section about the zero pair-creation operations:

**The upward zero pair-creation operation $[C(3)\pi]0^{\uparrow}$, which is a maximal crossing of the origin germ in the Interactive Unfolding of $EA_2$, is realized in the Interactive Unfolding of $EA_3$ by a maximal crossing of the bi-valent line $g$ on the right side-sheet.**

**The downward zero pair-creation operation $[C(3)\pi]0^{\downarrow}$, which is a maximal crossing of the origin germ in the Interactive Unfolding of $EA_{-2}$, is realized in the Interactive Unfolding of $EA_3$ by a maximal crossing of the bi-valent line $g$ on the left side-sheet.**

Let us now see how the zero pair-creation operations are realized in the Interactive Unfolding of the Dual $EA_{-3}$. Clearly, since the singularity-configuration in the domain volume in the Interactive Unfolding of $EA_3$ was $0m^-0$, the singularity-configuration in the domain volume in the Interactive Unfolding of $EA_{-3}$ must be $0M^+0$. Consequently, whereas the upward zero pair-creation operation $[C(3)\pi]0^{\uparrow}$ was applied across the bi-valent line $g$ on the *right* side-sheet in the Interactive Unfolding of $EA_3$, it must be applied across the bi-valent line $g$ on the *left* side-sheet in the Interactive Unfolding of $EA_{-3}$. To see this, recall first the inverted structure of the separatrix for the Interactive Unfolding of $EA_{-3}$, as illustrated for example in Fig 11.8 (page393). The bi-valent line $g$ on the left side-sheet is the zero-separatrix cuspidal edge on that sheet. It is *above* the remainder of the zero-separatrix. Thus, Fig 11.42 shows a cross-section through the full separatrix for $EA_{-3}$, where the cross-section is given by the plane $a_2 = c$ for a negative constant $c$. The three successive functions from the left-most function, across the left cusp-point, into the central enclosed volume, illustrate the $[C(3)\pi]0^{\uparrow}$ operation:

$$[C(3)\pi]0^{\uparrow} \ : \ 0M^+0 \quad \longrightarrow \quad 0M^+0m^-0M^+0 \tag{11.6}$$

where the upward arrow on the operation $[C(3)\pi]0^{\uparrow}$ implies that it must be applied to the left 0 in the domain-triple $0M^+0$. This contrasts with the application of this operation in the Interactive Unfolding of $EA_3$, where it was applied to the right 0 in the domain-triple $0m^-0$. In both cases however, it sends the 0 to the sequence $0M^+0m^-0$, in accord with the definition of $[C(3)\pi]0^{\uparrow}$ in expression (4.5) on page 111.

In contrast, the three successive functions, in Fig 11.42, from the right-most function, across the right cusp-point, into the central enclosed volume, illustrate the $[C(3)\pi]0^{\downarrow}$ operation:

$$[C(3)\pi]0^{\downarrow} \ : \ 0M^+0 \quad \longrightarrow \quad 0M^+0m^-0M^+0 \tag{11.7}$$

where the downward arrow on the operation $[C(3)\pi]0^{\downarrow}$ implies that it must be applied to the *right* 0 in the domain-triple $0M^+0$. This contrasts with the application of this operation in the Interactive Unfolding of $EA_3$, where it was applied to the *left* 0 in the domain-triple $0m^-0$. In both cases, however, it sent the 0 to the triple $0m^-0M^+0$, in accord with the definition of $[C(3)\pi]0^{\downarrow}$ in expression (4.6) on page 111.

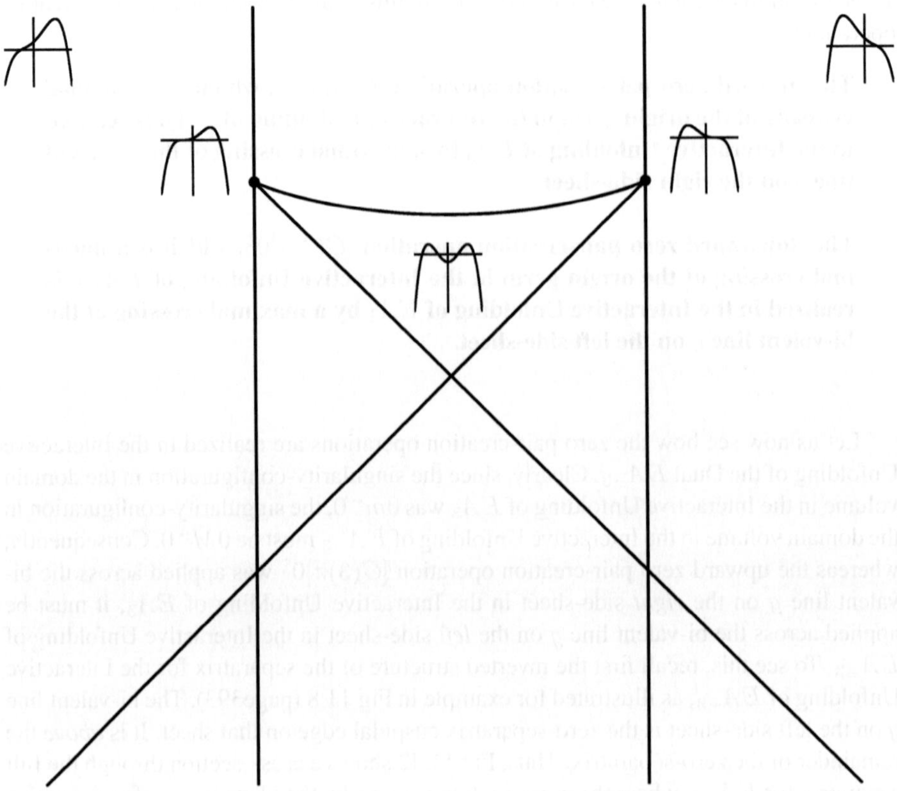

**Fig. 11.42.** The zero pair-creation operations $[C(3)\pi]0^{\uparrow}$ and $[C(3)\pi]0^{\downarrow}$ in the Interactive Unfolding of the Dual $EA_{-3}$.

Thus, from the previous two paragraphs, we see that, in the Interactive Unfolding of $EA_{-3}$, each of the points on the left bi-valent line $g$ is an example of the origin of the Interactive Unfolding of $EA_2$; and each of the points on the right bi-valent line $g$ is an example of the origin of the Interactive Unfolding of the Dual $EA_{-2}$. Therefore, we conclude this:

**The upward zero pair-creation operation $[C(3)\pi]0^{\uparrow}$, which is a maximal crossing of the origin germ in the Interactive Unfolding of $EA_2$, is realized in the Interactive Unfolding of $EA_{-3}$ by a maximal crossing of the bi-valent line $g$ on the left side-sheet.**

**The downward zero pair-creation operation $[C(3)\pi]0^{\downarrow}$, which is a maximal crossing of the origin germ in the Interactive Unfolding of $EA_{-2}$, is realized in the Interactive Unfolding of $EA_{-3}$ by a maximal crossing of the bi-valent line $g$ on the right side-sheet.**

We can now understand the extent of the "vertical planar neighborhood" illustrated in Fig 11.39 (page 433). Recall that this neighborhood is equivalent to a neighborhood of the origin of the control space for $EA_2$, and possesses the full separatrix structure for $EA_2$ and the full set of operations in the Process Grammar operation-diagram of the Interactive Unfolding of $EA_2$, as illustrated in Fig 11.38 (page 430). What we can see is that this neighborhood can be extended vertically both upward and downward to infinity, as well as rightward to infinity and leftward to the left side-sheet (but not including the sheet). The neighborhood therefore becomes the entire infinite half-plane terminating next to the left side-sheet. In carrying out this extension, the only other separatrix line that is added to the neighborhood is the cross-section of the opposite diagonal sheet. The role of this sheet is that crossing it has the effect of lowering or raising the *anchor-extremum* with respect to the $t$-axis in the function; i.e., it does *not* affect the two extrema produced by pair-creation across the right side-sheet. In this way the entire half-plane corresponds to the control space of the Interactive Unfolding of $EA_2$, and its Process Grammar operation-diagram is the same as that for $EA_2$, except that one adds the continuation operation for the opposite diagonal line when considering also the anchor-extremum.

The corresponding arguments apply for the "half-plane neighborhood" of a point on the *left* bi-valent line $g$ in the Interactive Unfolding of $EA_3$, such a neighborhood being equivalent to the Interactive Unfolding of $EA_{-2}$, again except for the extra diagonal zero-separatrix line for the anchor-extremum. And the corresponding arguments apply in the Interactive Unfolding of $EA_{-3}$, where left and right switch with respect to $EA_2$ and $EA_{-2}$.

Finally, in all these unfoldings, let us contrast the condition of *upward* versus *downward* in the pair-creation operations. We saw that the three *upward* pair-creation operations are given by crossing the critical-point separatrix in the Interactive Unfolding of $EA_2$, as illustrated in Fig 11.38 (page 430). This means that the three upward operations correspond to crossing the *right* side-sheet in the Interactive Unfolding of $EA_3$. Furthermore, they correspond to crossing the *left* side-sheet in the Interactive Unfolding of the Dual $EA_{-3}$. In contrast, we have seen that the three *downward* pair-creation operations are given by crossing the critical-point separatrix in the Interactive Unfolding of the Dual $EA_{-2}$. This means that the three downward operations correspond to crossing the *left* side-sheet in the Interactive Unfolding of $EA_3$. Furthermore, they correspond to crossing the *right* side-sheet in the Interactive Unfolding of the Dual $EA_{-3}$.

A convenient way of understanding this is seen by returning to the hierarchy shown in Fig 4.15 page 120. The top node is the pair-creation operator $\pi$; and this has two cases, the upward pair-creation operator $\pi^{\uparrow}$, and the downward pair-creation operator $\pi^{\downarrow}$. The *upward* operator $\pi^{\uparrow}$ corresponds to crossing the critical-point separatrix for $EA_2$. Therefore it corresponds to crossing the *right* side-sheet of the critical-point separatrix for $EA_3$, and the *left* side-sheet of the critical-point separatrix for $EA_{-3}$. In contrast, the *downward* operator $\pi^{\downarrow}$ corresponds to crossing the critical-point separatrix for $EA_{-2}$. Therefore it corresponds to crossing the *left* side-sheet of the critical-point separatrix for $EA_3$, and the *right* side-sheet of the critical-point separatrix for $EA_{-3}$.

## 11.11 The Pair-creation Operations in the Interactive Unfoldings of $EA_3$ and $EA_{-3}$

This chapter has shown that all the pair-creation operations in the Process Grammar, as well as all their applications, are realized within the Interactive Unfoldings of $EA_3$ and $EA_{-3}$. We can summarize their realization as follows:

**REALIZING THE PAIR-CREATION OPERATIONS**
**IN THE**
**INTERACTIVE UNFOLDINGS OF $EA_3$ AND $EA_{-3}$**

**Positive & Negative Pair-Creation Operations**

**The positive and negative pair-creation operations (4 operations), which each have 3 applications (i.e., a total of 12 applications), are realized by crossing the 12 bi-valent bounded regions of the critical-point separatrices for $EA_3$ and $EA_{-3}$.**

**The anchor-extremum of an operation changes from positive to negative in moving its transition point, within a side-sheet, across the bi-valent line $e$.**

**The operation itself changes from positive to negative in moving its transition point, within a side-sheet, across the bi-valent line $g$.**

**Zero Pair-Creation Operations**

**In between the positive operation and negative operation on a side-sheet, there is a zero operation, given by a maximal crossing through the bi-valent line $g$.**

**The 2 zero operations correspond to the 2 opposite lines $g$ within the same Interactive Unfolding, that of $EA_3$, or that of $EA_{-3}$.**

**The 2 applications of a zero operation correspond respectively to the line $g$ on one side-sheet for $EA_3$, and the line $g$ on the opposite side-sheet for $EA_{-3}$.**

**Upward and Downward Pair-Creation Operations**

**The 3 upward operations cross the right side-sheet for $EA_3$.**
**Furthermore, they cross the left side-sheet for $EA_{-3}$.**

**The 3 downward operations cross the left side-sheet for $EA_3$.**
**Furthermore, they cross the right side-sheet for $EA_{-3}$.**

# 12

# Operators $CC$ and $[C(4)B]$ in the Interactive Unfoldings of $EA_3$ and $EA_{-3}$

## 12.1 Introduction

Sections 5.10 – 5.12 introduced the Process Grammar operators which describe a process that opposes and breaks-through a compressive process, resulting in a zone-restricted penetrative process. We defined two alternative scenarios for this: the Two-Stage Scenario and the One-Stage Scenario. The present chapter will show that the operators used to describe these two scenarios are realized in the Interactive Unfoldings of $EA_3$ and $EA_{-3}$. In fact, they complete the set of operators that are realized in these Interactive Unfoldings.

**FUNDAMENTAL FACT OF INTERACTIVE SINGULARITY THEORY**

**A crucial aspect of Interactive Singularity Theory is that it gives the organization of relationships between the morphological operators, i.e. their Interactive Structures, in a way that matches the *causal* structure of the operators as defined by the correct process-inference rules.**

**In particular, we will see, in this chapter, that a process which opposes and breaks-through a compressive process, resulting in a zone-restricted penetrative process, is given by the Interactive Structures in the Interactive Unfoldings of $EA_3$ and $EA_{-3}$ in a way that matches the causal structure of those situations.**

**In contrast to Interactive Singularity Theory, the conventional singularity theories of shape fail to match these fundamentally important causal structures in biological morphology and manufacturing design.**

## 12.2 Two-Stage Scenario

We will first briefly recall the Two-Stage Scenario from sections 5.10 – 5.11, and then go on to show how it is realized in the Interactive Unfoldings of $EA_3$ and $EA_{-3}$. The two successive stages of the scenario are

**Stage 1: A penetrative process breaking-through a compressive process that it opposes.**

**Stage 2: Restricting the zone-of-influence of that opposing penetrative process.**

For example, consider the case where the opposed compressive process is that of the positive compressive extremum $m^+$. The two successive stages are illustrated by the two successive transitions in Fig 12.1. That is, as we can see, in the transition from the first to the second shape, the downward compressive process, at the top of the first shape, is opposed by an upward force which breaks-through that compressive process and emerges as the protrusion given by the bold upward arrow in the second shape. Next, in the transition from the second to the third shape, the zone-of-influence of that protruding process is restricted, on each side, by the two indentations that appear in the third shape.

Recall that the Process Grammar operations that give these two stages are these: Apply, to the top $m^+$ extremum on the first shape, the Process Grammar operation

$$Bm^+ : \ m^+ \longrightarrow m^+M^+m^+ \tag{12.1}$$

which we call, *breaking-through of a protrusion*. The result is the second shape in Fig 12.1, where the top squashing in the first shape has now bifurcated into two copies, one on each side of the upward protrusion that has broken-through in the second shape. By our *causal* theory of the $B$ operator (section 3.24), this protruding process *caused* the bifurcation. Notice, in Fig 12.1, the breaking-through protrusion has bifurcated the top shield in the first shape into what our theory calls a *double-shield* in the second shape. It is always the case that the $Bm^+$ operation creates a double-shield.

Now, to restrict the zone-of-influence of the protruding process that has broken-through, we pinch the double-shield; that is, we *continue* the two side $m^+$ processes, on the second shape, till they indent, as shown in the third shape. That is, we apply, to

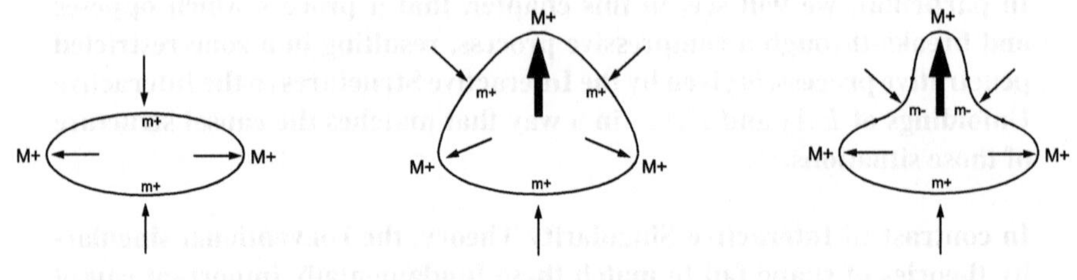


**Fig. 12.1.** Illustration of the Two-Stage Scenario with a starting $m^+$ extremum.

each of these two side $m^+$, the Process Grammar operation $Cm^+ : m^+ \longrightarrow 0m^-0$; i.e., *squashing continues till it indents*. Thus the codomain in expression (12.1), above, undergoes the following transition:

$$CCm^+ \;:\; m^+M^+m^+ \;\longrightarrow\; \overbrace{0m^-0}\,M^+\,\overbrace{0m^-0} \qquad (12.2)$$

where the over-braces in the final string indicate the effect of the simultaneous uses of the $Cm^+$ operation on the two side $m^+$ extrema. The Process Grammar operator $CC$ that defines this transition is called the 2-fold pinching operator (Type 1). It is two simultaneous applications of the continuation operator $C$. In the current situation, it restricts the zone-of-influence of the protruding process that broke-through the compressive process, and it does so by pinching the double-shield that was created by the breaking-through penetrative process. The full definition of $CC$ was given on page 191.

Putting together expressions (12.1) and (12.2) above, we have seen that, in the case of a starting $m^+$, the Two-Stage Scenario is given by

$$m^+ \;\longrightarrow\; m^+M^+m^+ \;\longrightarrow\; \overbrace{0m^-0}\;M^+\;\overbrace{0m^-0} \qquad (12.3)$$

Notice that the final string of *seven* singularities has grown out of the initial string of *only one* singularity $m^+$. These seven singularities consist of three extrema and four zeros. The existence of three extrema results from Stage 1, which is a 3-fold extremum-bifurcation; and the creation of four zeros is achieved by Stage 2, which is two simultaneous 2-fold zero-bifurcations. Thus, in terms of the singularities involved, the two stages can be described as extremum-bifurcation followed by zero-bifurcation.

The dual of the situation, we have been describing, starts with a $M^-$ extremum, as illustrated in Fig 12.2. The upward resistance process, at the $M^-$ in the first shape, is opposed and bifurcated by the downward indenting process shown in the second shape, resulting in the *double-bay* in the second shape. Then the two side resistance processes, of the double-bay, undergo continuation till they both protrude as shown in the third shape. This *restricts* the *zone-of influence* of the indenting process that broke-through. The two successive transitions are given by

$$BM^- \;:\; M^- \;\longrightarrow\; M^-m^-M^- \qquad (12.4)$$

$$CCM^- \;:\; M^-m^-M^- \;\longrightarrow\; \overbrace{0M^+0}\;m^-\;\overbrace{0M^+0} \qquad (12.5)$$

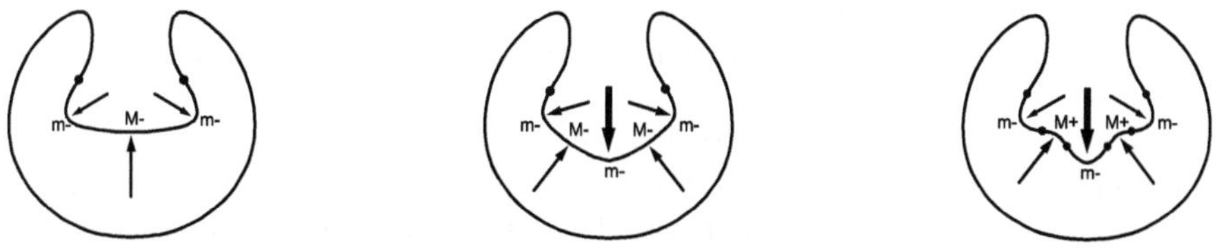


**Fig. 12.2.** Illustration of the Two-Stage Scenario with a starting $M^-$ extremum.

**Realizing the Two-Stage Scenario in the Interactive Unfoldings of $EA_3$ and $EA_{-3}$**

We will now see that the Two-Stage Scenario is realized in the Interactive Unfoldings of $EA_3$ and $EA_{-3}$ in an *extremely interesting way*. The argument will first be illustrated with the case of a process opposing a $m^+$ process, and this will be shown to be realized in the Interactive Unfolding of $EA_3$.

Since the two stages (1) breaking-through of a penetrative process, and (2) restricting the zone-of-influence of that process, are realized respectively by (1) extremum-bifurcation and (2) zero-bifurcation, they must be realized respectively by (1) crossing the critical-point separatrix and (2) crossing the zero-separatrix. Fig 12.3 shows these two separatrices, on the left and right respectively, so that we can use these to visualize the two successive stages.

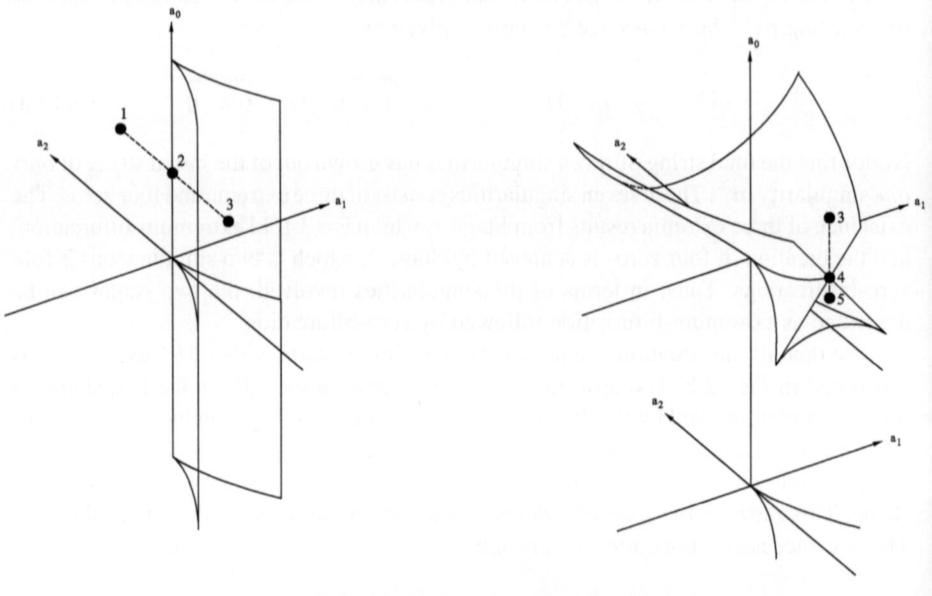


**Fig. 12.3.** The Two-Stage Scenario in the Interactive Unfolding of $EA_3$.

Begin by considering the first stage, i.e., the operation $Bm^+$, breaking-through of a protrusion. We saw in section 10.2.2 that the $Bm^+$ operation is modeled by a sequence of three points as indicated by dots 1, 2, 3, in the left diagram in Fig 12.3. All three points are above the zero-separatrix shown in the right diagram. The operation $Bm^+$ goes from point 1 to 3 (left diagram) through point 2, which is a cusp-point on the critical-point separatrix.

For later considerations, it is worthwhile reminding ourselves of the qualitative structure of the functions at points 1, 2, 3. The three successive points are curvature functions that have the qualitative structure illustrated by the three successive functions in Fig 12.4. We see that the first function has the singularity-configuration $m^+$ and the last function

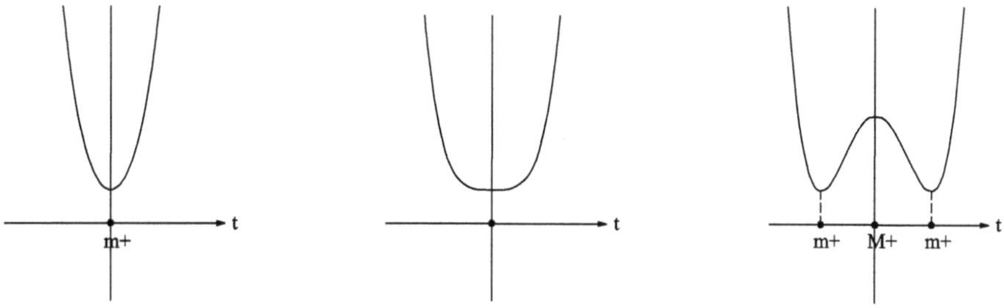


**Fig. 12.4.** Stage 1: Functions corresponding to points 1,2,3, in Fig 12.3.

has the singularity-configuration $m^+M^+m^+$. The middle function is the transition-function, and its singularity-configuration is a minimum which is 3-fold degenerate as a critical-point, but non-zero. The sequence of three functions therefore accords with the operation $Bm^+$, which defines Stage 1 in the Two-Stage Scenario.

Now consider Stage 2 in the Two-Stage Scenario: i.e., pinching (zone-of-influence-restriction). This is given by the Process Grammar operation $CCm^+$, which is two simultaneous uses of the operation $Cm^+$. Our crucial proposal is as follows: Using the right-hand diagram in Fig 12.3 to illustrate, Stage 2 is given by the transition from point 3 to 5 going through point 4, which is a point on the *intersection curve* of the two diagonal surfaces, i.e., the two minima surfaces of the zero-separatrix. The reason is this: Crossing either of these two surfaces downwards represents a use of the $Cm^+$ operation. Therefore, crossing the intersection curve downwards means crossing *both* of these two surfaces *simultaneously*, and thus is two *simultaneous* uses of $Cm^+$.

Notice that the final function, corresponding to point 5, is inside the *enclosed volume* of the zero-separatrix (the central triangular volume). It is therefore a *maximally-bifurcated* function in the Interactive Unfolding of $EA_3$; that is, all possible critical points and zeros have undergone bifurcation, and no further bifurcations are possible in the control space.

Now, the three successive points, 3, 4, 5, are curvature functions that have the qualitative structures illustrated by the three functions in Fig 12.5. The first function, which corresponds to point 3, has the singularity-configuration $m^+M^+m^+$. In the second

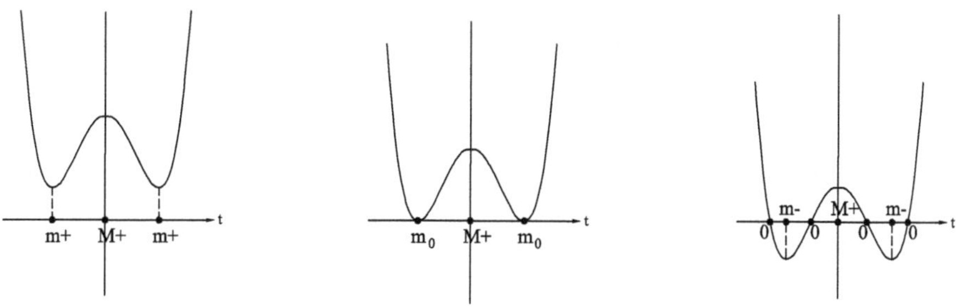


**Fig. 12.5.** Stage 2: Functions corresponding to points 3,4,5, in Fig 12.3.

function, which corresponds to point 4 (i.e., on the *intersection curve*), the two minima have descended and are now both zero. In fact, each of these two minima is now a 2-fold degenerate zero. Then, by further lowering, we get the third function, which corresponds to point 5 (i.e., in the *enclosed volume*). Here, the two 2-fold degenerate zeros of the middle function have bifurcated each into two zeros, giving a total of four zeros, as shown. However, the points have not bifurcated as critical points.

Now note that the Two-Stage Scenario in the *dual* situation, i.e., starting with extremum $M^-$, is simply the corresponding pair of trajectories with respect to the inverted separatrices in the Interactive Unfolding of $EA_{-3}$.

As a result of the above discussion, we conclude the following concerning the operator $CC$:

**ACTION OF OPERATOR $CC$ IN THE INTERACTIVE UNFOLDINGS OF $EA_3$ AND $EA_{-3}$**

**The pinching operator $CC$ (order 1) is realized by a maximal crossing of the intersection curve of the two minima zero-separatrix surfaces of the Interactive Unfolding of $EA_3$; and is also realized by a maximal crossing of the intersection curve of the two maxima zero-separatrix surfaces of the Interactive Unfolding of $EA_{-3}$.**

**In the Interactive Unfolding of $EA_3$ the operator is realized as the operation $CCm^+$, and in the Interactive Unfolding of $EA_{-3}$ it is realized as the operation $CCM^-$.**

## 12.3 One-Stage Scenario

We will first briefly recall the One-Stage Scenario from section 5.12, and then show how it is realized in the Interactive Unfoldings of $EA_3$ and $EA_{-3}$.

To recall the main properties of the One-Stage Scenario, consider the case where the initial extremum is $m^+$. The Two-Stage scenario goes through the three successive shapes illustrated in Fig 12.1 (page 442). In contrast, the One-Stage Scenario goes directly from the first to the third shape in Fig 12.1, without going through the intervening shape; i.e., as illustrated in Fig 12.6.

An important thing to observe is that, in *both* the Two-Stage Scenario and the One-Stage Scenario, the *final state* exhibits *three* penetrative processes: One of them is the process that opposed and broke-through the initial compressive process. Our theory calls this the *opposing penetrative process*. In Fig 12.1 and Fig 12.6, the opposing penetrative process is the upward protruding process seen in the final state. The other two penetrative processes seen in the final state are the two penetrative processes that appear at the two sides of the opposing penetrative process. Our theory calls these the *restricting*

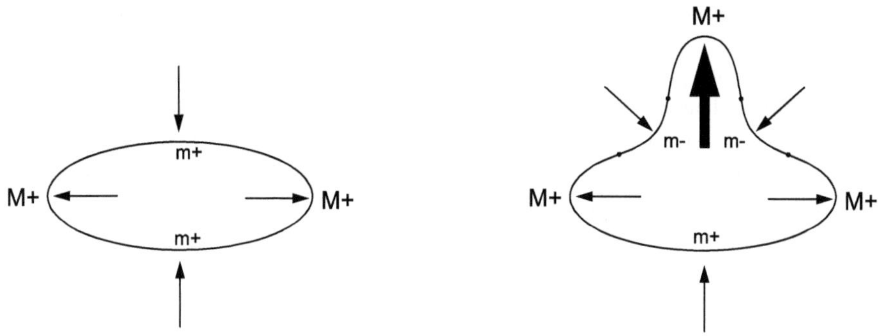


**Fig. 12.6.** The One-Stage Scenario.

*penetrative processes*, because they restrict the zone-of-influence of the opposing penetrative process. In Fig 12.1 and Fig 12.6, the restricting penetrative processes are the two indenting processes at the sides of the upward protruding process in the final state.

In the Two-Stage Scenario, the two restricting penetrative processes are continuations of the two compressive processes that have bifurcated from the initial compressive process. In contrast, in the One-Stage Scenario, the two restricting penetrative processes have bifurcated, *in their penetrative state*, from the application-point of the opposing penetrative process. That is, in the One-Stage Scenario, the zone-of-influence of the opposing penetrative process is *always* restricted, rather than starting to be restricted *later* (as occurs in the Two-Stage Scenario). Furthermore, in the One-Stage Scenario, the restricting penetrative processes do not initially appear at a finite distance from the center (as occurs in the Two-Stage Scenario), meaning that, in its formation, the zone-of-influence, and its boundary, radiate outwards from the central extremum.

Thus, in the One-Stage Scenario shown in Fig 12.6, the two indentation extrema $m^-$, shown in the right shape, must have bifurcated out from the application-point of the opposing protruding process, at the same time that the protrusion extremum, $M^+$, was created. And furthermore, the four zeros of curvature, between the positive and negative curvature regions, must have simultaneously bifurcated out from the application-point.

This means that, in this scenario, the top $m^+$, in the first shape of Fig 12.6, must undergo a simultaneous 7-fold bifurcation

$$m^+ \longrightarrow 0m^-0M^+0m^-0 \tag{12.6}$$

into three extrema and four zeros shown in the top of the second shape. The Process Grammar operator which gives this transition is labeled $[C(4)B]$, because the transition can be viewed as the *simultaneous* application of two operators, the 4-fold zero-bifurcation operator $C(4)$ and the 3-fold extremum-bifurcation operator $B$, to the *same* application-point. In expression (12.6), the operator is realized as the operation $[C(4)B]m^+$; and the dual is obviously $[C(4)B]M^-$. The operator is given the name, *3-fold radial-undulation operator*, which is a term that was fully explained in Comment 5.3 (page 194).

We also claimed that this operator, $[C(4)B]$, for the One-Stage Scenario, can be viewed as a coalescence of the two operators, $CC$ and $B$, for the Two-Stage Scenario, in the following sense: Whereas $CC$ bifurcates the four zeros as two pairs simultaneously from *two* separate points, in contrast, $C(4)$ bifurcates the four zeros simultaneously from *one* point. Thus we can understand the two separate application-points of the $CC$ operator as having coalesced into the one application-point of $C(4)$. Furthermore, this coalescence simultaneously pulls the other stage $B$ of the Two-Stage Scenario into the single stage of the One-Stage Scenario.

Note that the operator $[C(4)B]$ was defined in the bold print on page 195.

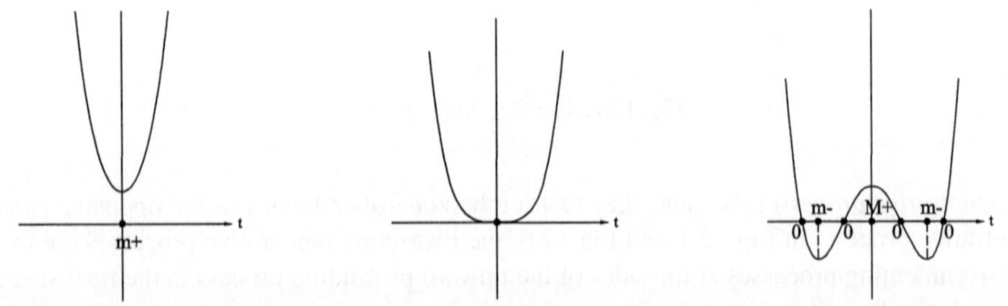


**Fig. 12.7.** The qualitative succession of curvature functions in the One-Stage Scenario, illustrated in the case of the initial compressive extremum being positive.

In the case of the initial compressive extremum being positive, the qualitative structure of the operator is illustrated by the three successive curvature functions shown in Fig 12.7. We see that the first function has the singularity-configuration $m^+$; and the last function has the 7-fold singularity-configuration $0m^-0M^+0m^-0$. Since we are describing the One-Stage Scenario, the seven singularities of the last function must simultaneously coalesce, backwards in time, to the single zero in the transition-function (the middle function); i.e., a point that is a 4-fold degenerate zero and a 3-fold degenerate critical point.

## Realizing the One-Stage Scenario in the Interactive Unfoldings of $EA_3$ and $EA_{-3}$

We are now going to model the operator $[C(4)B]$ in terms of the Interactive Unfoldings of $EA_3$ and $EA_{-3}$. Our proposal is that the operator is realized by going from the first shape to the last shape via the *germ* at the origin of the control space; i.e., via the actual *interactive singularity* from which the entire unfolding is produced.

To demonstrate this, let us first consider the Interactive Unfolding of $EA_3$. The three successive functions, which in Fig 12.7 qualitatively illustrate the operation $[C(4)B]m^+$, correspond to the three successive points, 1, 2, 3, shown in the control space for $EA_3$ in Fig 12.8. That is, the domain of the operation is given by point 1, which is any point in the volume behind the critical-point separatrix but above the zero-separatrix.

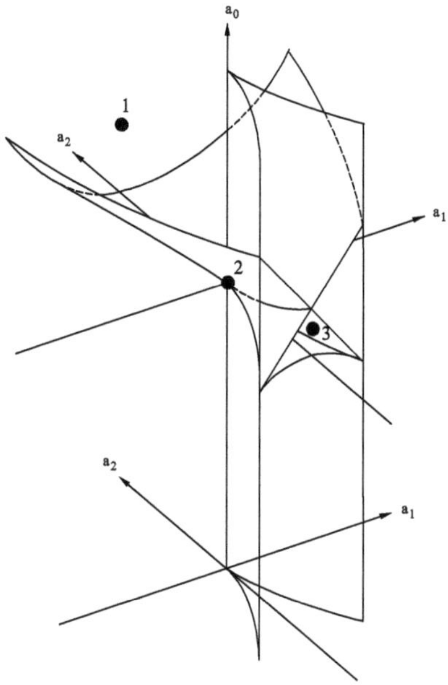


**Fig. 12.8.** The realization of the Process Grammar operation $[C(4)B]m^+$ in the Interactive Unfolding of $EA_3$.

This is because the singularity-configuration of any function in this volume is a positive minimum, $m^+$, which is the domain of the operation $[C(4)B]m^+$. Next the transition-state of the operation is given by point 2 (in Fig 12.8), which is the germ $t^4$ at the origin of the control space, because this is the only function in the control space that has a singularity which is *both* 3-fold degenerate as a critical point and 4-fold degenerate as a zero. That is:

**The singularity in the transition-state is the *interactive* singularity $EA_n$ as opposed to the conventional singularity $A_n$ which other singularity theories would use.**

Finally, the codomain of the operation is given by point 3 (in Fig 12.8), which is any point in the *enclosed volume*, because the singularity-configuration of any function in this volume is $0m^-0M^+0m^-0$, which is the *3-fold undulation*.

Thus the operation $[C(4)B]m^+$ goes from point 1 *directly* to the origin, and then *directly* into the enclosed volume. An essential fact is that the *interactive* singularity $EA_3$ given by the germ at the origin will instantly undergo the complete simultaneous 7-fold bifurcation, into three critical points and four zeros, the moment one moves off the germ into the enclosed volume.

Notice also that, by crossing the germ in this way, one is crossing *two* separatrices simultaneously. That is, we have the following crucial statement:

**The 3-fold radial-undulation operator, $[C(4)B]$, is labeled by** ***two operators*****, $C(4)$ and $B$, because it simultaneously crosses** ***two separatrices*****: the zero-separatrix and the critical-point separatrix.**

Correspondingly, consider now the operation $[C(4)B]M^-$. It is given by the transition

$$M^- \longrightarrow 0M^+0m^-0M^+0 \tag{12.7}$$

It is realized in the Interactive Unfolding of the Dual $EA_{-3}$, by going through the sequence indicated by the numbers 1, 2, 3, in Fig 12.9; i.e., by going from any point in the volume behind the critical-point separatrix, but below the zero-separatrix, *directly* to the origin, and then *directly* into the enclosed volume.

Generally, considering the two realizations $[C(4)B]m^+$ and $[C(4)B]M^-$ of the operator $[C(4)B]$, an important concept is described in the following set of statements:

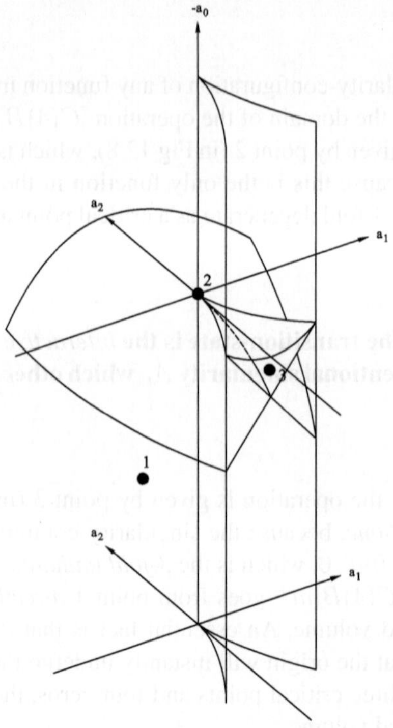


**Fig. 12.9.** The realization of the Process Grammar operation $[C(4)B]M^-$ in the Interactive Unfolding of the Dual $EA_{-3}$.

**ACTION OF OPERATOR $[C(4)B]$ IN THE INTERACTIVE UNFOLDINGS OF $EA_3$ AND $EA_{-3}$**

**The Process Grammar operator $[C(4)B]$ corresponds, in the control spaces for $EA_3$ and $EA_{-3}$, to the 7-fold bifurcation (3-fold critical-point bifurcation and 4-fold zero-bifurcation) obtained by a maximal crossing of the germ, from the volume whose functions have the minimal-length singularity-configuration (only a compressive extremum), into the volume whose functions have the maximal-length singularity-configuration.**

**That is: The $[C(4)B]$ operator is the maximal possible bifurcation in the Interactive Unfoldings of $EA_3$ and $EA_{-3}$.**

**In the Interactive Unfolding of $EA_3$, the operator $[C(4)B]$ is realized as the operation $[C(4)B]m^+$.**

**In the Interactive Unfolding of $EA_{-3}$, the operator $[C(4)B]$ is realized as the operation $[C(4)B]M^-$.**

Related to this, a crucial fact, whose consequence is necessary to understand is this:

**FURTHER INADEQUACY OF CATASTROPHE THEORY**

**An additional major inadequacy of Catastrophe Theory is that it fails to model the important morphological situation of a zone-restricted penetrative process breaking-through a compressive process. The reason is this:**

**An important fact is that the Process Grammar operation $[C(4)B]m^+$ starts with a minimum that is *positive*, goes through a transition-state which is a 3-fold degenerate critical point that has value *zero*, and ends with a state which has a maximum that is *positive* and two minima that are *negative*. Correspondingly, the Process Grammar operation $[C(4)B]M^-$ starts with a maximum that is *negative*, goes through a transition-state which is a 3-fold degenerate critical point that has value *zero*, and ends with a state which has a minimum that is *negative* and two maxima that are *positive*.**

**Catastrophe Theory does not make the distinction between positive, zero, and negative values of critical-points. Therefore:**

**Catastrophe Theory fails to have what our theory defines as the *maximal bifurcation* of the interactive singularity $EA_3$ and the *maximal bifurcation* of the interactive singularity $EA_{-3}$, and therefore fails to recognize the existence of these important morphological transitions.**

## THE COALESCENCE THAT PRODUCES THE OPERATOR $[C(4)B]$

**On page 196, we claimed that one can regard the operator $[C(4)B]$ as the coalescence of the two successive stages, $B$ and $CC$, of the Two-Stage Scenario, in the following sense, given by our theory of the *causal* structure of the One-Stage Scenario: Whereas, in the Two-Stage Scenario, the initial compressive process undergoes continuation after it has bifurcated into two copies, in contrast, in the One-Stage Scenario, the continuation of the initial compressive process occurs simultaneously to its bifurcation. That is, in this case, the term coalescence means that the continuation of the initial compressive process has coalesced with the bifurcation of that process.**

**The present chapter shows how this is given by the *Interactive Structure* in the Interactive Unfoldings of $EA_3$ and $EA_{-3}$, as follows:**

**Whereas, in the Two-Stage Scenario, the two successive stages are the crossing of the critical-point separatrix followed by the crossing of the zero-separatrix at its intersection curve, in contrast, in the One-Stage Scenario, the single stage crosses the critical-point separatrix and the zero-separatrix *simultaneously*, and does so at the only point where the cuspidal edge of the critical-point separatrix is coincident with the zero-separatrix.**

**Thus, in coalescing the operators $B$ and $CC$, one slides the transition-function for $CC$ along the intersection curve of the zero-separatrix, to the origin, thus causing the *two* 2-fold degenerate zeros of the transition-function for $CC$ to become the *single* 4-fold degenerate zero of the transition-function for $C(4)$.**

**Simultaneously, one slides the transition-function for $B$ along the vertical cuspidal edge of the critical-point separatrix, to the origin, therefore changing the function's critical point from non-zero value to zero value.**

**Furthermore, to ensure that the operator crosses *both* the critical-point separatrix and the zero-separatrix simultaneously, the entire path from the transition-function for $B$ to the transition-function for $CC$ is reduced to the single point that is the transition-function for $[C(4)B]$.**

## 12.4 Fundamental Fact of Interactive Singularity Theory

**FUNDAMENTAL FACT**
**OF**
**INTERACTIVE SINGULARITY THEORY**

**A fundamental fact of Interactive Singularity Theory is that it gives the organization of relationships between morphological transitions, i.e. their *Interactive Structures*, in a way that matches the *causal* structure of the transitions as defined by the correct process-inference rules.**

**As a powerful example, in this chapter we have seen the following:**

**The case of the Two-Stage Scenario of a process that opposes and breaks-through a *squashing* process, resulting in a *zone-restricted protruding* process, is given by the following *Interactive Structure* in the Interactive Unfolding of $EA_3$: The first stage is the crossing of the critical-point separatrix at a cusp-point above the zero-separatrix, and the second stage is the crossing of the zero-separatrix through the intersection curve of its two minima sheets.**

**The case of the One-Stage Scenario of a process that opposes and breaks-through a squashing process, resulting in a zone-restricted protruding process, is given by the following *Interactive Structure* in the Interactive Unfolding of $EA_3$: The transition is given by crossing the point in the full separatrix where the intersection curve of the two minima sheets of the zero-separatrix coalesces with the cuspidal edge of the critical-point separatrix.**

**The corresponding Interactive Structures in the Interactive Unfolding of $EA_{-3}$ give the Two-Stage Scenario and One-Stage Scenario of a process that opposes and breaks-through a *resistance* process, resulting in a *zone-restricted indenting* process.**

**All these cases use Interactive Structures because they use relationships between Process Grammar operations (succession in the Two-Stage Scenario and coalescence in the One-Stage Scenario) where the organization of the relationships between the operations is given by the structural relationship between the critical-point separatrix and zero-separatrix.**

**A fundamental fact is that these Interactive Structures match the *causal* structures of these scenarios as defined by the correct process-inference rules.**

**In contrast to Interactive Singularity Theory, the conventional singularity theories of shape fail to match these fundamentally important causal structures in biological morphology and manufacturing design.**

## 12.5 $[C(4)B]$ as Part-Formation

Let us return to the concept stated on page 451 that the $[C(4)B]$ operator is the maximal bifurcation in the Interactive Unfoldings of $EA_3$ and $EA_{-3}$. This relates to our theory of parts in Chapter 5. Our argument has lead to the following conclusion:

**OPERATOR $[C(4)B]$ AS PART-FORMATION**

**The operator $[C(4)B]$ produces the *maximally bifurcated parts* in the Interactive Unfoldings of $EA_3$ and $EA_{-3}$, by maximal bifurcation.**

**Our theory claims that this is an important morphological transition in biological morphology and manufacturing design, for the following reason:**

**In the case of $[C(4)B]$, that is, a zone-restricted penetrative process breaking-through a compressive process, the following is a crucial *causal* fact:**

**As we have said, in this scenario, the initial compressive process continues *immediately*, on being opposed, due to the coalescence.**

**This means that, in the One-Stage Scenario, the compressive process has an *immediate* role of strongly opposing the process that opposes it.**

**Therefore the compressive process forces the central compressive extremum to undergo additional flattening, in fact to zero curvature.**

**After the flattening created by this strong oppositional role of the compressive process, this oppositional role is bifurcated, and continues further, thus creating the zone-of-influence restriction.**

**In biological morphology and manufacturing design, this strong oppositional role can have been created by DESIGN INTENT.**

## 12.6 Quantitative Example of $[C(4)B]m^+$

This section gives a quantitative example of the operator $[C(4)B]$, by showing specific functions from the control space for $EA_3$. The following three figures, Figs 12.10, 12.11, 12.12, correspond to points 1,2,3 in Fig 12.8.

The text now continues in the captions of the figures, which the reader is recommended to read.

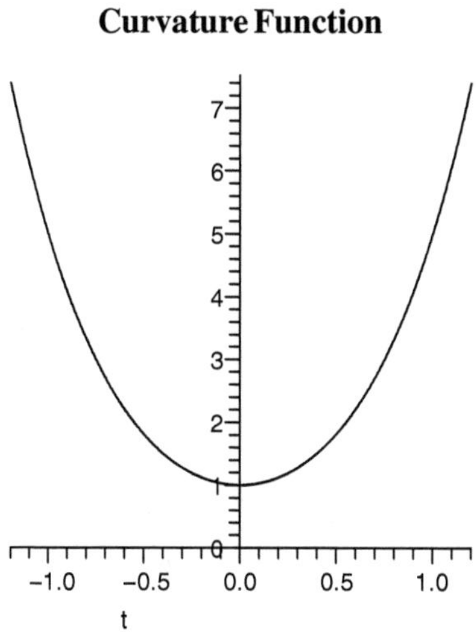


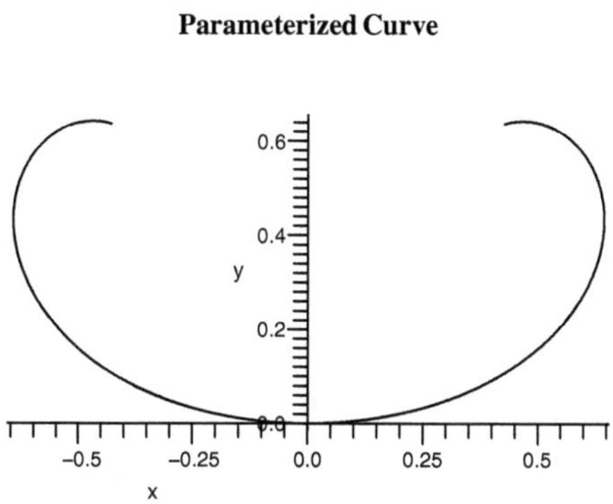


**Fig. 12.10.** Prior to Bifurcation: The upper figure shows the curvature function $F(t) = t^4 + 1 + 3t^2$ at the point $(a_0, a_1, a_2) = (1, 0, 3)$ in the 3D control space of the Interactive Unfolding of $EA_3$. This point is in the volume above the zero-separatrix and behind the critical-point separatrix, and therefore has the singularity-configuration $m^+$. The function clearly shows this, with its one singularity, the positive minimum $m^+$ of value 1. This value is the same as the position $a_0 = 1$ of the function in the 3D control space, because, in the control space, the function is located directly above the positive $a_2$-axis. The lower figure shows the unit-speed parameterized curve that results from this curvature function. The extremum $m^+$ is at the center of the curve, directly below the minimum in the function.

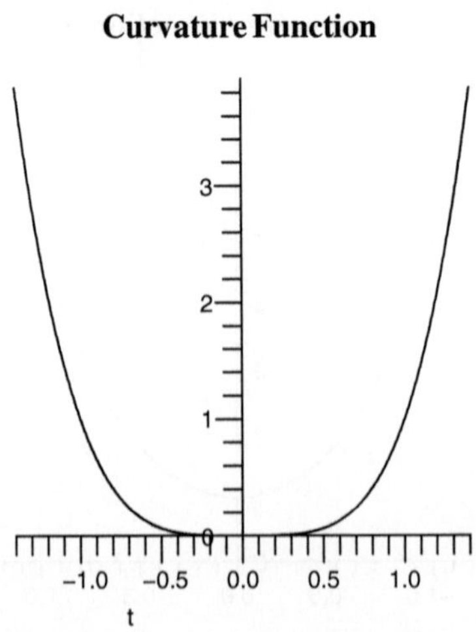


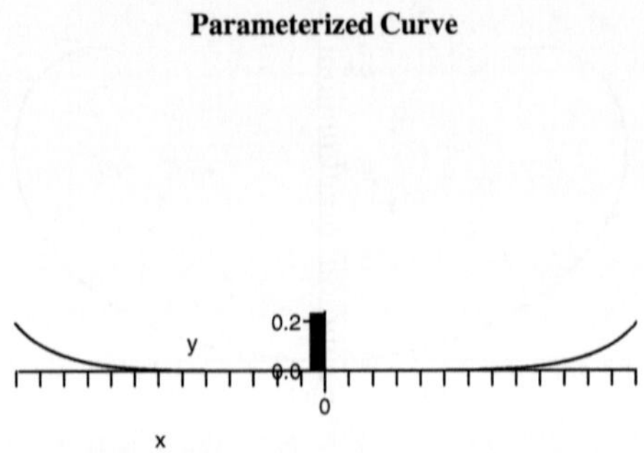


**Fig. 12.11.** The Germ: The upper figure shows the curvature function $F(t) = t^4$ at the origin $(a_0, a_1, a_2) = (0, 0, 0)$ of the 3D control space; i.e., the germ of the Interactive Unfolding of $EA_3$. The minimum of the function is now zero. Furthermore, it has become 3-fold degenerate as a critical point, and 4-fold degenerate as a zero. The lower figure shows the unit-speed parameterized curve that results from this curvature function. The central extremum on the parameterized curve is correspondingly flat (zero curvature), and is close to flatness over a large region, because of the considerable degeneracy of this point in the curvature function.

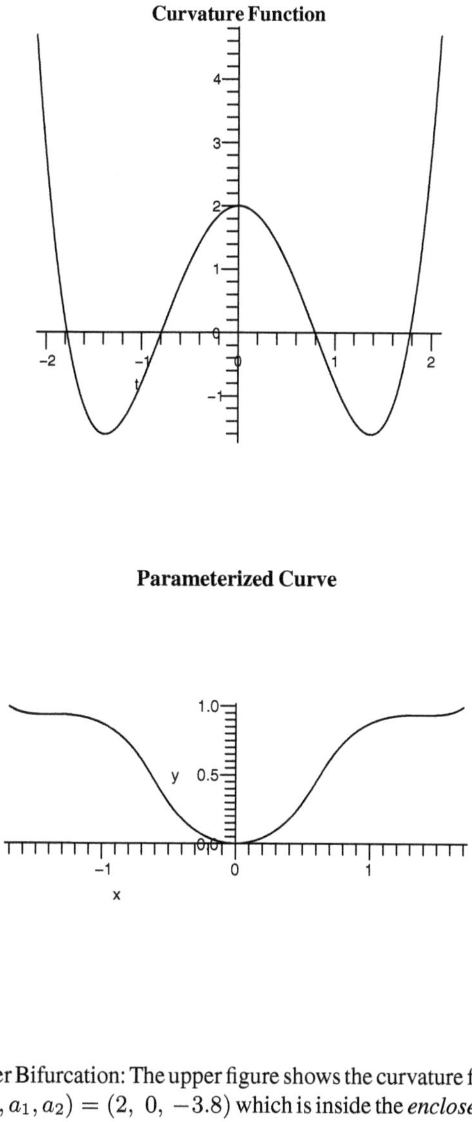


**Fig. 12.12.** After Bifurcation: The upper figure shows the curvature function $F(t) = t^4 + 2 - 3.8t^2$ at the point $(a_0, a_1, a_2) = (2,\ 0,\ -3.8)$ which is inside the *enclosed volume* of the zero-separatrix of the Interactive Unfolding of $EA_3$. The consequence is that the one singularity of the previous function has undergone a 7-fold bifurcation in the present function; i.e., a 3-fold critical-point bifurcation into non-degenerate negative minimum, positive maximum, negative minimum, and a 4-fold zero-bifurcation into 4 non-degenerate zeros. The singularity-configuration of the function is therefore $0m^-0M^+0m^-0$. That is, there has been a *maximal bifurcation.* Notice that the central extremum $M^+$, in the function, has value 2 which is the same as the position $a_0 = 2$ of the function in the 3D control space, because, in the control space, the function happens to be located directly above the negative $a_2$-axis. The lower figure shows the unit-speed parameterized curve that results from this curvature function. The 7 singularities in the curve are directly below the 7 singularities in the function. This 7-fold singularity-configuration is seen in the curve as a zone-restricted penetrative process; i.e., a central downward protrusion bounded on each side by an upward indentation.

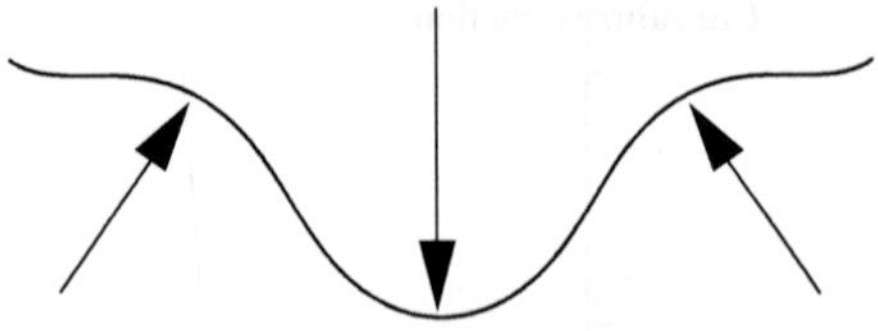

**Fig. 12.13.** This figure shows the parameterized curve of the previous figure, with the processes given by our process-inference rules. The central arrow is the protrusion process that has broken-through; and the two side arrows are the pinching processes that restrict its zone-of-influence.

## 12.7 Increasing the Pinching

This section illustrates the fact that going deeper into the enclosed volume of the zero-separatrix increases the pinching effect of the restricting penetrative extrema.

The three successive shapes in Fig 12.14 start with the last parameterized curve in the sequence just given, and not only move further into the enclosed volume, along the negative $a_2$-axis, but also move upward within the enclosed volume. We see that the protrusion successively increases, and the pinching successively creates a *neck*.

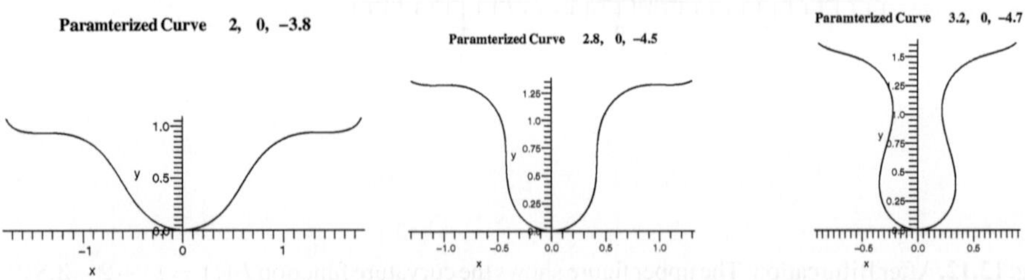


**Fig. 12.14.** As one goes further and upward in the enclosed volume, the protrusion lengthens and the pinching creates a neck. (The coordinates $(a_0, a_1, a_2)$, in the 3D control space, are shown above each shape.)

In contrast, the next figure, Fig 12.15, shows a sequence of six successive shapes along a straight line of *constant* height $a_0 = 2.9$, along the negative $a_2$-axis in the enclosed volume. The central protrusion in the shape therefore remains the same curvature 2.9, which equals that constant height $a_0 = 2.9$. However, the curvature of the restrictive indentations successively increases. No neck appears in the shape.

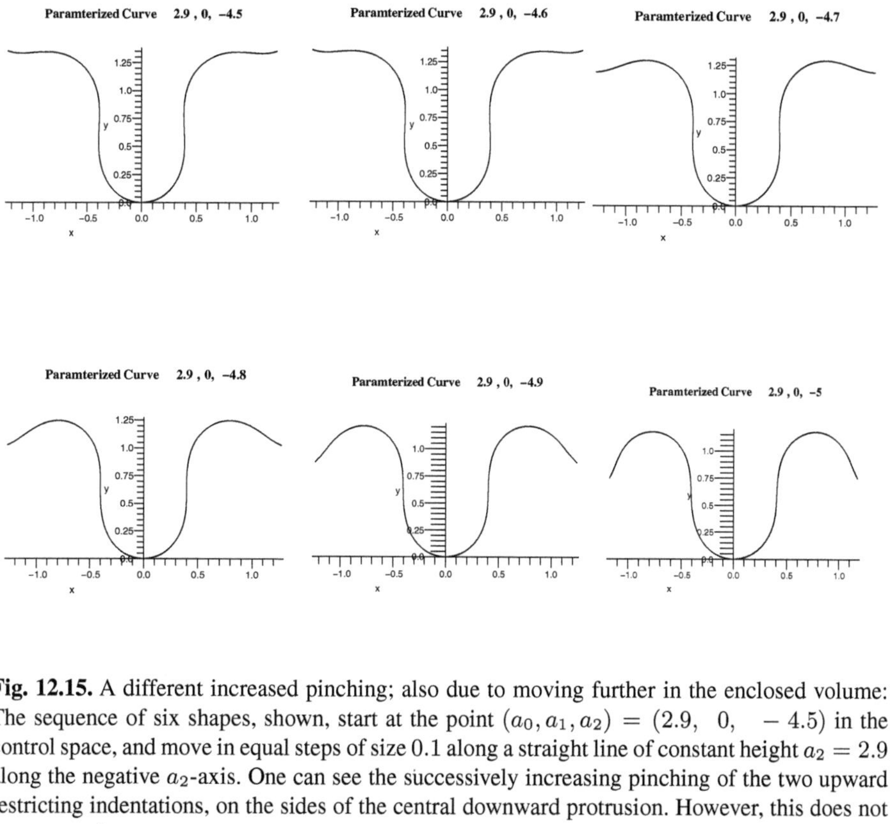


**Fig. 12.15.** A different increased pinching; also due to moving further in the enclosed volume: The sequence of six shapes, shown, start at the point $(a_0, a_1, a_2) = (2.9, \;\; 0, \;\; -4.5)$ in the control space, and move in equal steps of size $0.1$ along a straight line of constant height $a_2 = 2.9$ along the negative $a_2$-axis. One can see the successively increasing pinching of the two upward restricting indentations, on the sides of the central downward protrusion. However, this does not create a neck.

# 13

# Five-Fold Part-Formation Operators in the Symmetry-Restricted Interactive Unfoldings of $EA_5$ and $EA_{-5}$

## 13.1 Introduction

Chapter 6 elaborated a set of Process Grammar operators for the creation of 5-fold parts. These are listed as follows:

**5-FOLD PART-FORMATION OPERATORS**

**Symmetrically-paired pair-creation operators:**
$\pi^{\uparrow}\pi^{\downarrow},\ \ \pi^{\downarrow}\pi^{\uparrow},\ \ [C(3)\pi^{\uparrow}][C(3)\pi^{\downarrow}],\ \ [C(3)\pi^{\downarrow}][C(3)\pi^{\uparrow}]$

**Double-use of the 3-fold process-bifurcation operators:**
$B_{-3} \circ B_3,\ \ B_3 \circ B_{-3}$

**5-fold process-bifurcation operators:**
$B(5),\ \ [C(6)B(5)]$

**Continuation and pinching operators:**
$C,\ \ CC,\ \ CCC$

The present chapter will show that *all* these operators can be modeled within what we will call the Symmetry-Restricted Interactive Unfoldings of $EA_5$ and $EA_{-5}$. The germs in these cases are $t^6$ and $t^{-6}$ because they contain, respectively, the *complete* singularities $EA_5$ and $EA_{-5}$, each of which is simultaneously 5-fold degenerate as a critical point and 6-fold degenerate as a zero, which is the maximum degeneracy that we will require for creating the 5-fold parts. Since the operators listed above are symmetrically applied with respect to the initial extremum, we will consider, in the control space of the Interactive Unfoldings, the set of functions that are symmetric. This is what will be meant by the term *Symmetry-Restricted Interactive Unfoldings of* $EA_5$ *and* $EA_{-5}$.

## 13.2 Symmetry-Restricted Butterfly Catastrophe

Before developing the Symmetry-Restricted Interactive Unfolding of $EA_5$, we begin by reviewing what would be the corresponding catastrophe approach, as follows:

In Catastrophe Theory, the universal unfolding of the germ $t^6$ is the butterfly catastrophe, which has the following canonical form:

$$f(t;\ a_4, a_3, a_2, a_1) \quad = \quad \overbrace{t^6}^{germ} \quad + \ \overbrace{a_4t^4 + a_3t^3 + a_2t^2 + a_1t}^{perturbation}$$

For convenience of plotting, we have written the control variables in reverse order from Table 7.1 (page 244). Nevertheless, each variable $a_i$ still controls the term $t^i$.

Now, since the 5-fold part-formation operators developed in Chapter 6 are symmetric with respect to the initial extremum, we can remove the odd-power terms from the unfolding function of the butterfly catastrophe. That is, we will, correspondingly, consider the *symmetry-restricted* version of the above unfolding function, thus:

$$F(t;\ a_4, a_2) \quad = \quad \overbrace{t^6}^{germ} \quad + \ \overbrace{a_4t^4 + a_2t^2}^{perturbation} \qquad (13.1)$$

The control space is therefore 2-dimensional, given by coordinates $(a_4\ ,\ a_2)$. This unfolding function defines what one can call the *symmetry-restricted butterfly catastrophe*.

Fig 13.1 illustrates the fact that the two-dimensional control space $(a_4\ ,\ a_2)$ is divided into three Morse regions: Regions 1, 3, and 5, where the functions have respectively 1, 3, and 5, critical points. Being Morse regions, the functions within a region are qualitatively the same as each other, with respect to their critical-point structure; i.e., *right*-equivalent. A crucial fact to observe is that every function in the control space has at least one critical point. Also, the curve shown, in Fig 13.1, as separating Region 1 from Region 5, is the half-parabola $a_2 = \frac{a_4^2}{3}$ over the negative $a_4$-axis. As illustrated in the same figure, it will sometimes be helpful, in locating functions, to divide Region 1 into Regions 1a and 1b at the positive $a_2$-axis; and similarly to divide Region 3 into Regions 3a and 3b at the negative $a_2$-axis. However, the reader must remember that, only in the case of Regions 1, 3, and 5, will the term "Region" refer to a *full* Morse region.

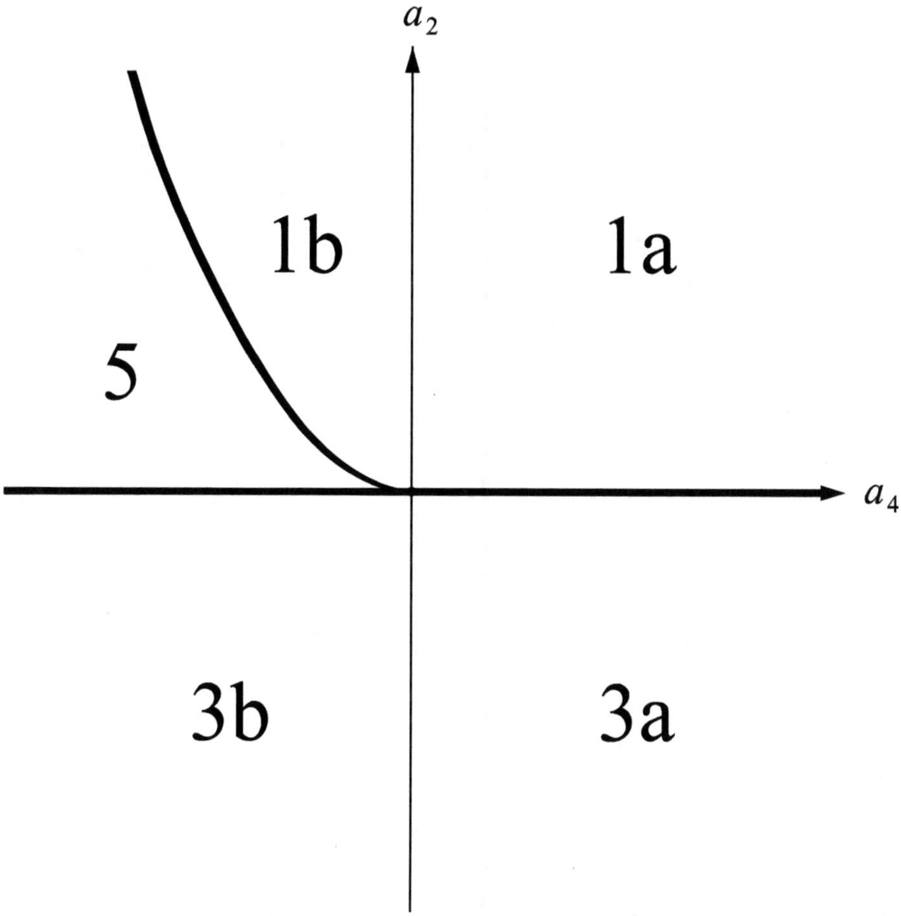


**Fig. 13.1.** The symmetry-restricted butterfly catastrophe: division of the control space.

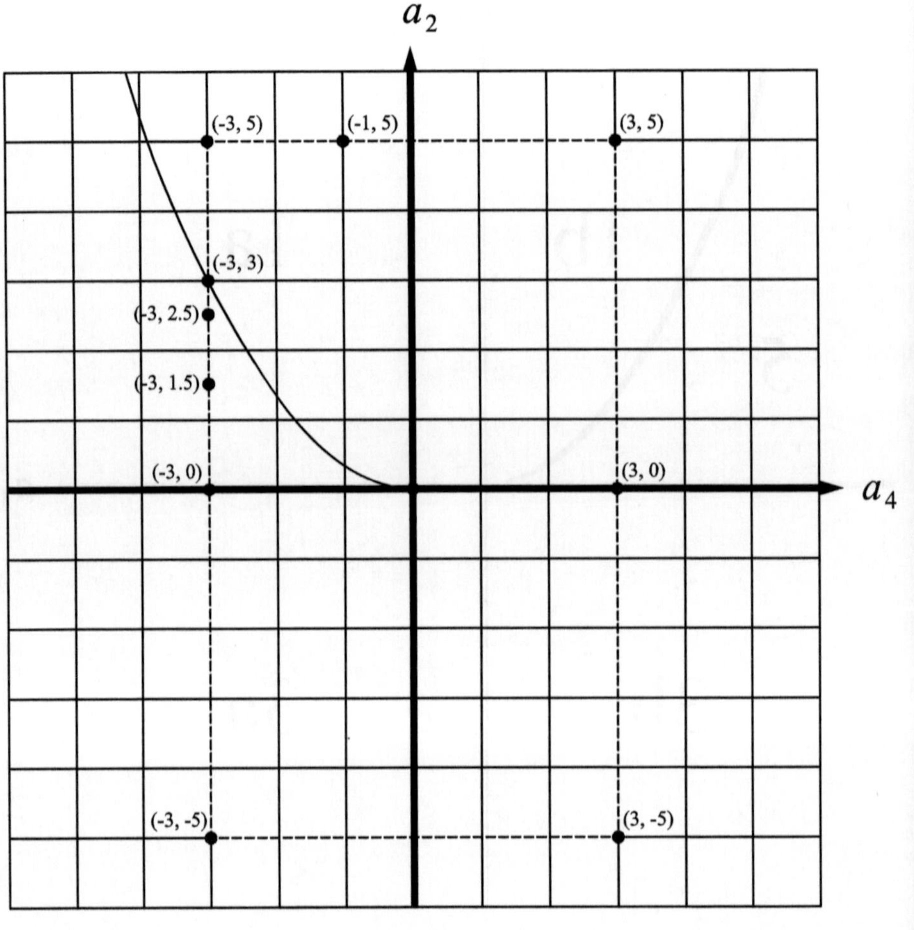


**Fig. 13.2.** Going around a rectangle in the control space of the symmetry-restricted butterfly catastrophe.

## Transitions between Regions

We will now examine how Catastrophe Theory views the functions as changing qualitatively as one goes from region to region; i.e., changes with respect to their critical points, not their zeros. To do so, we will go along the four sides of the rectangle shown in Fig 13.2, given by the corners $(a_4\ ,\ a_2) = (\pm3, \pm5)$. Since, throughout this route, we will be currently viewing the functions from the Catastrophe Theory point-of-view, the term *degeneracy* will mean degeneracy as a critical point, not as a zero.

**Region 1a $\longrightarrow$ Region 3a.**
This transition is illustrated in Fig 13.2 by going vertically down the right side of the rectangle; i.e., from position $(3, 5)$ to position $(3, 0)$ to position $(3, -5)$. The functions at these three successive positions are shown in Fig 13.3. The first function has a single non-degenerate minimum. When the function reaches the second position, which is within the positive $a_4$-axis of the control space in Fig 13.2, this minimum becomes degenerate as a critical point, as can be seen in the second graph in Fig 13.3. Then, as the function moves off this position within the positive $a_4$-axis, towards the third position $(3, -5)$ in the control space, the degenerate minimum in the function undergoes a 3-fold critical-point bifurcation into two non-degenerate minima and a central non-degenerate maximum, as illustrated in the third graph in Fig 13.3.

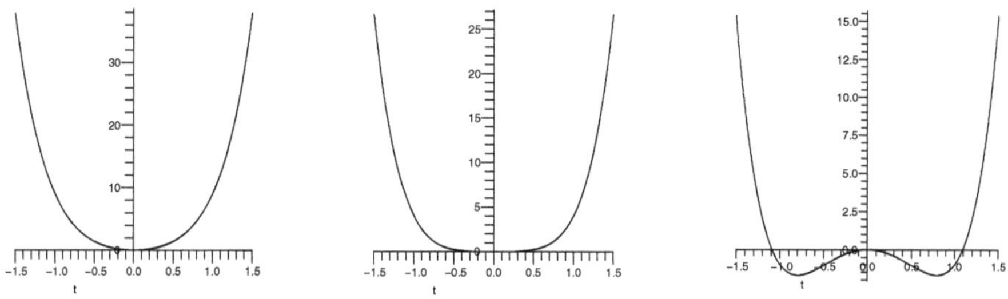

**Fig. 13.3.** Transition from the function at point $(3, 5)$ to point $(3, 0)$ to point $(3, -5)$ in the control space shown in Fig 13.2.

This qualitative progression, in the three successive functions, is entirely due to the inversion (from positive to negative) of the quadratic term $t^2$ in the unfolding function (13.1). We can therefore see that movement down this side of the rectangle corresponds to movement down the $a_2$-axis of the cusp catastrophe $A_3$. The same bifurcation happens along any downward crossing through the positive $a_4$-axis of the control space in Fig 13.2. Therefore *every* point on the positive $a_4$-axis corresponds to the cusp-point in the cusp catastrophe $A_3$.

**Region 1a $\longrightarrow$ Region 1b.**
This transition is illustrated in Fig 13.2 by going horizontally along the top side of the rectangle; i.e., from position $(3, 5)$ to position $(-1, 5)$ to position $(-3, 5)$. The functions at these three successive positions are shown in Fig 13.4. They each have one critical point: a non-degenerate minimum. This illustrates the fact that Regions 1a and 1b form, together, a single Morse region. Therefore, the movement we have just described, along the top of the rectangle shown in the control space, is simply a movement within the same Morse region. From the graphs in Fig 13.4, the reader can see that the only effect of the movement is that the sides of the graph undergo a flattening. This is the beginning of a rotating-tangent effect that will cause a qualitative change when one crosses the half-parabola, as will be seen in the next transition.

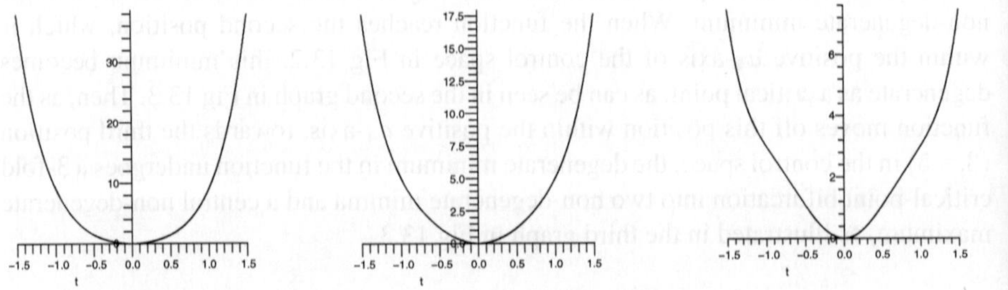

**Fig. 13.4.** Transition from the function at point $(3, 5)$ to point $(-1, 5)$ to point $(-3, 5)$ in the control space shown in Fig 13.2.

**Region 1b $\longrightarrow$ Region 5.**
This transition is illustrated in Fig 13.2 by going vertically down the left side of the rectangle; i.e., from position $(-3, 5)$, which is above the half-parabola, to position $(-3, 3)$, which is in the half-parabola, to position $(-3, 2.5)$, which is just below the half-parabola. The functions at these three successive positions are shown in Fig 13.5.

The first graph in Fig 13.5 has only one critical point: a non-degenerate minimum. Then, in the transition from the first graph to the second, a gradient in the left side of the first graph undergoes anti-clockwise rotation, while a gradient within the right side of the first graph undergoes a simultaneous clockwise rotation. The result, in the second graph, is that the rotating gradient on each side has reached the horizontal, producing a critical inflection-point on each side of the graph. Finally, in the transition to the third function, the two gradient-rotations have continued, so that the critical inflection-point, on the left side, has bifurcated into a minimum-maximum pair; and the critical inflection-point, on the right side, has bifurcated into a maximum-minimum pair; i.e., there is *pair-creation* on each side.

Earlier in the book, we saw that pair-creation into a minimum-maximum pair occurs by crossing the critical-point separatrix of the dual-fold catastrophe $A_{-2}$; and that pair-creation into a maximum-minimum pair occurs by crossing the critical-point separatrix

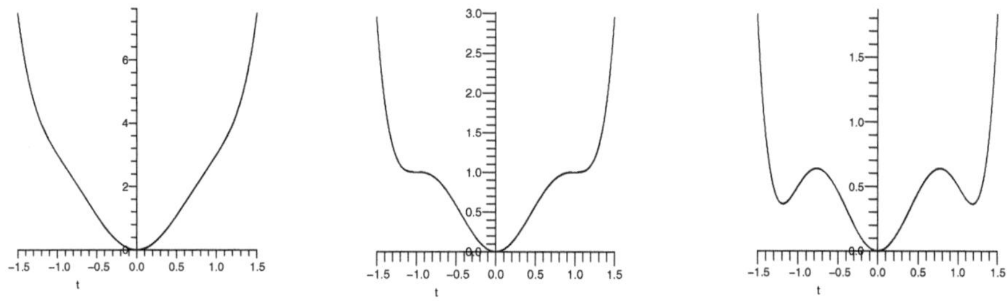


**Fig. 13.5.** Transition from the function at point $(-3, 5)$ to point $(-3, 3)$ to point $(-3,\ 2.5)$ in the control space shown in Fig 13.2.

of the fold catastrophe $A_2$. Thus, a point on the semi-parabola in the control space in Fig 13.2 corresponds to the simultaneous occurrence of (1) the critical-point separatrix of the dual-fold catastrophe (2) the critical-point separatrix of the fold catastrophe. That is, any point in the semi-parabola can be notated as $A_{-2}A_2$. In fact, the semi-parabola is the projection of the edges of two folds in the critical-point manifold of the butterfly catastrophe – one fold corresponding to the $A_{-2}$ catastrophe and the other fold corresponding to the $A_2$ catastrophe. Furthermore, these two folds, in the critical-point manifold, are symmetrically related to each other.

From this, we can see why, in the sequence of functions in Fig 13.5, the *symmetric double pair-creation* occurs as a result of crossing the semi-parabola in the control space of Fig 13.2. We can thus refer to the semi-parabola as the *symmetric double-fold curve*. Clearly, crossing this curve corresponds to our symmetrically-paired pair-creation operator $\pi^{\downarrow}\pi^{\uparrow}$, that is, the positive downward-upward operator, defined on page 204. However, there is a serious problem:

> **The positive downward-upward operator $\pi^{\downarrow}\pi^{\uparrow}$ of the Process Grammar does not allow the anchor-extremum (i.e., the minimum between the two application points) to have zero-value, as occurs in the catastrophe-theoretic functions in Fig 13.5. This is one of the reasons why we will have to create the Symmetry-Restricted *Interactive* Unfolding of $EA_5$.**

**Region 5 $\longrightarrow$ Region 3b.**

In this transition, we move further downward along the left side of the rectangle in Fig 13.2, crossing the negative $a_4$-axis.

In order to understand what happens, recall first that movement vertically down the control space corresponds to decreasing the coefficient $a_2$ of the $t^2$ term in the unfolding function. Consider the effect of this, within Region 5. The first two graphs in Fig 13.6 show the movement down from point $(-3,\ 2.5)$ to point $(-3,\ 1.5)$ in the control space. Both of these points are within Region 5, and therefore both their functions have five critical points.

However, from their graphs, we see that the effect of decreasing the $t^2$ term is to diminish the "amplitude" of the central minimum, while the "amplitude" of the two

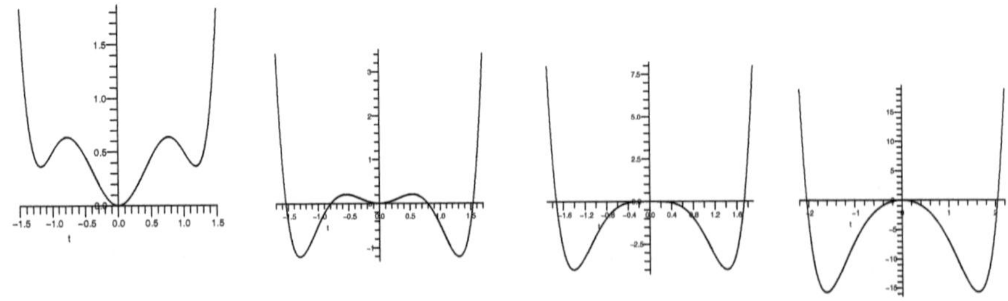

**Fig. 13.6.** Transition from the function at point $(-3,\ 2.5)$ to point $(-3,\ 1.5)$ to point $(-3,\ 0)$ to point $(-3, -5)$ in the control space shown in Fig 13.2.

side-minima increases. This trend continues as we further decrease the coefficient $a_2$ of the $t^2$ term. Thus, when the coefficient $a_2$ reaches zero, i.e., we reach the $a_4$-axis in the control space, the amplitude of the central minimum reaches zero size. As a result, the central minimum and the two maxima, of the function, coalesce to a single degenerate maximum – as we can see in the transition from the second to the third graph in Fig 13.6.

Then, as we move further downward in the control space, from the $a_4$-axis, most of the function remains qualitatively the same, the one qualitative change being that the central maximum becomes non-degenerate. The fourth graph in Fig 13.6 exhibits an example of the changed function, that at point $(-3, -5)$ in the control space.

It is extremely important now to view this same trajectory in the opposite direction; i.e., moving from Region 5 upward along the same line in the control space, crossing a point on the $a_4$-axis. Observe what happens in this movement: The single non-degenerate maximum of the fourth graph in Fig 13.6 becomes degenerate in the third graph, this graph being a point in the $a_4$-axis, and bifurcates into a non-degenerate minimum, maximum, minimum, in the second graph, this graph being a point above the $a_4$-axis.

Therefore, the upward crossing of the $a_4$-axis corresponds to a 3-fold bifurcation of a maximum. But, recall that this is what happens when one crosses the cusp-point of the *dual* cusp catastrophe $A_{-3}$. Thus, any point on the negative $a_4$-axis corresponds to the cusp-point of $A_{-3}$.

**Region 1a $\longrightarrow$ Region 5.**

The only way to go from Region 1a to Region 5, without going through another region, is via the origin $(a_4, a_2) = (0, 0)$ of the control space. An example of such a sequence of functions is the sequence of three functions shown in the *bottom row* of page 471, where the first function is in Region 1a, the second function is at the origin, and the third function is in Region 5.

By the unfolding equation (13.1) on page 462, the function at the origin is

$$f(t; 0, 0) = t^6$$

which is the *germ* for $A_5$. It has only one critical point: a 5-fold *degenerate* minimum.

Thus, at a starting point in Region 1a, the function contains only one critical point: a single *non-degenerate* minimum. Then moving onto the origin, this minimum becomes

5-fold degenerate. Finally moving from the origin into Region 5, this 5-fold degenerate minimum undergoes a simultaneous 5-fold bifurcation into three non-degenerate minima and two non-degenerate maxima.

**The Critical-Point Separatrix.**
From the above discussion, we can therefore see that the critical-point separatrix of the symmetry-restricted butterfly catastrophe is given by the bold lines (the half-parabola and $a_4$-axis) in Fig 13.7. The separatrix is divided into *four* components, listed below, where the functions within a component are qualitatively equivalent to each other. Since a function in the separatrix has at least one degenerate critical point, the function is locally equivalent, at the degenerate critical point, to the germ singularity of one of the cuspoid catastrophes $A_{\pm k}$. These equivalences are marked on the separatrix components in Fig 13.7, and are defined as follows:

1. Each point on half-parabola = Simultaneous symmetric occurrence of separatrix-point in dual-fold and in fold catastrophe, $A_{-2}A_2$.
2. Each point on positive $a_4$-axis = Cusp-point of cusp catastrophe $A_3$.
3. Each point on negative $a_4$-axis = Cusp-point of dual cusp catastrophe $A_{-3}$.
4. Origin point = $A_5$.

Furthermore, the separatrix divides the control plane into open regions where the points within a single region are Morse functions that are qualitatively equivalent to each other with respect to their critical-point structure.

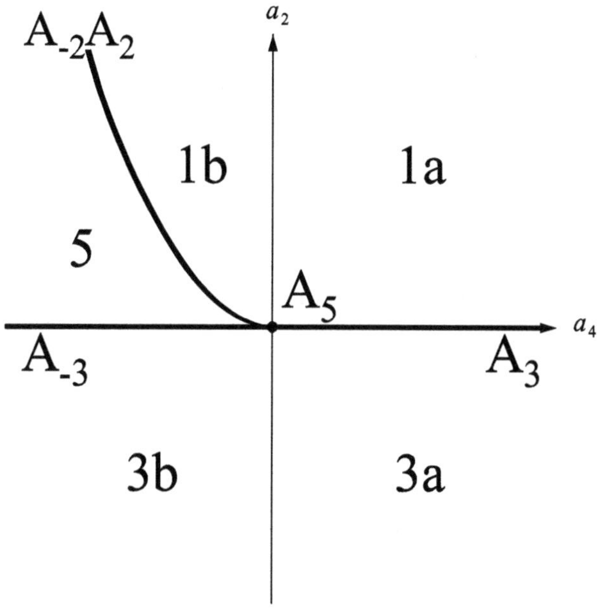


**Fig. 13.7.** The critical-point separatrix of the symmetry-restricted butterfly catastrophe.

## 13.3 Inadequacy of Modeling 5-Fold Part-Formation in Catastrophe Theory

In Chapter 6, we gave three scenarios for 5-fold part-formation. They were:

(1) Symmetrically-paired pair-creation around a central extremum.
(2) Two successive 3-fold extremum-bifurcations from the central extremum.
(3) One simultaneous 5-fold extremum-bifurcation from the central extremum.

Recall that a fundamental condition of the Process Grammar is that the extrema which appear in the domain and co-domain of extremum-bifurcations cannot be zeros, because zero extrema, being structurally unstable, must be transition-states, and therefore zeros in the domain and co-domain must be *transversal* to the $t$-axis. Observe that violating this fundamental condition makes the above three types of extremum-bifurcation correspond, respectively, to three routes in the symmetry-restricted butterfly catastrophe. All three routes go from Region 1 to Region 5 in the control space. They are, respectively, the following:

**3 ALTERNATIVE ROUTES**

**(1) Region 1 $\longrightarrow$ 5 across the half-parabola. This simultaneously bifurcates a minimum-maximum pair and a maximum-minimum pair, respectively, in the left and right sides of the initial minimum (simultaneous dual-fold and fold catastrophes $A_{-2}A_2$).**

**(2) Region 1 $\longrightarrow$ 3 $\longrightarrow$ 5, where the first stage crosses the positive $a_4$-axis and the second stage crosses the negative $a_4$-axis. In the first stage, the initial minimum undergoes a 3-fold bifurcation into a central maximum and two outer minima (cusp catastrophe $A_3$). In the second stage, this central maximum undergoes a 3-fold bifurcation into a central minimum and two outer maxima (dual cusp catastrophe $A_{-3}$).**

**(3) Region 1 $\longrightarrow$ 5 via the origin. The initial minimum undergoes a simultaneous 5-fold bifurcation into three minima alternating with two maxima. ($A_5$).**

The functions along these three routes are illustrated, respectively, in Figs 13.8 – 13.10. The important thing to observe from these functions is this: There is a serious problem in trying to use the *three routes*, in the symmetry-restricted butterfly catastrophe, to model the *three scenarios*, in the Process Grammar. The problem is this: As illustrated in Figs 13.8 – 13.10, *all* the functions along the three routes in the symmetry-restricted butterfly catastrophe have critical-points of zero value.

From this, our crucial conclusion is that each of these functions must be located on a *zero-separatrix*. However, Catastrophe Theory is not interested in zero-separatrices.

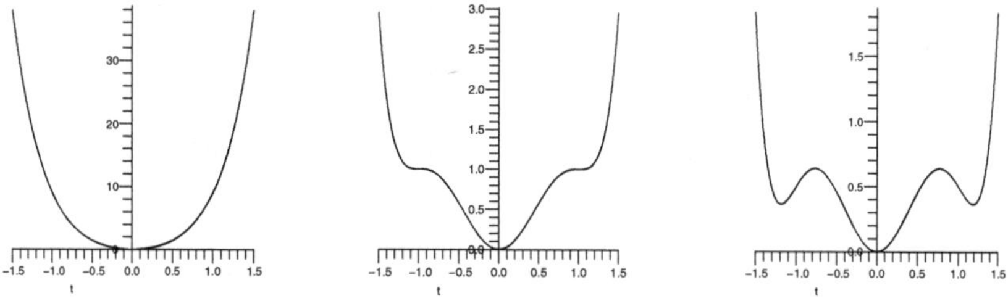


**Fig. 13.8.** Route 1 in the symmetry-restricted butterfly catastrophe: Two simultaneous symmetrically related 2-fold extremum-bifurcations (dual-fold and fold catastrophes). The middle function is the intervening non-Morse function.

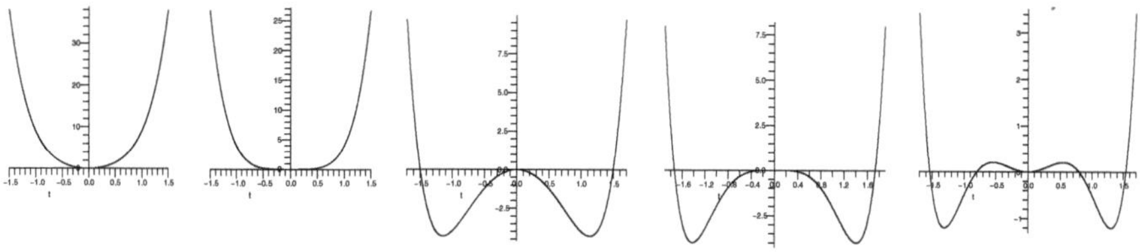

**Fig. 13.9.** Route 2 in the symmetry-restricted butterfly catastrophe: 3-fold extremum-bifurcation (cusp catastrophe) followed by another 3-fold extremum-bifurcation (dual-cusp catastrophe). The second and fourth functions are the intervening non-Morse functions.

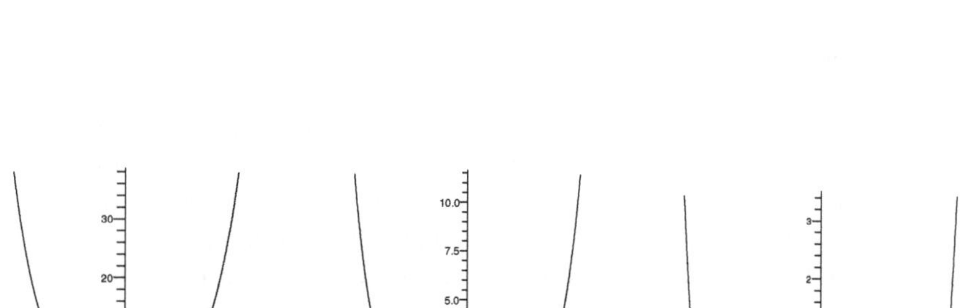


**Fig. 13.10.** Route 3 in the symmetry-restricted butterfly catastrophe: Simultaneous 5-fold extremum-bifurcation. The middle function is the intervening non-Morse function, the germ.

To handle this structure, we need to create the Symmetry-Restricted *Interactive* Unfolding of $t^6$ as the *complete* singularity $EA_5$. Before doing so, let us list the different types of zero-separatrix, whose presence can be inferred from the functions in the catastrophe.

1. From Fig 13.8, we infer that the positive $a_2$ half-plane must be part of the *minima* zero-separatrix.

2. From Fig 13.9, we infer that, in addition, the positive $a_4$-axis must be part of the *minima* zero-separatrix. We also infer that the negative $a_2$ half-plane, and the negative $a_4$-axis, must be part of the *maxima* zero-separatrix.

3. From Fig 13.10, we infer that, in addition, the origin must be part of the *minima* zero-separatrix.

In addition, the functions in an earlier figure, Fig 13.6 on page 468, imply the presence of more zero-separatrices: Recall that the first two functions in that figure correspond to points that are *both* within Region 5. Most crucially they are a move *downwards* in this region from point $(-3,\ 2.5)$ to point $(-3,\ 1.5)$ in the control space, as shown in Fig 13.2 page 464. What we observe, in these first two functions (Fig 13.6 page 468), is that the two outer minima are initially positive, and then become negative. They must therefore cross a zero-separatrix for *two* minima. This leads to the inference of additional zero-separatrix structure, as follows:

4. From Fig 13.6 (page 468), we infer that, between the half-parabola and the negative $a_4$-axis, there must be a curve that represents a zero-separatrix of *two simultaneous minima*.

Now observe this: From statements 1-3 above, we infer that the entire control plane of the symmetry-restricted butterfly catastrophe must be a zero-separatrix sheet. And from statement 4, we infer that there must also be another zero-separatrix sheet cutting through that sheet.

In relation to this, consider the following: A fundamental fact is that the process-continuation operators of the Process Grammar cross the zero-separatrix sheets. Therefore, the operators must move within the enlarged space of the Symmetry-Restricted Interactive Unfolding of $EA_5$. Indeed, within this enlarged space, we will discover that the zero-separatrix has even more sheets to be crossed than those listed above.

Furthermore, this enlarged space is required also because the process-bifurcation operators cannot be modeled in the symmetry-restricted butterfly catastrophe, since, in all three routes, in the catastrophe space, the domain extrema or anchor extrema of the bifurcations have zero value. For example, the space does contain the distinction between the application of a process-bifurcation operator to a positive minimum, i.e., a squashing, and its application to a negative minimum, i.e., an indentation. Furthermore, by noting the zero extrema, in the functions in the canonical unfolding of the symmetry-restricted butterfly catastrophe, we see that again our enlarged space must realize the following important fact, which we have stated previously:

**In the Process Grammar, the process-continuation operators must organize the process-bifurcation operators.**

That is, the zero extrema, seen in the catastrophe, must, in the Interactive Unfolding, be transitions in process-continuation operators that relate the positive and negative versions of the extrema; and must therefore be such that the process-continuation operators relate the positive and negative process-bifurcation operations; i.e., what we have called the *Interactive* Structure.

In particular, we saw in Chapter 6 that the relevant process-continuation operators, for 5-fold part-formation, are the ordinary continuation operator $C$, the 2-fold pinching operator $CC$, and the 3-fold pinching operator $CCC$. These must therefore have an organizing role with respect to the process-bifurcation operators that are responsible for 5-fold part-formation. Thus, to accommodate both the process-continuation and process-bifurcation operators, we must now invent the Symmetry-Restricted Interactive Unfolding of $EA_5$. This will allow us to show that the Symmetry-Restricted Interactive Unfoldings of $EA_5$ and $EA_{-5}$ realize the entire set of Process Grammar operations listed in Chapter 6 for 5-fold part-formation.

## 13.4 Symmetry-Restricted Interactive Unfolding of $EA_5$

Using our usual method of creating the unfolding function for an Interactive Unfolding, the unfolding function for the Symmetry-Restricted Interactive Unfolding of $EA_5$ is obtained by taking the unfolding function for the catastrophe version, that is, $F(t; a_4, a_2) = t^6 + a_4t^4 + a_2t^2$, and adding the control parameter $a_0$, thus:

$$F(t; a_4, a_2, a_0) \quad = \quad t^6 + a_4t^4 + a_2t^2 + a_0 \qquad (13.2)$$

Therefore we obtain a larger control space, $(a_4, a_2, a_0)$, which can be viewed as the 2D control space, shown in Fig 13.7 (page 469), extended in the perpendicular direction by the extra parameter $a_0$, which will reveal far more complex structures.

We will now construct the zero-separatrix in this 3D control space. Recall, from section 8.5, that Method 1 for constructing the zero-separatrix was by first constructing the critical-*value* surface of the catastrophe version of the germ unfolding, and then inverting it. This method will be used again here, because it gives insights into the structure of the separatrix.

First observe that the critical-value surface exists in the 3D space $(a_4, a_2, F)$, where $F$ gives the *values* of critical points in the functions located at the points in the catastrophe 2D control-plane $(a_4, a_2)$.

We will now build up the critical-value surface in a succession of stages. Begin by recalling our conclusion, in section 13.3, that one of the sheets in the surface must be the entire $(a_4, a_2)$ plane at F-level 0. The reason is that, because we are considering the symmetry-restricted butterfly catastrophe, any function $f(t)$ in that plane has a critical point at its origin $t = 0$ that has value $f(0) = 0$. Thus the $(a_4, a_2, 0)$ plane will be called the $F = 0$ sheet of the critical-value surface. Viewing this plane as Fig 13.7 (page 469), together with the example functions in Figs 13.8 – 13.10, we observe the following facts, which were noted in the previous section:

**$F = 0$ SHEET**
**OF CRITICAL-VALUE SURFACE**

**1. For all functions in the positive $a_2$ half-plane, and the positive $a_4$-axis (as well as the germ), the origin critical-point is a zero *minimum.***

**2. For all functions in the negative $a_2$ half-plane, and the negative $a_4$-axis, the origin critical-point is a zero *maximum.***

Now let us add, to this, the other sheets that belong to the critical-value surface. In order to discover and understand these additional sheets, we will go along three successive lines $a_2 = constant$ in the catastrophe control plane $(a_4, a_2)$.

### Line $a_2 = -3$

Begin by going along the line $a_2 = -3$ in the catastrophe control plane as shown in Fig 13.11. Let us examine the functions along this line. For example, consider the functions at positions $a_4 = -4$ and $a_4 = +4$, on this line (i.e., at the dots shown).

These two functions are drawn in Fig 13.12. The functions exhibit the qualitative structure of any function along this line: two negative minima $m^-$ of the same $F$-value as each other (in the function), and a central maximum of $F$-value zero. Because the $F$-values of the two minima are equal, we will say that this $F$-value corresponds to a *double-minimum*.

Now observe the following: The two functions are drawn to scale. Therefore what we see is that, in going from the right function to the left function, the $F$-value of the double-minimum has plunged downward enormously. However the $F$-value of the maximum has remained zero.

These facts are captured in Fig 13.13. Note that the vertical axis in this figure is the $F$-axis, and the horizontal axis is the $a_4$-axis. The curve shown is marked $a_2 = -3$ $mm$ because it gives the value of the double-minima in the functions along the line $a_2 = -3$ in the control space in Fig 13.11. We observe the enormous downward plunge of the double-minima, in going from right-to-left in the diagram.

Also note that the horizontal axis $F = 0$, in Fig 13.13, represents the zero-value of the maximum in all the functions along the line $a_2 = -3$.

Thus, pick any vertical line in Fig 13.13. Consider this line to be located at a point on the line $a_2 = -3$ in the control space. Therefore, this vertical line corresponds to the $F$-axis of the function at that point in the control space. This function has a double-minimum where the vertical line intersects the curve labeled $a_2 = -3$ $mm$, and a maximum where the vertical line intersects the $F = 0$ axis.

Therefore, we conclude the following: In Fig 13.13, the curve marked $a_2 = -3$ $mm$ and the $F = 0$ axis, represent the slice through the *critical-value surface* along the line $a_2 = -3$ in the 2D control space. That is, the critical-value surface has two sheets over this line.

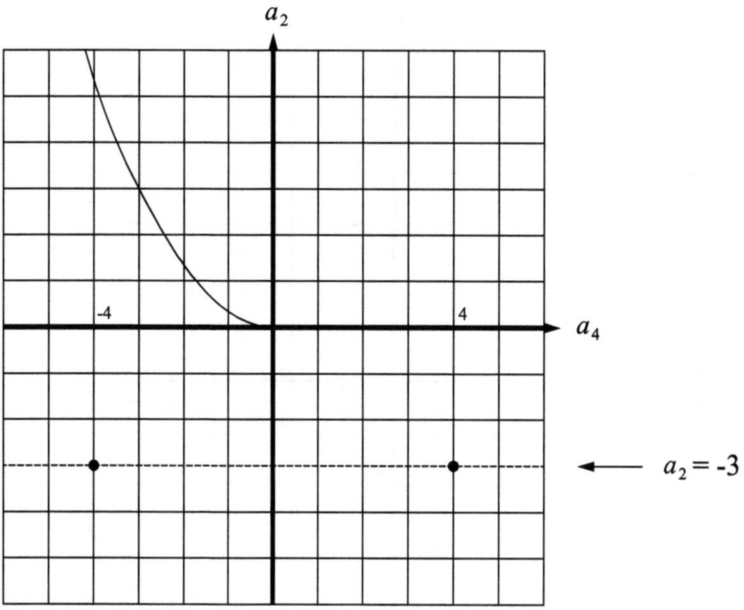


**Fig. 13.11.** We will be moving along the line $a_2 = -3$ in the catastrophe control plane.

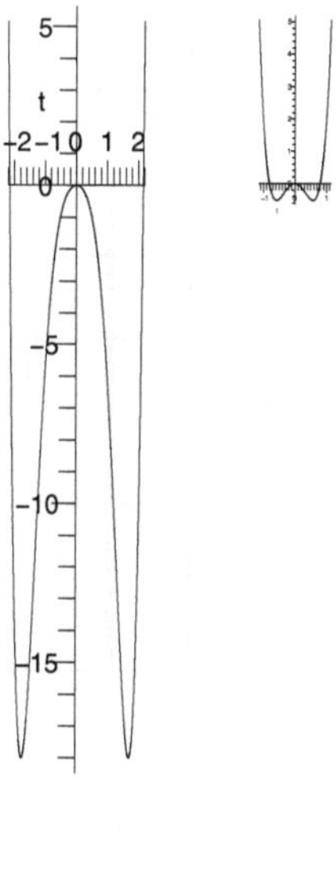


**Fig. 13.12.** The functions at the two points marked in Fig 13.11.

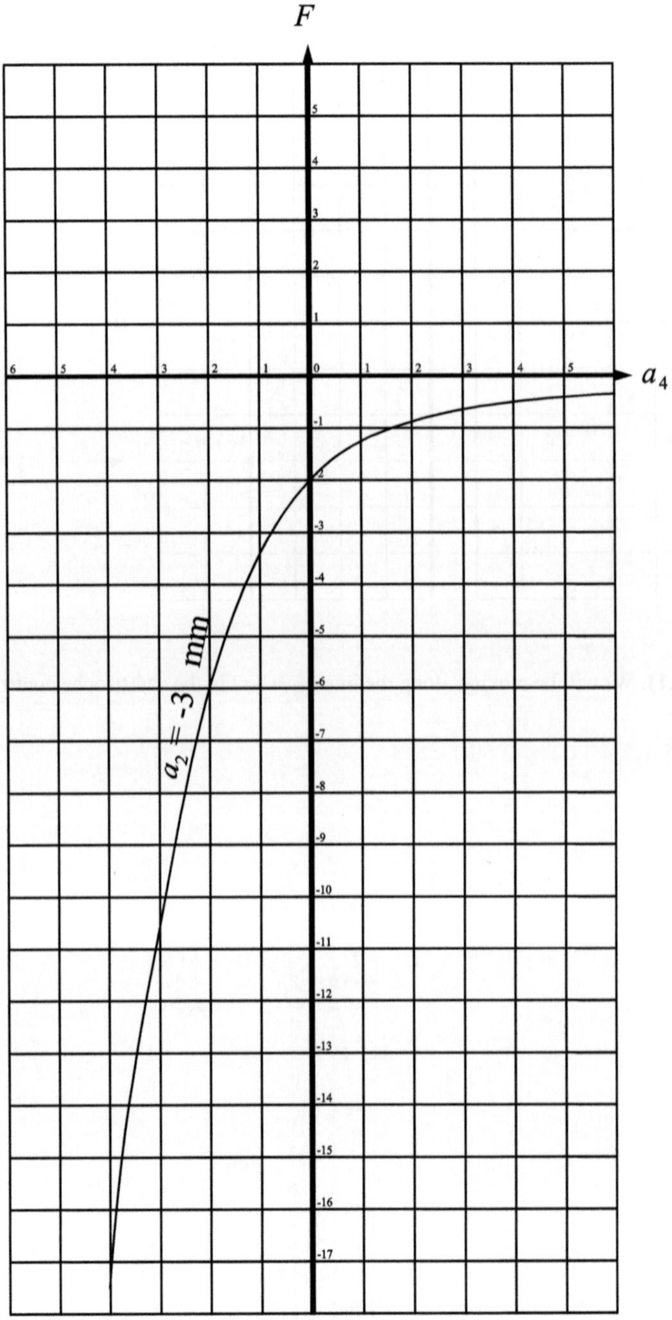


**Fig. 13.13.** The curve, marked $a_2 = -3\ mm$, gives the $F$-values of the double-minima in the functions along the line $a_2 = -3$ in the catastrophe control plane $(a_4, a_2)$. Note that all these functions also have a maximum at the $F = 0$ value. So imagine that each function, corresponding to a vertical line in this figure, has two minima where the vertical line intersects the curve, and a maximum where the vertical line intersects the $F = 0$ axis in this figure.

**Line $a_2 = 0$**

Now let us consider the line $a_2 = 0$ in the catastrophe control space, as illustrated in Fig 13.14. We will examine the functions along this line. For example, consider the functions at the three successive positions $a_4 = 4, 0, -4$, on this line (i.e., at the dots shown).

These three positions correspond respectively to the three successive functions drawn in Fig 13.15. The right-most function has only one critical point: a minimum which has zero $F$-value. The middle function, which is the germ, also has only one critical point: a minimum which has zero $F$-value. The left-most function has three critical points: a single maximum which has zero $F$-value, and two minima which have the same negative $F$-value as each other (i.e., a double-minimum).

These facts are captured in Fig 13.16 in the following curve, which represents the critical-*value* structure: Going from right-to-left, the critical-value curve starts as the positive $a_4$-axis, which represents the single minimum of zero $F$-value, in each of the functions along this axis. Then the critical-value curve continues as the origin, which represents the single minimum of zero $F$-value, within the germ. Then, immediately on leaving the origin, the critical-value curve bifurcates into two branches: (1) the negative $a_4$-axis, which represents the single maximum of zero $F$-value, in each of the functions along this axis; and (2) the downward curve labeled $a_2 = 0 \ mm$, which represents, within each of the same functions, the double-minimum, which again plunges downward. Notice that the point at which the critical-value curve bifurcates is a cusp.

Within the same figure, Fig 13.16, we also show the curve $a_2 = -3 \ mm$ that was previously shown in Fig 13.13.

A valuable way to understand Fig 13.16 is as follows: Imagine that the $a_2$-axis is perpendicular to the figure (at the origin), and that the viewer is on the negative $a_2$-axis, facing the figure. Then the catastrophe 2D control space $(a_4, a_2)$ is, of course, the horizontal plane that is perpendicular to the figure, at the level of the $a_4$-axis shown. This horizontal plane is one sheet of the critical-value surface; i.e., the $F = 0$ sheet. In this 3D view, the curve $a_2 = -3 \ mm$ is nearer the viewer and the curve $a_2 = 0 \ mm$ is further into depth from the viewer. The two curves are within a *second sheet* of the critical-value surface. Therefore, this second sheet rises up, as one goes into depth away from the viewer, and hits the horizontal $F = 0$ sheet at the positive $a_4$-axis and the origin.

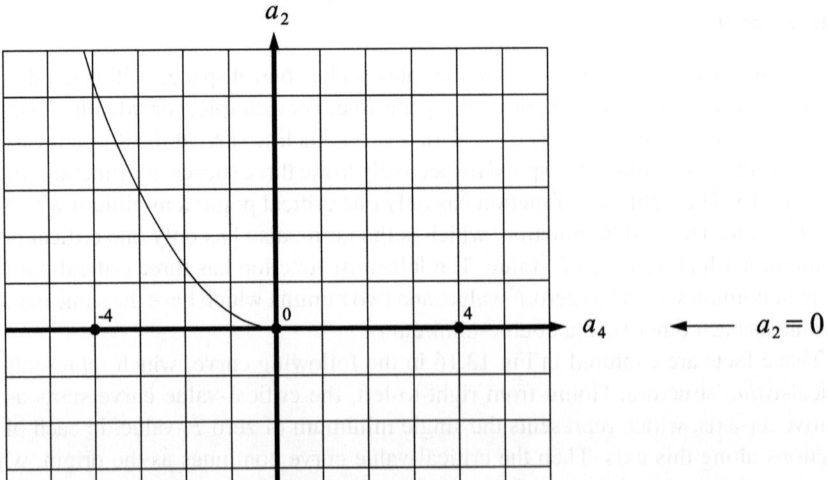


**Fig. 13.14.** We will be moving along the line $a_2 = 0$ in the catastrophe control plane.

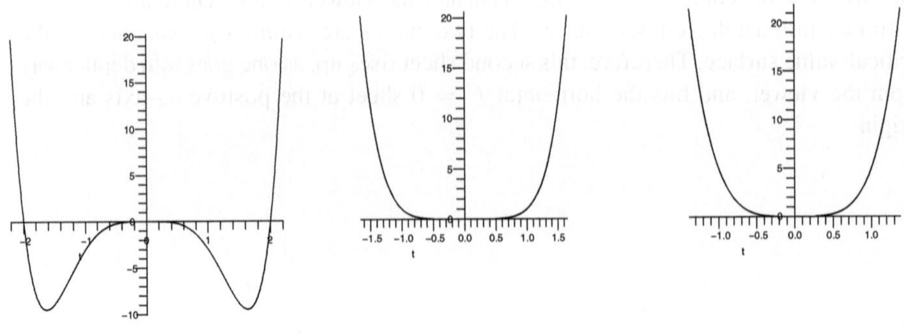

**Fig. 13.15.** The functions at the three successive points marked in Fig 13.14.

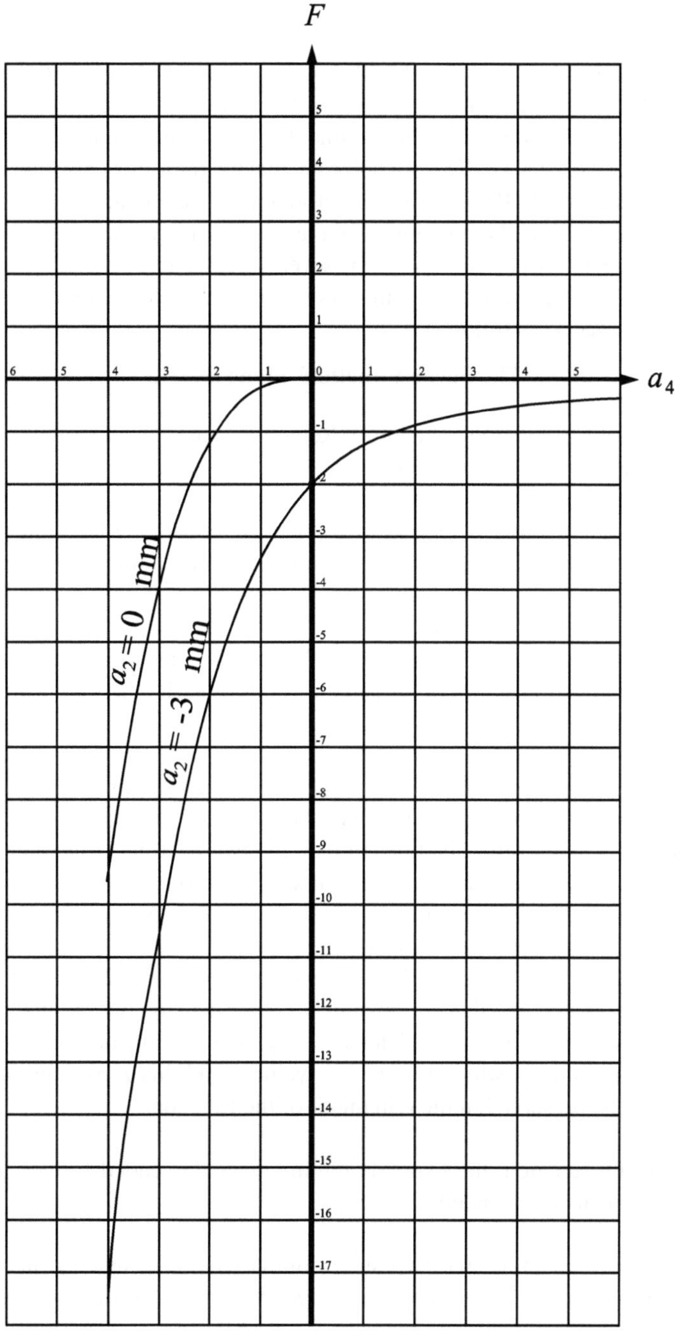


**Fig. 13.16.** This figure adds, to Fig 13.13, the curve $a_2 = 0\ mm$ that gives the $F$-values of the double-minima in the functions along the negative $a_4$-axis. These functions also have a maximum of value zero. The functions on the positive $a_4$-axis, and the origin, all have a minimum of value zero.

### Line $a_2 = 2$

Now let us consider the line $a_2 = 2$ in the $(a_4, a_2)$-control plane, as illustrated in Fig 13.17. We will examine the functions along this line. For example, consider the functions at the seven successive dots shown on the line. From right-to-left, they are at positions, $a_4$ = 3, -2, -2.24, -2.6, -2.8, -2.828, -3.4. Observe that the right-most position is in Region 1a. The next position is in Region 1b, just before the half-parabola. The next position is *on* the half-parabola. And the remaining four are successively after the half-parabola.

These seven positions correspond to the seven successive functions which are drawn in Fig 13.18. It is worth going along this row, from right-to-left, carefully studying the bifurcations that occur, as follows:

**The Seven Functions along $a_2 = 2$**

1. The right-most function, which is in Region 1a, has only one critical point: a minimum which has zero $F$-value.

2. The second function, which is in Region 1b, shows the beginning of rotation in the sides, but still has only one critical point: a minimum which has zero $F$-value.

3. The third function, which is on the half-parabola, has three critical points: two critical inflection-points of the same $F$-value as each other, and a minimum which has zero $F$-value.

4. The fourth function has five critical points, because a minimum-maximum pair and a maximum-minimum pair have been created, respectively, on the left side and right side of the function. In fact, the two maxima, which have been created, are positive and the same $F$-value as each other; and therefore constitute a *double-maximum*, which has a single $F$-value. Also, the two minima, which have been created, are positive and the same $F$-value as each other; and therefore constitute a *double-minimum*, which has a single $F$-value. Observe that the two created maxima are obviously *higher* than the two created minima – a fact which will be important later. Also note that the function has preserved the central minimum, which has zero $F$-value – and this zero minimum will continue to be preserved in the remainder of the functions.

5. In the fifth function, the two side minima have descended further, and the maxima have also descended.

6. In the sixth function, the two side minima have now reached the $t$-axis; which means that there are now *3 zero critical points* (all minima), whereas before there was only one.

7. In the seventh function, the two side minima have moved below the $t$-axis, which means that their zero values, in the previous function, have each undergone a 2-fold zero-bifurcation. There are now therefore 5 zeros.

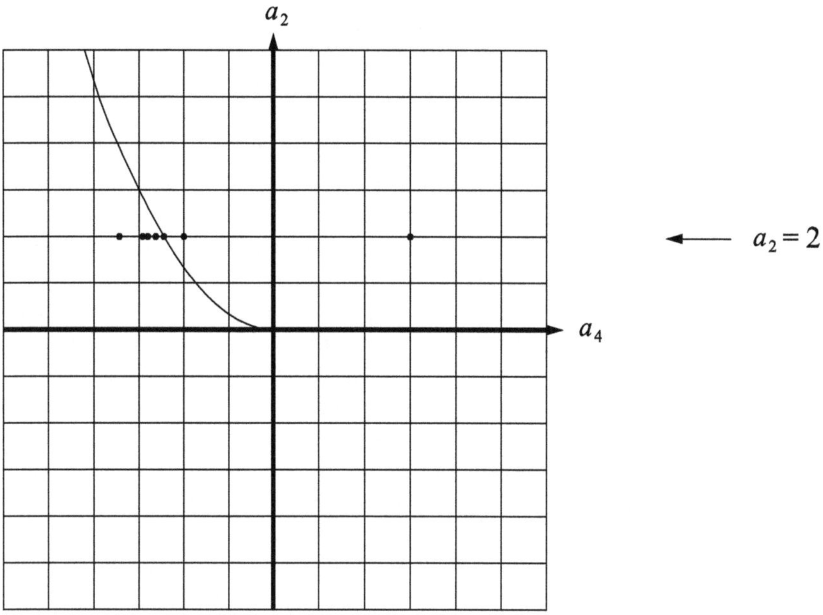


**Fig. 13.17.** We will be moving along the line $a_2 = 2$ in the catastrophe control plane.

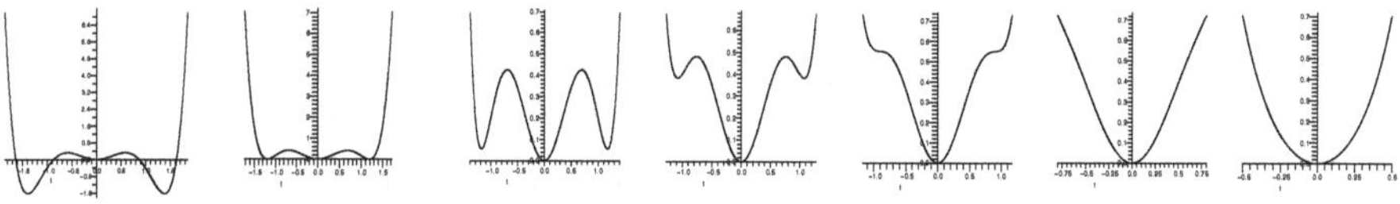

**Fig. 13.18.** The functions at the seven successive points marked in Fig 13.17.

These facts are captured in Fig 13.19 in the following critical-value curves added to the previous cases: Going from right-to-left, the first added curve is the entire $a_4$-axis which, in the current situation, has the following interpretation: Since the seven functions are all on the line $a_2 = 2$ in the control-plane, we are currently viewing the $a_4$-axis, in the figure, as having the $a_2$-value 2. It is therefore in the positive-$a_2$ half-plane, of the control space, which the reader will recall, from page 474, is a sheet of minima in the critical-value surface. Thus, for these seven functions, the $a_4$-axis, in the figure, represents the single minimum of zero $F$-value, that occurs in all those functions.

Now the seven functions correspond to *seven vertical lines* at the corresponding positions on the $a_4$-axis ($a_4$ = 3, -2, -2.24, -2.6, -2.8, -2.828, -3.4), and it is worth studying these vertical lines, as follows: The right-most function corresponds to a vertical line in the right half of the figure. Its only critical point has value where this vertical line crosses the $a_4$-axis; i.e., value zero. The same is true of the next function. However, the third function, that on the half-parabola (in the control space), corresponds to the vertical line (in Fig 13.19) which hits the cusp-point (in the upper-left quadrant) from which two curves bifurcate. The upper curve is labeled $a_2 = 2\ MM$ and the lower curve is labeled $a_2 = 2\ mm$. The reason is as follows: The cusp-point corresponds to the fact that the function located on the half-parabola contains two critical inflection-points, from which two maxima and two minima are going to bifurcate. Furthermore, the two maxima are going to be the same value as each other; i.e., they will be a double-maximum $MM$ yielding a single critical value. And also, the two minima are going to be the same value as each other; i.e., they will be a double-minimum $mm$ yielding a single critical value.

Thus, to model this bifurcation of four extrema, consider the fourth function. It is positioned in the catastrophe control plane *after* the half-parabola as shown in Fig 13.17. Therefore, as we observe from Fig 13.19, its vertical line must be *after the cusp-point* and must cross both curves, $a_2 = 2\ MM$ and $a_2 = 2\ mm$. The interpretation of this crossing is that the function has two maxima of the $F$-value where the vertical line crosses the curve $a_2 = 2\ MM$, and two minima of the $F$-value where the vertical line crosses the curve $a_2 = 2\ mm$. Notice that the $a_2 = 2\ MM$ curve is higher than the $a_2 = 2\ mm$ curve, which follows from the fact that the bifurcated maxima are obviously higher than the bifurcated minima. Furthermore, observe that the same vertical line also crosses the $a_4$-axis because the function also has a central minimum that has zero $F$-value.

Now go on to the fifth function. It corresponds to a vertical line a little further to the left. As can be seen from the two curves, $a_2 = 2\ MM$ and $a_2 = 2\ mm$, which it crosses, both the double-maxima and the double-minima have descended, but are still positive.

Next, the sixth function corresponds to a vertical line a little further to the left, where the $a_2 = 2\ mm$ curve *crosses the $a_4$-axis*. This means that the two side minima have descended to *zero* $F$-value, and the function therefore has 3 zero minima. The double-maximum $MM$ curve has again descended but has remained positive.

The seventh function corresponds to a vertical line a little further leftward, where the $a_2 = 2\ mm$ curve *has gone below the $a_4$-axis*. This means that the outer two side minima of the function have descended below the value zero. The double-maximum $MM$ curve

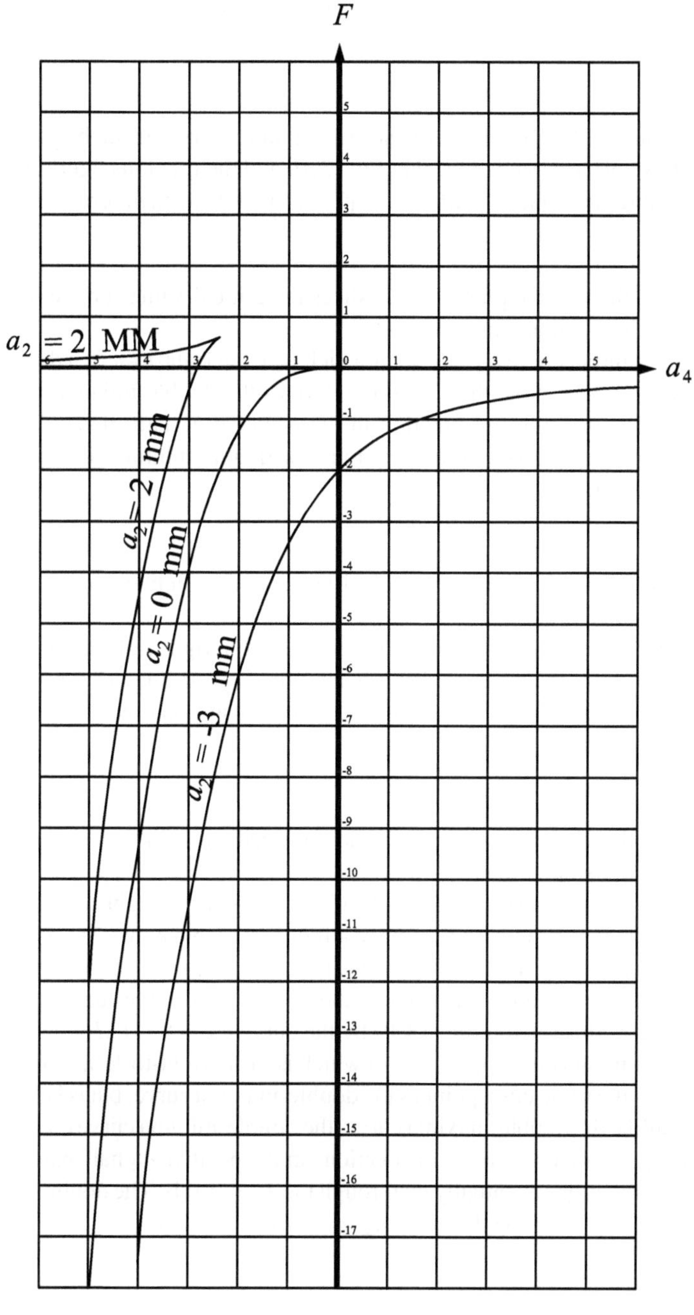


**Fig. 13.19.** This figure adds, to Fig 13.16, three curves corresponding to the functions along the line $a_2 = 2$. One is the $a_4$-axis consisting entirely of zero minima. The other two curves start at a *cusp-point*, which is over a point on the half-parabola in the 2D control space. The two curves are (1) the curve $a_2 = 2\ MM$, which is the $F$-value of the double-maxima emanating from the two critical inflection-points in that starting function, and (2) the curve $a_2 = 2\ mm$, which is the $F$-value of the double-minima also emanating from those two critical inflection-points.

has again descended but has remained positive, indicating that the two maxima have remained positive.

Now notice that Fig 13.19 also shows the critical-value curves from the previous figures. From the above discussion, we see that Fig 13.19 and the previous figures have revealed that there are three qualitatively different types of slices through the critical-value surface:

**Type 1**
For negative $c$, the vertical plane $a_2 = c$ slices the critical-value surface in two curves:
(i) The $F = 0$ line in that plane.
(ii) A double-minima curve $a_2 = c\ mm$ which is, from right-to-left, a downward curve below the $F = 0$ line. Note that, at no point, does the double-minima curve touch the $F = 0$ line, because, for any function in the negative $a_2$ half-plane, the minima are always below the $F = 0$ axis in the function. The double-minima curve is illustrated in Fig 13.19 by the curve $a_2 = -3\ mm$.

**Type 2**
The vertical plane $a_2 = 0$ slices the critical-value surface in two curves:
(i) The $F = 0$ line in that plane.
(ii) A double-minima curve $a_2 = 0\ mm$ which is, from right-to-left, a downward curve that starts at the origin and goes down below the $F = 0$ line. Therefore the double-minima curve and the $F = 0$ line are joined at a cusp-point at the origin. The curve $a_2 = 0\ mm$ is illustrated in Fig 13.19.

**Type 3**
For positive $c$, the vertical plane $a_2 = c$ slices the critical-value surface in three curves:
(i) The $F = 0$ line in that plane.
(ii) A double-maxima curve $a_2 = c\ MM$ which is, from right-to-left, a downward curve starting at a cusp-point in the upper-left quadrant. Note that, at no point, does the double-maxima curve reach the $F = 0$ line, because, for any function in the positive $a_2$ half-plane, the maxima are always above the $F = 0$ axis in the function. The double-maxima curve is illustrated in Fig 13.19 by the curve $a_2 = 2\ MM$.
(iii) A double-minima curve $a_2 = c\ mm$ which is, from right-to-left, a downward curve that starts from the same cusp-point as the double-maxima curve, but is below that curve. Note, that, unlike the double-maxima curve, the double-minima curve crosses the $F = 0$ line, because, for any line $a_2 = c$ of functions in the positive $a_2$ half-plane, the double-minima in the functions eventually go through the $F = 0$ axis. The double-minima curve is illustrated in Fig 13.19 by the curve $a_2 = 2\ mm$.

The next figure, Fig 13.20, takes the previous figure and adds extra examples of Type 3 slices; i.e., extra vertical slices $a_2 = c$ with positive $c$. We see that the cusp-point on each slice rises successively and moves left-ward.

Again, with the assumption that the viewer is on the negative $a_2$-axis, facing the graph, the curves, from right-to-left are successively moving deeper into space away from the viewer. In particular, as one goes from right-to-left, each successive pair of curves, joined at a cusp-point, is successively deeper in space. Thus, the cusp-points start at the origin and go successively deeper into space.

Concerning the succession of cusp-points, it is important to notice the following: As stated earlier, a pair of curves joined at a cusp-point in the upper-left quadrant represents the bifurcating double-minima and double-maxima from two inflection-points within a function on the *half-parabola*. Therefore, *the line of cusp-points must be over the half-parabola* in the control plane.

In conclusion, we see that the *critical-value surface consists of three sheets*: (1) the $F = 0$ sheet; (2) the double-minima sheet; and (3) the double-maxima sheet. From the above discussion, we see the following properties concerning these three sheets: The double-minima sheet is below the $F = 0$ sheet in the negative $a_2$ half-plane, and moves upwards as one moves in the increasing $a_2$-direction (i.e., away from the viewer), and reaches the $F = 0$ sheet at the line $a_2 = 0$, where, at the origin, it forms a cusp with the $a_4$-axis (which is within the $F = 0$ sheet). This cusp then continues upwards and leftwards, where the lower and upper sheets of the cusp are respectively (1) the continuation of the double-minima sheet, and (2) the double-maxima sheet, which starts at the negative $a_4$-axis.

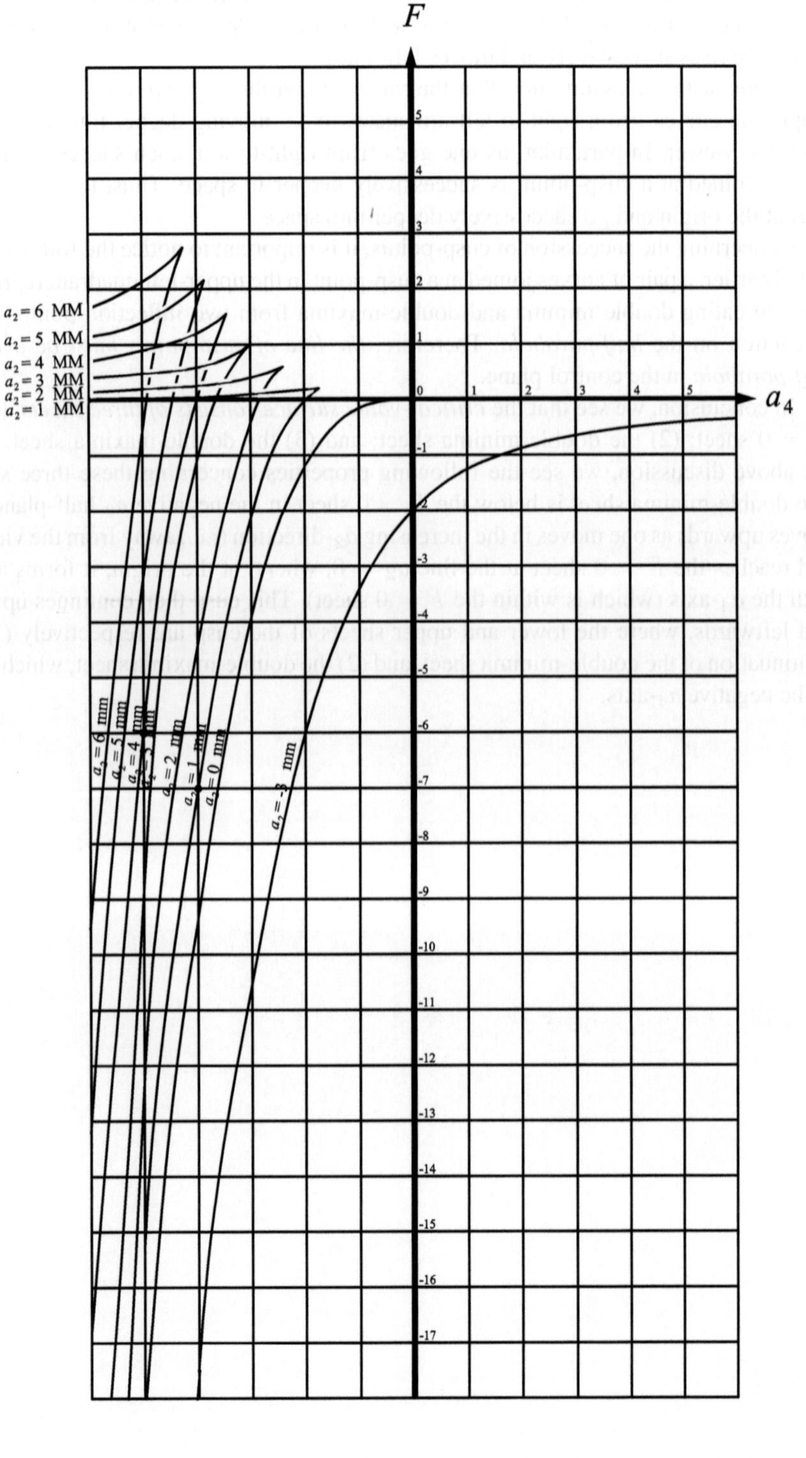


**Fig. 13.20.** The critical-value curves for individual values of $a_2$. The curves marked $mm$ are double-minima; and the curves marked $MM$ are double-maxima.

## The Zero-Separatrix

From the above information, we can now obtain the zero-separatrix of the Symmetry-Restricted Interactive Unfolding of $EA_5$. By Method 1, in section 8.5, we simply invert the critical-value surface in the $F$-direction, and replace the inverted $F$-axis by the control parameter $a_0$.

Thus, by inverting the critical-value curves from Fig 13.20, we obtain the corresponding zero-separatrix curves, shown in Fig 13.21. Again, with the assumption that the viewer is on the negative $a_2$-axis, facing the graph, the curves, from right-to-left are successively moving deeper into space away from the viewer. In particular, as one goes from right-to-left, each successive pair of curves, joined at a cusp-point, is successively deeper in space. Thus, the cusp-points start at the origin and go successively deeper into space.

The figure after this, Fig 13.22, exhibits the zero-separatrix *three dimensionally*. Both the 2D and 3D representations will be useful to us, as we will see later.

The next page of text describes the three sheets of the zero-separatrix in detail, and the reader should compare it with the 3D representation in Fig 13.22. It is important for the reader to fully absorb this description in detail:

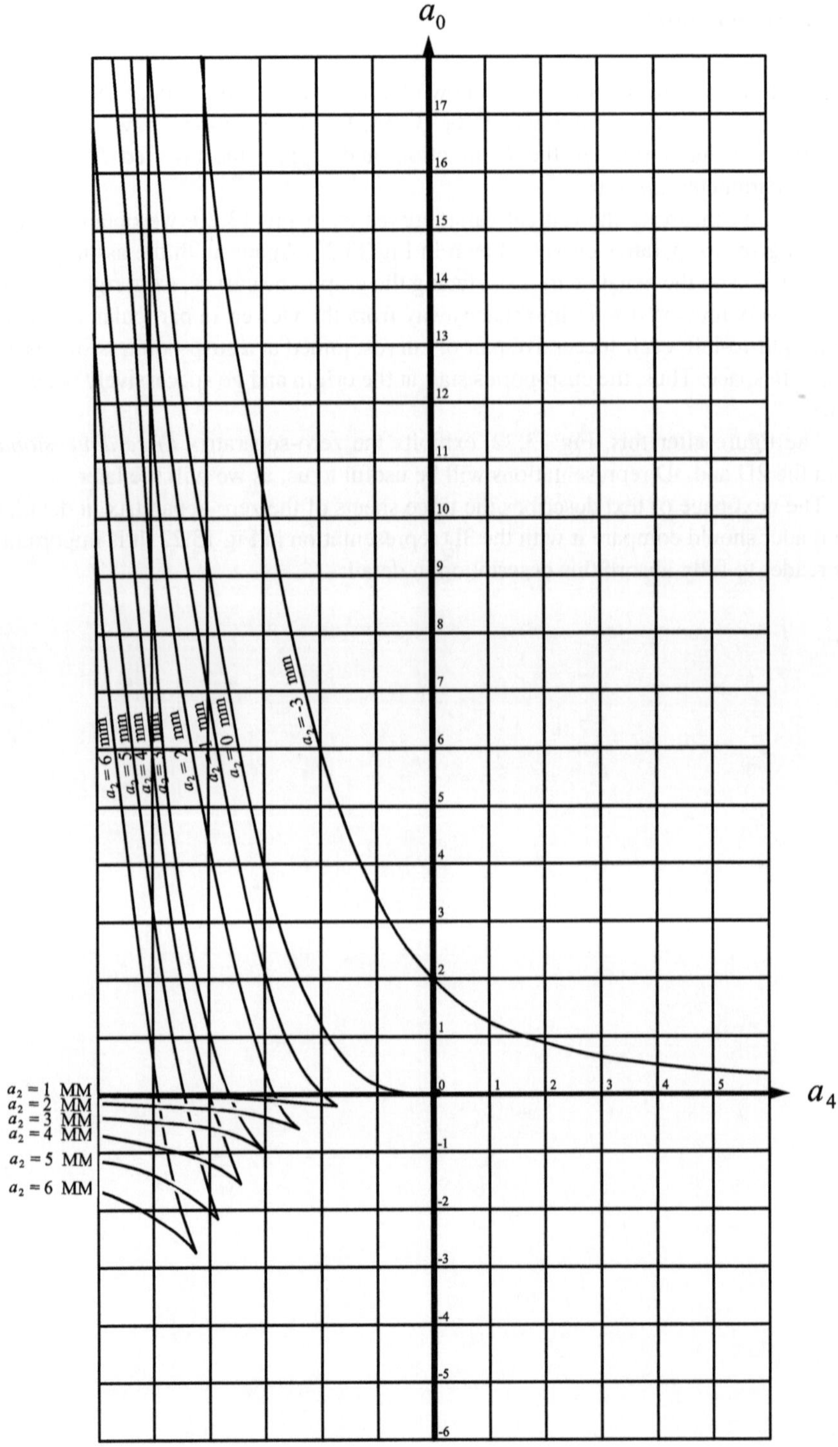


**Fig. 13.21.** The zero-separatrix curves for the Symmetry-Restricted Interactive Unfolding of $EA_5$.

**THE 3 SHEETS OF THE ZERO-SEPARATRIX**

**The zero-separatrix, of the Symmetry-Restricted Interactive Unfolding of $EA_5$, consists of three sheets:**

**1. The $a_0 = 0$ Plane.**
**This corresponds to the 2D control space $(a_4, a_2)$ of the symmetry-restricted butterfly catastrophe, but is within the 3D control space of the Symmetry-Restricted Interactive Unfolding of $EA_5$. In the positive $a_2$ half-plane, and the positive $a_4$-axis, this sheet represents functions whose central *minimum* is zero. In the negative $a_2$ half-plane, and the negative $a_4$-axis, this sheet represents functions whose central *maximum* is zero.**

**2. The Double-Minima Sheet.**
**This corresponds to Regions 3 and 5 of the catastrophe control space $(a_4, a_2)$. Therefore its boundary corresponds to the positive $a_4$-axis and the half-parabola; i.e., it does not exist in Regions 1a and 1b. The sheet is *above* the $a_0 = 0$ plane over *all* of Region 3, and moves downwards as one moves in the increasing $a_2$-direction. It then terminates at the positive $a_4$-axis in the $a_0 = 0$ plane, and forms a cusp with the negative $a_4$-axis where it still remains above the $a_0 = 0$ plane. Then, continuing in the increasing $a_2$-direction, the sheet remains partially above the $a_0 = 0$ plane, but cuts through that plane at a curve between the negative $a_4$-axis and the half-parabola, and therefore descends below the $a_0 = 0$ plane, and terminates at a line of cusp-points. This line of cusp-points is directly under the half-parabola, and has descended from the initial cusp-point at the origin.**

**3. The Double-Maxima Sheet.**
**This sheet is always under the $a_0 = 0$ plane, and corresponds to Region 5 of that plane; i.e., its boundary corresponds to the half-parabola and the negative $a_4$-axis in that plane. The sheet is always joined to the double-minima sheet at the line of cusp-points, i.e., under the half-parabola. The sheet moves downwards from the negative $a_4$-axis and also moves downwards along the half-parabola from the origin.**

The following additional comments can further help the reader visually interpret Fig 13.22. The curve of six dots at the far left, starting from the end of the negative $a_4$-axis and curving downwards, is itself in the side-plane of the figure; i.e., directly under the left-most line of the central plane. The curve of cusp-points, starting at the origin and going downwards, ends at the *back plane* of the figure – i.e., the lowest cusp-point shown is in this back plane. The figure shows the curve of cusp-points as directly over the half-parabola in the *bottom* $(a_4, a_2)$ plane, where the half-parabola is the only curve shown in the bottom plane. In fact, as stated above, the curve of cusp-points is directly

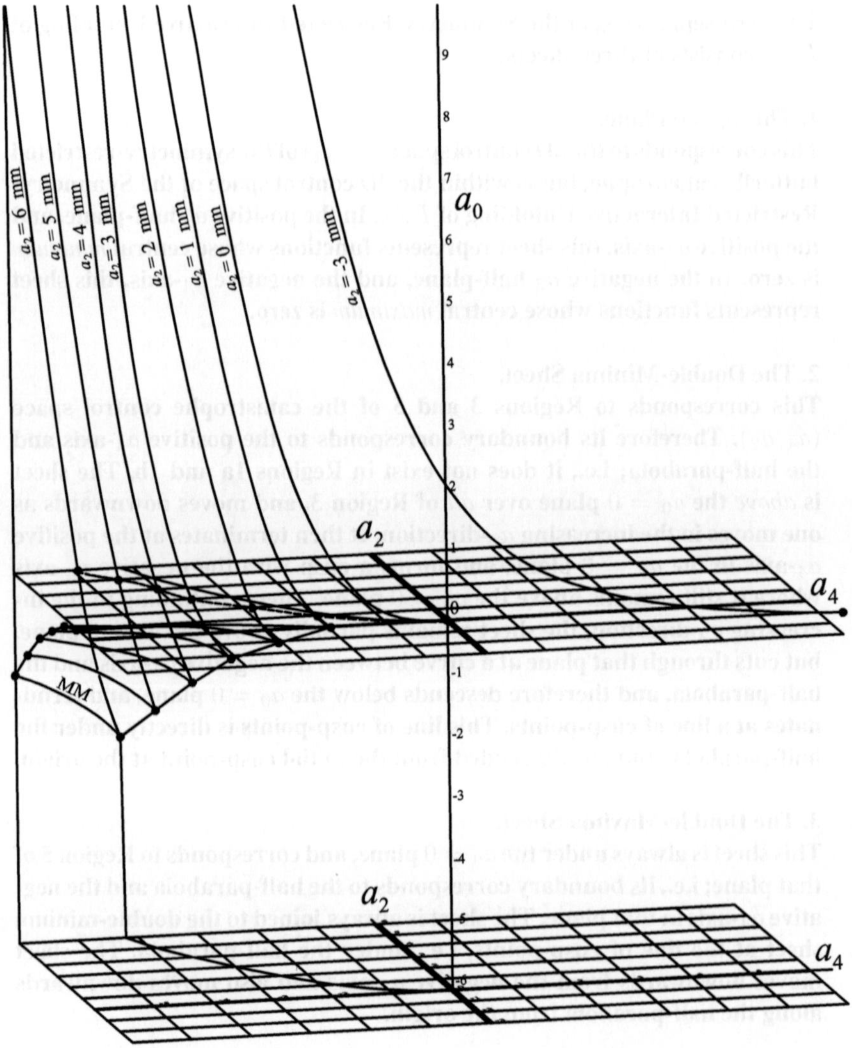


**Fig. 13.22.** The zero-separatrix, of the Symmetry-Restricted Interactive Unfolding of $EA_5$, consists of three sheets: (1) the double-minima sheet $mm$, (2) the double-maxima sheet $MM$, and (3) the $a_0 = 0$ plane.

*below* the half-parabola in the *central* $a_0 = 0$ plane, where the half-parabola is the *solid* curve shown in that plane. In contrast, the *dashed* curve, shown in the $a_0 = 0$ plane, is where the $mm$-sheet *crosses* that plane.

Note the following: Each pair of curves, consisting of a $mm$-curve with its corresponding $MM$-curve joined at its cusp-point, is within one of the vertical parallel planes of constant $a_2$. The figure shows six such pairs of curves, and they are in the six parallel planes through the six dots on the left side-curve.

While the zero-separatrix was derived from the critical-value surface by inversion, it is important to understand that the two structures define two very different mathematical objects. To see this, we successively contrast them as follows:

**Critical-Value Surface:**
The control space is only a *two*-dimensional *subspace* $(a_4, a_2)$ of the entire 3D space $(a_4, a_2, F)$ required to contain the surface.

**Zero-Separatrix:**
The control space *is* the entire *three*-dimensional space $(a_4, a_2, a_0)$.

**Critical-Value Surface:**
Given a point $(a_4, a_2)$ in the horizontal plane, the entire vertical line through that point corresponds to a *single* function $F(t; a_4, a_2)$.

**Zero-Separatrix:**
Given a point $(a_4, a_2)$ in the horizontal plane, the vertical line through that point corresponds to an *entire family* of functions; i.e., the set of functions obtained by taking the graph of a function $F(t; a_4, a_2)$ and translating it along its $F$-axis by all possible amounts.

**Critical-Value Surface:**
A single point in the vertical line through $(a_4, a_2)$ represents only one possible value of the function $F(t; a_4, a_2)$.

**Zero-Separatrix:**
A single point in the vertical line through $(a_4, a_2)$ represents an entire function.

### The Full Separatrix

Having constructed the zero-separatrix of the Symmetry-Restricted Interactive Unfolding of $EA_5$, we now construct the *full* separatrix. This is done by adding the critical-point separatrix of the Symmetry-Restricted Interactive Unfolding of $EA_5$. Clearly, the latter separatrix is given by sweeping, along the $a_0$-dimension, the separatrix of the catastrophe; i.e., the $a_4$-axis and the half-parabola which were illustrated in Fig 13.7 (page 469). Therefore, the *full* separatrix of the Symmetry-Restricted Interactive Unfolding

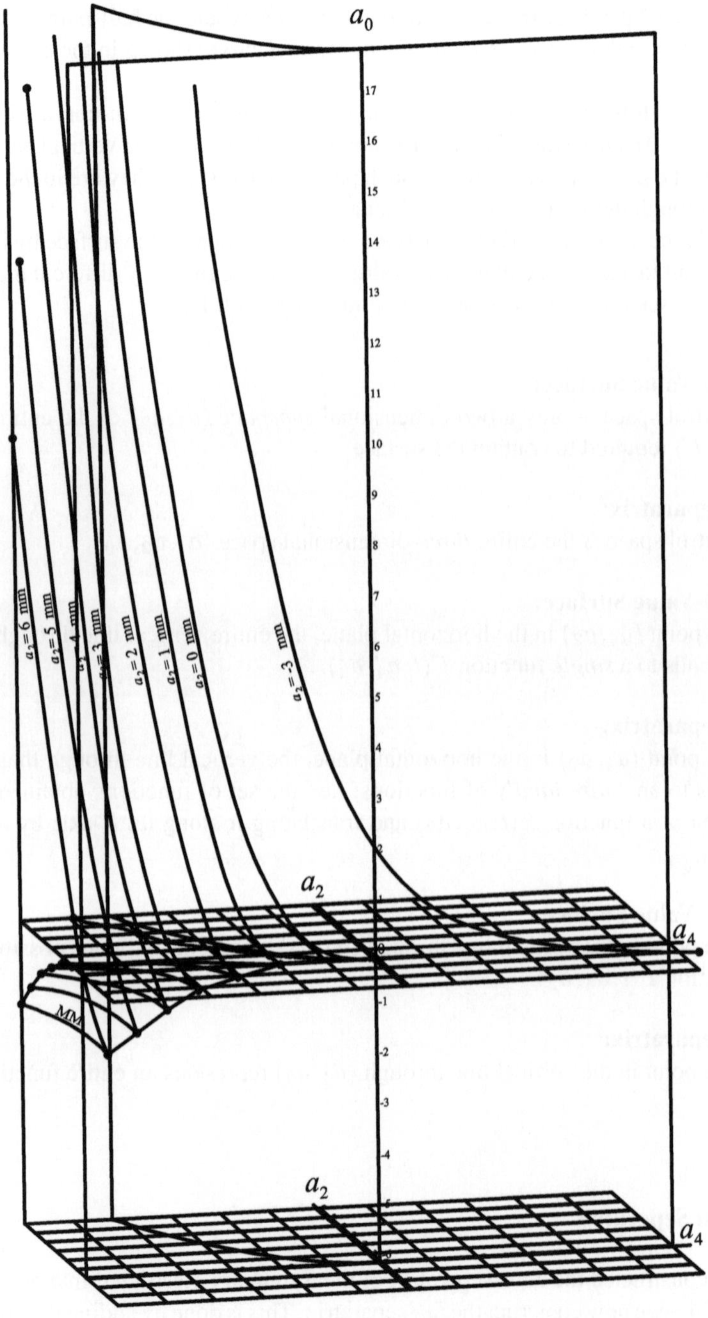


**Fig. 13.23.** The full separatrix (critical-point separatrix and zero-separatrix) of the Symmetry-Restricted Interactive Unfolding of $EA_5$.

of $EA_5$ is as illustrated in Fig 13.23. It consists of the zero-separatrix, together with the critical-point separatrix. Observe, within this figure, that the critical-point separatrix consists of the vertical plane, shown along the $a_4$-axis, and the vertical half-parabola sheet, shown as curving *behind* that vertical plane, away from the $a_0$-axis.

Note, of course, that the full separatrix of the Symmetry-Restricted Interactive Unfolding of the Dual $EA_{-5}$ will be the vertical inversion of the full separatrix of the Symmetry-Restricted Interactive Unfolding of $EA_5$.

## 13.5 Scenario 1

Now recall that Chapter 6 gave three alternative scenarios for 5-fold part-formation. These were summarized on page 210. The present section, and the following sections, will go successively through those scenarios, and show that they can all be modeled in the Symmetry-Restricted Interactive Unfolding of $EA_5$. We will subsequently see that the continuation operators, $C$, $CC$, and $CCC$, can also be modeled in the Symmetry-Restricted Interactive Unfolding of $EA_5$.

Let us begin with Scenario 1, which is the simultaneous use of the pair-creation operators $\pi^{\uparrow}$ and $\pi^{\downarrow}$ around the central extremum. In the case where that central extremum, i.e., the anchor-extremum, is a minimum, this scenario is given by the three *downward-upward* operations, listed on page 204. These operations were illustrated by the four figures on page 205, where the first two figures showed the two applications of the first operation; and the other two figures showed the single applications of the other two operations.

We will now see that all three operations, and their corresponding four applications, can be modeled in the Symmetry-Restricted Interactive Unfolding of $EA_5$. In the 3D control space, as illustrated in Fig 13.23, all three operations are modeled by crossing the vertical half-parabola sheet (at the back of the figure), from Region 1 to Region 5. The difference between the four applications, of the three operations, is *where* the crossings occur on the half-parabola sheet.

To make the distinctions between the four applications, we should first note the following fundamental fact: *the line of cusp-points in the zero-separatrix is also on the half-parabola sheet.* We will call this line the **double $g$-line**, for important reasons that will become clear later.

**Positive Downward-Upward Operation: First Application**

This is the use of Scenario 1 to create a *triple-shield*, as illustrated in the top diagram on p205. It is given by:

$$\pi^{\downarrow}\pi^{\uparrow}\emptyset^{+} \ : \ \emptyset^{+} \ m^{+} \ \emptyset^{+} \ \longrightarrow \ \overbrace{m^{+}M^{+}} \ m^{+} \ \overbrace{M^{+}m^{+}}$$

In the 3D control space, Fig 13.23, the initial $m^{+}$ corresponds to a function anywhere in Region 1 of the $(a_4, a_2)$ plane, but *above* that central plane, which is a sheet in the zero-separatrix. Then, in the operation, the function crosses the vertical half-parabola sheet (at the back), above the zero-separatrix, to a function that is also above the zero-separatrix.

**Positive Downward-Upward Operation: Second Application**
This is the use of Scenario 1 to *pinch* an indenting process in its two positive tails, as illustrated in the second diagram on p205. It is given by:

$$\pi^{\downarrow}\pi^{\uparrow}\emptyset^{+} \;:\; \emptyset^{+}\ 0m^{-}0\ \emptyset^{+} \;\longrightarrow\; \overbrace{m^{+}M^{+}}\ 0m^{-}0\ \overbrace{M^{+}m^{+}}$$

In the 3D control space, Fig 13.23, the initial $0m^{-}0$ corresponds to a function in Region 1 of the $(a_4, a_2)$ plane, but *below* that central plane. Then, in the operation, the function crosses the vertical half-parabola sheet, below the central plane but above the line of cusp-points, the double $g$-line, to a function that is also below the central plane but above the $mm$-sheet. This is because, in the operation, the two created minima are positive.

**Negative Downward-Upward Operation**
This is the use of Scenario 1 to create a *complete double-bay*, as illustrated in the third diagram on p205. It is given by:

$$\pi^{\downarrow}\pi^{\uparrow}\emptyset^{-} \;:\; \emptyset^{-}\ m^{-}\ \emptyset^{-} \;\longrightarrow\; \overbrace{m^{-}M^{-}}\ m^{-}\ \overbrace{M^{-}m^{-}}$$

In the 3D control space, Fig 13.23, the initial $m^{-}$ is given by a function in Region 1 of the $(a_4, a_2)$ plane, again below that central plane. Then, in the operation, the function crosses the vertical half-parabola sheet, entirely below the zero-separatrix, to a function that is also entirely below the zero-separatrix, i.e., below both the double-minima sheet and the double-maxima sheet. This is because, in the operation, the created minima and maxima are *all* negative.

**Zero Downward-Upward Operation**
This is the use of Scenario 1 to create a *pinched complete double-bay*, as illustrated in the fourth diagram on p205. It is given by:

$$[C(3)\pi^{\downarrow}][C(3)\pi^{\uparrow}]0 \;:\; 0^{\downarrow}\ m^{-}\ 0^{\uparrow} \longrightarrow \overbrace{0m^{-}0M^{+}0}\ m^{-}\ \overbrace{0M^{+}0m^{-}0}$$

In the 3D control space, Fig 13.23, the initial $m^{-}$ is given by a function in Region 1 of the $(a_4, a_2)$ plane, again below that central plane. Then, in the operation, the function crosses the vertical half-parabola sheet, *through the line of cusp-points*, i.e., the double $g$-line, into what we will call the **enclosed volume**: that below the central plane and *between* the $mm$-sheet and $MM$-sheet. This is because, in the operation, the two created minima are *negative*, and the two created maxima are *positive*. Observe the crucial fact that any function in the enclosed volume is maximally bifurcated in the Symmetry-Restricted Interactive Unfolding of $EA_5$; that is, there are no further critical-point bifurcations and zero-bifurcations that the function can undergo in the Symmetry-Restricted Interactive Unfolding of $EA_5$.

## 13.6 How Scenario 1 Relates the Symmetry-Restricted Interactive Unfolding of $EA_5$ to the Interactive Unfolding of $EA_3$

We will now see that there is an interesting way in which Scenario 1, in the Symmetry-Restricted Interactive Unfolding of $EA_5$, can be viewed in terms of the Interactive Unfolding of $EA_3$.

To illustrate, go back to Fig 11.6 (page 386), which showed the functions obtained by using the non-zero pair-creation operators to cross the side-sheets of the critical-point separatrix of the Interactive Unfolding of $EA_3$.

Consider the top pair of functions, as follows: The left top function is obtained by crossing the left side-sheet, by applying the downward pair-creation operator $\pi^{\downarrow}$ to the left side of the positive minimum in the initial function. The right top function is obtained by crossing the right side-sheet, by applying the upward pair-creation operator $\pi^{\uparrow}$ to the right side of the positive minimum in the initial function. Observe therefore that both these cases use the same anchor-extremum, a positive minimum. Therefore, if one applied these two operations simultaneously, one would get the downward pair-creation acting on the left, and, simultaneously, the upward pair-creation acting on the right, of the positive minimum.

Exactly the corresponding argument applies to the middle two functions in Fig 11.6 (page 386). The only change is that the anchor-extremum minimum has changed from positive to negative. Also, exactly the corresponding argument applies to the bottom two functions. Here the only change is that both the downward and upward pair-creation operations have changed from positive to negative, i.e., their application-points have changed from positive to negative.

Thus to summarize: The consistent fact is that crossing the left side-sheet corresponds to the downward pair-creation operator, and crossing the right side-sheet corresponds to the upward pair-creation operator. And, if we applied these two operators *simultaneously*, to the same initial function, we would get the downward pair-creation acting on the left, and, simultaneously, the upward pair-creation acting on the right, of the anchor-extremum minimum.

But this is exactly what Scenario 1 does in the Symmetry-Restricted Interactive Unfolding of $EA_5$. That is, it symmetrically pairs the downward pair-creation operator and the upward pair-creation operator respectively on the left and right side of the anchor-extremum minimum.

Thus the reader can imagine the half-parabola sheet in the control space of the Symmetry-Restricted Interactive Unfolding of $EA_5$, as the simultaneous occurrence of the left side-sheet and right side-sheet in the control space of the Interactive Unfolding of $EA_3$.

Furthermore, the *regions* on these sheets also correspond, as follows: Recall the top two functions in Fig 11.6 (page 386) are obtained by crossing the $e$-bounded region; i.e., the region of each side-sheet *above* the bi-valent line $e$ on the sheet. In the Symmetry-Restricted Interactive Unfolding of $EA_5$ this corresponds to crossing the half-parabola sheet *above* the central sheet; i.e., the first application of the positive downward-upward operation.

Then, in moving down to the middle pair of functions in Fig 11.6 (page 386), we have crossed the bi-valent line $e$, which means that we have changed the anchor-extremum from positive to negative, but kept the application-points as positive. Thus the two middle functions are obtained by crossing the *e&g*-bounded region; i.e., the region of each side-sheet *between* the bi-valent lines $e$ and $g$. In the Symmetry-Restricted Interactive Unfolding of $EA_5$, this corresponds to crossing the half-parabola sheet *below* the central plane, but above the line of cusp-points; i.e., the second application of the positive downward-upward operation. Note that the intersection of the central plane with the half-parabola sheet, in the Symmetry-Restricted Interactive Unfolding of $EA_5$, corresponds to the bivalent-lines $e$, in the Interactive Unfolding of $EA_3$. This is because, in the Symmetry-Restricted Interactive Unfolding of $EA_5$, moving the transition-point of the operation down the half-parabola sheet, across the central plane, has the effect of changing the anchor-extremum from positive to negative, but keeping the application-points positive.

Furthermore, the line of cusp-points, i.e., the double $g$-line, in the half-parabola sheet, in the Symmetry-Restricted Interactive Unfolding of $EA_5$, corresponds to the bivalent-lines $g$ in the Interactive Unfolding of $EA_3$, for the following reasons: Recall that an entire side-sheet in the Interactive Unfolding of $EA_3$ is the set of functions that have a critical inflection-point on one side of the anchor-extremum. Correspondingly, the entire half-parabola sheet in the Symmetry-Restricted Interactive Unfolding of $EA_5$ is the set of functions with *two* critical inflections-points, one on each side of the anchor-extremum. Recall also that the line $g$ on a side-sheet in the Interactive Unfolding of $EA_3$ is the set of functions where the critical inflection-point has zero $F$-value. Correspondingly, the double $g$-line in the Symmetry-Restricted Interactive Unfolding of $EA_5$ is the set of functions where the two critical inflection-points have zero $F$-value.

Thus, in the Interactive Unfolding of $EA_3$, when the transition-point, of pair-creation, crosses the $g$-line downward on a side-sheet, this has the effect of changing a positive pair-creation operation to a negative one; i.e., changing the application-point from positive to negative. Correspondingly, in the Symmetry-Restricted Interactive Unfolding of $EA_5$, when the transition-point, of symmetrical-paired pair-creation, crosses the double $g$-line downward on the half-parabola sheet, this has the effect of changing a positive downward-upward operation to a negative one; i.e., changing the *two* application-points from positive to negative.

Now return to Fig 11.6 (page 386). While the figure showed the three pairs of functions corresponding respectively to crossing the three bounded regions, it did not show the corresponding pair of functions within the central enclosed volume – i.e., the functions obtained by maximally crossing the $g$-line on each side-sheet. These two crossings (left and right) are the two *zero* pair-creation operations; that is, the downward zero pair-creation operation $[C(3)\pi^{\downarrow}]$ which crosses the left side-sheet at the $g$-line, and the upward zero pair-creation operation $[C(3)\pi^{\uparrow}]$ which crosses the right side-sheet at the $g$-line.

Notice that, if one applied these two operations simultaneously, one would get these two zero pair-creation operations acting simultaneously on the left and right sides of the negative minimum (anchor-extremum) in the function.

But this is exactly what the zero downward-upward operation $[C(3)\pi^{\downarrow}][C(3)\pi^{\uparrow}]$ does, in the Symmetry-Restricted Interactive Unfolding of $EA_5$, by crossing the double $g$-line into the enclosed volume in that space.

What the above discussion has therefore shown is this: The four types of crossings we have seen on a side-sheet in the Interactive Unfolding of $EA_3$, that is, two positive pair-creations, a negative pair-creation, and a zero pair-creation, correspond to the four types of crossings we see on the half-parabola sheet in the Symmetry-Restricted Interactive Unfolding of $EA_5$; that is, two positive downward-upward operation applications, the negative downward-upward operation, and the zero downward-upward operation. Furthermore, the regions and lines on a side-sheet in the Interactive Unfolding of $EA_3$ correspond to the regions and lines on the half-parabola sheet in the Symmetry-Restricted Interactive Unfolding of $EA_5$. In fact, the regions and lines on the half-parabola sheet in the Symmetry-Restricted Interactive Unfolding of $EA_5$ are the simultaneous symmetrical occurrence of the regions and lines of the two side-sheets in the Interactive Unfolding of $EA_3$.

With respect to this, observe also that, at the $g$-line on a side-sheet in the Interactive Unfolding of $EA_3$, two sheets of the zero-separatrix meet as a cusp-structure. The two sheets are (1) a *minima* sheet, which is the *upper* sheet of the cusp-structure, and (2) a *maxima* sheet, which is the *lower* sheet of the cusp-structure. Correspondingly, at the double $g$-line on the half-parabola sheet in the Symmetry-Restricted Interactive Unfolding of $EA_5$, two sheets of the zero-separatrix meet as a cusp-structure. The two sheets are (1) a *double-minima* sheet, which is the *upper* sheet of the cusp-structure, and (2) a *double-maxima* sheet, which is the *lower* sheet of the cusp-structure. Thus, the above discussion has shown:

> **The same surface-structure exists with respect to the half-parabola sheet in the Symmetry-Restricted Interactive Unfolding of $EA_5$ and a side-sheet in the Interactive Unfolding of $EA_3$. Furthermore, that surface-structure in the Symmetry-Restricted Interactive Unfolding of $EA_5$ represents functions whose singularity-configurations are the symmetric-doubling of the singularity-configurations of the functions represented by the corresponding surface-structure in the Interactive Unfolding of $EA_3$.**

Now, an important basis for this are the following facts that were established in section 11.10: A vertical slice through the right side-sheet of the Interactive Unfolding of $EA_3$ corresponds to the control plane of the Interactive Unfolding of $EA_2$. And, similarly, a vertical slice through the left side-sheet of the Interactive Unfolding of $EA_3$ corresponds to the control plane of the Interactive Unfolding of the Dual $EA_{-2}$. For example, Fig 11.38 (page 430) showed the Process Grammar operation-diagram for the Interactive Unfolding of $EA_2$. Recall that, as illustrated in this diagram, the full separatrix structure of the Interactive Unfolding of $EA_2$ consists of (1) the cusp-structure, which is the zero-separatrix, and (2) the vertical axis, which is the critical-point separatrix. This figure, giving the organization of the Interactive Unfolding of $EA_2$, has the same organization as a vertical slice through the right side-sheet of the Interactive Unfolding of $EA_3$. The only difference is that the functions in the Interactive Unfolding of $EA_2$ do not contain an *anchor-extremum*, whereas the functions in the Interactive Unfolding of $EA_3$ do. Thus there is nothing in the Interactive Unfolding of $EA_2$ that corresponds to the $e$-line in the Interactive Unfolding of $EA_3$, because crossing the $e$-line gives the change of the anchor-extremum from positive to negative. Nevertheless, the remainder of the structure is the same.

Therefore, we can see that a vertical slice through the half-parabola sheet in the Symmetry-Restricted Interactive Unfolding of $EA_5$ has the same structure as the Interactive Unfolding of $EA_2$ (except that the component given by the central plane in the Symmetry-Restricted Interactive Unfolding of $EA_5$ is omitted in $EA_2$ since it corresponds to the $e$-line). Thus, in that slice in the Symmetry-Restricted Interactive Unfolding of $EA_5$, we have the same cusp-shape of the zero-separatrix, and the same vertical line of the critical-point separatrix. Furthermore, that slice in the Symmetry-Restricted Interactive Unfolding of $EA_5$ represents functions whose singularity-configurations are the symmetric-doubling of the singularity-configurations of the functions in the Interactive Unfolding of $EA_2$. That is, the functions in that slice in the Symmetry-Restricted Interactive Unfolding of $EA_5$ contain the symmetrical pairing of the functions in the Interactive Unfoldings of $EA_2$ and $EA_{-2}$.

In this and the last section, we have concentrated on the *downward-upward* operations that realize Scenario 1. Clearly, the dual statements are true for the *upward-downward* operations that realize Scenario 1. These latter operations were defined on p203. The crucial fact is that these operations are necessarily applied to the sides of maxima $M^{\pm}$; that is, the anchor-extremum is a maximum. Therefore, the operations are modeled in the Symmetry-Restricted Interactive Unfolding of the Dual $EA_{-5}$. Thus, analogously to the above arguments, we can set up a correspondence between the half-parabola sheet in the Symmetry-Restricted Interactive Unfolding of $EA_{-5}$ and the side-sheets in the Interactive Unfolding of $EA_{-3}$, which are again given by the Interactive Unfoldings of $EA_2$ and $EA_{-2}$. However, in this case, the Interactive Unfoldings of $EA_2$ and $EA_{-2}$ are applied respectively to the left and right sides of the functions, whereas in the downward-upward case they were applied respectively to the right and left sides.

## 13.7 Scenario 2

Let us now turn to Scenario 2, that was defined for 5-fold part-formation. This is the successive double-application of the 3-fold bifurcation operator $B$. In section 6.2, we saw that this has four cases: $B_{-3} \circ B_3$ applied to $m^+$ and $m^-$; and $B_3 \circ B_{-3}$ applied to $M^-$ and $M^+$. They were illustrated respectively with four diagrams, which are now shown again *together* on p499.

We are now going to understand these four cases in terms of the Symmetry-Restricted Interactive Unfoldings of $EA_5$ and $EA_{-5}$. The first two cases are movements within the control space of the Symmetry-Restricted Interactive Unfolding of $EA_5$. To illustrate this, consider the Process Grammar operator-diagram in Fig 13.28 (p500), which shows a neighborhood of the origin within a $(a_4, a_2)$-slice, where the slice is above or below the central plane, and the neighborhood (within the slice) does not reach any position where the zero-separatrix crosses the slice; i.e., the neighborhood is therefore entirely above or below the zero-separatrix. Scenario 2 can be seen as the application of $B_3$

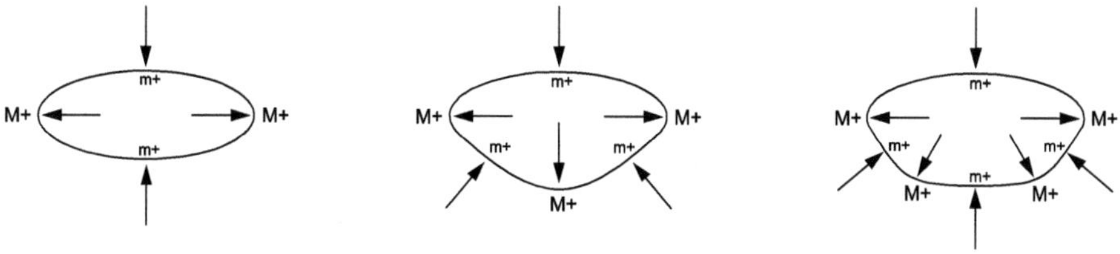


**Fig. 13.24.** $B_{-3} \circ B_3 m^+$ creating a *triple-shield*.

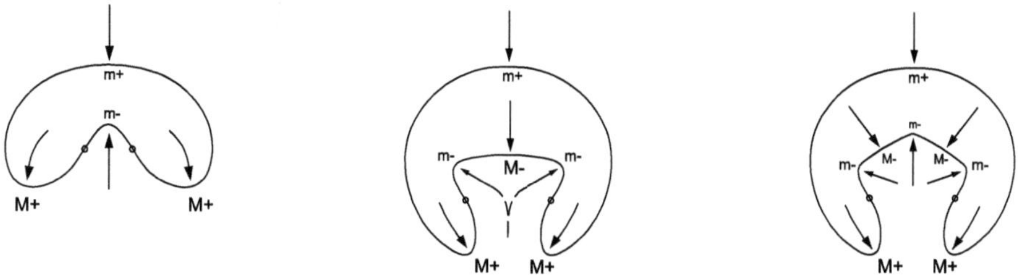


**Fig. 13.25.** $B_{-3} \circ B_3 m^-$ creating a *complete double-bay*.

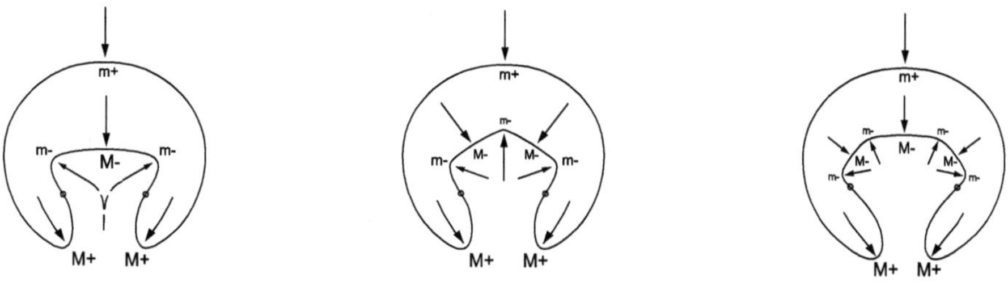


**Fig. 13.26.** $B_3 \circ B_{-3} M^-$ creating a *triple-bay*.

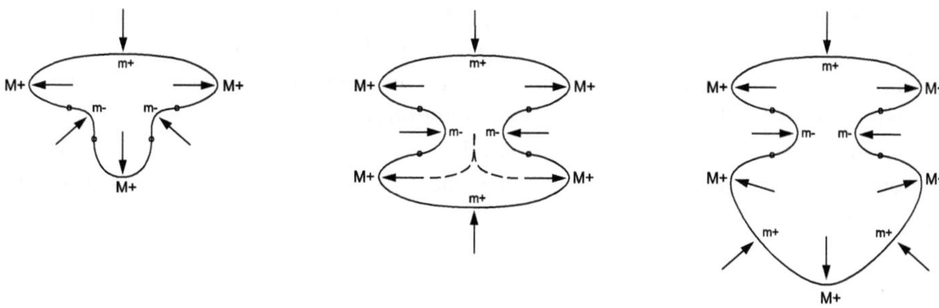


**Fig. 13.27.** $B_3 \circ B_{-3} M^+$ creating a *complete double-shield*.

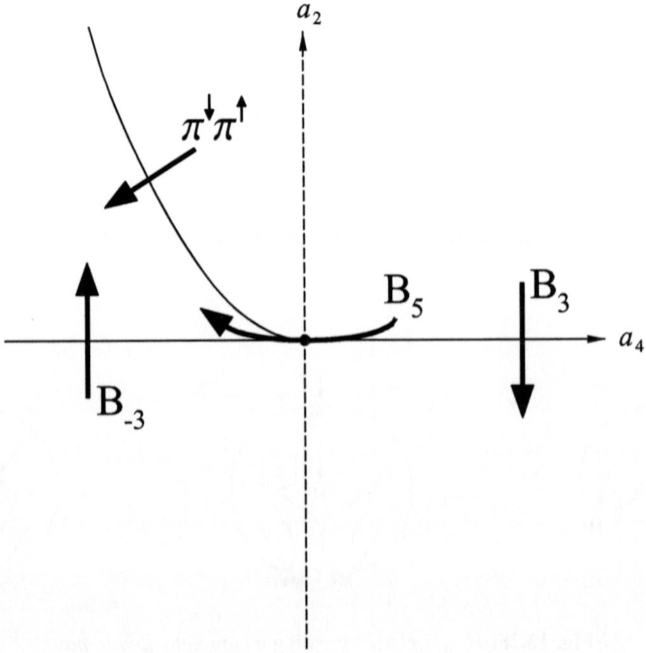


**Fig. 13.28.** The Process Grammar operator-diagram in a $(a_4, a_2)$-planar region that is entirely above or below the zero-separatrix.

followed by $B_{-3}$, in such a neighborhood. That is, the scenario is the transition from Region 1 to Region 3 to Region 5.

Thus, consider the sequence illustrated in the top figure on p499. According to section 6.2, this illustrates the application of

$$B_3m^+ \ : \ m^+ \ \longrightarrow \ m^+M^+m^+$$

followed by

$$B_{-3}M^+ \ : \ M^+ \ \longrightarrow \ M^+m^+M^+$$

Thus, looking at the 3D control space in Fig 13.23 page 492, we see that the initial $m^+$ is given by a function in Region 1 of the $(a_4, a_2)$ control plane but above the central plane. Then, under the first operation, the function crosses the vertical plane over the positive $a_4$-axis, but above the central plane, to a function above the $mm$-sheet! Then, under the second operation, the function crosses the vertical plane over the negative $a_4$-axis, but entirely above all three zero-separatrix sheets, to a function that is also above the zero-separatrix sheets.

Now consider the sequence illustrated by the figure on the second level in p499. This takes exactly the same route in the 2D control plane, that is, $B_3$ followed by $B_{-3}$, but this time the route is *below* all the zero-separatrix sheets.

The sequences illustrated by the figures on the third and fourth levels in p499 are simply duals of those on the first and second levels, and are therefore the dual routes, $B_{-3}$ followed by $B_3$, in the Symmetry-Restricted Interactive Unfolding of the Dual $EA_{-5}$.

## 13.8 Scenario 3

Let us now turn to Scenario 3, that was defined for 5-fold part-formation. In this scenario, all five extrema bifurcate out *simultaneously* from a single central extremum. Section 6.2 defined the four operations that realize this bifurcation, when the created 5-fold part has curvature of a single sign. The four operations are: $B_5$ applied to $m^+$ and $m^-$, and $B_{-5}$ applied to $M^-$ and $M^+$. In section 6.2, these were illustrated respectively with four diagrams which are shown together here on p502.

We will now understand these four operations in terms of the Symmetry-Restricted Interactive Unfoldings of $EA_5$ and $EA_{-5}$. The first two operations are movements within the control space of the Symmetry-Restricted Interactive Unfolding of $EA_5$. To illustrate this, return to the Process Grammar operator diagram in Fig 13.28 (p500), which shows a neighborhood of the origin within a $(a_4, a_2)$-slice, where the slice is above or below the central plane, and the neighborhood (within the slice) is above or below the zero-separatrix. Scenario 3, that is $B_5$, is a transition, within such a neighborhood, from Region 1 to 5 via the 2D origin in the slice.

Observe that $B_5$ can be regarded as a coalescence of $B_3$ and $B_{-3}$, in the following sense: The coalescence is obtained by taking a path $B_{-3} \circ B_3$, and contracting the transition-points of $B_{-3}$ and $B_3$, together with the path between those transition-points, to the 2D origin in that neighborhood.

Now, the obvious fact is that this origin is a point on the $a_0$-axis of the 3D control space illustrated in Fig 13.23 (p492). However, the deeper fact is that the $a_0$-axis is the vertical line of cusp-points where the vertical half-parabola sheet pulls away from the vertical plane through the $a_4$-axis. Such cusp-points will be called the *vertical cusp-points*. All of them are points that are only on the critical-point separatrix, except the 3D origin which is also on the zero-separatrix. The latter will be dealt with later. Thus, each positive and negative vertical cusp-point has a neighborhood as illustrated in Fig 13.28 (p500).

Thus, consider the shape-transition illustrated by the figure at the top level of p502. This is the formation of a *triple-shield* by a simultaneous 5-fold bifurcation of a starting $m^+$ extremum. According to section 6.2, this is given by the operation

$$B_5 m^+ \; : \; m^+ \;\longrightarrow\; m^+ M^+ m^+ M^+ m^+$$

In terms of the 3D control space in Fig 13.23 (p492), the initial $m^+$ is given by a function in Region 1 of the $(a_4, a_2)$ control plane, but above the central plane. Then, in

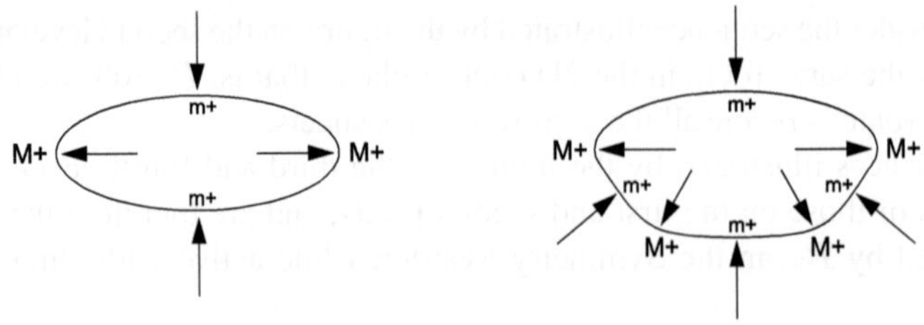


**Fig. 13.29.** Formation of a triple-shield using $B_5 m^+$.

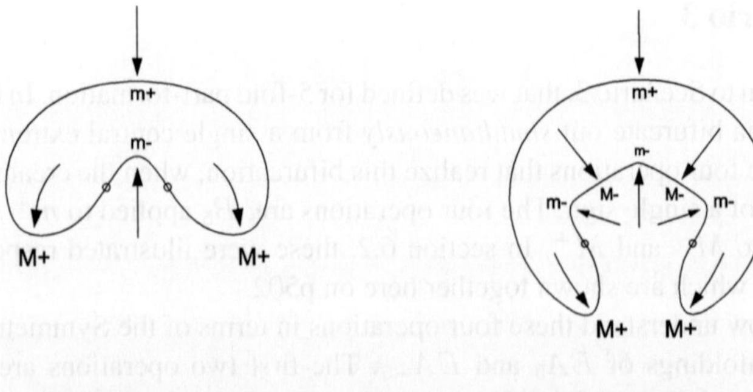


**Fig. 13.30.** Formation of a complete double-bay using $B_5 m^-$.

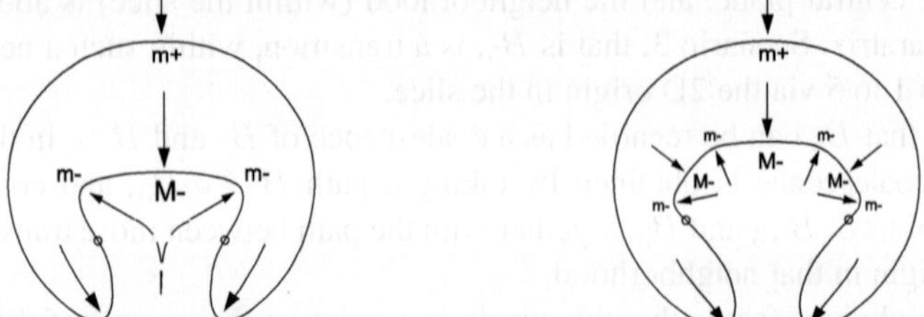


**Fig. 13.31.** Formation of a triple-bay using $B_{-5} M^-$.

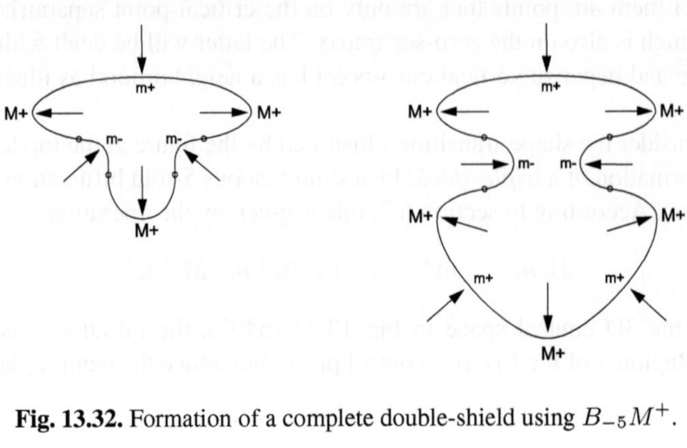


**Fig. 13.32.** Formation of a complete double-shield using $B_{-5} M^+$.

the operation $B_5$, the function crosses a vertical cusp-point above the central plane to a function in Region 5 above all the zero-separatrix surfaces.

Next, consider the transition illustrated by the figure on the second level in p502. This is the formation of a *complete double-bay* by a simultaneous 5-fold bifurcation of a starting $m^-$ extremum. According to section 6.2, this is given by the operation

$$B_5 m^- \ : \ m^- \ \longrightarrow \ m^- M^- m^- M^- m^-$$

In terms of the 3D control space in Fig 13.23 (p492), the initial $m^-$ is given by a function in Region 1 of the $(a_4, a_2)$ control plane, but below the central plane. Then, in the operation $B_5$, the function crosses a vertical cusp-point below the central plane to a function in Region 5 below all the zero-separatrix surfaces.

The shape-transitions illustrated by the figures on the third and fourth levels in p502 are the creation, respectively, of a *triple-bay* and a *complete double-shield*, each by a simultaneous 5-fold bifurcation. These are simply duals of the transitions on the first and second levels, and are therefore the dual routes, $B_{-5}M^-$ and $B_{-5}M^+$, in the Symmetry-Restricted Interactive Unfolding of the Dual $EA_{-5}$

## 13.9 Symmetric Continuations on a Triple-Compressive Part

Recall that we called a 5-fold part that has three compressive processes, a *triple-compressive part.* The two cases are a *triple-shield* and a *triple-bay* . In section 6.3, we defined the continuation operators that continued these compressive processes symmetrically with respect to the central extremum. These operators are the ordinary continuation operator $C$, the 2-fold pinching operator $CC$, and the 3-fold pinching operator $CCC$. We will now show that these operators are modeled in a very interesting way within the Symmetry-Restricted Interactive Unfoldings of $EA_5$ and $EA_{-5}$. As usual, we will illustrate the argument with the Symmetry-Restricted Interactive Unfolding of $EA_5$, as follows:

To help our explanation, it is worth considering Fig 13.33, which takes a vertical planar slice through the 3D control space along a line $a_2 = c$ for some positive constant $c$. In this slice, the dashed vertical line, on the far right, corresponds to the $a_0$-axis. All the other lines in the figure, i.e., the solid lines, show the planar slice through the full separatrix. In this set of lines, the vertical solid line is where the slice cuts through the half-parabola sheet; i.e., the critical-point separatrix. The remaining solid lines (the $mm$, $MM$, and horizontal line) are where the slice cuts through the zero-separatrix.

Observe that, in this slice, Region 5 begins at the vertical solid line, and is the half-plane left-ward from that line. As can be seen from the figure, the zero-separatrix divides Region 5 as follows:

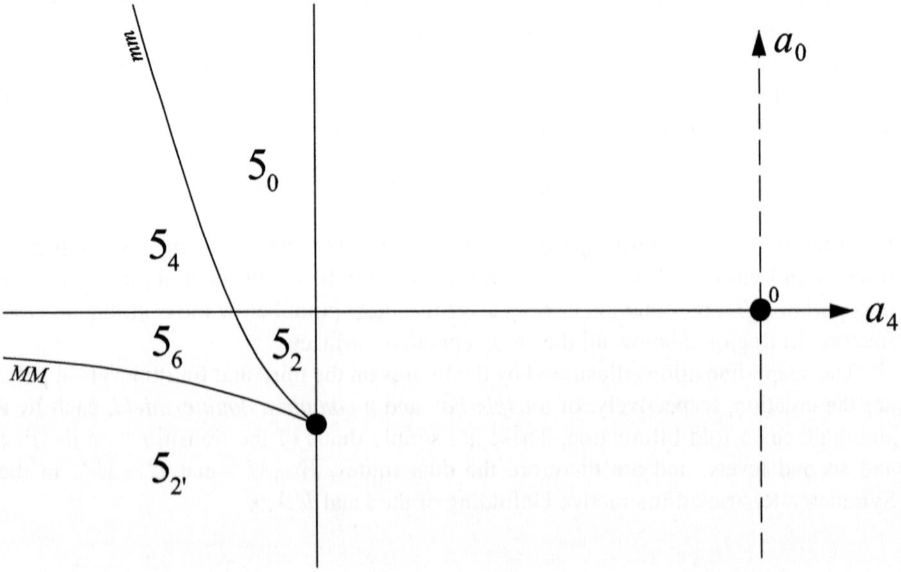


**Fig. 13.33.** A vertical planar slice through the full separatrix of the Symmetry-Restricted Interactive Unfolding of $EA_5$, along the line $a_2 = c$ for some positive constant $c$. The regions $5_i$ are defined as the regions whose functions have 5 critical points and $i$ zeros.

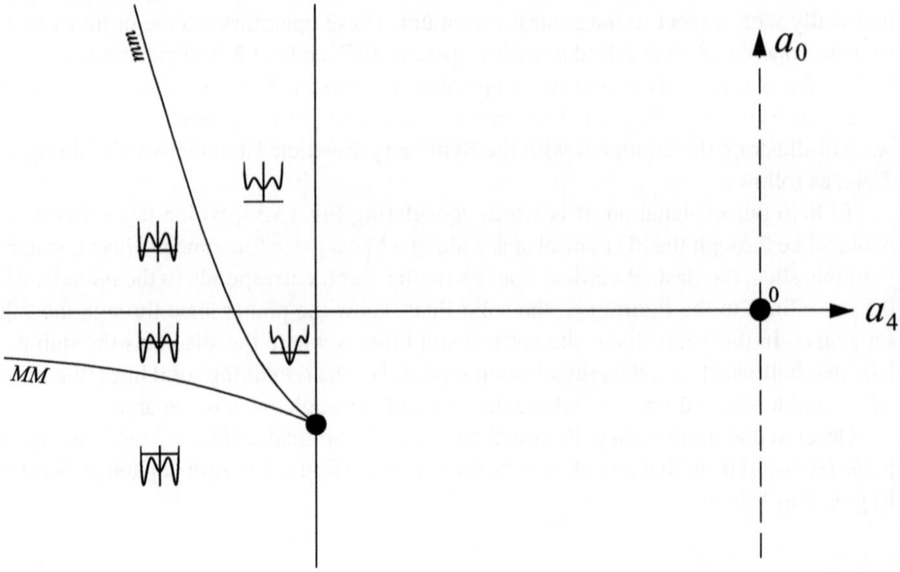


**Fig. 13.34.** Representative functions in the regions $5_i$ of the previous figure.

**As illustrated in Fig 13.33 and Fig 13.34:**
**A vertical planar slice $a_2 = c$ (positive $c$) through Region 5 is divided into five regions by the zero-separatrix.**
**The regions are labeled:**
**$5_i$ = functions with 5 critical points and $i$ zeros.**
**The regions are:**

$5_0$ = functions in which all critical points are positive; i.e., *no* zeros.

$5_2$ = functions in which only the central minimum is negative; i.e., *two* zeros around that minimum.

$5_4$ = functions in which only the two side minima are negative; i.e., *four* zeros around those two minimum.

$5_6$ = functions in which all three minima, but not the maxima, are negative; i.e., *six* zeros around those three minimum.

$5_{2'}$ = functions in which all critical points are negative; i.e., *two* zeros around the entire set of critical points.

Fig 13.33 shows the labeling of the regions $5_i$ in relation to the zero-separatrix; and Fig 13.34 shows representative functions of those regions.

Now, in section 6.3, our study of the *continuation of processes*, on the 5-fold part we call a triple-shield, began by considering the creation of a triple-shield from a central compressive extremum, as illustrated again here in Fig 13.35, in the transition from the bottom $m^+$ extremum of the ellipse, the left shape, to the triple-shield at the bottom of the right shape. The triple-shield has, of course, three compressive extrema, $m^+$, corresponding to three squashing processes.

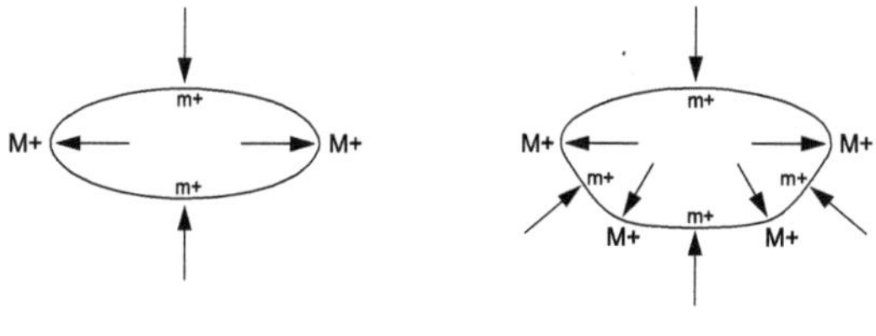


**Fig. 13.35.** Creation of a 5-fold part, a triple-shield.

Now observe the following: The bottom extremum $m^+$ in the first shape corresponds, in Fig 13.33, to a function in the space to the *right* of the critical-point separatrix (vertical solid line), and *above* the horizontal sheet of the zero-separatrix (horizontal line). And the triple-shield corresponds to a function in the region labeled $5_0$ in Fig 13.33; that is, to the left of the critical-point separatrix (vertical solid line) and above both the double-minima separatrix $mm$ and the horizontal sheet of the zero-separatrix (horizontal line). A representative function in this region, $5_0$, is shown in Fig 13.34.

Now, in section 6.3, we observed that the three squashing processes, in the triple-shield (Fig 13.35), can *continue* till they indent. Since, in each of the three alternative scenarios that created the triple-shield, the three squashing processes were created *symmetrically* with respect to the starting extremum in the first shape, it is likely that the continuations of the squashing processes would be symmetrically organized.

The first symmetrical continuation we defined was simply the continuation of the central squashing process of the triple-shield till it indents, as illustrated in the transition from the first to the second shape in Fig 13.36. This is given by the Process Grammar operation

$$Cm^+ \ : \ m^+ \longrightarrow 0m^-0$$

We can now model this in the Symmetry-Restricted Interactive Unfolding of $EA_5$ as follows: As observed above, the triple-shield corresponds to a function in Region $5_0$ in Fig 13.33. Then the operation $Cm^+$, applied to the central minimum, is modeled by the following transition between regions:

$5_0 \longrightarrow 5_2$ through the horizontal central sheet of the zero-separatrix.

From Fig 13.33, the reader can see that, although Region $5_2$ is below the central sheet, it is above the double-minima sheet $mm$. The functions corresponding to the transition we have just defined, from Region $5_0$ to Region $5_2$, are shown in Fig 13.34. We can see that the central minimum, within the function, moves downwards through the function's own $t$-axis. This, of course, accords with the definition of $Cm^+$.

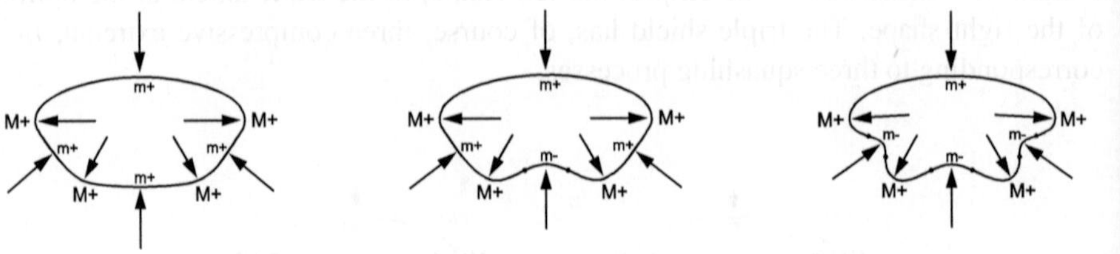


**Fig. 13.36.** Continuation $C$ followed by 2-fold pinching $CC$.

In section 6.3, we observed that this symmetrical continuation in the triple-shield, i.e., the application of $Cm^+$ to the central extremum, can be followed by a further symmetrical continuation: The two side squashing processes can continue simultaneously till they indent, as illustrated in the transition from the second shape to the third shape in Fig 13.36. This is an example of the 2-fold *pinching* operator $CC$, thus:

$$CCm^+ : \; m^+ \, X \, m^+ \quad \longrightarrow \quad \overbrace{0m^-0} \; X \; \overbrace{0m^-0}$$

where, in the present case, $X$ is the sequence $M^+0m^-0M^+$ between the two side $m^+$ extrema in the second shape in Fig 13.36.

We will now model this second transition in the Symmetry-Restricted Interactive Unfolding of $EA_5$. As illustrated in Fig 13.33, this transition is modeled by the following transition between regions:

$5_2 \longrightarrow 5_6$ through the double-minima sheet $mm$ of the zero-separatrix.

It is important to emphasize the fact that Region $5_6$ is not only below the double-minima sheet $mm$, but also below the horizontal central sheet, and above the double-maxima sheet $MM$. Thus Region $5_6$ is what we have called the *enclosed volume* in the Symmetry-Restricted Interactive Unfolding of $EA_5$. The crucial fact is that any function in the enclosed volume is maximally bifurcated in the Symmetry-Restricted Interactive Unfolding of $EA_5$; that is, there are no further critical-point bifurcations and zero-bifurcations that the function can undergo in the Symmetry-Restricted Interactive Unfolding of $EA_5$. The functions corresponding to the transition we have just defined, from Region $5_2$ to Region $5_6$, are shown in Fig 13.34. We can see that the two side minima, within the function, move downwards through the function's own $t$-axis. This, of course, accords with the definition of the 2-fold pinching operation $CCm^+$ given above. Notice that, because the maximally bifurcated function has three negative minima and two positive maxima, it necessarily has six zeros.

Thus to recapitulate: The sequence of operations, illustrated by the sequence of shapes in Fig 13.36, is modeled by the following sequence of regions

$$5_0 \longrightarrow 5_2 \longrightarrow 5_6$$

in the Symmetry-Restricted Interactive Unfolding of $EA_5$. That is, the sequence goes first through the horizontal central sheet of the zero-separatrix, and then through the double-minima sheet $mm$ of that separatrix.

Now, keeping the symmetrical actions of the continuation operators, with respect to the central extremum, section 6.3 also applied the sequence just given, $C$ followed by $CC$, in the reverse order, as illustrated in Fig 13.37. Let us consider how this is modeled in the Symmetry-Restricted Interactive Unfolding of $EA_5$.

Begin by considering the first transition, $CC$. Here, the two side squashing processes of the triple-shield, in the first shape (Fig 13.37), continue simultaneously till they indent in the second shape.

We can now model this first transition in the Symmetry-Restricted Interactive Unfolding of $EA_5$, thus: As illustrated in Fig 13.33, this transition is modeled by the following transition between regions:

$5_0 \longrightarrow 5_4$ through the double-minima sheet $mm$ of the zero-separatrix.

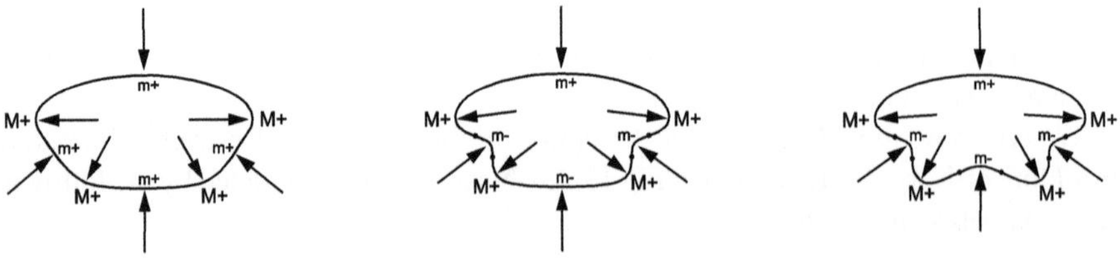


**Fig. 13.37.** 2-fold pinching $CC$ followed by continuation $C$.

It is important to emphasize the fact that, although Region $5_4$ is *below* the double-minima sheet, it is *above* the horizontal central sheet. The functions corresponding to the transition we have just defined, from Region $5_0$ to Region $5_4$, are shown in Fig 13.34. We can see that the two side minima, within the function, move downwards through the function's own $t$-axis. This, of course, accords with the definition of the 2-fold pinching $CC$. Notice that, because the resulting function has two negative minima, and no other negative critical-points, it necessarily has four zeros.

The second transition in the sequence shown in Fig 13.37 is the application of continuation operator $C$; that is, the continuation of the bottom central squashing process, in the second shape, till it indents, in the third shape.

We can now model this second transition in the Symmetry-Restricted Interactive Unfolding of $EA_5$, thus: As illustrated in Fig 13.33, this transition is modeled by the following transition between regions:

$5_4 \longrightarrow 5_6$ through the horizontal central sheet of the zero-separatrix.

Thus, the final region is again Region $5_6$, the enclosed volume. The functions corresponding to the transition we have just defined, Region $5_4$ to Region $5_6$, are shown in Fig 13.34. We can see that the central minimum, within the function, moves downwards through the function's own $t$-axis. This, of course, accords with the definition of the operation $Cm^+$. Since this operation induces a 2-fold zero-bifurcation, we can see that it adds two zeros to the existing four zeros in the function, thus giving the total of six zeros of a maximally bifurcated function.

Thus to recapitulate: The sequence of operations, illustrated by the sequence of shapes in Fig 13.37, is modeled by the following sequence of regions

$$5_0 \longrightarrow 5_4 \longrightarrow 5_6$$

in the Symmetry-Restricted Interactive Unfolding of $EA_5$ That is, the sequence goes first through the double-minima sheet $mm$ of the zero-separatrix, and then through the horizontal central sheet of the zero-separatrix.

In the present section, so far, we have considered two alternative scenarios, $C$ followed by $CC$, and $CC$ followed by $C$, by which all three squashing processes in the triple-shield were continued till they indented.

In modeling these two scenarios in the Symmetry-Restricted Interactive Unfolding of $EA_5$, we have seen that both begin in Region $5_0$ and end in Region $5_6$. The difference is that the first scenario goes via Region $5_2$, and the second scenario goes via Region $5_4$. Thus, the first scenario proceeds by first crossing the horizontal central sheet, and then crossing the double-minima sheet $mm$. In contrast, the second scenario proceeds by first crossing the double-minima sheet $mm$, and then crossing the horizontal central sheet.

In Section 6.3, we argued that there is one more symmetrical case. In this, *all three* squashing processes continue *simultaneously* till they indent, as illustrated in the transition from the left to the right shape in Fig 13.38. We called this *3-fold pinching*, and defined it by the Process Grammar operator, $CCC$, which, in the present example, is realized by the following operation

$$CCCm^+ \ : \ m^+M^+m^+M^+m^+ \ \longrightarrow \ \overbrace{0m^-0}\ M^+\ \overbrace{0m^-0}\ M^+\ \overbrace{0m^-0}$$

where the three over-braces, in the codomain, indicate the three simultaneous uses of the continuation operator $C$.

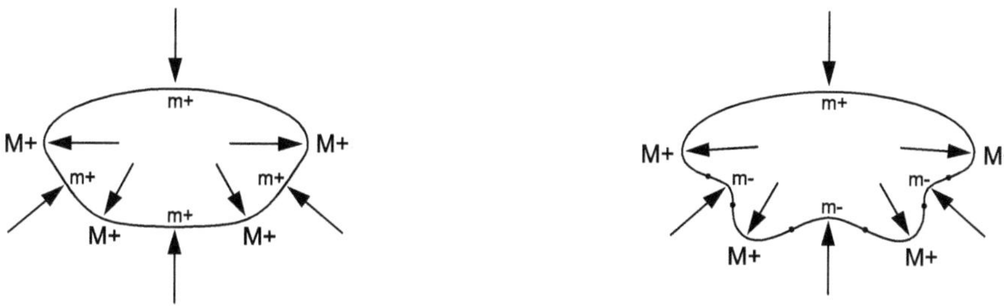


**Fig. 13.38.** 3-fold pinching.

In fact, in the present context, we shall see that it is more appropriate to understand $CCC$ as the simultaneous occurrence of the 2-fold pinching operator $CC$ and the 1-fold continuation operator $C$. The reason is as follows:

Our model for the operator $CCC$ in the Symmetry-Restricted Interactive Unfolding of $EA_5$ is this: The first thing to observe is that the transition defined by the operation $CCCm^+$ goes between the same starting domain and final codomain as the shape-sequences considered earlier in this section. That is, as illustrated in Fig 13.33, the transition is modeled by starting in Region $5_0$ and ending in Region $5_6$.

The next thing to observe is that, since, in the operation $CCCm^+$, all three positive minima $m^+$ undergo their 2-fold zero-bifurcation at the same time, this must mean that, in the transition-function of the operation $CCCm^+$, all three minima must be zero at the same time. In the Symmetry-Restricted Interactive Unfolding of $EA_5$, this can only be the case if the transition-function is on the intersection of the double-minima sheet $mm$ and the horizontal central sheet. This intersection is shown as the dashed curve in the 3D control space as illustrated in Fig 13.22 (page 490). Of course, this gives the intersection point seen in the 2D slice in Fig 13.33.

Thus the operation $CCCm^+$ is modeled in the Symmetry-Restricted Interactive Unfolding of $EA_5$ as the following transition between regions:

$5_0 \longrightarrow 5_6$ through the *intersection* between the double-minima sheet and the horizontal central sheet of the zero-separatrix.

The way the reader should understand this is as follows:

**Since we have seen that the 3-fold pinching operation $CCC$ applied to the triple-shield in the Symmetry-Restricted Interactive Unfolding of $EA_5$ is given by a downward crossing of the *intersection curve* between the double-minima sheet and the horizontal central sheet of the zero-separatrix, this means a crossing of *both* sheets *downwards*.**

**With respect to this, observe the following:**

**Recall that we have seen that crossing the horizontal central sheet downwards means that the function is undergoing the operation $C$ on its central minimum.**

**Also, recall that we have seen that crossing the double-minima sheet $mm$ downwards means that the function is undergoing the operation $CC$ on its two side minima.**

**Therefore, since downward crossing the *intersection curve* means downward crossing the two sheets simultaneously, this means that the two operators, $C$ and $CC$, are being applied simultaneously.**

**Thus we see that the 3-fold pinching operator $CCC$ applied to a triple-shield is modeled, in the Symmetry-Restricted Interactive Unfolding of $EA_5$, as decomposed into two operators, $C$ and $CC$, because the Symmetry-Restricted Interactive Unfolding of $EA_5$ describes this 3-fold pinching as the simultaneous crossing of *two different zero-separatrix sheets*.**

The next page states some of the main results that were given in this section.

## OPERATORS $C, CC, CCC$ IN THE SYMMETRY-RESTRICTED INTERACTIVE UNFOLDING OF $EA_5$ APPLIED SYMMETRICALLY TO A TRIPLE-SHIELD

**A** ***triple-shield*** **is located in Region 5 of the Symmetry-Restricted Interactive Unfolding of $EA_5$** ***above the zero-separatrix.***

**The application of the three operators, $C$, $CC$, $CCC$, symmetrically to the triple-shield, correspond respectively to the three following crossings of the zero-separatrix:**

**Operator $C$**
**This operator, which continues the central compressive process of the triple-shield, corresponds, in the Symmetry-Restricted Interactive Unfolding of $EA_5$, to a downward crossing (from above the zero-separatrix in Region 5) through the horizontal central sheet within Region 5.**

**Operator $CC$**
**This operator, which continues the two side compressive processes of the triple-shield, corresponds, in the Symmetry-Restricted Interactive Unfolding of $EA_5$, to a downward crossing (from above the zero-separatrix in Region 5) through the double-minima sheet within Region 5.**

**Operator $CCC$**
**This operator, which continues the three compressive processes of the triple-shield, corresponds, in the Symmetry-Restricted Interactive Unfolding of $EA_5$, to a maximal downward crossing (from above the zero-separatrix in Region 5) through the intersection curve between the double-minima sheet and the horizontal central sheet.**

**Clearly, the dual statements hold for the continuation operators, $C$, $CC$, $CCC$, in the Symmetry-Restricted Interactive Unfolding of the Dual $EA_{-5}$.**

Finally, we note that other uses of these continuation operators, within the Symmetry-Restricted Interactive Unfoldings of $EA_5$ and $EA_{-5}$, will be given later this chapter.

## 13.10 Five-fold Radial-Undulation Operator $[C(6)B(5)]$

In the previous section, we started by considering 5-fold part-formation in which the bottom compressive extremum $m^+$ of the ellipse in Fig 13.35 (page 505) underwent a transition to a triple-shield. Next we considered a set of scenarios that continued the three squashing processes of the triple-shield till they all eventually indented. Observe that the result was that the triple-shield was converted into what we call a *5-fold undulation*. This is a configuration of 6 zeros alternating with 5 penetrative extrema of alternating opposite signs, $M^+$ and $m^-$; i.e., implying *penetrative processes of alternating opposite directions*.

Therefore, the entire shape development, starting with the bottom compressive extremum $m^+$ of the ellipse, and ending with the 5-fold undulation, is as shown in Fig 13.39.

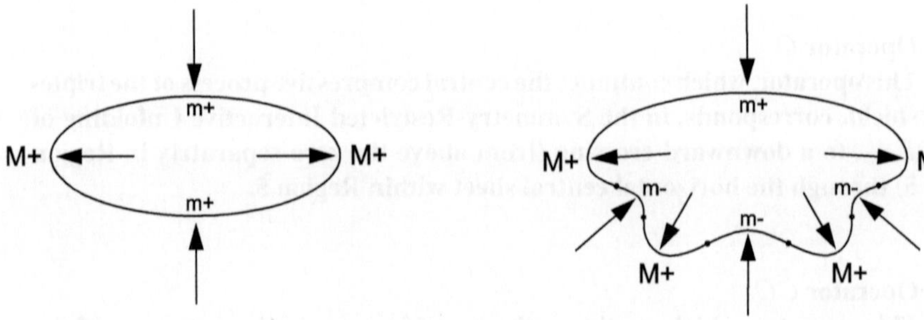


**Fig. 13.39.** Transition from a compressive extremum $m^+$ to a 5-fold undulation.

What we want to do now is consider the scenario in which one goes *directly* from the left shape to the right shape in Fig 13.39, without going through the intervening stages considered in the previous section. That is, the bottom $m^+$ of the ellipse must bifurcate *simultaneously* into the six zeros and five extrema of the undulation in the right shape, i.e., it must undergo the following 11-fold bifurcation:

$$m^+ \longrightarrow 0m^-0M^+0m^-0M^+0m^-0 \qquad (13.3)$$

In section 6.4, we defined the Process Grammar operator that gives this transition as $[C(6)B(5)]$, because the transition can be viewed as the *simultaneous* application of two operators, the 6-fold zero-bifurcation operator $C(6)$, and the 5-fold extremum-bifurcation operator $B(5)$, to the *same* application-point. In expression (13.3), the operator is realized as the operation $[C(6)B(5)]m^+$; and the dual is obviously $[C(6)B(5)]M^-$. The operator was given the name, *5-fold radial-undulation operator*, which is a term that was fully explained in Definition 4.15 (page 122).

Note that the operator $[C(6)B(5)]$ was defined in the bold print on page 221.

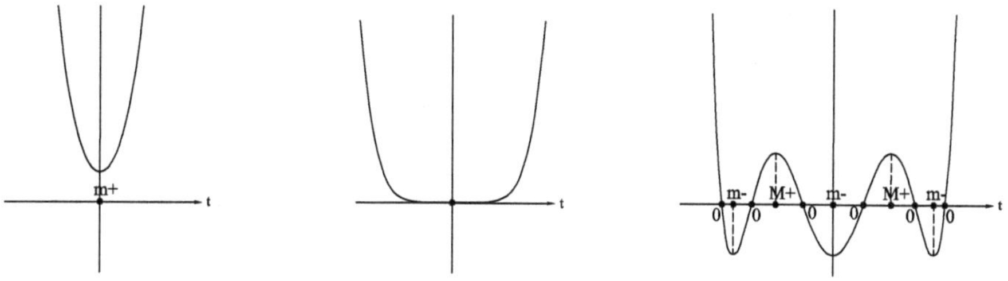


**Fig. 13.40.** The qualitative structure of the successive curvature functions in the operation $[C(6)B(5)]m^+$.

The qualitative structure of the operation $[C(6)B(5)]m^+$ is illustrated by the three successive curvature functions in Fig 13.40. We see that the first function has the singularity-configuration $m^+$; and the last function has the 11-fold singularity-configuration of the 5-fold undulation. The middle function is the transition-function of the operation. The eleven singularities of the last function must simultaneously coalesce, backwards in time, to the single degenerate zero in the transition-function.

We are now going to model the operator $[C(6)B(5)]$ in terms of the Symmetry-Restricted Interactive Unfoldings of $EA_5$ and $EA_{-5}$. We will see that the operator is the *maximal crossing* of the *germ* at the origin of the 3D control space. However, it is important to carefully understand this in terms of the particular *volumes* involved in the control space, as follows:

Let us begin by considering the Symmetry-Restricted Interactive Unfolding of $EA_5$. Clearly, the first function in Fig 13.40 belongs to the volume *above* Region 1 of the horizontal central plane of the control space. That is, using Fig 13.23 (page 492), this volume is *behind* the critical-point separatrix, i.e., the central *vertical* plane and the half-parabola sheet; and also above the central *horizontal* plane.

Next, the transition-function is the germ $t^6$ at the origin of this space, because the germ is the only function in the control space that has a singularity which is *both* 5-fold degenerate as a critical point and 6-fold degenerate as a zero; i.e., it is 11-fold degenerate in the particular way that is required of the transition-function of the operation $[C(6)B(5)]m^+$.

However, we must now carefully understand how the operation then moves in the control space, from its transition-function, the origin, to the final function, thus:

First, we know that the final function must be in the *enclosed volume*. In the previous section, we represented this volume by Region $F_6$ in the vertical slice shown in Fig 13.33 (page 504). Observe the crucial fact that Region $F_6$ is below *both* the double-minima sheet and the central horizontal plane, but above the double-maxima sheet.

Now, the vertical slice shown in this figure is along the line $a_2 = c$ for positive $c$. If we successively reduce $c$ to zero, then the $MM$-curve, shown in this slice, will successively rise to become the negative $a_4$-axis, at $c = 0$; where the $mm$-curve will form a cusp with the negative $a_4$-axis.

Correspondingly, Fig 13.21 (page 488) shows that, as we move the $a_2 = c$ slice, from the value $c = 0$ in the *positive* $a_2$-direction, Region $5_6$ is initially non-existent (when $c = 0$), and successively expands downward. That is, this figure shows the *successively expanding* $5_6$ *Regions* on the successive slices.

This means that when, under the operation $[C(6)B(5)]m^+$, one leaves the transition-function (the origin), to enter the enclosed volume, one must move *downwards*, i.e., below the horizontal central plane, and simultaneously move in the positive $a_2$ direction.

This enables us to fully understand the actual movement in the 3D structure in Fig 13.23 (page 492). That is, one moves from the origin *directly* into the volume that is below *both* the horizontal central plane and the double-minima sheet $mm$, but above the double-maxima sheet $MM$. It is this volume that we call the *enclosed volume*. The volume can be intuitively understood as a "bent cone" whose peak is the origin.

Thus, putting together the entire transition involved in the operation $[C(6)B(5)]m^+$, we see that the operation goes from a point *above* the Region 1 of the horizontal central plane, *directly* to the origin, which is the "peak" of the enclosed volume, and then *directly* from this "peak" into the enclosed volume itself (i.e., into the "bent cone"). An essential fact is that, by moving from the origin *directly* into the enclosed volume, the 11-fold degenerate singularity, within the germ at the origin, will *immediately* undergo the complete simultaneous 11-fold bifurcation, into five critical points and six zeros, the moment one moves off the origin into the enclosed volume.

Notice also that, in crossing the germ in this way, one is crossing *two* separatrices simultaneously:

> **The 5-fold radial-undulation operator, $[C(6)B(5)]$, is labeled by** *two operators*, **$C(6)$ and $B(5)$, because it simultaneously crosses** *two separatrices*: **the zero-separatrix and the critical-point separatrix.**

This should be carefully understood as follows: The operation $[C(6)B(5)]m^+$ crosses the *zero-separatrix* by going through the origin into the enclosed volume, because the enclosed volume is bounded entirely by the surfaces of the zero-separatrix, as illustrated in Fig 13.22 (page 490), which is a figure that shows *only* the zero-separatrix. And the same operation crosses the *critical-point separatrix* by going through the vertical line of cusps which connect the two sheets of the critical-point separatrix: the half-parabola sheet, and the vertical central plane along the $a_4$-axis. Now observe the following:

## ACTION OF OPERATOR $[C(6)B(5)]$ IN THE SYMMETRY-RESTRICTED INTERACTIVE UNFOLDINGS OF $EA_5$ AND $EA_{-5}$

**The Process Grammar operator $[C(6)B(5)]$ corresponds, in the control spaces of the Symmetry-Restricted Interactive Unfoldings of $EA_5$ and $EA_{-5}$, to the 11-fold bifurcation (5-fold critical-point bifurcation and 6-fold zero-bifurcation) obtained by a maximal crossing of the germ, from the volume whose functions have the** ***minimal-length singularity-configuration*** **(only a compressive extremum), into the volume whose functions have the** ***maximal-length singularity-configuration.***

**That is: The $[C(6)B(5)]$ operator is the maximal possible bifurcation in the Symmetry-Restricted Interactive Unfoldings of $EA_5$ and $EA_{-5}$.**

**In the Symmetry-Restricted Interactive Unfolding of $EA_5$, the operator is realized as the operation $[C(6)B(5)]m^+$; and in the Symmetry-Restricted Interactive Unfolding of $EA_{-5}$, it is realized as the operation $[C(6)B(5)]M^-$.**

The fact that the $[C(6)B(5)]$ operator is the maximal possible bifurcation, in the Symmetry-Restricted Interactive Unfoldings of $EA_5$ and $EA_{-5}$, relates to our *theory of parts* in Chapter 5. That is, the above argument has lead to the following conclusion:

## OPERATOR $[C(6)B(5)]$ AS PART-FORMATION

**The operator $[C(6)B(5)]$ produces the** ***maximally bifurcated parts*** **in the Symmetry-Restricted Interactive Unfoldings of $EA_5$ and $EA_{-5}$, by** ***maximal bifurcation.***

## 13.11 Double-Continuation of Double-Compressive Parts

Recall, from section 6.5, the following issue: When we studied the continuation operators, $C$, $CC$, $CCC$, in section 6.3, we applied them to a 5-fold part that has a *triple-compressive* structure (i.e., a triple-shield or a triple-bay). An important fact is that all three operators, $C$, $CC$, $CCC$, can be applied *symmetrically* to such a part.

However, in section 6.5, we observed that the operator $CC$ can be applied also to a 5-fold part that has a *complete double-compressive* structure (i.e., a complete double-bay or a complete double-shield).

Note that the application of the three operators, $C, CC, CCC$, to a triple-compressive part, was modeled in the Symmetry-Restricted Interactive Unfolding of $EA_5$, in section 13.9. In the present section, we show how the application of the operator $CC$, to a complete double-compressive part, is also modeled in the Symmetry-Restricted Interactive Unfolding of $EA_5$.

It is important to understand the issues involved. To do so, recall first that our application of the continuation operators, $C$, $CC$, $CCC$, to a triple-shield, was preceded by the formation of that 5-fold part (the triple-shield) from a single *compressive* extremum, a *positive minimum* $m^+$, as illustrated in Fig 13.35 (page 505).

In the present case, our application of the operator $CC$, to a complete double-bay, will be preceded by the formation of that 5-fold part (the complete double-bay) from a single *penetrative* extremum, a *negative minimum* $m^-$. This is illustrated in Fig 13.41. The left shape shows the indentation process inferred from the initial $m^-$ in the center of the shape. The transition from the left shape to the right shape is given by applying any one of the Three Scenarios, $\pi^{\downarrow}\pi^{\uparrow}\emptyset^-$, $B_{-3} \circ B_3 m^-$, or $B_5 m^-$. The result is the *complete double-bay* shown in the right shape.

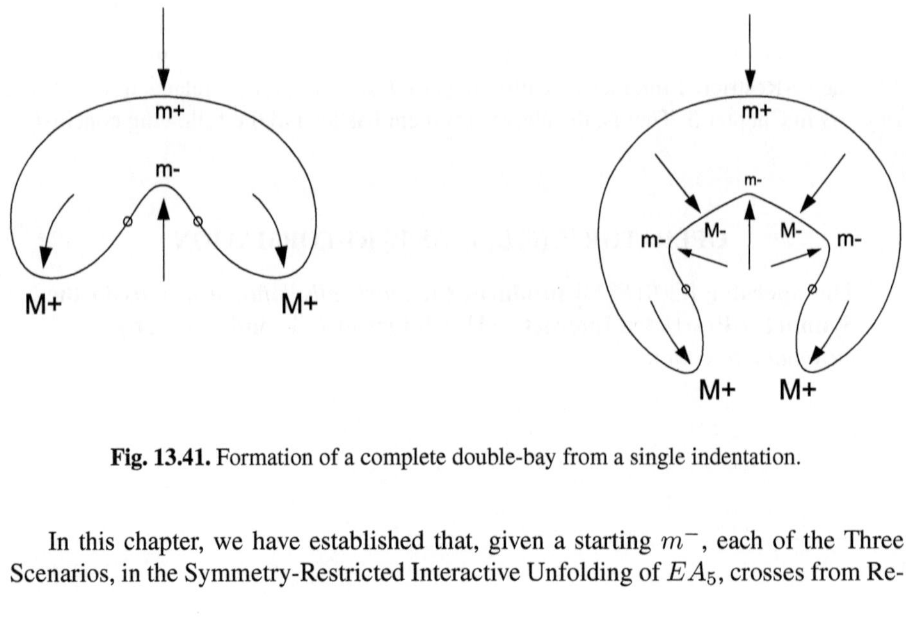


**Fig. 13.41.** Formation of a complete double-bay from a single indentation.

In this chapter, we have established that, given a starting $m^-$, each of the Three Scenarios, in the Symmetry-Restricted Interactive Unfolding of $EA_5$, crosses from Region 1 to Region 5, entirely under the zero-separatrix, via particular components of the critical-point separatrix.

The resulting shape, the complete double-bay, is in Region $5_{2'}$ in Fig 13.33 (page 504).

It is important for the reader to compare the creation of a 5-fold part in the previous case, i.e., *the triple-shield*, see Fig 13.35 (page 505), with the creation of a 5-fold part in the present case, i.e., *the complete double-bay*, see Fig 13.41. Both start with a single *minimum*. In the previous case, the minimum was *compressive*. In the present case, the minimum is *penetrative*. In the 3D control space of the Symmetry-Restricted Interactive Unfolding of $EA_5$, this is modeled by the fact that, in the previous case, the function is in Region 1 *above* the horizontal central plane; whereas, in the present case, the function is in Region 1 *below* the horizontal central plane.

Therefore the consequence, of applying any one of the Three Scenarios, is this: In the previous case, the resulting 5-fold part consists of *three* compressive extrema and *two* penetrative extrema. In the present case, the resulting 5-fold part consists of *two* compressive extrema and *three* penetrative extrema.

Thus, we conclude this: In the previous case, the resulting 5-fold part, the triple-shield, corresponds to a function in Region $5_0$, illustrated in Fig 13.33 (page 504). In contrast, in the present case, the resulting 5-fold part, the complete double-bay, corresponds to a function in Region $5_{2'}$, also illustrated in Fig 13.33 (page 504).

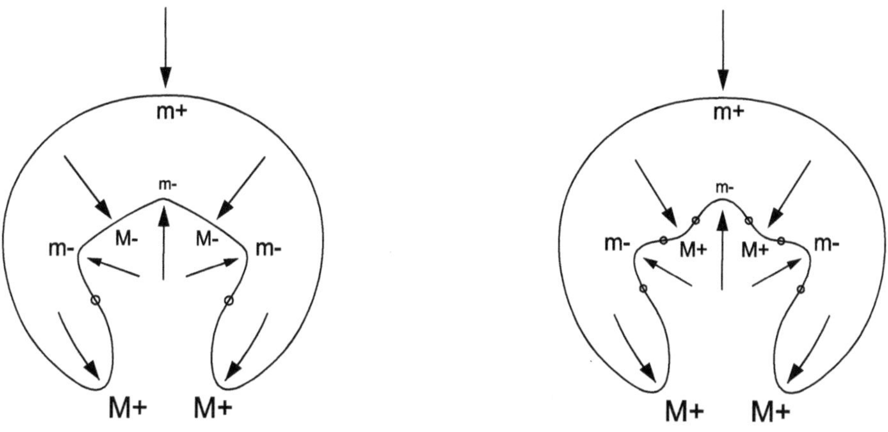


**Fig. 13.42.** The operation $CCM^-$ gives the transition from a complete double-bay to a pinched complete double-bay.

Therefore, an important fact is that, in the previous case, all three operators $C$, $CC$, $CCC$, can be applied symmetrically to the compressive extrema. However, in the present case, only the operator $CC$ can be applied symmetrically to the compressive extrema. The latter application is illustrated in Fig 13.42. The first shape in this figure is the 5-fold part that was produced in the above transition, Fig 13.41; that is, the complete double-bay. We now apply the 2-fold pinching operator $CC$ to the two side $M^-$ extrema in the complete double-bay; thus making their two arrows continue till they protrude, as shown in the right shape in Fig 13.42. That is, the transition is given by the operation:

$$CCM^- \ : \ M^- \, m^- \, M^- \ \rightarrow \ \overbrace{0M^+0} \, m^- \, \overbrace{0M^+0}$$

Observe that the operation has the effect of *pinching* the complete double-bay. Therefore, the result is what our theory calls the *pinched complete double-bay*.

We will now model this transition in the Symmetry-Restricted Interactive Unfolding of $EA_5$. As illustrated in Fig 13.33 (page 504), this is modeled by the following transition between regions:

$5_{2'} \longrightarrow 5_6$ through the double-maxima sheet $MM$ of the zero-separatrix.

The functions corresponding to this transition are shown in Fig 13.34. That is, as seen in the figure, for any function in Region $5_{2'}$, the critical points are *all* negative. This is because Region $5_{2'}$ is *entirely* below the zero separatrix. Then, in the transition to Region $5_6$, the two maxima in the function move upwards through the function's own $t$-axis. Since the three minima remain negative, this produces a maximally-bifurcated function; i.e., three negative minima and two positive maxima.

The crucial fact is that, whereas the resulting configuration in Region $5_6$ was previously produced by continuing the *three* compressive processes in a *triple-shield*, it is produced in the present situation by continuing the *two* compressive processes in a *complete double-bay*.

**The difference between these two histories, the creation of a triple-shield that gets pinched and the creation of a complete double-bay that gets pinched, is given by *interactions* of crossing the critical-point separatrix and the zero-separatrix, i.e., the *Interactive Structure* as defined by our *Interactive Singularity Theory*. Furthermore, it uses the crucial fact that Interactive Singularity Theory matches the *causal* structure defined by our process-inference rules which use the PISA axes; that is, it distinguishes the fact that the triple-shield history began with a *compressive* extremum, whereas the complete double-bay history began with a *penetrative* extremum.**

## 13.12 All Continuation Operators in the Symmetry-Restricted Interactive Unfolding of $EA_5$

In sections 13.9 and 13.11, the continuation operators, $C$, $CC$, $CCC$, were applied to *5-fold parts*. Therefore, in the 3D control space of the Symmetry-Restricted Interactive Unfolding of $EA_5$, they were applied over *Region 5* (i.e., between the regions labeled $5_i$). However, the operators $C$ and $CC$ are also applicable over *other regions* of the control space, where, in those other regions, they also go, respectively, through the horizontal central sheet and the double-minima sheet of the zero-separatrix.

## ALL CONTINUATION OPERATORS $C, CC, CCC$ IN THE SYMMETRY-RESTRICTED INTERACTIVE UNFOLDING OF $EA_5$

**The following gives the entire set of applications of the operators, $C$, $CC$, $CCC$, in the Symmetry-Restricted Interactive Unfolding of $EA_5$, i.e.,** *not only to 5-fold parts.* **The crucial fact is that these operators relate to the three sheets of the zero-separatrix, which were described in the bold print on page 489. The reader will find page 489 helpful in understanding the following:**

**The operators applied to minima:**

**Operator $C$**
**When applied to a minimum, this operator corresponds to a downward crossing of the horizontal central sheet in the positive $a_2$ half-plane.**

**Operator $CC$**
**When applied to minima, this operator corresponds to a downward crossing of the double-minima sheet, which exists everywhere across Regions 3 and 5 of the $a_0 = 0$ plane.**

**Operator $CCC$**
**When applied to minima, this operator corresponds to a maximal downward crossing of the intersection curve between the double-minima sheet and the horizontal central sheet.**

**The operators applied to maxima:**

**Operator $C$**
**When applied to a maximum, this operator corresponds to an upward crossing of the horizontal central sheet in the negative $a_2$ half-plane; i.e., Region 3 of the $a_0 = 0$ plane.**

**Operator $CC$**
**When applied to maxima, this operator corresponds to an upward crossing of the double-maxima sheet, which exists across Region 5 of the $a_0 = 0$ plane.**

**Operator $CCC$**
**When applied to maxima, this operator does not occur in the Symmetry-Restricted Interactive Unfolding of $EA_5$.**

Clearly, the dual statements hold for the operators $C$, $CC$, $CCC$ in the Symmetry-Restricted Interactive Unfolding of $EA_{-5}$.

## 13.13 Quantitative Example of the Operation $\pi^{\downarrow}\pi^{\uparrow}\emptyset^{+}$ Applied with $m^{+}$ Anchor-Extremum

On page 204, we listed the alternative operations that realize the downward-upward operator $\pi^{\downarrow}\pi^{\uparrow}$. The first one listed was the positive operation $\pi^{\downarrow}\pi^{\uparrow}\emptyset^{+}$. When applied to the sides of a squashing process, i.e., with anchor-extremum $m^{+}$, the operation is

$$\pi^{\downarrow}\pi^{\uparrow}\emptyset^{+} \ : \ \emptyset^{+} \ m^{+} \ \emptyset^{+} \ \longrightarrow \ \overbrace{m^{+}M^{+}} \ m^{+} \ \overbrace{M^{+}m^{+}}$$

This was illustrated in the top figure on page 205.

The present section will give a quantitative example of this operation in the Symmetry-Restricted Interactive Unfolding of $EA_5$. As usual, we will preserve the curvature of the initial extremum. Previously, for the 3-fold bifurcation operator $B$ and the pair-creation operator $\pi$ in the Interactive Unfolding of $EA_3$, this meant adding a varying curvature factor to the unfolding function, because, in being pushed sideways, the initial extremum undergoes a change in its value. However, for the operator $\pi^{\downarrow}\pi^{\uparrow}$ in the Symmetry-Restricted Interactive Unfolding of $EA_5$, the initial extremum remains at the center of the function, and therefore remains the same value – thus requiring no varying curvature factor.

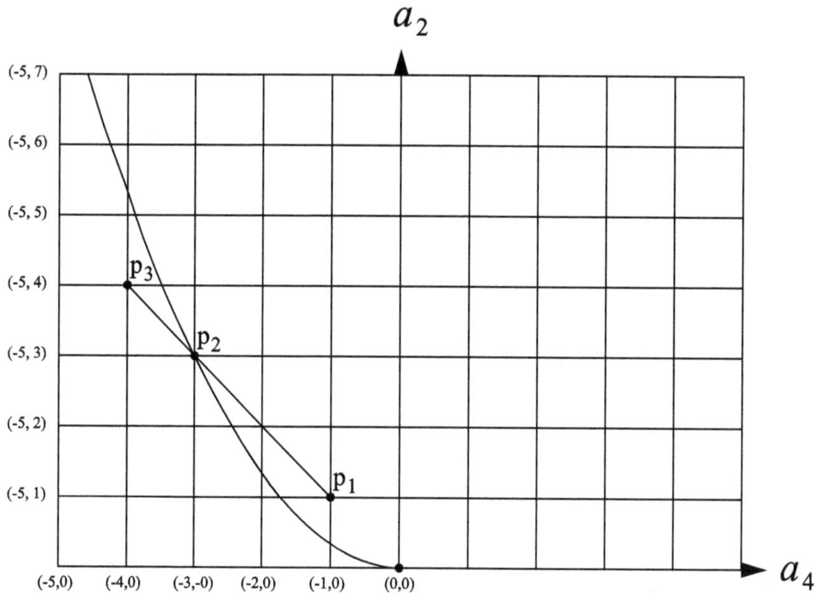


**Fig. 13.43.** The $(a_4, a_2)$-coordinates of the three points chosen, in the next three figures, to illustrate the $\pi^{\downarrow}\pi^{\uparrow}\emptyset^{+}$ operation.

**The Chosen Path:**

The downward-upward operator $\pi^{\downarrow}\pi^{\uparrow}$ goes from Region 1 to 5, by crossing the half-parabola sheet in the 3D control space. In the present case, this crossing must be *above* the zero-separatrix. The path, which we shall take to realize this, will be a sequence of three points whose $(a_4, a_2)$-coordinates will be the following, as illustrated in Fig 13.43:

$p_1 = (-1,\ 1)$
$p_2 = (-3,\ 3)$ on half-parabola
$p_3 = (-4,\ 4)$

The preserved curvature of the central extremum will be chosen to be $0.1$, which will therefore be the third coordinate $a_0$ for each of these points.

We will, as generally in Interactive Singularity Theory, consider the functions at these three points to be *curvature functions*. We will correspondingly see how the bifurcation properties, discussed earlier, determine properties of the resulting *parameterized curves*. The three functions, with their resulting parameterized curves, are shown respectively in the three figures, Fig 13.44 – 13.46. The reader should now read the captions of these figures, because they contain essential information.

**Curvature Function**

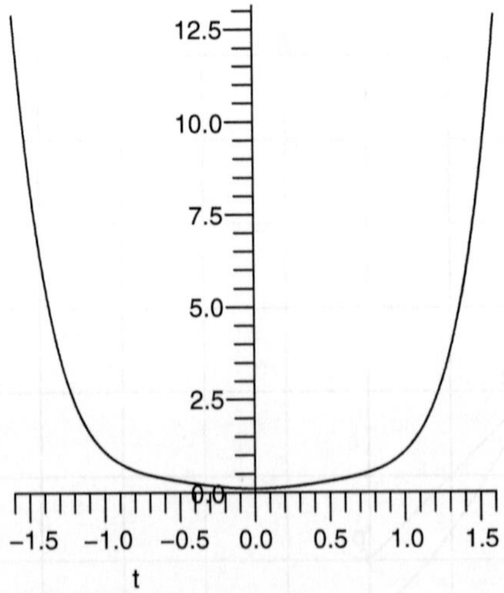


**Parameterized Curve**




**Fig. 13.44.** Pre-Bifurcation: The upper figure shows the curvature function $f(t) = t^6 - t^4 + t^2 + 0.1$ at the position $p_1 = (a_4,\ a_2,\ a_0) = (-1,\ 1,\ 0.1)$ in the 3D control space of the Symmetry-Restricted Interactive Unfolding of $EA_5$. Notice that the function shows the diagonal flattening of the sides of the graph, typical of functions just prior to the half-parabola. Also notice that the minimum of the function is the positive value 0.1, and is therefore an extremum of type $m^+$. The unit-speed parameterized curve, determined by the function, is shown below, with the $m^+$ extremum at the center.

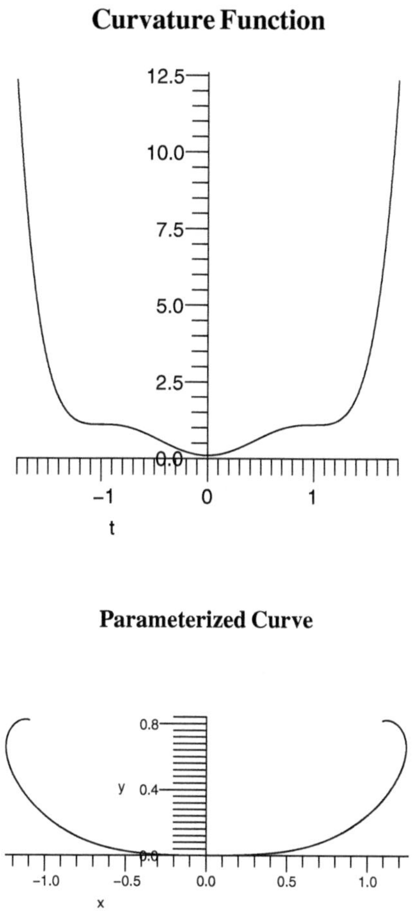


**Fig. 13.45.** Bifurcation Point: The upper figure shows the curvature function $f(t) = t^6 - 3t^4 + 3t^2 + 0.1$ at the point $p_2 = (a_4,\ a_2,\ a_0) = (-3,\ 3,\ 0.1)$ on the half-parabola sheet in the 3D control space in the Symmetry-Restricted Interactive Unfolding of $EA_5$. The lower figure shows its corresponding unit-speed parameterized curve. The function has preserved the central minimum $m^{+}$, but the neighborhood of the minimum has risen in curvature, thus causing greater bending in the corresponding regions in the curve below. Most importantly, because the function is on the half-parabola sheet of the critical-point separatrix, a critical inflection-point has appeared in each side of the function. This is the transition in the operation $\pi^{\downarrow}\pi^{\uparrow}\emptyset^{+}$. Each of the two critical inflection-points is responsible for producing an almost circular region below it in the parameterized curve.

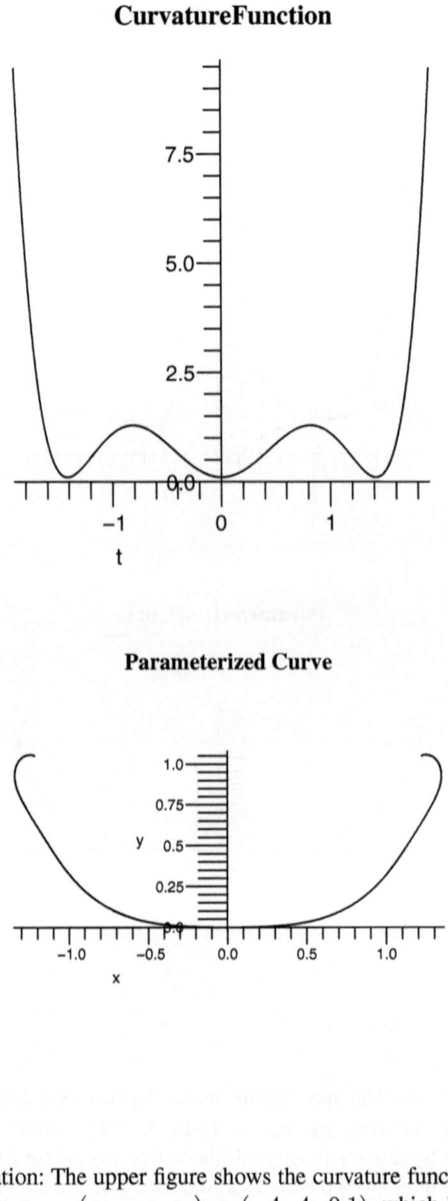


**Fig. 13.46.** Post-Bifurcation: The upper figure shows the curvature function $f(t) = t^6 - 4t^4 + 4t^2 + 0.1$ at the position $p_3 = (a_4,\ a_2,\ a_0) = (-4,\ 4,\ 0.1)$, which is a point *after* the half-parabola sheet in the 3D control space of the Symmetry-Restricted Interactive Unfolding of $EA_5$. Each of the two critical inflection-points in the previous function has undergone a bifurcation into a maximum and minimum. Consequently, there are now five extrema, where previously there was only one. All five extrema are positive, and therefore the function is above the zero-separatrix in the control space. The lower figure shows the resulting unit-speed parameterized curve. Observe that the parameterized curve has three compressive extrema corresponding to the three positive minima in the function. Also, on the parameterized curve, the three compressive extrema are separated by two penetrative extrema corresponding to the two positive maxima in the function.

The next figure, Fig 13.47, shows the three successive parameterized curves from Figs 13.44 – 13.46. We will now understand this progression in terms of the forces inferred by the Process Grammar. By the Symmetry-Curvature Duality Theorem, the first shape is created by a single upward force, against the single extremum at the center. Since the object (solid) is on the upper side of the curve, this force must be a *squashing*. In the second shape, a rounding has appeared on each side, due to the initial presence of new forces pushing diagonally on each side. In the third shape, these side forces have created a bifurcation, giving a compressive extremum on each side, as well as penetrative extrema that separate those compressive extrema from the bottom compressive extremum.

**Fig. 13.47.** The shape progression in the previous three figures.

In our theory of parts, the final configuration is classified as a *5-fold part*. Its five extrema are $m^{+}M^{+}m^{+}M^{+}m^{+}$, with the set of inferred forces shown in Fig 13.48. This means that there are two symmetrical pairings: (1) the pairing of the two side $m^{+}$ squashing forces, and (2) the pairing of the two $M^{+}$ protruding forces between them and the central squashing force.

Given that the four created extrema were produced by the downward-upward symmetrically-paired pair-creation operator $\pi^{\downarrow}\pi^{\uparrow}\emptyset^{+}$, the pair $m^{+}M^{+}$ on the left were created by the pair-creation operator $\pi^{\downarrow}$ and the pair $M^{+}m^{+}$ on the right were created by the pair-creation operator $\pi^{\uparrow}$. Recall, from Chapter 4, that, according to our theory, a pair-creation operator is used to simultaneously realize *two design intents* away from an extremum, i.e., on a spiral of a curve.

Therefore, the symmetrical-pairing of pair-creation operators realizes the *pairing of pairs of design intents*. In the present case, this is given as follows:

(1) In the top-right shape of Fig 13.49, the pairing shown is that of the processes at the two created compressive extrema. Therefore, in this case, the symmetrically-paired design intents have been the creation of two additional shields, at the sides of the initial shield.

(2) In the bottom-right shape of Fig 13.49, the pairing shown is that of the processes at the two created penetrative extrema. Therefore, in this case, the symmetrically-paired design intents have been the separation of the three resulting shields.

**FUNDAMENTAL STATEMENT**

**According to our theory, the symmetrical-pairing of design intents is fundamentally important in biological morphology and manufacturing design.**

**This accords with the Maximization of Transfer principle of our New Foundations to Geometry.**

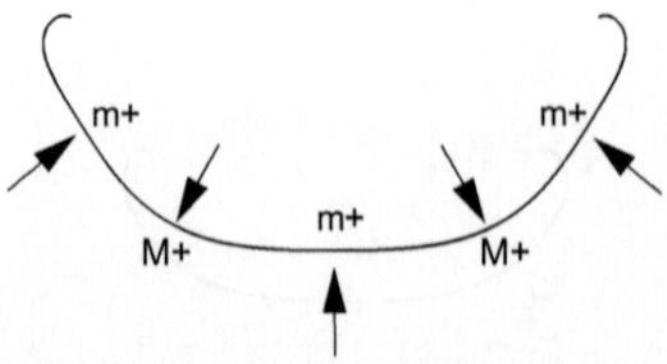


**Fig. 13.48.** The five extrema in the final shape, together with their inferred forces.

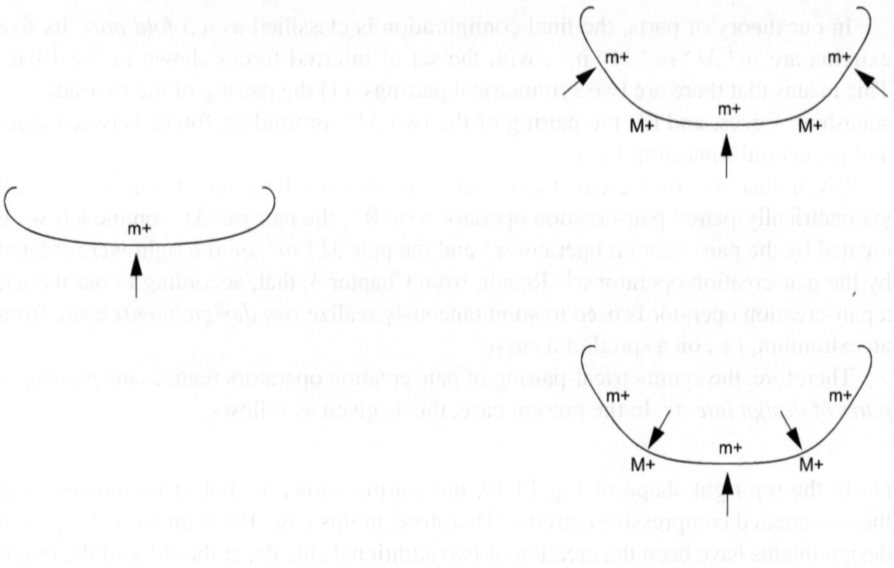


**Fig. 13.49.** From the five extrema, the symmetrical pairings of design intents.

## 13.14 Quantitative Example of Adding the Continuation Operation $Cm^+$ to the Previous Operation $\pi^{\downarrow}\pi^{\uparrow}\emptyset^+$

To give a quantitative example of the *Interactive Structure* of the Symmetry-Restricted Interactive Unfolding of $EA_5$, we will now apply continuation to the central process of the triple-shield created in the previous section.

When one applies the Process Grammar operation

$$Cm^+ : m^+ \longrightarrow 0m^-0$$

to the central $m^+$ of the final part given in the last section, its squashing process will continue till it indents. This is achieved by crossing the central plane of the zero-separatrix in the Symmetry-Restricted Interactive Unfolding of $EA_5$, to a function that is *between* that plane and the double-minima $mm$-surface. That is, using the map of regions shown in Fig 13.33 (page 504), we undergo the following transition between regions:

$$5_0 \longrightarrow 5_2$$

An example of the resulting function, and its parameterized curve, is shown in Fig 13.50. The text now continues in the caption of this figure.

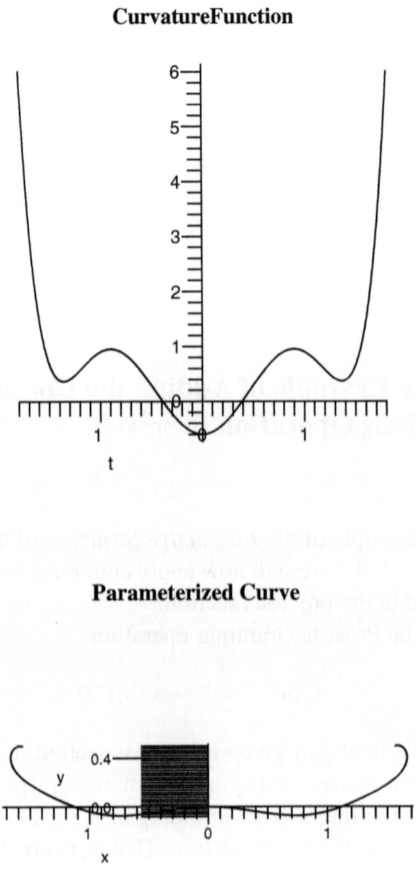


**Fig. 13.50.** Applying the continuation operation $Cm^+$ to the central positive minimum: The upper figure shows the resulting curvature function $f(t) = t^6 - 4t^4 + 4.5t^2 - 0.6$, at the position $p_4 = (a_4, a_2, a_0) = (-4, 4.5, -0.6)$ in the 3D control space of the Symmetry-Restricted Interactive Unfolding of $EA_5$. Due to the operation, the central minimum has been lowered through the $t$-axis, creating the sequence $0m^-0$. The new negative minimum $m^-$, in the function, corresponds to the upward indentation shown directly below it on the parameterized curve. The two side positive minima $m^+$ in the function correspond to the two side compressed regions directly below them on the parameterized curve. And, the two positive maxima $M^+$ in the function correspond to the two downward protrusions directly below them on the parameterized curve.

**Fig. 13.51.** The central continuation emphasizes the zone-of-influence restriction role of the two protrusions.

**DESIGN INTENT**

**Fig 13.51 shows the transition carried out by the $Cm^+$ operation applied to the central compressive extremum.**

**With respect to the continuation, this defines the design intent of the two $M^+$ forces as restricting the zone-of-influence of the central upward force.**

## 13.15 Quantitative Example of the Operation $\pi^{\downarrow}\pi^{\uparrow}\emptyset^-$

The second downward-upward operation listed on p204 is $\pi^{\downarrow}\pi^{\uparrow}\emptyset^-$. Its anchor-extremum is $m^-$, and therefore the operation is

$$\pi^{\downarrow}\pi^{\uparrow}\emptyset^- \;:\; \emptyset^- \; m^- \; \emptyset^- \;\longrightarrow\; \overbrace{m^- M^-} \; m^- \; \overbrace{M^- m^-}$$

This was illustrated by the figure on the third level in p205. It produces a complete double-bay.

The present section will give a quantitative example of this operation in the Symmetry-Restricted Interactive Unfolding of $EA_5$.

The downward-upward operator $\pi^{\downarrow}\pi^{\uparrow}$ goes from Region 1 to 5, by crossing the half-parabola sheet in the 3D control space. In the present case, this crossing must be entirely *below* the zero-separatrix. The path, which we shall take to realize this, will be a

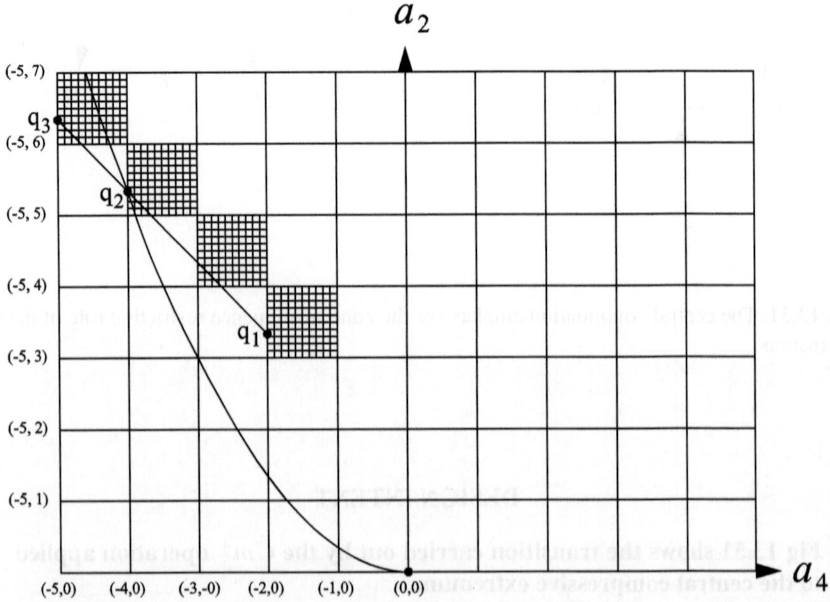


**Fig. 13.52.** The $(a_4, a_2)$-coordinates of the three points chosen, in the next three figures, to illustrate the $\pi^{\downarrow}\pi^{\uparrow}\emptyset^{-}$ operation.

sequence of three points whose $(a_4, a_2)$-coordinates will be the following, as illustrated in Fig 13.52:

$$\begin{aligned} q_1 &= (-2,\ 3.3) \\ q_2 &= (-4,\ 5.3) \quad \text{on half-parabola} \\ q_3 &= (-5,\ 6.3) \end{aligned}$$

The preserved curvature of the anchor-extremum will be chosen to be $-2.5$, which will therefore be the third coordinate $a_0$ for each of these points.

We will, as generally in Interactive Singularity Theory, consider the functions at these three points to be *curvature functions*. We will correspondingly see how the bifurcation properties, discussed earlier, determine properties of the resulting *parameterized curves*. The three functions, with their resulting parameterized curves, are shown respectively in the three figures, Fig 13.53 – 13.55. The reader should now read the captions of these figures, because they contain essential information.

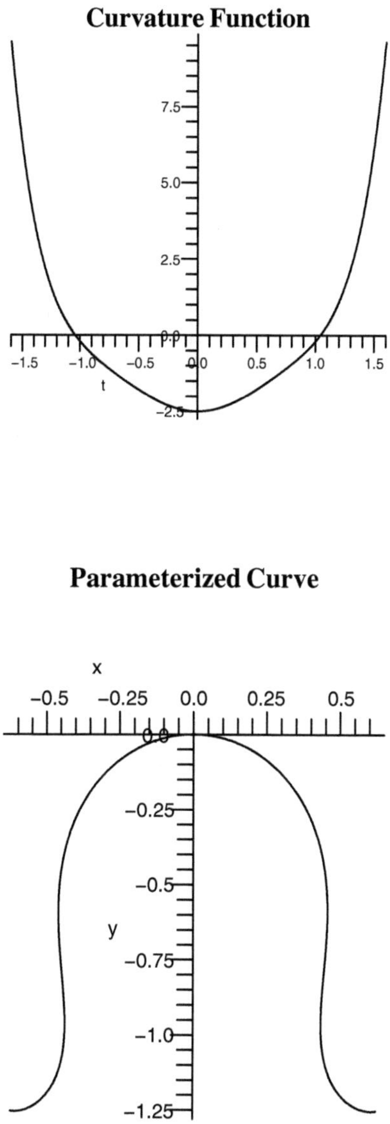


**Fig. 13.53.** Pre-Bifurcation: The upper figure shows the curvature function $f(t) = t^6 - 2t^4 + 3.3t^2 - 2.5$ at the position $q_1 = (a_4,\ a_2,\ a_0) = (-2,\ 3.3,\ -2.5)$ in the 3D control space of the Symmetry-Restricted Interactive Unfolding of $EA_5$. Notice that the function has the diagonal flattening of the sides of the graph, typical of functions just prior to the half-parabola. Also notice that the minimum of the function is the negative value $-2.5$, and is therefore an extremum of type $m^-$. The unit-speed parameterized curve, determined by the function, is shown below, with the $m^-$ extremum at the center.

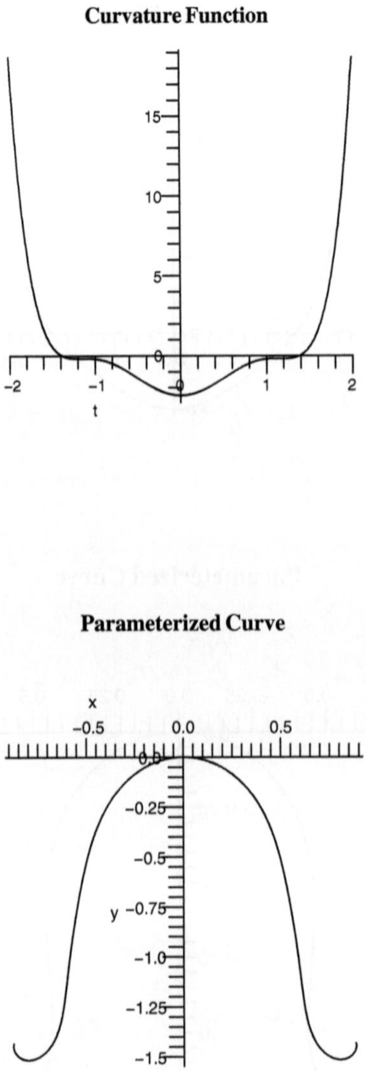


**Fig. 13.54.** Bifurcation Point: The upper figure shows the curvature function $f(t) = t^6 - 4t^4 + 5.3t^2 - 2.5$ at the point $q_2 = (a_4, a_2, a_0) = (-4, 5.3, -2.5)$ which is on the half-parabola sheet in the 3D control space of the Symmetry-Restricted Interactive Unfolding of $EA_5$, but below the zero-separatrix. The lower figure shows its corresponding unit-speed parameterized curve. The function has preserved the central minimum $m^-$ with the preserved curvature $-2.5$. However, the neighborhood of the minimum has risen almost to zero, thus causing greater flattening on each side of the parameterized curve below. Most importantly, because the function is on the half-parabola sheet of the critical-point separatrix, a critical inflection-point has appeared in each side of the function. Although these two critical inflection-points are near zero value in the curvature function, they are in fact negative, which ensures that the function is below the zero-separatrix in the control space of the Symmetry-Restricted Interactive Unfolding of $EA_5$.

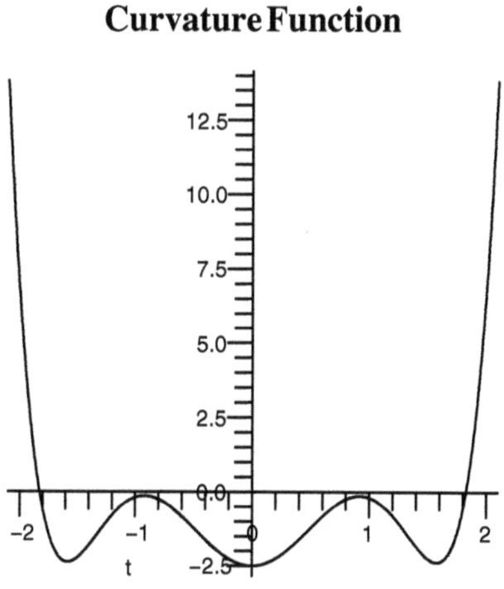


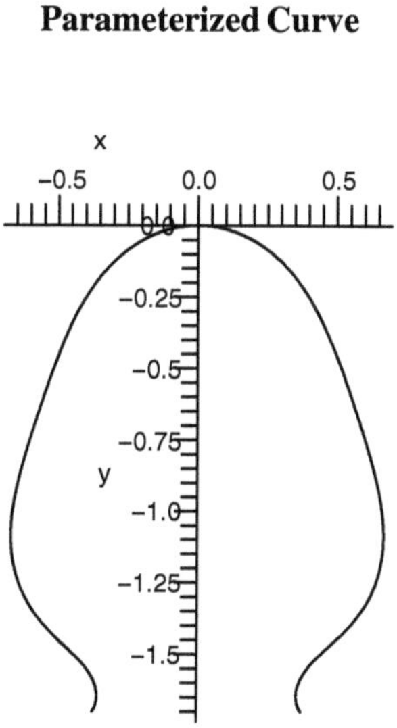


**Fig. 13.55.** Post-Bifurcation: The upper figure shows the curvature function $f(t) = t^6 - 5t^4 + 6.3t^2 - 2.5$ at the position $q_3 = (a_4,\ a_2,\ a_0) = (-5,\ 6.3\ -2.5)$ which is a point *after* the half-parabola sheet in the 3D control space of the Symmetry-Restricted Interactive Unfolding of $EA_5$. Each of the two critical inflection-points in the previous function has undergone a bifurcation into a maximum and minimum. Consequently, there are now five extrema, where previously there was only one. All five extrema are negative, and therefore the function is below the zero-separatrix in the control space. Thus, the resulting parameterized curve, shown below, has *three* penetrative extrema (on the top and two sides) corresponding to the three negative minima $m^-$ in the function; and these are separated by the two compressive extrema corresponding to the two negative maxima $M^-$ in the function.

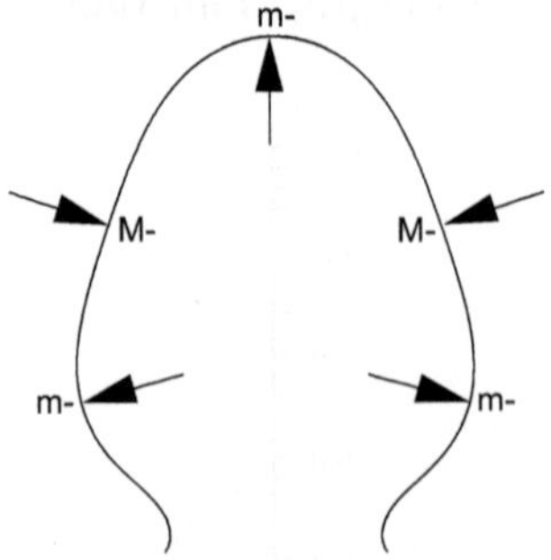


**Fig. 13.56.** The complete double-bay that has resulted from the operation $\pi^{\downarrow}\pi^{\uparrow}\emptyset^{-}$.

Fig 13.56 shows the final shape (in the sequence of three just given) together with the distribution of the five extrema and the associated process-arrows. The solid surrounds the curve, and the empty space is within the curve. The shape is an example of what our theory calls a *complete double-bay*. We see the three *indenting* arrows corresponding to the three $m^{-}$ extrema, and the two *resisting* arrows corresponding to the two $M^{-}$ extrema.

In the present case, the complete double-bay was created by the application of pair-creation operators *away* from the central extremum $m^{-}$, and such that these operators were symmetrically related.

## 13.16 Quantitative Example of Adding the 2-Fold Pinching Operation $CCM^{-}$ to the Previous Operation $\pi^{\downarrow}\pi^{\uparrow}\emptyset^{-}$

Next, let us continue the processes at the two side compressive extrema, $M^{-}$, in Fig 13.56, till they penetrate the internal space. This is a use of the 2-fold pinching operation of Type 1, $CCM^{-}$, defined on p191. That is:

$$CCM^{-} \;:\; M^{-}\, m^{-}\, M^{-} \;\rightarrow\; \overbrace{0M^{+}0}\, m^{-}\, \overbrace{0M^{+}0}$$

The path we shall take, to illustrate this behavior, will be the following sequence of two points, where the first point is the final function in the sequence on the previous three pages, and the second point is the result of applying the $CCM^{-}$ operation:

$$\begin{aligned} q_3 &= (a_4,\, a_2,\, a_0) = (-5,\; 6.3,\; -2.5) \\ q_4 &= (a_4,\, a_2,\, a_0) = (-6,\; 9,\; -2.9) \end{aligned}$$

The positions of these two points in the $(a_4, a_2)$-plane are shown in Fig 13.57. Note that, while the $CCM^{-}$ operation is upward in the sense that it goes from below to above the $MM$-sheet, the crossing need not be upward relative to the $a_0$-axis. Thus, consider

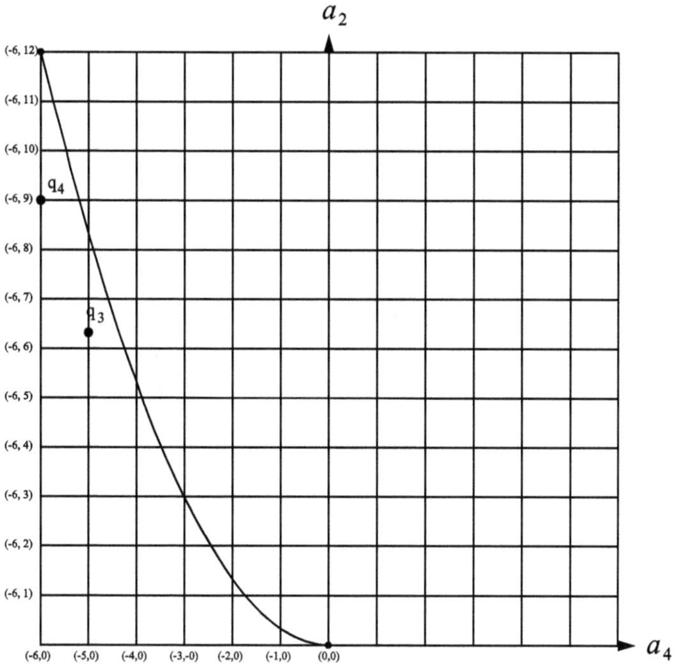


**Fig. 13.57.** Two successive points chosen to illustrate the 2-fold Type 1 pinching behavior of the $CCM^-$ operator on the side extrema.

the 3D structure of the separatrix, as illustrated in Fig 13.23 (page 492). We see that the $MM$-sheet slopes downward. Therefore, crossing this sheet upward, in the sense of going from below to above it, can actually be achieved by pulling the function sideways through the sheet. The present transition $q_3 \to q_4$ is an example of this sideways action, as can be seen in Fig 13.57. The way to understand this figure is as follows: The plane shown is the horizontal central plane of the 3D control space. The curve shown is the half-parabola in that plane. The curve of cusp-points, where the $mm$-sheet and $MM$-sheet join, is along this half-parabola, but below it in the 3D space, going downward from the origin. Correspondingly, as one moves from point $q_3$ to point $q_4$, the $MM$-sheet curves downward. In fact, due to this downward curving of the $MM$-sheet, point $q_3$ is below the $MM$-sheet; and point $q_4$ is above the $MM$-sheet. Therefore, the transition $q_3 \to q_4$ has gone upward through the $MM$-sheet; i.e., in the sense of going from below to above it. Most crucially, this means that point $q_4$ is in the *enclosed volume*; i.e., between all three sheets of the zero-separatrix.

Now let us illustrate the *quantitative* effect of applying the pinching operation $CCM^-$, by going from point $q_3$ to point $q_4$, as we have just discussed. In Fig 13.55 (page 533), we showed the function $q_3$ and its parameterized curve. The figure we now give on the next page, Fig 13.58, shows the function $q_4$ and its parameterized curve. Again, our text continues in the figure caption.

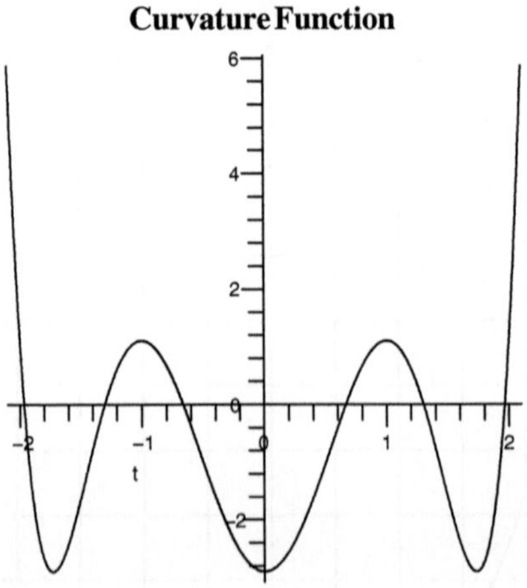


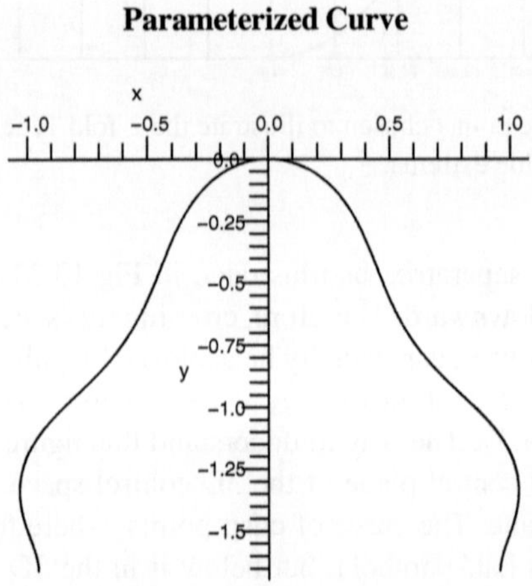


**Fig. 13.58.** Applying the $CCM^+$ operation: The upper figure shows the curvature function $f(t) = t^6 - 6t^4 + 9t^2 - 2.9$ at the position $q_4 = (a_4,\ a_2,\ a_0) = (-6,\ 9,\ -2.9)$ in the 3D control space of the Symmetry-Restricted Interactive Unfolding of $EA_5$. The two negative maxima in the previous function (p533) have now been raised through the $t$-axis to become positive maxima $M^+$. This means that, in the control space of the Symmetry-Restricted Interactive Unfolding of $EA_5$, the function has been pulled, from below the zero-separatrix, through the $MM$-sheet. The result is that the processes at the two compressive extrema, at the sides of the previous parameterized curve, have continued till they have become penetrative.

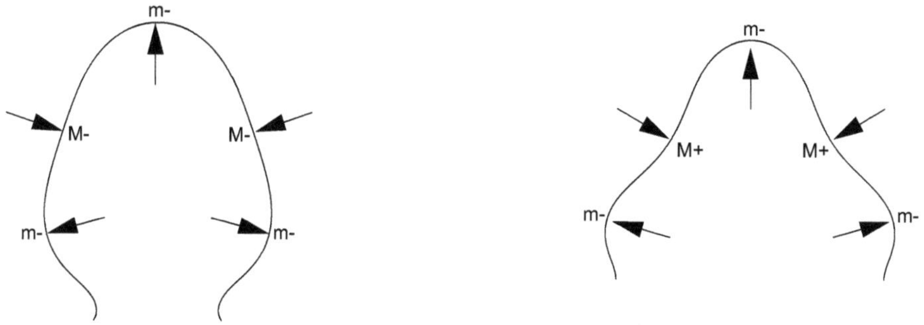


**Fig. 13.59.** Application of the 2-fold pinching operation $CCM^-$.

Fig 13.59 shows the $CCM^-$ shape transition that has just been carried out. The reader can clearly see that this is a 2-fold pinching of Type 1, rather than Type 2. That is, the two side compressive forces have been continued till they penetrated the internal space. This involved a 2-fold zero-bifurcation on each side, and no critical-point bifurcation. That is, the first shape is already in Region 5 defined by the $(a_4, a_2)$-plane; i.e., already across the half-parabola sheet. Therefore, what the transition does is make a movement within Region 5 (thus no critical-point bifurcation) but across the zero-separatrix (i.e., resulting in zero-bifurcation). Note that, to get into Region 5, a critical-point bifurcation was required in an earlier stage in the history.

**Given each of the full quantitative histories we have described, the relationship between the operators in the history is given by the *Interactive Structure* defined by Interactive Singularity Theory, and a fundamentally important fact is that Interactive Singularity Theory matches the *causal* structure defined by the correct process-inference rules, which conventional singularity theories of shape completely fail to model.**

# References


1. Blum, H., (1973). Biological shape and visual science. *Journal of Theoretical Biology*, **38**, 205-287.
2. Brady, J.M., (1983). Criteria for representations of shape. In: A.Rosenfeld & J. Beck, editors. *Human and Machine Vision: Vol 1.* Hillsdale, NJ: Erlbaum.
3. Bryant, S.V., French, V., & Bryant, P.J. (1981). Distal regeneration and symmetry. *Science*, **212**, 993-1002.
4. Costa, L. da F. (2000). Robust skeletonization through exact Euclidean distance transform and its application to neuromorphometry. *Real-Time Imaging*, **6**, 415-431.
5. De Sa, R., Radice, G., & Kerckhove, M. (1999). Timing and 3-D analysis of musculoskeletal development. NSF Final Report. Univesity of Richmond.
6. Deguchi, K. & Furukawa, R. (2001). Curvature-based diagram description of deforming contour and its application to X-ray Cardiac Image Diagnosis. *Interdisciplinary Information Sciences*, **7 (2)**, 197-208.
7. French, V., Bryant, P.J., & Bryant, S.V. (1976). Pattern regulation in epmorphic fields. *Science*, **193**, 969-994.
8. Katz, R.A. (2002) Form Metrics for Interactive Rendering via Figural Models of Perception. PhD Thesis, University of Northern Carolina.
9. Larsen, T.W. (1993). Proces grammatik og proces historie for 2D objekter. DAIMI IR-115, Aarhus Univ.Report
10. Lee, J.P. (1991). *Scientific Visualization with Glyphs and Shape Grammars.* Master's Thesis, School for Visual Arts, New York.
11. Leyton, M. (1984) Perceptual organization as nested control. *Biological Cybernetics,* **51**, 141-153.
12. Leyton, M. (1986a) Principles of information structure common to six levels of the human cognitive system. *Information Sciences,* **38**, 1-120. Entire journal issue.
13. Leyton, M. (1986b) A theory of information structure I: General principles. *Journal of Mathematical Psychology,* **30**, 103-160.
14. Leyton, M. (1986c) A theory of information structure II: A theory of perceptual organization *Journal of Mathematical Psychology,* **30**, 257-305.
15. Leyton, M. (1987a) Nested structures of control: An intuitive view. *Computer Vision, Graphics, and Image Processing,* **37**, 20-53.
16. Leyton, M. (1987b) Symmetry-curvature duality. *Computer Vision, Graphics, and Image Processing,* **38**, 327-341.
17. Leyton, M. (1987d) A Limitation Theorem for the Differential Prototypification of Shape. *Journal of Mathematical Psychology,* **31**, 307-320.

18. Leyton, M. (1988) A Process-Grammar for Shape. *Artificial Intelligence,* **34**, 213-247.
19. Leyton, M. (1989) Inferring Causal-History from Shape. *Cognitive Science,* **13**, 357-387.
20. Leyton, M. (1992). *Symmetry, Causality, Mind.* Cambridge, Mass: MIT Press.
21. Leyton, M. (1999) New foundations for perception. In Lepore, E. (Editor). *Invitation to Cognitive Science.* Blackwell, Oxford. p121 - 171.
22. Leyton, M. (2001). *A Generative Theory of Shape.* Berlin: Springer-Verlag.
23. Leyton, M. (2004) Musical works are maximal memory stores. *This paper is viewable on the web on Leyton's home page*, and is published in *Perspectives in Mathematical and Computer-Aided Music Theory.* G. Mazzola & T. Noll (Editors). Osnabruck: Osnabruck Music Publishing.
24. Leyton, M. (2006) Interoperability and objects. *International Journal of Product Lifecycle Management,* **1 (2)**, 98-128.
25. Leyton, M. (2006). *Shape as Memory: A Geometric Theory of Architecture.* Zurich: Birkhauser.
26. Leyton, M. (2006). *The Structure of Paintings.* Vienna: Springer-Verlag.
27. Leyton, M. (2011). *Versal Transitions in the Creation of Parabolic Loops and Ridge Loops on a Surface.* Tech Report 2010-2: The International Institute of Interoperability Science.
28. Lin W-C., Liang C-C. & Chen C-T. (1989). A computational model for process-grammar. *Artificial Intelligence*, **38** , 207-224.
29. Lopéz, A.M. (1997). Ridge/valley-like structures: creases, separatrices, and drainage patterns. http://www.dai.ed.ac.uk/CVonline/LOCAL_COPIES/LOPEZ/evonline.html
30. Meyer, B. (1997). *Object-Oriented Software Construction.* New Jersey: Prentice Hall.
31. Milios, E.E. (1989). Shape matching using curvature processes. *Computer Vision, Graphics, and Image Processing*, **47**, 203-226.
32. Ogniewicz, R.L. (1995). Automatic medial axis pruning by mapping characteristics of boundaries evolving under Euclidean geometric heat flow onto Voronoi skeletons. Harvard Robotics Laboratory Technical Report No. 95-4.
33. Parvin, B., Peng, C., Johnston, W. & Maestre, M. (1994). Tracking of Tubular Objects for Scientific Applications. *IEEE Conf. on Computer Vision and Pattern Recognition*, 295-30.
34. Pernot, J-P., Guillet, S., Leon, J-C., Falcidieno, B., & Giannini, F. (2003). Interactive Operators for free form features manipulation. In SIAM conference on CADG, Seattle, 2003.
35. Pernot, J-P., Guillet, S., Leon, J-C., Falcidieno, B., & Giannini, F. (2005). Un modèle de description de formes gauches et opérateurs de manipulation associeś. Tech report Laboratoire 3S Grenoble and CNR Genova.
36. Pizer, S.M., Fritsch, D.S., Yushkevich, P.A., Johnson, V.E., & Chaney, E.L. (1996). Segmentation, registration, and measurement of shape variation via image object shape. *IEEE Transactions on Medical Imaging*, **18(10)**, 851-865.
37. Richards, W., Koenderink, J.J., & Hoffman, D.D. (1987). Inferring three-dimensional shapes from two-dimensional silhouettes. *Journal of the Optical Society of America A*, **4**, 1168-1175.
38. Shemlon, S. (1994). *The Elastic String Model of Non-Rigid Evolving Contours and its Applications in Computer Vision.* PhD Thesis, Rutgers University.
39. Thom, R. (1972). *Structural Stabilité et Morphogénèse.* New York: Benjamin. English edition: Thom, R (1975) *Structural Stability and Morphogenesis*. Translated by D. H. Fowler. New York: Benjamin-Addison Wesley.
40. Torres, R da S, & Falcão, A. X. (2007). Contour salience descriptors for effective image retrieval and analysis. *Image and Vision Computing*, **25**, 3–13.
41. Zou, J. J. & Hong Yan, H. (2001). Cartoon image vectorization based on shape subdivision. *Computer Graphics International*, 131-145.

# Index

MIX
Papier aus verantwortungsvollen Quellen
Paper from responsible sources
FSC® C105338

If you have any concerns about our products,
you can contact us on
**ProductSafety@springernature.com**

In case Publisher is established outside the EU,
the EU authorized representative is:
**Springer Nature Customer Service Center GmbH**
**Europaplatz 3, 69115 Heidelberg, Germany**

Printed by Libri Plureos GmbH
in Hamburg, Germany